Pharmacokinetics and Pharmacodynamics of Novel Drug Delivery Systems: From Basic Concepts to Applications

Sankalp A. Gharat • Munira M. Momin •
Tabassum Khan
Editors

Pharmacokinetics and Pharmacodynamics of Novel Drug Delivery Systems: From Basic Concepts to Applications

A Machine-Generated Literature Overview

Springer

Editors
Sankalp A. Gharat
Department of Pharmaceutics
SVKM's Dr. Bhanuben Nanavati College of Pharmacy
Mumbai, Maharashtra, India

Munira M. Momin
Department of Pharmaceutics
SVKM's Dr. Bhanuben Nanavati College of Pharmacy
Mumbai, Maharashtra, India

Tabassum Khan
Department of Pharmaceutical Chemistry and Quality Assurance
SVKM's Dr. Bhanuben Nanavati College of Pharmacy
Mumbai, Maharashtra, India

ISBN 978-981-99-7857-1 ISBN 978-981-99-7858-8 (eBook)
https://doi.org/10.1007/978-981-99-7858-8

This Springer imprint is published by the registered company Springer Nature Singapore Pte Ltd.
The registered company address is: 152 Beach Road, #21-01/04 Gateway East, Singapore 189721, Singapore

Preface

The science of drug delivery has witnessed remarkable advancements in the ever-evolving era of pharmaceuticals and healthcare that promise to revolutionize clinical therapeutics. Among these breakthroughs, nanoparticles and novel drug delivery systems (NDDS) have emerged as game-changers, offering enhanced drug efficacy, reduced side effects, and improved patient outcomes. This book, *Pharmacokinetics and Pharmacodynamics of Novel Drug Delivery Systems: From Basic Concepts to Applications*, delves deep into the intricate world of these cutting-edge technologies. It seeks to unravel the mysteries of how nanoparticles interact with the human body, the dynamics of drug release, and the impact of these innovations on therapeutic response and clinical outcomes.

Nanoparticles represent a promising frontier in pharmaceuticals. Their ability to transport, protect, and release drugs with precision is nothing short of astounding. The intricate interplay between nanoparticles and biological systems, the optimization of drug loading and release kinetics, and the quest for enhanced therapeutic efficacy demand a comprehensive understanding. Beyond nanoparticles, novel drug delivery systems encompass a wide array of ingenious techniques and formulations. From liposomes to microneedles, from implants to inhalation systems, each chapter explores the unique characteristics, challenges, and opportunities associated with these innovative approaches to drug delivery.

This book is an effort to provide a comprehensive understanding of the nuances of NDDS and its impact on the fascinating science of PK & PD. Each chapter is a carefully crafted exploration of a specific aspect of nanoparticle-based drug delivery or a novel delivery system, offering a holistic view of the subject.

The book is intended for researchers, clinicians, students, and anyone with an interest in the future of drug delivery. Whether you are a research scientist looking to strengthen your understanding or a student embarking on a journey of discovery, this book offers something for everyone. It is our hope that this book will serve as a beacon, guiding you through the complexities and possibilities of this field. As you embark on this voyage, we invite you to explore, question, and envision the future of pharmaceuticals with us.

Welcome to the world of *Pharmacokinetics and Pharmacodynamics of Nanoparticles and Novel Drug Delivery Systems*.

Munira M. Momin

Disclaimer

Auto-summaries can be generated by either an abstractive or extractive auto-summarization:

- An extraction-based summarizer identifies the most important sentences of a text and uses the original sentences to create the summary.
- An abstraction-based summarizer creates new text based on deep learning. New phrases are created to summarize the content.

The auto-summaries you will find in this book have been generated via an extractive summarization approach.

Each chapter was carefully edited by Prof. Dr. Munira Momin, Dr. Tabassum Khan, and Prof. Sankalp Gharat. The editors selected the papers which were then auto-summarized. The editors have not edited the auto-summaries due to the extraction-based approach and have not changed the original sentences. You will find the editors' reviews and guidance on the auto-summaries in their chapter introductions.

In machine-generated books, editors are defined as those who curate the content for the book by selecting the papers to be auto-summarized and by organizing the output into a meaningful order. Next to the thoughtful curation of the papers, editors should guide the readers through the auto-summaries and make transparent why they selected the papers.

The ultimate goal is to provide a current literature review of Springer Nature publications on a given topic in order to support readers in overcoming information overload and to help them dive into a topic faster; to identify interdisciplinary overlaps; and to present papers which might not have been on the readers' radar.

Please note that the selected papers are not used to train an LLM while the auto-summaries are created.

Contents

About the Editors

Sankalp A. Gharat Prof. Sankalp Gharat is presently working as an assistant professor in the Department of Pharmaceutics at SVKM's Dr. Bhanuben Nanavati College of Pharmacy (BNCP), Mumbai, India. He has earlier served as assistant professor in the Department of Pharmaceutics (2018–2019) at Humera Khan College of Pharmacy, Mumbai, India. Before joining academics, he was associated with Macleods Pharmaceuticals R&D Centre, as a research scientist in the formulation and development of topical (ANDA) conventional and novel products in the Dermatological Section for regulated market (US and EU market). He has worked as a research associate fellow (2014–2016) in research project on cancer management sponsored by the University of Mumbai. His research interests include formulation development of novel drug delivery and targeted systems based on nanotechnology for cancer, rheumatoid arthritis, psoriasis, and ophthalmic use. In his area of research, he has several national and international papers published in peer-reviewed journals in collaboration with Laurentian University, Bhabha Atomic Research Centre, and Ajman University of United Arab Emirates. He has been awarded with title "SPDS India Young Researcher 2024" by Society for Pharmaceutical Dissolution Science (SPDS) and "Promising Young Faculty Award 2023" by SVKM's Dr. BNCP. He has secured first place for two consecutive years at the prestigious 15th and 16th AVISHKAR Research Convention—University of Mumbai, in Medicine and Pharmacy Category (2020–2021 and 2021–2022). Prof. Sankalp is a recipient of "AICTE Scholarship for

2014–2016" and "Meyer Organic Academic Excellence Award 2016" by Meyer Organics Pvt Ltd for his academic performance in bachelor's program. He has delivered several research talks and interacted with students of various pharmaceutical institutions. He is actively involved in training students for conducting cell culture-based experiments for toxicity testing of formulations and phytochemicals. He is a member of professional societies like the Association of Pharmaceutical Teachers of India (APTI) and Controlled Release Society (CRS-Indian Chapter).

Munira M. Momin Prof. Munira Momin is a professor and principal of SVKM's Dr. Bhanuben Nanavati College of Pharmacy, Mumbai, India. She has earlier served as professor in the Department of Pharmaceutics (2012–2015) at Oriental College of Pharmacy, Navi Mumbai, India, as associate professor (2005–2011) at LJ Institute of Pharmacy, Ahmedabad, India, and as assistant professor (2003–2005) at LM College of Pharmacy, Ahmedabad, and SK Patel College of Pharmaceutical Education and Research, Mehsana, India. Her research interest is in targeted drug delivery and bioengineered polymeric nanoparticles with a specialization in brain-targeted nanosystems. She has been conferred with prestigious Nehru-Fulbright Academic Excellence in Higher Education seminar for the year 2019. She is a recipient of topmost influential principal award (pharmacy category) for the year 2022 for her contribution in administration and research and bringing positive changes among students by setting up social sensitivity cell at her institute. Dr. Momin is a recipient of ACG-Best Research Paper Award of IDMA during the year 2020–2021 and ML Khurana Memorial Best Research Paper Award by IPA for the papers published during 2008–2009. She also has FH Jani Gold Medal for securing first merit during BPharm program. She has published more than 80 scientific articles in peer-reviewed international journals and authored or co-authored numerous books and book chapters. She has five granted Indian patents to her credit. She is a member of many professional societies, i.e., Indian Pharmaceutical Association (IPA), Association of Pharmaceutical Teachers of India (APTI), Society for Pharmaceutical Dissolution Science (SPDS), and Controlled Release society (CRS-Indian Chapter). She is a scientific chair of Disso-India, an annual conference arranged by SPDS.

Tabassum Khan Dr. Tabassum Khan is Professor, Department of Pharmaceutical Chemistry and Quality Assurance at SVKM's Dr. Bhanuben Nanavati College of Pharmacy, Mumbai, India. She is a recipient of the prestigious Erasmus+ Teaching Mobility Fellowship, ERASMUS Foundation, European Union Programme 2023; Shastri Indo-Canadian Institute, Shastri Mobility Programme Fellowship Award 2018; P.D. Sethi National Award for Best Research Paper on HPTLC, 2017, and the Indian Pharmaceutical Association B. Pharm Merit Award, 1990. She is on the reviewer board for science fund of the Republic of Serbia, 2023 and 2021; National Science Centre Poland, 2022; New Frontier Seed Grant Program, Network for Canadian Oral Health Research, Nova Scotia, Canada, November 2019 and 2018; Shastri Indo Canadian Institute, MHRD, India, under the Collaborative Research Category, October 2019; and Medical Research Council, UK, under the South Africa (Newton)–UK Antibiotic Accelerator 2019 program, 2019. Her thrust areas of research include medicinal chemistry, natural product chemistry, and nanotechnology-based drug development. She has 2 Indian patents, 78 publications, and 8 book chapters to her credit. She is life member of the Indian Pharmaceutical Association and Society of Pharmaceutical Education and Research and member of the International Federation of Pharmacists and American Association of Pharmaceutical Scientists.

Chapter 1
Introduction to Pharmacokinetics and Pharmacodynamic Studies of Novel Drug Delivery Systems

Sankalp A. Gharat, Munira M. Momin, and Tabassum Khan

1.1 Introduction

Nanoformulations have evolved to circumvent the limitations of conventional drug delivery systems. Novel Drug Delivery Systems (NDDS) have emerged as innovative approaches that enhance the efficacy and safety of drug therapies. NDDS includes a diverse range of nanotechnologies designed to optimize the delivery of therapeutic agents to their intended targets within the body. The need for nanoformulations arises from the limitations of traditional drug administration methods, that often results in suboptimal therapeutic outcomes and reduced patient compliance [1]. The history of NDDS is an interesting narrative of human ingenuity and scientific progress; the first controlled-release polymer device was developed in 1964. Bangham discovered the liposome in 1965; albumin-based nanoparticles were reported in 1972; liposome-based formulations were developed in 1973; the first micelle was developed and approved in 1983; the USFDA approved the first controlled formulation in 1989; and the first polyethylene glycol (PEG)-protein conjugate hit the market in 1990 [2].

The term "NDDS" covers various macro-micro-nano strategies and platforms that modulate the formulation for release, and targeting of drugs. These systems are designed to specific disease states, and patient populations, offering personalized

S. A. Gharat · M. M. Momin (✉)
Department of Pharmaceutics, SVKM's Dr. Bhanuben Nanavati College of Pharmacy, Mumbai, Maharashtra, India
e-mail: sankalp.gharat@bncp.ac.in; munira.momin@bncp.ac.in

T. Khan
Department of Pharmaceutical Chemistry and Quality Assurance, SVKM's Dr. Bhanuben Nanavati College of Pharmacy, Mumbai, Maharashtra, India
e-mail: tabassum.khan@bncp.ac.in

S. A. Gharat et al. (eds.), *Pharmacokinetics and Pharmacodynamics of Novel Drug Delivery Systems: From Basic Concepts to Applications*,
https://doi.org/10.1007/978-981-99-7858-8_1

and optimized treatment approaches. Drug products containing nanomaterials possess distinct characteristics due to potential alteration of their chemical, physical, or biological properties in comparison to conventional pharmaceuticals, significantly impacting the quality, safety, and effectiveness of the product [3]. Nanomaterial-based drug formulations may exhibit a different pharmacokinetic profile as compared to traditional formulations containing the same drug. Once a drug loaded nanoparticle enters the bloodstream, it has the capacity to engage with specialized immune cells known as macrophages. These macrophages engulf the nanoparticles and facilitate its targeted delivery to specific locations that are typically challenging to access for conventional formulations [4]. In another scenario, a pharmaceutical formulated as a nanomaterial incorporates a unique "shielding" mechanism designed to prevent interactions with immune cells, enabling the drug to circulate in the bloodstream for extended periods, ultimately reaching its intended destination, such as tumor tissues. This capability to selectively target specific areas while avoiding others substantially reduces the risk of side effects and off-target toxicity, resulting in enhanced therapeutic efficacy [5–7].

Nanomaterials find their primary application in the treatment of conditions like cancer and microbial infections [8–10]. However, it is a challenging process to comprehend the transit of these nanoparticles within the biological system. Understanding PK (Pharmacokinetics) and PD (Pharmacodynamics) of nanoparticles is vital for optimizing drug delivery, ensuring efficacy, and minimizing potential side effects when using these nanoscale carriers for targeted therapies. Additionally, PK/PD studies guide the design of nanoparticle-based drug delivery systems, contributing to safer and more efficient healthcare solutions [11]. Before delving into PK/PD of nanoparticles, it is essential to have a comprehensive understanding of the current global scenario of NDDS.

The global NDDS market was estimated to be valued at approximately US$12.7 billion in year 2022. Projections indicate that it is poised to reach a revised market size of around US$61 billion by the year 2030, demonstrating a compounded annual growth rate (CAGR) of 21.6% over the analysis period spanning from 2022 to 2030. The NDDS market in the United States is estimated to be valued at approximately US$4.8 billion in 2022. China, as the world's second-largest economy, is expected to reach a projected market size of around US$7.4 billion by 2030, exhibiting a CAGR of 26.4% during the analysis period from 2022 to 2030. Noteworthy growth is also anticipated in other geographic markets, including Japan and Canada, which are forecast to experience growth rates of approximately 18.7% and 19.8%, respectively, over the period spanning from 2022 to 2030. In Europe, Germany is expected to grow at a CAGR of approximately 20.2% [12].

"Nanotechnology—Over a Decade of Progress and Innovation": A report by the USFDA, issued in July 2020 illustrates the increasing trend in Investigational New Drug (IND), New Drug Application (NDA), Abbreviated New Drug Application (ANDA) submissions of human drug products to the USFDA that involve nanomaterials from 1970 to 2019 [13], as shown in Fig. 1.1.

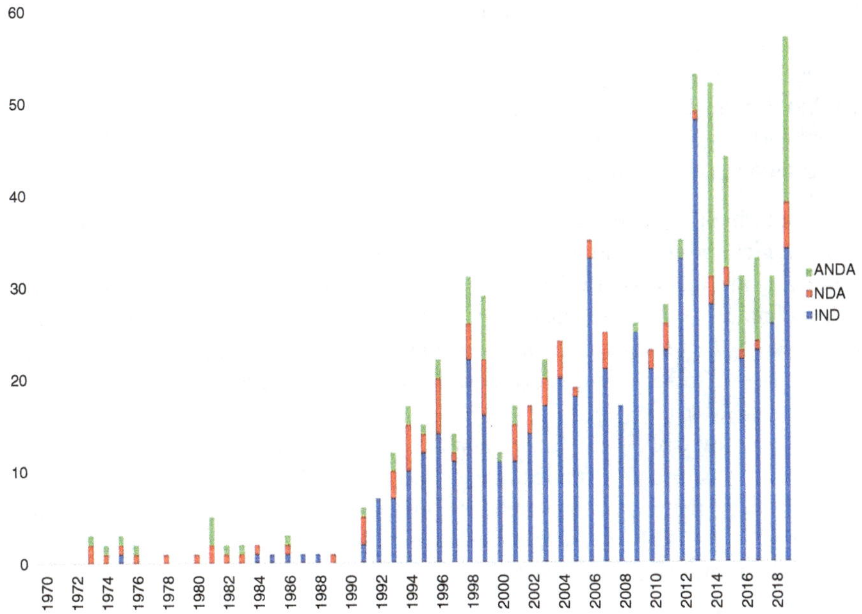

Fig. 1.1 Nanomaterials based applications submitted to the United States Food and Drug Administration (USFDA) in the period of 1970–2019 (Figure adapted from USFDA website data freely available in public domain: "Nanotechnology—Over a Decade of Progress and Innovation: A report by the USFDA") [13]

The growth of the NDDS market is driven by a combination of medical, technological, regulatory, and economic factors, with a focus on improving patient outcomes and treatment options [14], including:

Chronic Diseases: The rising prevalence of chronic diseases, such as cancer, diabetes, and cardiovascular diseases, creates a demand for advanced drug delivery systems to improve treatment outcomes.

Biopharmaceuticals: The growth of biopharmaceuticals, including monoclonal antibodies and gene therapies, has spurred the development of specialized drug delivery technologies to enhance their delivery and effectiveness.

Personalized Medicine: The shift towards patient-centric healthcare and personalized medicine is fueling the demand for tailored drug delivery systems that meet individual patient needs.

Population: The aging population in many countries is driving the need for innovative drug delivery systems to address age-related health conditions.

Technological Advancements: Advances in nanotechnology, materials science, and microfabrication have opened up new possibilities for creating novel drug delivery platforms.

Regulatory Support: Favourable regulatory policies and incentives for innovative drug delivery systems encourage companies to invest in research and development.

Investment and Research: Significant investments in research and development by pharmaceutical companies, startups, and government initiatives are fueling innovation in drug delivery technologies.

Global Pandemic: The COVID-19 pandemic has accelerated research into vaccine delivery systems, including nanoparticle-based platforms, which will likely have lasting impacts on pharmaceutical industry [15].

As nanotechnology continues to evolve, the range and impact of marketed nanotechnological products are expected to grow, shaping the future of industries and improving the quality of life for people around the world. Over the past few decades, nanotechnology has transitioned from laboratory research to translational applications, leading to the emergence of a wide range of nanotechnological products in the global market. Table 1.1 summarises various approved nanotechnology based marketed formulations.

The translation of NDDS from the laboratory to commercialization is a complex and multifaceted process. While these technologies hold the potential to revolutionise healthcare, there are several challenges that companies and researchers must address to successfully commercialise these innovations. Several factors contribute to these translational gaps are as follows [19]:

(a) **Complexity of nanostructures:** Nano formulations involve complex structures with precise control over size, shape, surface properties, and drug loading. Scaling up these complex structures while maintaining their integrity and functionality is highly challenging.
(b) **Regulatory hurdles:** Regulatory agencies like the USFDA and EMA (European Medicines Agency) have stringent requirements for approving new drug formulations, especially those involving nanoparticles. Demonstrating safety and efficacy, characterizing the product's quality, and ensuring consistent manufacturing processes is time-consuming and expensive.
(c) **Biocompatibility and toxicity:** Nanoformulations must be carefully evaluated for their biocompatibility and potential toxicity. This requires extensive preclinical studies to understand their effects in different biological systems.
(d) **Stability and shelf life:** Maintaining the stability of nano formulations over time is crucial for commercial viability. Ensuring that nanoparticles remain stable, both in storage and during administration, is challenging.
(e) **Scalability:** Transitioning from small-scale laboratory production to large-scale commercial manufacturing is tough. Achieving the same level of precision and quality control at scale is a significant hurdle.
(f) **Cost-effectiveness:** The production of nanoparticles often involves expensive and specialized equipment and materials. Reducing the cost of manufacturing while maintaining quality is significant barrier to translation.
(g) **Targeted delivery and efficacy:** While nano formulations offer the promise of targeted drug delivery, ensuring that the nanoparticles effectively reach their intended target and produce the desired therapeutic effect is complex and requires further optimization.

Table 1.1 List of approved marketed nano formulations [16–18]

Sr. no	Formulation type	Name of the product	Name of drug	Manufacturer	Use
1	Liposomal formulation	Doxil® (Caelyx™)	Doxorubicin	Janssen Pharmaceuticals	Multiple myeloma, ovarian neoplasms, breast neoplasms, Kaposi sarcoma
		Onivyde®	Irinotecan	PharmaEngine	Metastatic pancreatic cancer
		Vyxeos®	Daunorubicin and Cytarabine	Jazz Pharmaceuticals	Treatment of certain types of acute myeloid leukemia
		DepoDur®	Morphine	Pacira Pharmaceuticals Ltd.	Post-operative pain management
		AmBisome®	Amphotericin B	Gilead Sciences International Ltd.	Severe systemic and deep mycoses and visceral leishmaniasis in immunocompetent patients
		DaunoXome®	Daunorubicin	Galen Limited	Treatment of advanced HIV-associated Kaposi's sarcoma
		Myocet®	Doxorubicin	Cephalon Europe	Breast neoplasms
		DepoCyte®	Cytarabine	Pacira Pharmaceuticals Ltd.	Meningeal neoplasms
		Marqibo®	Vincristine	Talon Therapeutics	Philadelphia chromosome-negative (Ph-) acute lymphoblastic leukemia
		Visudyne®	Verteporfin	Bausch & Lomb Incorporated	Degenerative myopia, age related macular degeneration
		Mepact®	Mifamurtide	IDM Pharma SAS	Osteosarcoma
		Epaxal®	Hepatitis A vaccine	Crucell Italy	Active immunisation against hepatitis A

(continued)

Table 1.1 (continued)

Sr. no	Formulation type	Name of the product	Name of drug	Manufacturer	Use
2	Polymeric nanoparticles	Abraxane®	Paclitaxel (Albumin-bound nanoparticles)	Celgene Corporation (now part of Bristol Myers Squibb)	Metastatic breast cancer
		Gebexol-PM®	Paclitaxel (Methoxy-PEG-poly[D,L-lactide] taxol)	Lupin Ltd	Breast cancer, pancreatic cancer, and non-small cell lung cancer
		Oncaspar®	(PEG-L-asparaginase)	Servier IP UK Ltd	Acute lymphoblastic leukaemia
		Neulasta®	PEG-filgrastim	Amgen Inc	Used to prevent neutropenia
		Rapamune®	Sirolimus (also known as Rapamycin) (Colloidal dispersion of nanocrystals stabilized with poloxamer)	Pfizer	Prophylaxis of organ rejection in renal transplant
		Emend®	Aprepitant (Colloidal dispersion of nanocrystals)	Merck Sharp & Dohme Corp	Nausea and vomiting
		Cholib®	Fenofibrate/Simvastatin (Colloidal dispersion of nanocrystals)	Abott Healthcare Products Ltd	Dyslipidemias
3	Metallic nanoparticles	Feraheme®	Ferumoxytol (superparamagnetic iron oxide nanoparticles)	AMAG Pharmaceuticals	Iron replacement therapy for iron-deficiency anemia
		NanoTherm®	Iron oxide nanoparticle	MagForce AG	Multiple myeloma
		Hensify® (NBTXR3)	Hafnium oxide nanoparticles stimulated with external radiation to enhance tumor cell death via electron production	Nanobiotix	Locally advanced squamous cell carcinoma
4	Lipidic nano formulations	Intralipid®	Lipid-based nanoemulsion of essential fatty acids and vitamins	Fresenius Kabi	Emergency management of local anaesthetics inadvertently administered intravenously
		Onpattro® (Patisiran) ALN-TTR02	Lipid nanoparticle RNAi for the knockdown of disease-causing TTR protein	Alnylam Pharmaceuticals	Transthyretin (TTR)-mediated amyloidosis

(h) **Clinical trials:** Conducting clinical trials to demonstrate the safety and efficacy of nano formulations is time-consuming and expensive. Moreover, recruiting patients for such trials is tough, particularly if the formulation targets a rare disease or a niche market.
(i) **Market adoption:** Even if a nanoformulation successfully traverses the regulatory process and demonstrates clinical efficacy, market adoption may be slow owing to factors such as physician acceptance and competition with established therapeutic modalities.
(j) **Intellectual property:** Protecting intellectual property associated with nanoformulations can be challenging and hinders their translation.

To bridge the translational gap for nanoformulations, collaboration among researchers, clinicians, regulatory agencies, and industry stakeholders is essential. Additionally, continued investment in research, development, and the establishment of standardized protocols for evaluation and manufacturing can help overcome some of these challenges.

1.2 Regulatory Guidelines for NDDS in Pharmaceutical Industry

Food products, cosmetics, medical devices, and medications have incorporated nanotechnology over the course of several decades. There is a wide range of pharmaceuticals incorporating nanomaterials that are under the jurisdiction of the USFDA's Centre for Drug Evaluation and Research (CDER). The number of authorised drug products utilising nanomaterials has steadily increased, which includes generic medications as well as experimental novel pharmaceuticals, new drug applications, and abbreviated new drug applications. Since the early 1970s, more than 60 applications have been accepted, and the demand continues to grow [20]. Current nanotechnology research is concentrated on the exploration of pivotal processes and material attributes that possess the potential to influence product quality, while comprehending quality within the broader framework of effectiveness and safety. The stability of liposomes, is also a significant aspect under scrutiny, given its direct bearing on product quality. Additionally, there is a focus on assessing the extent to which manufacturing deviations, such as temperature fluctuations or drying procedures, can affect particle size, distribution within the body, and overall stability.

Manufacturers are obligated to meticulously opt for and execute suitable quality control measures to enable the detection of any potential variations in the nanoproduct. The Office of Testing and Research (OTR) focuses their attention on assessing the drug's performance through the utilization of advanced analytical techniques. These assessments include examinations conducted both within the body (*in-vivo*) and in external settings (*in-vitro*). The primary objectives are to unravel the mechanisms governing the release of the drug from nanocarriers and to establish a correlation between *in-vivo* and *in-vitro* findings [21].

In 2014, the formation of the Nanotechnology Risk Assessment Working Group aimed to evaluate the implications of nanotechnology on drug products. Comprising of experts from various domains within the CDER, this working group is actively engaged in establishing standards for nanomaterials applied in drug development, thereby fostering technological progress. The collective findings of this working group indicate that, for the most part, current assessment procedures suffice for the evaluation of drugs incorporating nanomaterials. However, this is contingent upon the drug applicant's diligent execution of appropriate studies early in the developmental phase and the implementation of a formulation control strategy to ensure consistent clinical outcomes. The CDER has dedicated substantial efforts over recent years to comprehend the attributes of nanomaterials when utilized in drug products. This aims to establish a regulatory framework that can effectively evaluate the influence of these unique physical properties on the safety and efficacy of such nanotechnology-based products. The initial phase of this work concentrated on exploring the role of zinc oxide and titanium dioxide nanomaterials in sunscreens. CDER conducted studies to investigate the potential penetration of titanium dioxide nanomaterials into normal skin, demonstrating in pig model that such penetration did not occur beyond the dermis. In more recent times, CDER's ongoing research initiatives encompass a broader scope, including the characterization and safety assessment of nanomaterials in drug products [22].

The USFDA has released a series of guidance documents concerning the utilization of nanotechnology in products that fall under FDA regulation. These guidance documents are being issued as a part of the USFDA's ongoing efforts to implement the recommendations outlined in the FDA's 2007 Nanotechnology Task Force Report. It is important to note that while these guidance documents do not establish or grant any rights to individuals or impose binding obligations on the FDA or the public, they do serve as a reflection of the FDA's current perspective on the subject matter. Some of the guidance documents that address the use of nanotechnology or nanomaterials in products regulated by the FDA are discussed in Table 1.2:

The existing regulatory framework in Europe actively supports the advancement of novel nanomedicines and has demonstrated its efficacy in evaluating marketing authorization applications for such products. A substantial commitment is being made both at the European and international levels to ensure that regulatory science progresses in sync with the evolving knowledge of nanotechnology. In 2007, the EMA, in collaboration with experts from regulatory agencies in the United States, Japan, Canada, and Australia, participated in a global initiative led by the FDA. This initiative aimed to establish the defining characteristics of medicines utilizing nanotechnology. The consortium of regulatory bodies engaged in discussion and information sharing on guidelines as well as scientific and legislative efforts in their respective regions. This exchange of information and experience facilitated valuable insights, drawing from analogous frameworks such as cosmetics, medical devices, and consumer products. Under the chairmanship of the EMA, this group also formulated a working descriptor for "nano-medicines intended for internal use." This descriptor served as a common reference point to enable global regulatory authorities to exchange experiences and deliberate on their respective

Table 1.2 Summary of USFDA's Final Guidance to Industry on use of nanotechnology

Sr. no	Final Guidance for Industry	Summary	Ref
1	Guidance for Industry: Assessing the Effects of Significant Manufacturing Process Changes, Including Emerging Technologies, on the Safety and Regulatory Status of Food Ingredients and Food Contact Substances, Including Food Ingredients that Are Color Additives.	This document offers guidance to food ingredient and food contact substance (FCS) manufacturers, as well as end users of food ingredients categorized as color additives. The guidance outlines the FDA's current perspective on evaluating the impact of substantial changes in manufacturing processes, such as the incorporation of nanotechnology, on the safety and regulatory status of a food substance.	[23]
2	Guidance for Industry: Considering Whether an FDA-Regulated Product Involves the Application of Nanotechnology	This guidance document establishes a comprehensive framework for the FDA's regulatory approach to nanotechnology products. It is intended to serve as a reference for manufacturers, suppliers, importers, and various stakeholders. Within this framework, it highlights two critical factors that should be taken into account when determining whether a product, subject to FDA regulation, incorporates nanotechnology: a) Whether a material or end product has been intentionally engineered to possess at least one external dimension, internal structure, or surface feature falling within the nanoscale range, which spans approximately 1 nanometer (nm) to 100 nanometers (nm). b) Whether a material or end product has been purposely designed to exhibit specific properties or phenomena, encompassing physical, chemical, or biological effects, which can be attributed to its dimensions. This evaluation applies even if these dimensions extend beyond the nanoscale range, up to 1 μm (1000 nm).	[24]

(continued)

Table 1.2 (continued)

Sr. no	Final Guidance for Industry	Summary	Ref
3	Final Guidance for Industry: Liposome Drug Products: Chemistry, Manufacturing, and Controls; Human Pharmacokinetics and Bioavailability; and Labeling Documentation.	This guidance document outlines the specific information that should be included in a new drug application (NDA) or abbreviated new drug application (ANDA) for a liposome drug product, which is subject to review by the Center for Drug Evaluation and Research (CDER). The discussion within this guidance encompasses the following key areas related to liposome drug products: a) Chemistry, Manufacturing, and Controls (CMC): This section provides guidance on the submission of information pertaining to the chemistry, manufacturing, and quality control of liposome drug products. b) Human Pharmacokinetics and Bioavailability (or Bioequivalence, in the case of an ANDA): This part offers recommendations regarding the submission of data on human pharmacokinetics and bioavailability, which are crucial aspects of evaluating the effectiveness and safety of liposome drug products. c) Labeling in NDAs and ANDAs: This section addresses the labeling requirements and recommendations for liposome drug products included in NDAs and ANDAs. The guidance primarily focuses on the distinct technical considerations associated with liposome drug products.	[25]
4	Final Guidance for Industry: Safety of Nanomaterials in Cosmetic Products	This guidance document offers industry professionals and various stakeholders, including academia and other regulatory bodies, insights into the FDA's current perspective on the safety evaluation of nanomaterials present in cosmetic products. Its primary aim is to aid industry players and stakeholders in the identification of potential safety concerns associated with nanomaterials in cosmetic formulations and in the establishment of a structured approach for their evaluation. Furthermore, this guidance document includes contact information for manufacturers and sponsors interested in engaging with the FDA to discuss safety-related matters concerning the utilization of specific nanomaterials in cosmetic products. This serves as a channel for dialogue and consultation on safety considerations in these cases.	[26]

5	Final Guidance for Industry: Use of Nanomaterials in Food for Animals	This guidance document outlines the current stance of the Food and Drug Administration (FDA) regarding the use of nanomaterials and the application of nanotechnology in animal food. Its purpose is to assist industry professionals and other stakeholders in recognizing potential concerns related to the safety and regulatory status of animal food that either contains nanomaterials or involves the utilization of nanotechnology. In this context, "animal food" refers to food intended for consumption by animals.	[27]
6	Final Guidance for Industry: Drug Products, Including Biological Products, that Contain Nanomaterials	This guidance document offers direction regarding the development of human drug products, including biological products, that incorporate nanomaterials within the final dosage form. It centers on pertinent considerations concerning the FDA's regulation of such drug products in accordance with the Federal Food, Drug, & Cosmetic Act (FD&C Act) and Public Health Service Act (PHS Act). The guidance is intended to provide recommendations to applicants and sponsors of investigational, premarket, and postmarket submissions for these specific products.	[28]

approaches to regulating this emerging field. It ensured that regulatory science advancements were collaborative and open to input from various stakeholders, including the work of ISO 199 TC. Building upon the collaborative work of this international expert group, the EMA organized the First International Scientific Workshop on Nanomedicines on September 2–3, 2010. The workshop brought together approximately 200 participants from Europe and around the world, representing 27 countries, including Australia, Canada, India, Japan, and the United States. These participants engaged in discussions concerning the advantages and challenges associated with the application of nanotechnologies in medicine [24, 29].

1.3 Need of Pharmacokinetic and Pharmacodynamic Studies (PK/PD) of NDDS

Understanding pharmacokinetics and pharmacodynamics play a significant role in overcoming the translational challenges of NDDS. PK/PD studies are essential in evaluating the impact of these nano systems, which aim to enhance drug efficacy and safety through controlled and targeted release as shown in Fig. 1.2.

PK is the study of how the body interacts with a drug after administration. It involves the processes of absorption, distribution, metabolism, and excretion (ADME) that a drug undergoes within the body. In the context of novel drug delivery systems, understanding pharmacokinetics becomes crucial for optimizing drug release profiles, extending drug circulation times, and achieving targeted delivery. Various routes of administration, including oral, parenteral, transdermal, and inhalation, offer distinct pharmacokinetic profiles that influence the design and selection of delivery systems. By studying the PK of novel drug delivery systems, researchers can assess their bioavailability, tissue distribution, and elimination kinetics. This knowledge helps optimize dosage regimens, predict drug concentrations at the target site, and determine the duration of therapeutic effect. On the other hand, PD refers to the study of a drug's mechanism of action and its effects on the body. It encompasses receptor binding, signal transduction, and the resulting physiological responses. Novel drug delivery systems often aim to enhance pharmacodynamic outcomes by providing controlled and sustained drug release, localized drug action, or modulation of drug release in response to specific physiological cues. PD studies of NDDS provide insight into their mechanism of action, therapeutic efficacy, and potential side effects [11].

PK/PD studies of NDDS provide valuable insights into the behavior of drugs and nanoparticles in the body, their interactions with biological systems, and their therapeutic effects. Understanding PK/PD of nanoformulations helps in [30]:

(a) **Optimizing drug delivery:** PK studies help determine the ADME of nanoformulations. This information guides the designing of nanoparticles to enhance drug delivery. For example, understanding the drug's PK profile can help modify the nanoparticles to improve drug release kinetics, bioavailability, and tissue-specific targeting.

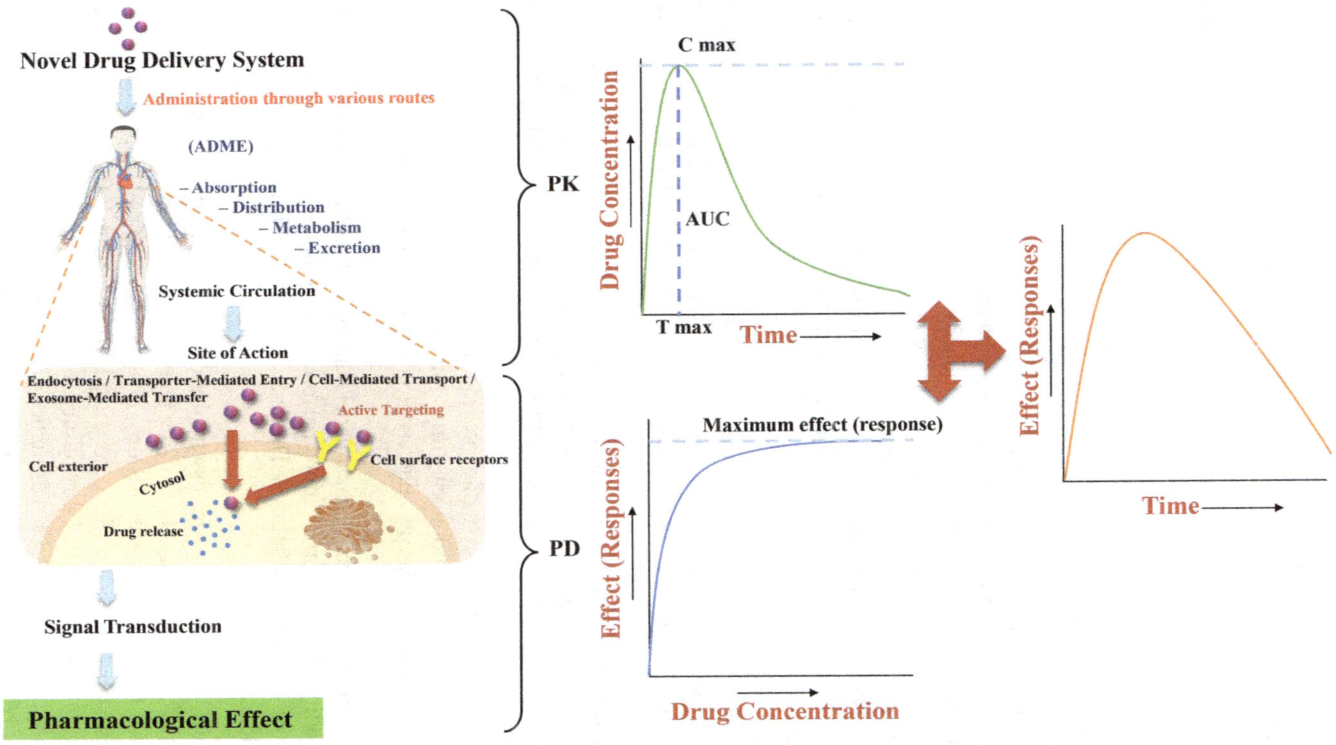

Fig. 1.2 Pharmacokinetics (PK) and pharmacokinetics (PD) of nanoparticles; (*AUC* area under the curve, *Cmax* maximum drug concentration, *Tmax* time taken by the drug to reach maximum concentration)

(b) **Dosing regimen:** PD studies elucidates the relationship between drug concentration and therapeutic effect. This knowledge helps to establish the appropriate dosing regimen of nanoformulations, ensuring that the drug concentration remains within the therapeutic window while minimizing side effects.
(c) **Minimizing toxicity:** PK studies identify potential toxicity associated with nanoformulations, such as accumulation of nanoparticles in specific tissues. Understanding the distribution of nanoparticles in the body and their clearance rates helps design safer formulations.
(d) **Predicting clinical efficacy:** PD studies help establish a dose-response relationship, enabling researchers to predict the clinical efficacy of nanoformulations. This information is crucial for designing effective clinical trials and determining appropriate endpoints.
(e) **Patient stratification:** PK/PD data helps identify patient populations that are most likely to benefit from nanoformulations. This personalized medicine approach improves the chances of success in clinical trials and enhances the value proposition of the nanoformulation.
(f) **Biomarker development:** PK/PD studies aids in the identification of biomarkers that correlate with therapeutic response or toxicity. These biomarkers serve as valuable tools for patient selection, monitoring treatment efficacy, and predicting treatment outcomes.
(g) **Regulatory approval:** Regulatory agencies require comprehensive PK/PD data as part of the drug approval process. A thorough understanding of the behavior of the drug in the body can facilitate and streamline the approval process for nanoformulations.
(h) **Dose optimization:** PK/PD modeling and simulations helps optimize dosing regimens, reducing the risk of underdosing or overdosing patients. This is particularly important for nanoformulations, where drug release and distribution is complex.
(i) **Long-term safety:** PK studies provide insights into the long-term safety of nanoformulations by monitoring drug and nanoparticle persistence in the body. This information helps in addressing concerns about potential accumulative toxicity.
(j) **Translational success:** A robust understanding of PK/PD can improve the predictability of a nanoformulations performance in humans, reducing the likelihood of translational failures. This can save time and resources in the drug development process.

In summary, PK/PD studies are essential tools for characterizing the behavior of nanoformulations *in-vivo* and optimizing their design for clinical use. They provide crucial data for regulatory submissions, dose selection, patient stratification, and overall translational success. Incorporating PK/PD insights into the development process can enhance the probability of successful translation and commercialization of nanoformulations. While PK/PD studies offer numerous advantages in the development of NDDS, they also introduce unique challenges. Some challenges specific to the PK/PD aspects of NDDS are as follows [31–34]:

(a) **Complex pharmacokinetics:** The complex nature of NDDS leads to more complex PK profiles. Factors such as drug release rates, particle size, surface characteristics, and carrier interactions influence drug absorption, distribution, and elimination.
(b) **Variability in drug release:** NDDS often exhibit controlled or sustained drug release profiles, leading to variability in drug exposure. This variability requires careful consideration and adjustments to ensure consistent therapeutic effects.
(c) **Non-linear pharmacokinetics:** Some NDDS exhibit non-linear PK, where changes in dose do not lead to proportional changes in drug concentration. Understanding and predicting these non-linear relationships can be challenging.
(d) **Tissue distribution and accumulation:** Targeted NDDS alters the tissue distribution pattern of drugs. Achieving desired drug accumulation at target sites while minimizing off-target accumulation can be complex and depends on factors like particle size, surface modification, and targeting ligands.
(e) **Stability and degradation:** NDDS have significant impact on the drug stability and its susceptibility towards degradation. Factors like pH, temperature, and interaction with carrier materials can impact drug stability and subsequently affect PK.
(f) **Interactions with biological carriers:** NDDS often interact with biological barriers, such as cell membranes and the blood-brain barrier and these interactions influence the ADME of encapsulated drugs. Biological responses to NDDS are further more complex and multifaceted, influenced by factors beyond drug release, such as carrier-cell interactions and immune responses.
(g) **Biocompatibility and immunogenicity:** Some NDDS triggers immune responses or adverse reactions, affecting both PK and PD. Evaluating the biocompatibility and potential immunogenicity of these materials is important.
(h) **Dosing strategies:** Determining appropriate dosing strategies for NDDS is challenging. Balancing the release profile with therapeutic efficacy and safety considerations requires careful optimization.
(i) **PK/PD modeling:** Developing accurate PK/PD models for NDDS is complex due to the numerous variables involved. Specialized modeling approaches are often needed to accurately predict the drug behavior.
(j) **Targeting and non-targeting effects**: NDDS offer targeted delivery, however unintended interactions with off-target tissues can occur, affecting both PK and PD responses.
(k) **Patient variability:** The impact of NDDS on PK/PD can vary among individuals due to genetic, physiological, and disease-related factors. Individualized treatment approaches may be needed.
(l) **In-vivo validation:** NDDS require in-vivo validation to assess PK/PD behavior accurately. Designing and conducting these studies can be resource-intensive.
(m) **Long-term effects:** Understanding the long-term PK/PD effects of NDDS, especially in chronic diseases, requires extensive research to ensure safety and sustained therapeutic benefits.

Addressing these challenges requires interdisciplinary collaboration among formulation scientists, pharmacologists, material scientists, clinicians, and regulatory experts. Combining in vitro studies, preclinical assessment, advanced modeling techniques, and clinical trials is essential to comprehensively understand and optimize the PK/PD behavior of NDDS, ensuring their safe and effective translation from the lab to clinical practice.

1.4 Conclusion

The advent of novel drug delivery systems has revolutionized the pharmaceutical industry by offering improved therapeutic outcomes and patient comfort. As NDDS continues to evolve, they hold promise of transforming the future of medicine, making treatments more effective, accessible to individual health conditions. The study of PK/PD of nanoparticles represents a dynamic and rapidly evolving field with immense potential for improving drug delivery and patient outcomes. The subsequent chapters delve deeper into PK/PD of specific delivery systems, and their applications.

Bibliography

1. Gharat SA, Momin MM, Bhavsar C (2016) Oral squamous cell carcinoma: current treatment strategies and nanotechnology-based approaches for prevention and therapy. Crit Rev Ther Drug Carrier Syst 33(4)
2. Harrington KJ, Lewanski C, Northcote AD, Whittaker J, Peters AM, Vile RG, Stewart JS (2001) Phase II study of pegylated liposomal doxorubicin (Caelyx™) as induction chemotherapy for patients with squamous cell cancer of the head and neck. Eur J Cancer 37(16):2015–2022
3. Ezike TC, Okpala US, Onoja UL, Nwike PC, Ezeako EC, Okpara JO, Okoroafor CC, Eze SC, Kalu OL, Odoh EC, Nwadike U (2023) Advances in drug delivery systems, challenges and future directions. Heliyon
4. Hu G, Guo M, Xu J, Wu F, Fan J, Huang Q, Yang G, Lv Z, Wang X, Jin Y (2019) Nanoparticles targeting macrophages as potential clinical therapeutic agents against cancer and inflammation. Front Immunol 10:1998
5. Suk JS, Xu Q, Kim N, Hanes J, Ensign LM (2016) PEGylation as a strategy for improving nanoparticle-based drug and gene delivery. Adv Drug Deliv Rev (1):28–51
6. Shi L, Zhang J, Zhao M, Tang S, Cheng X, Zhang W, Li W, Liu X, Peng H, Wang Q (2021) Effects of polyethylene glycol on the surface of nanoparticles for targeted drug delivery. Nanoscale 13(24):10748–10764
7. Shen Z, Fisher A, Liu WK, Li Y (2018) PEGylated "stealth" nanoparticles and liposomes. In: Engineering of Biomaterials for Drug Delivery Systems. Woodhead Publishing, pp 1–26
8. Yao Y, Zhou Y, Liu L, Xu Y, Chen Q, Wang Y, Wu S, Deng Y, Zhang J, Shao A (2020) Nanoparticle-based drug delivery in cancer therapy and its role in overcoming drug resistance. Front Mol Biosci 7:193
9. Gharat S, Basudkar V, Momin M, Prabhu A (2023) Mucoadhesive oro-gel–containing chitosan lipidic nanoparticles for the management of oral squamous cell carcinoma. J Pharm Innov 20:1–8

10. Gao W, Thamphiwatana S, Angsantikul P, Zhang L (2014) Nanoparticle approaches against bacterial infections. Wiley Interdiscipl Rev Nanomed Nanobiotechnol 6(6):532–547
11. Liu E, Zhang M, Huang Y (2016) Pharmacokinetics and pharmacodynamics (PK/PD) of bionanomaterials. Biomed Nanomater 30:1–60
12. https://www.researchandmarkets.com/report/novel-drug-delivery-system#:~:text=Global%20Novel%20Drug%20Delivery%20Systems,the%20analysis%20period%202022%2D2030
13. https://www.fda.gov/media/140395/download
14. Weissig V, Elbayoumi T, Flühmann B, Barton A (2021) The growing field of nanomedicine and its relevance to pharmacy curricula. Am J Pharm Educ 85(8)
15. Carneiro DC, Sousa JD, Monteiro-Cunha JP (2021) The COVID-19 vaccine development: a pandemic paradigm. Virus Res 301:198454
16. Anselmo AC, Mitragotri S (2019) Nanoparticles in the clinic: an update. Bioeng Transl Med 4(3):e10143
17. https://www.fda.gov/drugs
18. https://www.ema.europa.eu/en
19. Hua S, De Matos MB, Metselaar JM, Storm G (2018) Current trends and challenges in the clinical translation of nanoparticulate nanomedicines: pathways for translational development and commercialization. Front Pharmacol 9:790
20. https://www.fda.gov/about-fda/fda-organization/center-drug-evaluation-and-research-cder
21. https://www.fda.gov/about-fda/center-drug-evaluation-and-research-cder/office-testing-and-research
22. Tyner KM, Zheng N, Choi S, Xu X, Zou P, Jiang W, Guo C, Cruz CN (2017) How has CDER prepared for the nano revolution? A review of risk assessment, regulatory research, and guidance activities. AAPS J 19:1071–1083
23. https://www.fda.gov/regulatory-information/search-fda-guidance-documents/guidance-industry-assessing-effects-significant-manufacturing-process-changes-including-emerging
24. https://www.fda.gov/regulatory-information/search-fda-guidance-documents/considering-whether-fda-regulated-product-involves-application-nanotechnology
25. https://www.fda.gov/media/70837/download
26. https://www.fda.gov/regulatory-information/search-fda-guidance-documents/guidance-industry-safety-nanomaterials-cosmetic-products
27. https://www.fda.gov/media/88828/download
28. https://www.fda.gov/regulatory-information/search-fda-guidance-documents/drug-products-including-biological-products-contain-nanomaterials-guidance-industry
29. European Medicines Agency's workshop on nanomedicines (2009). https://www.ema.europa.eu/en/events/european-medicines-agencys-workshop-nanomedicines
30. Zou H, Banerjee P, Leung SS, Yan X (2020) Application of pharmacokinetic-pharmacodynamic modeling in drug delivery: development and challenges. Front Pharmacol 11:997
31. Glassman PM, Muzykantov VR (2019) Pharmacokinetic and pharmacodynamic properties of drug delivery systems. J Pharmacol Exp Ther 370(3):570–580
32. Hafeez MN, Celia C, Petrikaite V (2021) Challenges towards targeted drug delivery in cancer nanomedicines. Processes 9(9):1527
33. Workman P (2002) Challenges of PK/PD measurements in modern drug development. Eur J Cancer 38(16):2189–2193
34. Desai N (2012) Challenges in development of nanoparticle-based therapeutics. AAPS J 14(2):282–295

Chapter 2
Absorption, Distribution, Metabolism and Excretion of Novel Drug Delivery Systems

Sankalp A. Gharat, Munira M. Momin, and Tabassum Khan

Introduction by the Editor

Nanotechnology has revolutionized various fields, from drug delivery to diagnosis, by enabling the designing and engineering of drug carriers at the nanoscale level. Nanoparticles, possess unique properties that make them highly valuable in numerous applications. However, understanding how these nanoparticles are absorbed, distributed, metabolized, and excreted within biological systems is crucial for ensuring their safe utilization [1]. Once inside the body, nanoparticles get distributed across different tissues and organs, facilitated by their unique physicochemical characteristics as depicted in Fig. 2.1. It is therefore important to understand the factors that affect the absorption, distribution, metabolism and excretion (ADME) of nanoparticles. The factors affecting ADME of nanoparticles are as follows [2–4]:

(a) Physicochemical properties of nanoparticles (size, shape, surface area to volume ratio, surface charge, surface coating, surface chemistry, surface topology, crystallinity, functionalization, PEGylation, ligand conjugation, composition of nanoparticles, density, porosity, solubility of nanoparticles, stability and aggregation potential of nanoparticles)
(b) Route of administration of nanoparticles and type of delivery system

S. A. Gharat · M. M. Momin (✉)
Department of Pharmaceutics, SVKM's Dr. Bhanuben Nanavati College of Pharmacy, Mumbai, Maharashtra, India
e-mail: sankalp.gharat@bncp.ac.in; munira.momin@bncp.ac.in

T. Khan
Department of Pharmaceutical Chemistry and Quality Assurance, SVKM's Dr. Bhanuben Nanavati College of Pharmacy, Mumbai, Maharashtra, India
e-mail: tabassum.khan@bncp.ac.in

S. A. Gharat et al. (eds.), *Pharmacokinetics and Pharmacodynamics of Novel Drug Delivery Systems: From Basic Concepts to Applications*,
https://doi.org/10.1007/978-981-99-7858-8_2

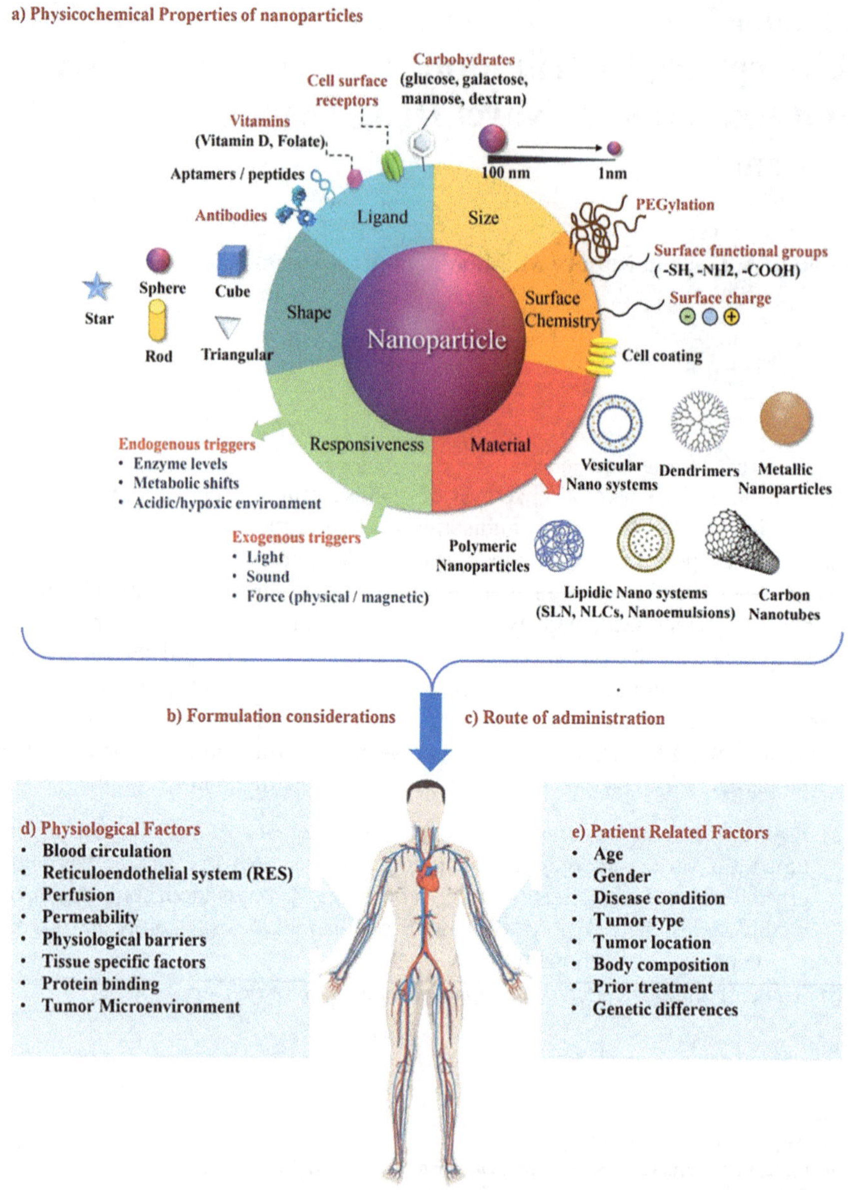

Fig. 2.1 Factors affecting absorption, distribution, metabolism, excretion of nanoparticles

(c) Physiological factors (Blood circulation, Reticuloendothelial system (RES) interaction, perfusion, permeability, physiological barriers, tissue specific factors, protein binding)
(d) Patient related factors (age, gender, disease type, tumor type and location, body composition and prior treatments)

The first step in the ADME process involves the absorption of nanoparticles into the biological system. The absorption mechanisms largely depend on the physicochemical properties of the nanoparticles and the target tissue. For instance, nanoparticles can be absorbed through various routes such as ingestion, inhalation, dermal, and injection. In the case of oral ingestion, nanoparticles pass through the gastrointestinal tract, wherein factors like size, surface charge, and surface chemistry influence their uptake. Similarly, inhalation exposes the respiratory system to airborne nanoparticles, raising concerns about their potential effects on lung health. The extent of nanoparticle absorption is further influenced by factors like mucosal barriers, particle stability, and the presence of specific transporters [5].

Once absorbed, nanoparticles are distributed throughout the body via the circulatory system. Their nano size allows them to travel through capillaries, enabling their access to various organs and tissues. The distribution of nanoparticles is influenced by blood flow, particle size, and surface characteristics. For example, nanoparticles with specific surface modifications can be designed to target certain tissues or cells, enhancing their accumulation in those areas [6]. This targeted distribution is particularly advantageous in medical applications, such as cancer treatment, where nanoparticles are able to deliver drugs directly to tumor cells, minimizing side effects on healthy tissues [7].

Once inside the cells or in contact with intracellular compartments, nanoparticles undergo various metabolic processes. The extent and nature of these processes depend on the composition and type of nanoparticles. For example, lipid-based nanoparticles are metabolized by cellular lipid pathways, while polymeric nanoparticles undergo enzymatic degradation [8, 9]. Metabolism involves the transformation of nanoparticles within the body, often through enzymatic processes. The metabolism significantly affects biological fate and potential toxicity of nanoparticles. Liver and spleen are the key organs involved in nanoparticle metabolism and the liver enzymes can modify nanoparticles, altering their physicochemical properties and influencing their clearance from the body [9]. Metabolism can also influence the interaction between nanoparticles and endogenous biological molecules, impacting their biological effects. However, the complexity of nanoparticle metabolism and the potential formation of metabolites with different properties raise challenges in predicting their behaviour within the body accurately [10].

Excretion is the final step in the ADME process, where nanoparticles and their metabolites are eliminated from the body. The primary routes of excretion include urine and faeces, with the kidneys and liver playing crucial roles. Renal excretion eliminates smaller nanoparticles, while hepatic excretion is more relevant for larger particles. Nanoparticles with hydrodynamic diameter <6 nm are freely filtered in the glomerulus, irrespectively of their molecular charge. However, filtration of nanoparticles between 6 and 8 nm is dependent on charge interactions between the nanoparticles and the negative charges of the glomerular basement membrane. Therefore, positively charged particles are more readily filtered than same-sized negatively charged particles. Due to size limitations, nanoparticles >8 nm do not undergo glomerular filtration, irrespective of its surface charge [11]. During hepatic clearance, the nanoparticles of hydrodynamic diameter <100 nm get endocytosed by hepatocytes [12]. The efficiency of excretion depends on factors like particle size, shape, surface properties, and potential aggregation [13]. In some cases, nanoparticles might be excreted more slowly, leading to long-term accumulation in organs and potential adverse effects [14].

Understanding the ADME of nanoparticles is essential for assessing their safety and optimizing their applications. However, several challenges exist like, the diverse physicochemical properties of nanoparticles make it difficult to establish generalized rules for their behaviour within biological systems. Additionally, the interactions between nanoparticles and biological components can be complex and unpredictable, leading to unexpected outcomes. To address these challenges, researchers employ advanced techniques such as in vivo imaging, cellular models, and computational simulations to gain insights into nanoparticle behaviour [15]. Regulatory bodies also play a vital role in ensuring the safe development and use of nanoparticle-based products by establishing guidelines and standards for testing and evaluation.

In conclusion, the absorption, distribution, metabolism, and excretion of nanoparticles within biological systems are intricate processes that depend on various factors, including nanoparticle properties and target tissues. As nanotechnology continues to advance, a comprehensive understanding of these processes becomes increasingly crucial. In-depth assessment of nanoparticle behaviour within the body will facilitate the development of safe and effective applications in medicine, electronics, and environmental remediation. It is imperative that researchers, regulators, and stakeholders collaborate to ensure the rational development of nanoparticles to maximize their benefits and minimize the potential risks. Understanding the ADME process of nanoparticles is crucial to assess their potential health risks and design safer nanomaterials for diverse applications. This chapter delves into the intricate processes of absorption, distribution, metabolism, and excretion (ADME) of nanoparticles.

2.1 Physicochemical Properties Affecting ADME of Novel Drug Delivery Systems

Introduction by the Editor

The physicochemical properties of nanoparticles play a pivotal role in influencing the absorption, distribution, metabolism, and excretion (ADME) of novel drug delivery systems. These properties determine the interaction of nanoparticles with biological systems, impacting their therapeutic efficacy and safety profiles. Nanoparticles in the range of 1–100 nm exploit the enhanced permeability and retention (EPR) effect, enabling passive accumulation in tumor tissues due to their leaky blood vessels. Smaller nanoparticles of diameter less than 10 nm avoid immune clearance and penetrate tissues more effectively, while larger ones with diameters of 150–200 nm face greater uptake by macrophages, impacting biodistribution. Surface charges on the nanoparticles largely affect their interaction with cells and proteins. Positively charged nanoparticles enhance cellular uptake due to electrostatic interaction, while negatively charged ones avoid recognition by macrophages, prolonging circulation time. Charge also influences protein corona formation, which can impact the in vivo behaviour of nanoparticles. Surface hydrophilicity and hydrophobicity affects interaction with biological fluids and cells. Hydrophilic coatings, such as polyethylene glycol (PEG), can reduce opsonization and extend circulation, improving biodistribution. Hydrophobic surfaces may promote the adsorption of hydrophobic drugs, affecting release kinetics. Surface functionalization with ligands or targeting moieties influences nanoparticle-cell interactions. Specific ligands can enhance cellular uptake by binding to receptors overexpressed on target cells, aiding in targeted drug delivery and reducing off-target effects. Furthermore, the composition and type of nanosystem have a significant impact on the ADME of nanoparticles. Biodegradable polymers, such as poly (lactic-co-glycolic acid) (PLGA), influence drug release kinetics and degradation rates, affecting both efficacy and toxicity. Inorganic nanoparticles like gold and iron oxide serve as contrast agents for imaging and provide localised hyperthermia for cancer treatment. Liposomal preparations undergo enzymatic degradation, releasing encapsulated drugs for therapeutic action. Surface modifications, like PEGylation, extend the circulation time by reducing recognition. The metabolism of liposomes involves their recognition by the reticuloendothelial system, primarily in the liver and spleen, leading to uptake and subsequent breakdown. In conclusion, the intricate interaction of nanoparticle physicochemical properties impacts their ADME profile. Designing nanoparticles of desired physicochemical properties can optimise their interaction with biological system, enhancing their ability to reach target sites, control drug release, and minimise systemic side effects, ultimately revolutionising the field of drug delivery leading to superior therapeutic outcomes.

Machine generated summaries

Disclaimer: The summaries in this chapter were generated from Springer Nature publications using extractive AI auto-summarization: An extraction-based summarizer aims to identify the most important sentences of a text using an algorithm and uses those original sentences to create the auto-summary (unlike generative AI). As the constituted sentences are machine selected, they may not fully reflect the body of the work, so we strongly advise that the original content is read and cited. The auto generated summaries were curated by the editor to meet Springer Nature publication standards.

To cite this content, please refer to the original papers.

Machine generated keywords: tumor, adult, peg, solid tumor, patient advanced, pegylation, advanced, age, nasal, cancer, nanoparticle, mucosa, normal, old, kinase, charge, pegylation, efficient, micelle, peptide, iontophoretic, loading, nasal, delivery, interest, tissue, iontophoresis, flux, preparation, nanoparticle

Effect of Charge and Molecular Weight on Transdermal Peptide Delivery by Iontophoresis [16] This is a machine-generated summary of:

Abla, Nada; Naik, Aarti; Guy, Richard H.; Kalia, Yogeshvar N.: Effect of Charge and Molecular Weight on Transdermal Peptide Delivery by Iontophoresis [16]

Published in: Pharmaceutical Research (2005)
Link to original: https://doi.org/10.1007/s11095-005-8110-2

If you want to cite the papers, please refer to the original.

For technical reasons we could not place the page where the original quote is coming from.

Abstract-Summary "The study was conducted to investigate the impact of charge and molecular weight (MW) on the iontophoretic delivery of a series of dipeptides."

"Increasing MW was compensated by additional charge; for example, Lys (MW = 147 Da, +1) and H-Lys-Lys-OH (MW = 275 Da, +2) had equivalent steady-state fluxes of 225 ± 48 and 218 ± 40 nmol cm^{-2} h^{-1}, respectively."

"For peptides with similar MW, e.g., H-Tyr-d-Arg-OH (MW = 337 Da, +1) and H-Tyr-d-Arg-NH_2 (MW = 336 Da, +2), the higher valence ion displayed greater flux (150 ± 26 vs. 237 ± 35 nmol cm^{-2} h^{-1})."

"The iontophoretic flux of zwitterionic dipeptides was less than that of acetaminophen and dependent on pH. For the series of dipeptides studied, flux is linearly correlated to the charge/MW ratio."

Introduction

"Peptides, which are usually charged at physiological pH, are poor candidates for passive transdermal delivery, but are very well suited for iontophoresis."

"As iontophoresis also enables tight control over the drug administration kinetics, it is especially useful for the delivery of peptides that exhibit input rate-dependent pharmacological activity [17, 18]."

"The work presented in this paper represents a systematic study of the factors governing transdermal iontophoretic peptide delivery, in particular, the dependence of iontophoretic transport on peptide charge and molecular weight (MW)."

"The main objectives were: (1) to determine whether increasing peptide charge could compensate increasing MW, (2) to evaluate the impact of increasing charge in a peptide of given MW, and (3) to investigate zwitterion iontophoresis and the effect of partial charge on transport."

"The effect of metabolism on peptide transport, particularly with respect to the role of peptide structure in determining hydrolytic susceptibility, and the site of enzymatic degradation in the skin, was also investigated."

Materials and Methods

"The buffers used were either (1) 30 mM $NaHCO_3$/133 mM NaCl, pH 7.4 (donor), 150 mM $NaHCO_3$/133 mM NaCl, pH 7.4 (receiver) for lysine, H-Lys-Lys-OH, H-His-Lys-OH, and H-Glu-ε-Lys-OH; or (2) phosphate-buffered saline (PBS), pH 7.4 (16.8 mM Na_2HPO_4/1.4 mM KH_2PO_4/136.9 mM NaCl) in all compartments, used with H-His-Lys-OH, H-Glu-ε-Lys-OH and tyrosine-containing dipeptides; or (3) citrate buffer (10 mM citric acid/133 mM NaCl, pH 4.5) for certain experiments with H-Tyr-His-OH and H-Tyr-Gln-OH."

"Although the peptide bond in the lysine-containing dipeptides, H-Lys-Lys-OH and H-His-Lys-OH, has a UV absorbance at ~210 nm, cleavage of this linkage during iontophoresis (see below), meant that transport had to be estimated from the free lysine concentrations in the receiver compartment."

Results and Discussion

"In a third iontophoretic study, free lysine and H-His-Lys-OH (10 mM in 30 mM $NaHCO_3$/133 mM NaCl, pH 7.4) were separately iontophoresed across porcine skin for 4 h. In both cases, only free lysine was detected in the receiver compartment at the end of the experiment; however, the receiver levels of lysine subsequent to dipeptide iontophoresis were threefold lower than those after lysine iontophoresis."

"The transport of two zwitterionic dipeptides (H-Tyr-Gln-OH and H-Glu-ε-Lys-OH), which contain both positive and negative charge centers at physiological pH, was also investigated."

"The total flux of H-Tyr-Gln-OH remained lower than that of acetaminophen because the dipeptide was now partially positively charged (formulation pH < pI of H-Tyr-Gln-OH) and not conducive to cathodal iontophoresis."

Conclusions

"At a given molecular weight, increasing peptide charge enhanced EM flux."

"The probability of interaction with the transport pathway will also increase with increasing molecular weight (and hence molecular volume) and this would obviously hinder delivery."

"Published investigations on the transport of higher molecular weight, multiply charged species are rare, although the delivery of cytochrome c (MW = 12.4 kDa, multiple charges) across excised human skin has been reported [19, 20]."

"The transport data for the histidine-containing dipeptides indicate the influence of neighboring amino acids and the local microenvironment on the acidity of the imidazole group; this may have implications for the delivery of peptides containing charged residues for which selection of the formulation pH might be critical to achieve the desired ionization state of the molecule."

Smart pH-sensitive nanoassemblies with cleavable PEGylation for tumor targeted drug delivery [21] This is a machine-generated summary of:

Zhao, Guanren; Long, Ling; Zhang, Lina; Peng, Mingli; Cui, Ting; Wen, Xiaoxun; Zhou, Xing; Sun, Lijun; Che, Ling: Smart pH-sensitive nanoassemblies with cleavable PEGylation for tumor targeted drug delivery [21]

Published in: Scientific Reports (2017)

Link to original: https://doi.org/10.1038/s41598-017-03111-2

If you want to cite the papers, please refer to the original.

For technical reasons we could not place the page where the original quote is coming from.

Abstract-Summary "DTX nanoassemblies driven by PEG-s-PEI thus formulated exhibited an excellent pH-sensitivity PEGylation cleavage performance at extracellular pH of tumor microenvironment, compared to normal tissues, thereby long circulated in blood but were highly phagocytosed by tumor cells."

Introduction

"Sensitively cleavable PEGylation of positively charged NPs, with PEG chains stretching in circulation and resident tissues, but cleaved in tumor microenvironment, may exhibit long blood circulation time with efficient phagocytosis by tumor cells [22–26]."

"PH sensitive cleavage of PEGylation was also succeed applied in gene delivery and liposome coating for long blood circulation time and efficient phagocytosis by tumor cells [27–29]."

"We discovered a one-pot and high efficient fabrication of polymer nanotherapeutics based on commercially available homopolymers (such as polyethyleneimine (PEI)) and small molecule drugs through multiple interactions mediated self-assembly, with high drug loading capacity and desirable therapeutic benefits, which showed great potential and advantages in clinical transformation as efficient oral nanocontainers for other hydrophobic drugs [30, 31]."

"Nanoparticles thus produced, with facile material synthesis, high drug loading capacity, desirable therapeutic benefits, low toxicity for intravenous application and pH-triggered deshielding of PEG, can serve as efficient and tumor environment targeting nanocontainers for anti-cancer drugs, and conducive to clinical transformation of PEGylation cleavable nanotherapeutics."

Results

"Based on computational results, self-assemblies of DTX, IND and PEG-s-PEI with various DTX/IND/PEG-s-PEIs weight ratios were conducted by dialysis against deionized water using methanol as the common solvent [32]."

"At pH 5.5 and 6.5, DTX/IND/PEG-s-PEI NAs formed significantly surface potential shifting in 120 min, while stably retained low surface charge for the whole 2 h time period at pH 7.4."

"Evaluation on in vitro antitumor activities by MTT assay showed the superior efficacy of assembled DTX/IND/PEI and DTX/IND/PEG-s-PEI NAs over raw DTX against various cancer cells including B16F10 and HepG2 in acidy and normal mediums."

"The cytotoxicity of raw DTX, DTX/IND/PEG-b-PEI and DTX/IND/PEI were significantly decreased in acidy medium, but the cytotoxicity of DTX/IND/PEG-s-PEI NAs was not effected [33]."

"Treatment with DTX/IND/PEG-s-PEI NAs dramatically inhibited the tumor volume."

Discussion

"We hypothesize that PEGylation cleavable nanotherapeutics with high drug loading capacity, long blood circulation time and efficient phagocytosis by tumor cells, can be easily constructed by combining cleavable PEGylation with this facile, convenient, cost-effective and easily scalable one-pot self-assembly strategy discovered by our previous studies."

"The anti-tumor activities of DTX/IND/PEG-s-PEI-2 NAs at low DTX concentrations in acidy medium was inhibited with no improvement than normal medium, even though the DTX accumulation in tumor cells was significantly increased."

"The anti-tumor activities of DTX/IND/PEG-s-PEI-2 NAs at high concentrations were higher in acidy medium than normal, and significant difference between IC_{50} at pH 6.5 and 7.4 was still observed for DTX/IND/PEG-s-PEI-2 NAs, which was contributed by the more DTX accumulated in tumor cells that offset the impact of acidic environment on chemo-sensitivity."

"A pH-sensitive PEGylation cleavable PEGylated PEI (PEG-s-PEI) was successfully synthesized, and DTX was highly efficiently packaged into NAs driven by multiple noncovalent interactions-mediated host-guest assembly of IND and PEG-s-PEI."

Methods

"Cells were incubated at 37 °C with 5% CO_2 for 24 h. Then the culture medium was replaced by 500 μL of fresh medium (pH 6.5 or pH 7.4) containing DTX/IND/PEG-s-PEI NAs (fabricated based on a formulation with DTX/IND/PEG-s-PEI weight ratio of 10:10:10), respectively."

"DTX/IND/PEG-s-PEI NAs, DTX/IND/PEG-b-PEI NAs and pristine DTX (solubilized in the aqueous solution containing Cremophor EL and ethanol) were intravenous injected (i.v.) at a dose of 10.0 mg/kg, respectively."

"Cells were then treated with the medium at various pH values (5.5, 6.5 or 7.4) containing DTX/IND/PEG-s-PEI, DTX/IND/PEG-b-PEI NAs, or pristine DTX (solubilized in the aqueous solution containing Cremophor EL and ethanol) at different doses for 24 h. The cell viability was quantified by MTT method."

New PTX-HS15/T80 Mixed Micelles: Cytotoxicity, Pharmacokinetics and Tissue Distribution [34] This is a machine-generated summary of:

Liu, Baoyu; Gao, Wei; Wu, Hui; Liu, Hong; Pan, Hongchun: New PTX-HS15/T80 Mixed Micelles: Cytotoxicity, Pharmacokinetics and Tissue Distribution [34]

Published in: AAPS PharmSciTech (2021)
Link to original: https://doi.org/10.1208/s12249-021-01929-8

If you want to cite the papers, please refer to the original.

For technical reasons we could not place the page where the original quote is coming from.

Abstract-Summary "Compared with single micelle, the new PTX-HS15/T80 mixed micelle system (PTX-HS15/T80 MMs) had achieved better results in solubilization, stability, and sensitization before."

"In vitro cytotoxicity test showed that the new PTX-HS15/T80 MMs had a stronger ability to inhibit the proliferation of cancer cells."

"The results of in vivo tissue distribution showed that, compared with the single micelle Taxol®, the new PTX-HS15/T80 MMs had good distribution characteristics in the lung (AUC $_{\text{(lung 0–4 H)}}$ increased about 26%) and low concentration in the heart (AUC $_{\text{(Heart 0–4 H)}}$ decreased about 10%)."

"The above results suggested that the new PTX-HS15/T80 MMs may have a certain therapeutic potential against lung cancer and reduce the toxic and side effects."

INTRODUCTION

"Taxol injection (Taxol®) is a single micelle preparation composed of Cremophor EL, but in clinical use, patients need to be given pretreatment such as corticosteroids or antihistamines [35], because Cremophor EL used in the preparation can induce the activation of human serum complement, release histamine, and induce asymptomatic arrhythmia."

"These limitations prompted researchers to seek a Cremophor EL-free PTX formulation to facilitate the development of new formulations for PTX, such as Lipusu®, a liposome composed of phosphatidylcholine and phosphatidylglycerol; Abraxane®, an albumin binding paclitaxel nanoparticle preparation; and polymer micelle-Genexol-PM®."

"It is of great significance to develop a low toxicity, high efficiency, and economic paclitaxel preparation."

"On the basis of keeping the advantages of small size, strong solubilization ability, safety and economy, and simple preparation, the mixed micelle is also characterized by the dense structure, enhanced solubilization effect, and high utilization rate of carrier materials [36]."

"[37], we developed a simple, safe, efficient, and economic new PTX-HS15/T80 MMs preparation through the combination of surfactants."

MATERIALS AND METHODS

"HeLa cells were inoculated in 96-well plate, each well contained 4000 HeLa cells, and the medium was 200 μL. After 24 h of culture, the culture medium containing different concentrations of test drug (PTX-HS15/T80 MMs or Taxol® injection) was replaced, and each concentration was set with 6 multiple pores."

"In the experimental group, PTX-HS15/T80 MMs was injected intravenously (diluted to 1.5 mg/mL with water for injection before use)."

"PTX-HS15/T80 MMs was injected into tail vein of experimental group."

"Before test, the samples were sonicated for 5 min, dissolved with 200 μL methanol, then 5 min of ultrasound, and vortexed for 3 min; after being centrifuged at 12000 rpm × 10 min, 20 μL of supernatant was taken for HPLC analysis."

RESULTS

"The assay method offered limits of quantification (LOQ) of 0.05 μg/mL in plasma and 0.10 μg/mL in tissue homogenate samples; the limit of detection (LOD) of PTX in plasma was 20 ng."

"All of the above results indicated that the HPLC-UV method for the determination of PTX in plasma and tissue met the requirements of the methodology."

"According to the area under the concentrate time curve, the concentrations of PTX in various tissues at the indicated times were found in the order of liver > ovarian > kidney > spleen > heart > lungs, respectively."

"At 0.5 H, 1 H, and 2 H, the PTX contents of mixed micelles were higher than that of Taxol®, and $AUC_{(lung\ 0\text{–}4\ H)}$ increased by about 26%, suggesting that the mixed micelles can improve the distribution of drugs in the lung tissues."

DISCUSSION

"The results of in vitro cytotoxicity test showed that the proliferation effects of single micelles and mixed micelles were significantly different, which showed that PTX-HS15/T80 MMs can more effectively inhibit the proliferation of cancer cells."

"It is reported that paclitaxel has certain cardiac toxicity [38], but the mixed micelles reduced the drug content in heart tissue ($AUC_{(heart\ 0\text{–}4\ H)}$ was decreased by about 10%), which was conducive to improving the safety of the preparation."

"Paclitaxel is a first-line drug for the clinical treatment of lung cancer and ovarian cancer, and the mixed micelles showed good lung tissue distribution characteristics and safety advantages."

"The anaphylaxis test results of previous studies [37] showed that compared with the single micelle Taxol®, the new PTX-HS15/T80 MMs had greater security with no sensitization, indicating that it may provide a good research direction and value for clinical application."

CONCLUSION

"Based on the theory of synergistic solubilization, we developed a new PTX-HS15/T80 MMs by the collaboration of HS15 and T80, which improved the solubilization ability of paclitaxel."

"This study further demonstrated that PTX-HS15/T80 MMs performed better in anticancer cell proliferation and lung tissue distribution characteristic than Taxol® which was composed of single micelles."

"All these results indicated that PTX-HS15/T80 MMs were superior in terms of safety and efficacy."

Bigger or Smaller? Size and Loading Effects on Nanoparticle Uptake Efficiency in the Nasal Mucosa [39] This is a machine-generated summary of:

Albarki, Mohammed A.; Donovan, Maureen D.: Bigger or Smaller? Size and Loading Effects on Nanoparticle Uptake Efficiency in the Nasal Mucosa [39]

Published in: AAPS PharmSciTech (2020)

Link to original: https://doi.org/10.1208/s12249-020-01837-3

If you want to cite the papers, please refer to the original.

For technical reasons we could not place the page where the original quote is coming from.

Abstract-Summary "To better understand the size dependence of nasal epithelial uptake, PLGA nanoparticles (60 nm or 125 nm) loaded with Nile Red were prepared, and their uptake into excised sections of bovine nasal respiratory or olfactory mucosa was measured for 30 or 60 min."

"Nanoparticles were present both in the epithelial cells and in the submucosal tissues, and greater numbers of the 60-nm particles were present in the submucosa than the epithelium, while greater numbers of the 125-nm particles remained in the epithelial cell layer."

"The amount of Nile Red recovered from the mucosal tissues after exposure to 125-nm nanoparticles was at least 2-fold greater than from the 60-nm nanoparticles, however, due to the higher (~9-fold) loading capacity of the larger particles."

"Well-designed nanoparticles with diameters >100 nm show good uptake into the nasal epithelium and are capable of transfer to the submucosal tissues, near the location of significant populations of blood and lymphatic vessels."

INTRODUCTION

"While drug administration via the nasal mucosa has been shown to be an effective method for the administration of small-molecule drugs for both topical and systemic actions, little is known about the ability of the nasal tissues to transfer nanoparticles beyond the mucosal surface, and even less is understood about the characteristics of nanoparticles that would provide effective and efficient delivery."

"Owing to their small size, nanoparticles may provide improved targeting and transport through the nasal mucosa, and drug-loaded nanoparticles may enhance the delivery of drugs or vaccines via the intranasal route."

"Identifying the optimal nanoparticle characteristics, including size, that control nanoparticle uptake and transfer in the nasal mucosa will provide an improved understanding of nanoparticle trafficking and contribute to the design of new, effective particulate delivery systems."

METHODS

"At the end of the incubation, the receiver solution was collected, the exposed area of the tissues was rinsed with Nanopure® water, and the exposed tissue region was trimmed free from the remaining tissues and transferred to a 15-ml polypropylene centrifuge tube containing 2 ml trypsin-EDTA (0.25%) solution in order to separate the outer epithelial cell layer from the underlying submucosal tissue."

"After a 2-h incubation with trypsin-EDTA, the remaining submucosal tissue was transferred into a separate 15-ml polypropylene tube containing 1 ml of the organic solvent, Cellosolve® acetate, in order to disrupt and solubilize the submucosal tissues and to dissolve the entrapped nanoparticles to allow for quantification of the Nile Red content."

"1 ml of Cellosolve® acetate was added to the receiver solution which was removed and placed in another 15-ml polypropylene tube in order to quantify any nanoparticles translocated from the mucosal tissue into the receiver solution."

RESULTS

"Nanoparticles were detected in both the epithelial and submucosal layers, and a very limited number (< 1% of the number measured in the tissues) of the nanoparticles traversed the full thickness of the tissue and were transferred into the receiver medium."

"Studies of the uptake of the larger, 125-nm-diameter, nanoparticles across the nasal tissues revealed that these nanoparticles were also able to translocate into both the nasal respiratory and olfactory mucosae."

"The number of nanoparticles transferred to the submucosal layer (in both tissue types) was also significantly higher (5–7-fold, $p < 0.035$) for the 60-nm particles."

"Due to the higher mass carrying capacity of the 125 nm particles, however, the total amount of Nile Red quantified in the nasal tissues was greater following incubation with the 125 nm nanoparticles as compared to the 60 nm nanoparticles."

DISCUSSION

"We have seen similar behaviors in our studies, but the increase in Nile Red concentrations within the tissues over time, especially within the submucosal tissue region, indicates that the PLGA nanoparticles were being continuously taken up by the nasal mucosal tissues and trafficked both within the epithelial cell layer and in the lamina propria."

"A greater number of smaller (60 nm) nanoparticles were translocated into both the nasal respiratory and olfactory mucosa, especially in the submucosal tissue regions, compared with the larger, 125-nm nanoparticles."

"Despite the higher absolute number of 60-nm nanoparticles present in the nasal tissues, their smaller size significantly limits their total drug-payload carrying capacity."

"The results from these studies show that the higher Nile Red carrying capacity of the 125-nm particles, approximately 9 times the mass loaded in the 60-nm PLGA nanoparticles despite similar loading efficiencies, compensates for the lower number of 125-nm particles transferred into the tissues."

CONCLUSIONS

"Both 60-nm and 125-nm PLGA nanoparticles were able to be internalized into excised nasal mucosal tissues in as little as 30 min, but their total uptake represented less than 5% of the available nanoparticle load."

"The limited overall uptake (< 5 %) of the PLGA nanoparticles by the nasal mucosa suggests a potential limitation to the development of efficient nanoparticle delivery systems, yet nanoparticle systems may enable targeted delivery or may reduce the impact of other limitations to nasal administration, including mucociliary clearance and mucosal metabolism."

"Uptake was observed to be size dependent, where the smaller diameter nanoparticles were transferred in higher numbers compared with the larger nanoparticles."

"Since the larger, 125-nm particles carried a greater amount of encapsulated Nile Red compared with the smaller nanoparticles, the 125-nm particles provided a greater total tissue exposure to Nile Red despite the reduced absolute number of nanoparticles present in the tissues."

2.2 Effect of Surface Modification on ADME of Nanoparticles

Introduction by the Editor

Surface modification of nanoparticles significantly impacts their absorption, distribution, metabolism, and excretion (ADME) profiles. By engineering the outer layer of nanoparticles, researchers modify their interactions with biological systems, enhancing their therapeutic efficacy and safety. Surface modifications influence nanoparticle interaction with biological barriers, such as the blood-brain barrier and the gastrointestinal tract. By attaching specific ligands or targeting moieties, nanoparticles gain access to otherwise restricted sites. Ligand-mediated targeting enables nanoparticles to bind to specific cell receptors, enhancing cellular uptake and directing drug delivery to desired tissues, reducing off-target effects. Positively charged nanoparticles interact more effectively with negatively charged cell membrane, improving cellular internalization. Conversely, neutral or negatively charged surfaces extend the circulation time by reducing recognition by the immune system. Hydrophilic or hydrophobic surface modifications influence drug release kinetics. Hydrophilic coatings sustain drug release, prolonging therapeutic effect, while hydrophobic modification promotes rapid drug release for immediate action. Moreover, surface modifications impact nanoparticle stability, preventing aggregation and maintaining their integrity in biological fluids. Surface modification also influences nanoparticle recognition by the immune system. PEGylation, for instance, reduces opsonization and clearance by the reticuloendothelial system, extending circulation times and increasing the chance of reaching target sites. Furthermore, modifications alter the physicochemical properties of nanoparticles, affecting their interaction with drugs, encapsulated payloads, and the biological milieu. In conclusion, the surface modification of nanoparticles has profound effects on their ADME profile. These modifications allow researchers to fine-tune nanoparticles for site-specific targeting, controlled drug release, improved stability, and reduced immune recognition. Balancing the benefits and potential challenges, surface modification is key to designing nanoparticles that effectively navigate biological barriers, enhancing their therapeutic potential and minimising adverse effects.

Machine generated summaries

Disclaimer: The summaries in this chapter were generated from Springer Nature publications using extractive AI auto-summarization: An extraction-based summarizer aims to identify the most important sentences of a text using an algorithm and uses those original sentences to create the auto-summary (unlike generative AI). As the constituted sentences are machine selected, they may not fully reflect the body of the work, so we strongly advise that the original content is read and cited. The auto generated summaries were curated by the editor to meet Springer Nature publication standards.

To cite this content, please refer to the original papers.

Machine generated keywords: peg, nasal, surface, nanoparticle, pegylation, intranasal, mucosa, particle, uptake, permeation, polymeric, treat, circulation time, nanostructure, liver

Nose-to-brain drug delivery mediated by polymeric nanoparticles: influence of PEG surface coating [40] This is a machine-generated summary of:

de Oliveira Junior, Edilson Ribeiro; Santos, Lílian Cristina Rosa; Salomão, Mariana Arraes; Nascimento, Thais Leite; de Almeida Ribeiro Oliveira, Gerlon; Lião, Luciano Morais; Lima, Eliana Martins: Nose-to-brain drug delivery mediated by polymeric nanoparticles: influence of PEG surface coating [40]

Published in: Drug Delivery and Translational Research (2020)

Link to original: https://doi.org/10.1007/s13346-020-00816-2

If you want to cite the papers, please refer to the original.

For technical reasons we could not place the page where the original quote is coming from.

Abstract-Summary "In order to overcome rapid nasal mucociliary clearance, low epithelial permeation, and local enzymatic degradation, we investigated the influence of PEGylation on nose-to-brain delivery of polycaprolactone (PCL) nanoparticles (PCL-NPs) encapsulating bexarotene, a potential neuroprotective compound."

"Upon incubation with artificial nasal mucus, only 5 and 10% of PCL-PEG coating were able to ensure NP stability and homogeneity in mucus."

"Rapid mucus-penetrating ability was observed for 98.8% of PCL-$PEG_{5\%}$ NPs and for 99.5% of PCL-$PEG_{10\%}$ NPs."

"Fluorescence tomography images evidenced higher translocation into the brain for PCL-$PEG_{5\%}$ NPs."

"Bexarotene loaded into PCL-$PEG_{5\%}$ NPs resulted in area under the curve in the brain (AUC_{brain}) 3 and 2-fold higher than that for the drug dispersion and for non-PEGylated NPs ($p < 0.05$), indicating that approximately 4% of the dose was directly delivered to the brain."

"Combined, these results indicate that PEGylation of PCL-NPs with PCL-$PEG_{5\%}$ is able to reduce NP interactions with the mucus, leading to a more efficient drug delivery to the brain following intranasal administration."

Introduction

"The nose-to-brain drug delivery enables an alternative and direct access from the nasal cavity to the cerebral tissue, bypassing BBB restrictions [41, 42]."

"In order to overcome these issues, nose-to-brain delivery mediated by nanocarriers has demonstrated to be a successful strategy to aid drug transport to the CNS after intranasal administration [43–45]."

"Polymeric nanoparticles (NPs) may shield drugs from degradation on the nasal mucosa, expanding the residence time of the formulation at the absorption site [46]."

"NPs may facilitate cellular uptake and brain drug accumulation when administered nasally [47–49]."

"As high amounts of PEG were used for coating NP surfaces, an increase in NP adhesion to the nasal mucosa was observed, instead of the expected mucus penetration."

"We provide a proof-of-concept that the mucus-penetrating ability of PEGylated NPs enhances the translocation of NPs into the brain following intranasal administration."

Materials and methods

"The density of PEG coverage (Γ), expressed by the number of PEG chains/100 nm^2, was calculated by comparing the concentration of PEG (in percentage by mass, wt% PEG) in the NPs to the particles surface area (S) and volume (V), using PCL density (ρ) of 1.145 cm^{-3}, as described by Bertrand and co-workers [50]."

"Following overnight culture, the cells were incubated with Cou6-labeled NPs with different amounts of PCL-PEG (0, 5, or 10%) for 1 h at 37 °C."

"Animals were divided in three groups (n = 4) and received, via the nasal route, 20 μL of IR-780-labeled NPs with different PCL-PEG contents (0, 5, or 10%)."

"For brain homogenates, 20 μL IS solution was added to the samples and the drug was extracted with 1 mL ethyl acetate."

Results and discussion

"The PEG hydrophilic coverage was important to allow the particle penetration on nasal mucus; however, the lower hydrophilicity and negative zeta potential of PCL-$PEG_{5\%}$ NPs increased the particle accumulation on the brain tissue."

"Our data suggest that the use of 5% of PCL-PEG seems more appropriate to increase nose-to-brain delivery, since this amount of PEG coating on NP surface promotes mucus penetration without affecting internalization by epithelial cells and, potentially, by neurons, during nose-to-brain transport."

"DTI values were 2.48 ± 0.4 and 5.56 ± 0.3 for BEX PCL-NPs and BEX PEG-$PCL_{5\%}$ NPs, respectively, evidencing that the PEGylated formulation increased the brain availability of the drug."

"All these data demonstrate that PEGylation with 5% of PCL-PEG was effective in promoting NP permeation through the mucus, overcoming mucociliary clearance in the nasal cavity and, consequently, increasing the amount of drug in the brain."

Conclusions

"We demonstrated for the first time the importance of PEGylation, as a mucus penetration enhancer, on the transport of NPs via nose-to-brain."

"Comparing different rates of PEG coating, we identified that 5% of the PEGylated polymer (PEG-PCL) seems to be an optimal concentration for nose-to-brain delivery of PCL-NPs."

"This percentage of PEG coating did not reduce the uptake of NPs by nasal epithelial cells involved in nose-to-brain delivery."

"The data suggest that the use of 5% of PEG coating facilitates the transport of PCL-NPs from the nasal cavity to the CNS."

Surface modification of paclitaxel-loaded tri-block copolymer PLGA-b-PEG-b-PLGA nanoparticles with protamine for liver cancer therapy [51] This is a machine-generated summary of:

Gao, Nansha; Chen, Zhihong; Xiao, Xiaojun; Ruan, Changshun; Mei, Lin; Liu, Zhigang; Zeng, Xiaowei: Surface modification of paclitaxel-loaded tri-block copolymer PLGA-b-PEG-b-PLGA nanoparticles with protamine for liver cancer therapy [51]

Published in: Journal of Nanoparticle Research (2015)

Link to original: https://doi.org/10.1007/s11051-015-3121-3

If you want to cite the papers, please refer to the original.

For technical reasons we could not place the page where the original quote is coming from.

Abstract-Summary "In order to enhance the therapeutic effect of chemotherapy on liver cancer, a biodegradable formulation of protamine-modified paclitaxel-loaded poly(lactide-co-glycolide)-b-poly(ethylene glycol)-b-poly(lactide-co-glycolide) (PLGA-b-PEG-b-PLGA) nanoparticles (PTX-loaded/protamine NPs) was prepared."

"PTX-loaded and PTX-loaded/protamine NPs were characterized in terms of size, size distribution, zeta potential, surface morphology, drug encapsulation efficiency, and drug release."

"PTX-loaded/protamine NPs exhibited significantly higher cytotoxicity than PTX-loaded NPs and Taxol® did."

"All the results suggested that surface modification of PTX-loaded PLGA-b-PEG-b-PLGA NPs with protamine boosted the therapeutic efficacy on liver cancer."

Introduction

"In order to reduce the adverse effects and to increase the therapeutic efficacy of chemotherapeutic drugs, various drug delivery carriers, such as prodrug conjugates, micelles, microspheres, and nanoparticles (NPs), have been developed (Chen and others [52]; Geever and others [53]; Yanagihara and others [54]; Zeng and others [55])."

"PEG can limit the interaction between vehicles and target cells and thus inhibit effective cellular uptake of the NPs (Gao and others [56]; Romberg and others [57])."

"The surface properties of drug-loaded NPs have been modified to surmount these drawbacks and to enhance cellular uptake (Gomez and others [58])."

"It is of great significance to analyze the characteristics of protamine-modified NPs, including drug release behavior, cellular uptake, cytotoxicity, applications, etc (Zeng and others [59])."

"We prepared and evaluated protamine-coated PLGA-b-PEG-b-PLGA NPs as superb PTX nanocarriers for treating liver cancer."

"Drug loading capacity, EE, morphology, in vitro drug release, cellular uptake, and cytotoxicity of the NPs were detected, with commercial Taxol® used for comparison."

Materials and methods

"PTX-loaded PLGA-b-PEG-b-PLGA NPs were suspended in this solution at a concentration of 9.5 mg/ml by sonication at 25 W power output for 60 s. Due to the zeta potential of PTX-loaded PLGA-b-PEG-b-PLGA NPs was negative, protamine was easily conjugated onto the surface of NPs."

"The cells were incubated with 250 μg/mL coumarin 6-loaded NPs and coumarin 6-loaded/protamine NPs at 37 °C for 4 h, washed with cold PBS three times, fixed by methanol for 20 min, stained with 4′,6-diamidino-2-phenylindole dihydrochloride (DAPI, Fluka, Buche, Switzerland) for 10 min, and washed twice with PBS."

"The cells were incubated with the PTX-loaded and PTX-loaded/protamine NPs suspensions and Taxol® at equivalent drug concentrations ranging from 0.25 to 25 mg/ml for 24, 48, and 72 h. The drug-free protamine-modified NPs with the same amount of NPs were also used as comparison."

Results and discussion

"A similar drug release curve of PTX-loaded/protamine NPs was observed."

"All these results confirmed that protamine raised the cellular uptake efficiency of drug-loaded NPs, which was conducive to the therapeutic effects on cancer."

"PTX-loaded and PTX-loaded/protamine NPs were sterilized by gamma ray to eliminate any contamination effect, and then suspended to culture HepG2 cells."

"Besides, PLGA-b-PEG-b-PLGA and protamine were nontoxic in cell culture because drug-free NPs at various concentrations barely showed cytotoxicity."

"PTX-loaded/protamine NPs were more cytotoxic against HepG2 cells than PTX-loaded NPs and commercial Taxol®, probably because protamine augmented the cellular uptake efficiency of PTX-loaded NPs."

"The NPs and Taxol®, especially protamine-modified PTX-loaded NPs, reduced cell viability with increasing drug dose and culture time."

"PTX-loaded/protamine NPs were highly cytotoxic for HepG2 cells, and PLGA-b-PEG-b-PLGA which formulated drug nanocarriers was safe and nontoxic."

Conclusion

"Biodegradable nanocarrier PTX-loaded/protamine NPs were successfully synthesized to formulate PTX for liver cancer therapy."

"The size of PTX-loaded/protamine NPs was, as detected by DLS, about 180 nm."

"Surface modification of PTX-loaded PLGA-b-PEG-b-PLGA NPs with protamine is a promising treatment protocol for liver cancer."

Effect of PEG Surface Conformation on Anticancer Activity and Blood Circulation of Nanoemulsions Loaded with Tocotrienol-Rich Fraction of Palm Oil [60] This is a machine-generated summary of:

Alayoubi, Alaadin; Alqahtani, Saeed; Kaddoumi, Amal; Nazzal, Sami: Effect of PEG Surface Conformation on Anticancer Activity and Blood Circulation of Nanoemulsions Loaded with Tocotrienol-Rich Fraction of Palm Oil [60]

Published in: The AAPS Journal (2013)

Link to original: https://doi.org/10.1208/s12248-013-9525-z

If you want to cite the papers, please refer to the original.

For technical reasons we could not place the page where the original quote is coming from.

Abstract-Summary "The objective of this study was to investigate the effect of surface grafted polyethylene glycol (PEG) on the properties of the nanoemulsions."

"The effect of PEG surface topography on the antiproliferative activity of nanoemulsions against mammary adenocarcinoma cells, their susceptibility to protein adsorption, and its effect on blood hemolysis and circulation time was investigated."

"Nanoemulsions PEGylated with poloxamer or PEG_{2000}-DSPE were stable under physical stress."

"Poloxamer nanoemulsion, however, displayed higher uptake and potency against MCF-7 tumor cells in 2D and 3D culture and increased hemolytic effect and susceptibility to IgG adsorption, which was reflected in its rapid clearance and short circulation half-life (1.7 h)."

"PEGylation with PEG_{2000}-DSPE led to a 7-fold increase in mean residence time (12.3 h) after IV injection in rats."

"Differences between the nanoemulsions were attributed to polymer imbibitions and the differences in PEG conformation and density on the surface of the droplets."

INTRODUCTION

"One of the most successful approaches is grafting the surface of nanoparticles with polyethylene glycol (PEG), also known as PEGylation, which was shown to help nanoparticles evade clearance by the RES and increase their residence time in the blood [61–63]."

"As with poloxamers, 1,2-distearoyl-sn-glycero-3-phosphoethanolamine-N-[amino(polyethylene glycol)2000] (PEG_{2000}-DSPE), a block copolymer of PEG_{2000} covalently linked to a distearoyl lipid tail, has been used in many nanoplatforms (e.g., liposomes, polymeric nanoparticles, solid lipid nanoparticles, and microemulsions) as a PEGylating agent to increase their in vivo circulation time."

"The specific objectives of this study were to compare between the two nanoemulsions with respect to their (a) physical stability under stress, (b) hemolytic effect, (c) susceptibility to protein adsorption, (d) cellular uptake and in vitro antiproliferative activity against mammary adenocarcinoma cells in 2D and 3D culture, and (e) blood circulation time and clearance after IV injection in tumor-free rats."

MATERIALS AND METHODS

"To evaluate the ability of poloxamer 188 and PEG_{2000}-DSPE to shield the nanoemulsion droplets from protein adsorption and consequently uptake by the mononuclear phagocyte system (MPS), droplet size of the nanoemulsions was monitored while incubated with increasing concentration of immunoglobulin IgG. Of each nanoemulsion, 100 μL was diluted with 900 μL of IgG stock solution to achieve a final IgG concentration of 0.5, 1, 2, and 4 mg/mL. Mixtures were incubated while rotating at 37°C in a gravity convection oven for 24 h. At the conclusion of the test, samples were collected and analyzed for droplet size and were visually inspected for signs of phase separation."

"After overnight incubation at 37°C in a 5% CO_2 environment, cells were treated with different concentrations of TRF in either poloxamer 188 or PEG_{2000}-DSPE stabilized nanoemulsions."

RESULTS AND DISCUSSION

"The hemolytic effect of the poloxamer nanoemulsion after 8 h could be attributed to the differences between the two formulations in the conformation of the PEG chains on the surface of the droplets."

"Differences in half-life between the two nanoemulsions could be explained by the difference in the PEG coverage on the surface of the droplets, which was previously found to contribute to the hemolytic effect of poloxamer nanoemulsion and its higher uptake in tumor cells."

"It could be concluded that PEG chains on the surface of the poloxamer nanoemulsion droplets assume a brush conformation ($D < R_F$), whereas they assume a dense brush confirmation on the surface of the PEG_{2000}-DSPE nanoemulsion droplets ($L \approx 2R_F$)."

CONCLUSION

"We compared TRF nanoemulsions made with poloxamer 188 as the PEGylating agent versus one PEGylated with PEG_{2000}-DSPE."

"While equimolar PEG concentration was used in both formulations, the poloxamer 188 nanoemulsion had higher uptake and potency against MCF-7 cell."

"It was inferior to the PEG_{2000}-DSPE nanoemulsions in eliciting a PEGylation effect."

"The $T_{1/2}$ of γ-T3 from the PEG_{2000}-DSPE nanoemulsion was 7-fold higher than the poloxamer nanoemulsion and the control group which could be explained by different PEG surface density and conformation on the surface of the nanoemulsion droplets and the overall polymer chain architecture."

Intranasal Surface-Modified Mosapride Citrate-Loaded Nanostructured Lipid Carriers (MOS-SMNLCs) for Treatment of Reflux Diseases: In vitro Optimization,

Pharmacodynamics, and Pharmacokinetic Studies [64] This is a machine-generated summary of:

Hammad, Reham Waheed; Sanad, Rania Abdel Baset; Abdelmalk, Nevine Shawky; Aziz, Randa Latif; Torad, Faisal A.: Intranasal Surface-Modified Mosapride Citrate-Loaded Nanostructured Lipid Carriers (MOS-SMNLCs) for Treatment of Reflux Diseases: In vitro Optimization, Pharmacodynamics, and Pharmacokinetic Studies [64]

Published in: AAPS PharmSciTech (2018)

Link to original: https://doi.org/10.1208/s12249-018-1142-9

If you want to cite the papers, please refer to the original.

For technical reasons we could not place the page where the original quote is coming from.

Abstract-Summary "Mosapride (MOS) is a safe prokinetic agent potentially used to treat GERD."

"This study aimed to formulate MOS nanostructured lipid carriers (MOS-NLCs) via the intranasal route to improve its bioavailability."

"Melt–emulsification low temperature–solidification technique using 2^3 full factorial design was adopted to formulate MOS-NLCs."

"Glycerol addition significantly reduced the particle sizes and improved %EE and %drug released."

"The optimized MOS surface-modified nanostructured lipid carriers (MOS-SMNLCs-F7)(stearic acid, 4% glycerol, 0.5% LuterolF127, 0.5% chitosan) showed low particle size 413.8 nm ± 11.46 nm and high %EE 90.19% ± 0.06% and a three-fold increase in permeation of MOS with respect to the drug suspension."

"MOS-SMNLCs (F7) was also evaluated for its bioavailability compared with drug suspension and commercial product."

"Statistical analysis revealed a significant increase in gastric emptying rate to be 21.54 ± 1.88 contractions/min compared with10.02 ± 0.62 contractions/min and 8.9 ± 0.72 contractions/min for drug suspension and oral marketed product respectively."

"Pharmacokinetic studies showed 2.44-fold rise in bioavailability as compared to MOS suspension and 4.54-fold as compared to the oral marketed product."

"MOS-SMNLCs could be considered a step forward towards enhancing the clinical efficacy of Mosapride."

INTRODUCTION

"MOS is reported to treat GERD through improving esophageal peristalsis amplitude and gastric emptying time."

"MOS reduced esophageal acid exposure time and decreasing number of reflux events [65]."

"NLCs were used in many previous studies to increase drug solubility, drug loading, and drug permeation [66, 67]."

"The use of mucoadhesive polymers in intranasal formulation provided a better opportunity for good nasal absorption due to increased retention time and prolonged contact with the nasal mucosa [68]."

"Mosapride has not been previously formulated as surface-modified nanostructured lipid carriers for intranasal administration."

"This work has considered the factors affecting the formulation of Mosapride surface-modified nanostructured lipid carriers (MOS-SMNLCs) intended to be deposited intranasally to enhance MOS efficacy and to increase patient compliance."

"Effect of Chitosan (as surface modifier) on mucoadhesion and nasal permeation was also studied."

MATERIALS AND METHODS

"Ltd., India, L-Alpha-lecithin (MW 750.00 g/mol) (surfactant) was purchased from Acros Organics, USA."

"A specified amount of the lipid mixture consisted of solid lipid (WE85 or stearic acid) and liquid lipid (isopropyl myristate) (70:30), loaded with MOS and L-alpha Lecithin as a surfactant that were melted above the melting temperature of solid lipid (70°C) in 3 mL ethanol."

"Co-surfactant (lutrol F127), glycerol, and double-distilled water were mixed at the same temperature and added to the melted lipid while stirring."

"Surface modification of MOS-NLCs was done according to Gartziandia and others 2015 and Cui and others 2006 [69, 70], where 0.3% and 0.5% (w/v) Chitosan solutions were prepared and drop wisely added to equal volumes of the prepared MOS-NLCs dispersion at room temperature for 1 h using magnetic stirring 500 rpm (Jenway, LTD, U.K.), followed by incubation overnight at 4°C."

CHARACTERIZATION OF THE PREPARED MOSAPRIDE NANOSTRUCTURED LIPID CARRIERS (MOS-NLCS) AND MOSAPRIDE SURFACE-MODIFIED NANOSTRUCTURED LIPID CARRIERS (MOS-SMNLCS)

"The drug suspension and optimized MOS-SMNLCs were uniformly spread out on the sheep nasal mucosa, placed between the donor and receptor compartments."

"The gastric emptying rate and the intestinal motility for each rabbit was investigated before administration to be used as a control and after intranasal administration of the optimized MOS-SMNLCs and was compared to drug suspension and commercially available oral tablet produced by the Western pharmaceutical company, Egypt."

"The rabbits were fasted overnight before oral administration of Mosapride® tablet or the intranasal application of the optimized MOS-SMNLCs formulation, drug suspension but had free access to water."

"Blood samples were withdrawn from a cannula introduced into marginal ear vein and evacuated into heparinized tubes immediately at 0.5, 1.0, 1.5, 2.0, 3, 4.0, 6.0, 8.0, and 24.0 h following the intranasal application of the optimized MOS-SMNLCs, drug suspension, and orally administrated Mosapride tablet."

RESULTS AND DISCUSSION

"The MOS-NLCs prepared using stearic acid as solid lipid showed higher EE% and %DL than those prepared using WE85."

"Increasing the concentration of Lutrol F127 to 1% significantly ($p < 0.05$) decrease EE% and % DL of MOS-NLCs, due to the partitioning of more drug molecules out rapidly as a result of the solubilizing and emulsifying effect of Lutrol F127."

"The optimized MOS-NLCs (F7) was prepared by stearic acid as solid lipid, 4% glycerol concentration, and 0.5% Lutrol F127."

"Chitosan modified the surface of MOS-NLCs (F7) through electrostatic attraction between negatively charged lipid and positively charged Chitosan [71], increasing particles sizes which were consistent with results of particle size analysis."

"The broad peak at about 65.73°C related to lipid [72], which interpreted the increased melting enthalpy of pure stearic acid from 9.1 to 10.11 J/g in the MOS-SMNLCs formulation, was due to the increase in the melting temperature."

CONCLUSION

"The MOS-SMNLCs formulation was successfully formulated using stearic acid as solid lipid, isopropyl myristate as liquid lipid, l-alpha lecithin as a surfactant, lutrol F127 as co-surfactant, and Chitosan as a surface modifier and optimized to attain higher bioavailability."

"The addition of glycerol during formulation was promising in improving the drug-loading capacity of the prepared MOS-NLCs."

"The pharmacodynamics and pharmacokinetics studies of the optimized MOS-SMNLCs formulation ensured higher bioavailability than drug suspension and oral marketed tablets."

2.3 Effect of Patient Related Factors on ADME of Novel Drugs and Novel Drug Delivery Systems

Introduction by the Editor

Patient-related factors play a crucial role in the absorption, distribution, metabolism, and excretion (ADME) of novel drugs, nanoparticles, and novel drug delivery systems. These factors encompass a diverse range of individual characteristics that significantly influence the pharmacokinetics and therapeutic outcomes of these agents. Age is a key factor contributing to physiological changes. Paediatric and geriatric patients have altered drug metabolism and clearance rates due to differences in enzyme activity and organ function. Genetic polymorphisms have a remarkable impact on drug-metabolising enzymes, transporters, and drug targets, leading to variable responses to therapies. Underlying medical conditions, such as liver or kidney diseases, alter the drug metabolism and elimination pathways, affecting drug efficacy and toxicity. Metabolic diseases, like diabetes and obesity, can modify drug distribution in adipose tissues and alter pharmacokinetics. Patient-specific factors like gender influence drug responses due to hormonal differences. Lifestyle choices, such as diet, smoking, and alcohol consumption, influence drug metabolism through interactions with enzymes and transporters. Co-administration of other medications plays a profound role and affects drug interaction and ADME profiles, leading to either enhanced or reduced therapeutic effects. In the context of nanoparticles and novel drug delivery systems, patient-related factors extend to the biocompatibility and immunogenicity of these materials. Individual immune responses trigger clearance mechanisms or adverse reactions to nanoparticles, affecting their biodistribution and safety. Pre-existing sensitivities or allergies influence patient's tolerability of certain nanomaterials. Understanding these patient-related factors is vital for personalised medicine. Modifying treatment regimens based on individual characteristics can optimise drug dosing, minimise adverse effects, and improve therapeutic outcomes. Advancement in pharmacogenomics and precision medicine aim to predict individual responses to treatment based on genetic makeup. Clinicians and researchers are increasingly recognising the importance of patient-related factors in designing effective drug therapies and targeted nanoparticle-based interventions for enhanced patient care.

Machine generated summaries

Disclaimer: The summaries in this chapter were generated from Springer Nature publications using extractive AI auto-summarization: An extraction-based summarizer aims to identify the most important sentences of a text using an algorithm and uses those original sentences to create the auto-summary (unlike generative AI). As the constituted sentences are machine selected, they may not fully reflect the body of the work, so we strongly advise that the original content is read and cited. The auto generated summaries were curated by the editor to meet Springer Nature publication standards.

To cite this content, please refer to the original papers.

Machine generated keywords: age, adult, old, year, prodrug, acute, phase, review, child, disorder, tablet, female, prior, subject, stem cell

Influence of age on pharmacokinetics of capecitabine and its metabolites in older adults with cancer: a pilot study [73] This is a machine-generated summary of:

Shafiei, Mohsen; Galettis, Peter; Beale, Philip; Reuter, Stephanie E.; Martin, Jennifer H.; McLachlan, Andrew J.; Blinman, Prunella: Influence of age on pharmacokinetics of capecitabine and its metabolites in older adults with cancer: a pilot study [73]

Published in: Cancer Chemotherapy and Pharmacology (2023)
Link to original: https://doi.org/10.1007/s00280-023-04552-5

If you want to cite the papers, please refer to the original.

For technical reasons we could not place the page where the original quote is coming from.

Abstract-Summary "Capecitabine is an oral chemotherapy prodrug of 5-fluorouracil (5-FU) with unpredictable toxicity, especially in older adults."

"The aim of this study was to evaluate the pharmacokinetics (PK) of capecitabine and its metabolites in younger adults (<70 years) and older adults (≥70 years) receiving capecitabine for solid cancer."

"A linear mixed-effect analysis of variance (ANOVA) model was used to compare dose-normalised log-transformed PK parameters between age groups."

"Of the total 26 participants, 58% were male with a median age of 67 years (range, 37–85) with 54% aged<70 years and 46% aged≥70 years."

"Participants aged≥70 years, compared to those aged<70 years, had a greater 5-FU exposure based on area under the concentration–time curve (AUC) of 17% (90% CI 103–134%; 0.893 vs. 0.762 mg h/L) and 14% increase in maximal concentration, C_{max} (90% CI 82.1–159%; 0.343 vs. 0.300 mg/L)."

"5-FU exposure was significantly increased in older adults compared to younger adults receiving equivalent doses of capecitabine, and is a possible cause for increased toxicity in older adults."

Background

"Changes in 5-FU pharmacokinetics (PK) due to physiological changes with ageing may be responsible for the excess toxicity of capecitabine in older adults."

"There are limited data on the PK of capecitabine and its metabolites (5-FU, 5-DFCR, 5-DFUR) in older adults with cancer with conflicting results among the few published studies."

"Both studies demonstrated significant differences in capecitabine clearance (CL/F) and volume of distribution (Vd/F) and rate of absorption (ka) among older adults (aged > 70 years)."

"The aim of this study was to investigate the PK of capecitabine and its metabolites (5-DFCR, 5-DFUR and 5-FU) in younger (< 70 years) and older (≥ 70 years) adults receiving treatment for breast or gastrointestinal (gastric, pancreas, colorectal, and biliary) cancer and to explore the correlation between PK of capecitabine and chemotherapy-related toxicity and geriatric assessment domains."

Methods

"Individual estimates of PK parameters were determined by empiric Bayesian estimation using the PK of the drug (i.e., model), individual patient factors (i.e., body surface area, estimated creatinine clearance, and serum alkaline phosphatase activity) and the measured drug and metabolite concentration(s)."

"Determined model parameters were then used to calculate the following PK parameters on day 14 for each treatment cycle, and these included area under the plasma concentration–time curve over the 12-h dosing interval (AUCτ), maximum plasma concentration over the 12-h dosing interval (Cmax), and time of maximum plasma concentration (Tmax)."

"A linear mixed-effect analysis of variance (ANOVA) model was used to compare dose-normalised Ln-transformed PK parameters between age groups."

"Linear regression and logistic regression analysis were used to determine the correlation between capecitabine and metabolite PK and domains of geriatric assessment, inflammatory markers, and toxicity."

Results

"All 26 participants were included in the PK analysis of concentration–time data for capecitabine and its metabolites (Online Appendix 2)."

"The mean dose-normalised 5-FU concentration–time profiles showed a 17% increase in total exposure (AUCτ 90% CI 103–134%) and 14% increase in maximal concentrations (Cmax5-FU 90% CI 82.1–159%) over the dosing interval in the older age group, compared to the younger group (Online Appendix 3)."

"5-DFUR Cmax values exhibited great variability, such that the 90% confidence intervals were 78.7–146%, extending beyond the standard limits; nonetheless, the mean ratio was approximately 100% and no differences found in 5-DFUR PK between older and younger patients."

"Mean predicted dose-normalised (to a capecitabine dose of 1500 mg) concentration–time profiles on Cycle 1, Day 14 for 5-DFUR and 5-FU, stratified by age group, are presented in Online Appendix 8."

"The geometric mean ratio of older/younger group PK data and associated 90% confidence intervals are presented in Online Appendix 9."

Discussion

"Older adults, compared with younger adults, who had standard dose capecitabine for breast or gastrointestinal cancer had a statistically significant higher exposure to 5-FU."

"The increased exposure to 5-FU among older adults was positively correlated with the TUG-score (a measure of functional ability), but not other geriatric assessment variables, rates of severe chemotherapy-related toxicity or inflammatory markers."

"Previous studies determining the effect of age on capecitabine PK have showed differences between older adults and younger adults."

"An association between the geriatric assessment variables and PK parameters would enable clinicians, following completion of a geriatric assessment of older adults commencing chemotherapy, to identify older adults at, for example, increased risk of severe chemotherapy toxicity, hospitalisation, and/ or mortality due to change in PK parameters (e.g., exposure and Cmax) of a chemotherapy agent and prescribe appropriate dose modifications to minimise these risks."

Conclusions

"Compared to younger adults, older adults having capecitabine chemotherapy at the standard dose have significantly increased exposure to 5-FU but not to the other metabolites of capecitabine."

"The clinical significance of these findings requires further investigation in a larger cohort to determine whether it contributes to excess toxicity and/or provides a rationale for dose modifications in older adults receiving capecitabine."

A Review of the Pharmacological and Clinical Profile of Newer Atypical Antipsychotics as Treatments for Bipolar Disorder: Considerations for Use in Older Patients [74] This is a machine-generated summary of:

Vasudev, Akshya; Chaudhari, Sumit; Sethi, Rickinder; Fu, Rachel; Sandieson, Rachel M.; Forester, Brent P.: A Review of the Pharmacological and Clinical Profile of Newer Atypical Antipsychotics as Treatments for Bipolar Disorder: Considerations for Use in Older Patients [74]

Published in: Drugs & Aging (2018)

Link to original: https://doi.org/10.1007/s40266-018-0579-6

If you want to cite the papers, please refer to the original.

For technical reasons we could not place the page where the original quote is coming from.

Abstract-Summary "Although older agents, including lithium and valproic acid, offer significant antimanic efficacy, as supported by a recent randomized controlled trial (RCT), there is growing interest in using atypical antipsychotics to treat bipolar disorder in older adults."

"Newer atypical antipsychotics are of interest based on their tolerability and efficacy in the general adult bipolar population."

"The aim of this review was to systematically examine efficacy and tolerability of newer atypical antipsychotics for older adult bipolar disorder (OABD)."

"We conducted a systematic search utilizing the MEDLINE, EMBASE, PsycINFO and Cochrane Library electronic databases, with the aim of identifying all RCTs comparing newer atypical antipsychotics approved by the US FDA since 2002 (including brexpiprazole, cariprazine, lurasidone, iloperidone, asenapine, paliperidone, and aripiprazole) with placebo or another comparator, in the treatment of any phase of bipolar disorder (including mania, depression or mixed episodes while used as an acute or maintenance treatment) in older adults (>65 years)."

"Two post hoc studies on lurasidone suggest its reasonable safety and efficacy profile in the acute and maintenance treatment of OABD; however, there are no pharmacoeconomic data on the use of lurasidone in the treatment of OABD."

"Research data from open-label studies on oral asenapine and aripiprazole as add-on therapy suggest that these two agents are adequately tolerated and improved symptoms of depression and mania in OABD; hence, there is an urgent need to conduct RCTs on these two agents."

Introduction

"While one of these guidelines does offer some insight into the management of OABD [75], it does not offer a detailed review of the safety and efficacy of newer atypical antipsychotics for the treatment of OABD."

"The first large randomized controlled trial (RCT) on the management of acute mania in late-life bipolar disorder patients, the GERI-BD study, has recently been published [76]."

"Some of these agents have been approved for the treatment of the acute and/or maintenance phase of bipolar illness in the general adult population, while some have also been approved for use in the child and adolescent bipolar population; however, none of these agents are currently approved for the treatment of OABD."

"When subsequently expanding our search to post hoc studies assessing data from previous RCTs with a reduced age threshold of ≥ 55 years, we identified two post hoc studies of lurasidone in the treatment of late-life bipolar disorder."

Newer Atypical Antipsychotics for Older Adult Bipolar Disorder

"In a phase II trial, patients aged 18–65 years treated with cariprazine 3–12 mg/day demonstrated significant improvement in manic symptoms using the Young Mania Rating Scale (YMRS), in addition to improvements on the Clinical Global Impression (CGI) scale [77]."

"This review did not identify any ongoing or published studies of iloperidone for the acute or maintenance phases of treatment of either adult or late-life bipolar disorder, and hence no clinical recommendations can be offered."

"We identified a 12-week, prospective, open-label study designed to assess the efficacy and tolerability of adjunctive asenapine (n = 15, initiated at 5 mg/day and titrated as tolerated) in a non-dementia older adult sample (≥ 60 years of age) with suboptimal previous response to other bipolar disorder treatments [78]."

"Study participants included 766 patients aged 18–65 years administered flexible-dose paliperidone (3–12 mg/day) or olanzapine (5–20 mg/day) during a 3-week acute treatment phase before entry into the maintenance phase [79]."

Conclusions

"Given the complex needs of this group of patients, newer treatment options are sorely needed."

"Over the last 15 years, at least seven new atypical antipsychotics have come onto the market for the treatment of schizophrenia; however, only a few have gained marketing approval for the additional indication of treatment of bipolar disorder."

"Our review of these seven newer atypical antipsychotics found no RCT studies of any of these agents in any phase of OABD."

"On expanding our scope of this review, we included two post hoc studies of lurasidone, and found a reasonable efficacy profile of this agent in the acute and maintenance treatment of this illness, and hence lurasidone can be recommended for routine use."

"There are no data to support the use brexpiprazole, cariprazine, iloperidone, and paliperidone in the treatment of OABD."

Treosulfan Pharmacokinetics and its Variability in Pediatric and Adult Patients Undergoing Conditioning Prior to Hematopoietic Stem Cell Transplantation: Current State of the Art, In-Depth Analysis, and Perspectives [80] This is a machine-generated summary of:

Romański, Michał; Wachowiak, Jacek; Główka, Franciszek K.: Treosulfan Pharmacokinetics and its Variability in Pediatric and Adult Patients Undergoing Conditioning Prior to Hematopoietic Stem Cell Transplantation: Current State of the Art, In-Depth Analysis, and Perspectives [80]

Published in: Clinical Pharmacokinetics (2018)
Link to original: https://doi.org/10.1007/s40262-018-0647-4

If you want to cite the papers, please refer to the original.

For technical reasons we could not place the page where the original quote is coming from.

Abstract-Summary "Treosulfan is a prodrug that undergoes a highly pH- and temperature-dependent nonenzymatic conversion to the monoepoxide {(2S,3S)-1,2-epoxy-3,4-butanediol 4-methanesulfonate [S,S-EBDM]} and diepoxide {(2S,3S)-1,2:3,4-diepoxybutane [S,S-DEB]}."

"Treosulfan is tested in clinical trials as an alternative to busulfan in conditioning prior to hematopoietic stem cell transplantation (HSCT)."

"The pharmacokinetics of treosulfan, together with its biologically active epoxides, is comprehensively reviewed for the first time, with the focus on conditioning prior to HSCT."

"A clinically important aspect of the formation rate-limited elimination of S,S-EBDM and S,S-DEB is described, including the correlation between the exposure of the prodrug and S,S-EBDM in children."

"The significance of the elimination half-life of treosulfan and its epoxides for successful conditioning prior to HSCT is also raised."

"The organ disposition of treosulfan and S,S-EBDM in rats is discussed in the context of the clinical toxicity and myeloablative activity of treosulfan versus busulfan."

"The review is intended to be helpful to pharmacists and doctors in the comprehension of the clinical pharmacokinetics of treosulfan."

Introduction

"Prospective, multicenter, clinical phase II and III trials are ongoing that directly compare treosulfan versus busulfan in conditioning prior to allogeneic HSCT in children (1 month–18 years of age) with nonmalignant diseases, and in adults (18–70 years of age) with acute myeloid leukemia and myelodysplastic syndromes."

"In another ongoing phase III trial, treosulfan-based conditioning is compared not only with busulfan-based conditioning but also with fractionated total body irradiation-based conditioning in children and adolescents undergoing allogeneic HSCT for acute lymphoblastic leukemia, i.e. with the most frequent indication for HSCT in the pediatric population."

"The data covered the kinetics of the nonenzymatic epoxy transformation of treosulfan, including the quantitative description of pH and temperature effect; the pharmacokinetics of S,S-EBDM in children; the discovery of the formation-rate pharmacokinetics of the epoxy transformers; the organ disposition of treosulfan and S,S-EBDM in rats; treosulfan population pharmacokinetics in humans; and studies on the association of treosulfan exposure with the toxicity and efficacy of HSCT conditioning [81–93]."

Pharmacokinetics of the Prodrug Treosulfan

"In the pharmacokinetic studies of treosulfan involving a classic two-compartment model, the so-called volume of drug distribution at steady state (V_{ss}) has been reported."

"The reason children > 1 year of age (average treosulfan Cl_{tot} 3–6 mL/min/kg) experience a similar AUC as adults (approximate Cl_{tot} 2.0–2.5 mL/min/kg) following the same drug dosing per m^2 of body surface area (BSA) is that the BW increases unproportionally faster with age than the BSA."

"A hypothesis might be offered that infants treated with high-dose treosulfan prior to HSCT are more prone to experience a lower blood pH, as well as other body fluids, than older children, and, consequently, the decreased Cl_f and Cl_{tot} of the prodrug."

"The age-related differences in the Cl_{tot} of treosulfan have found reflection in the stratification of the drug doses prior to HSCT."

Factors Affecting Interindividual Variability of Volume of Distribution at Steady State and Cl_{tot} of Treosulfan

"In the model developed by Mohanan and others [93], none of the covariates tested, i.e. age, BW, BSA, sex, liver size, liver fibrosis, ferritin levels, liver enzymes, and hemoglobin level, were found to explain the wide interindividual variability of treosulfan pharmacokinetics in thalassemia major patients aged 1.5–25 years."

"Despite age, renal function (creatinine clearance), and administration of diuretics not being identified as significantly influencing the V_{ss} and Cl_{tot} of treosulfan [90, 91], these factors should not be neglected in future population pharmacokinetic modeling in larger and more homogeneous cohorts of HSCT patients."

"Additional covariates worth testing in treosulfan pharmacokinetic modeling are blood pH, as a readily measurable marker of acid-base balance, body temperature, and the volume of fluids infused intravenously to patients."

"The volume of fluids administered to HSCT patients on a daily basis might contribute to the interindividual variability of the V_{ss} and/or Cl_{tot} of treosulfan, and also their differences between various clinical centers."

Pharmacokinetics of Biologically Active Epoxy Derivatives of Treosulfan

"When treosulfan was administered to rabbits, the $t_{½}$ of S,S-EBDM did not differ statistically from that of the prodrug (1.6 h); likewise in the previous study in humans."

"The $t_{½}$ of S,S-EBDM and S,S-DEB is the same as that of treosulfan, despite the levels of epoxides in the body being very low compared with the prodrug due to their high Cl_{tot} [85]."

"Additional confirmation for this phenomenon was provided by an observation that the organ elimination of S,S-EBDM in rats proceeded at a similar rate as that of treosulfan (lungs, muscle, and bone marrow), except the brain, from which the epoxide was eliminated faster [87]."

"A clinical importance of the above facts is that once the elimination of treosulfan is completed, S,S-EBDM and S,S-DEB are also eliminated from the patient's body."

"This is further supported by the immeasurable liver levels of S,S-EBDM in rats administered treosulfan, and the rapid elimination of the epoxy transformer from the lungs compared with the other organs [87]."

Clinically Relevant Studies on the Organ Disposition of Treosulfan and S,S-EBDM in Rats

"The liver/plasma, lungs/plasma, brain/plasma, and bone marrow/plasma AUC ratios obtained for treosulfan amounted to, on average, 0.96, 0.82, 0.10, and 0.82, respectively."

"The average rat lungs/plasma, brain/plasma and bone marrow/plasma AUC ratios for biologically active S,S-EBDM were 0.50, 0.35, and 0.75, respectively."

"After scaling the organ AUC results of S,S-EBDM to the concentrations observed in the plasma of HSCT patients, the clinical exposure of the lungs and brain to the epoxide was lower than to busulfan."

"The mean brain-to-plasma treosulfan AUC ratio in younger and older animals was 0.16 and 0.08, respectively, and the tissue/plasma AUC ratio obtained for S,S-EBDM was found to be 0.5 and 0.2, respectively."

"These results led to the conclusion that very young patients receiving high-dose treosulfan prior to HSCT may experience higher neurotoxicity than older patients due to the increased penetration of the prodrug and S,S-EBDM across the incompletely mature blood-brain barrier."

Perspectives for Therapeutic Drug Monitoring of Treosulfan

"Relatively high variability of treosulfan pharmacokinetics in pediatric patients may raise the need for implementing therapeutic drug monitoring and individual dose adjustment in this group."

"Van der Stoep and others [92] and Mohanan and others [93] recently published the first results of a relationship between the exposure of treosulfan and early toxicity, as well as clinical outcome, in children undergoing conditioning prior to HSCT."

"No association was found between treosulfan exposure and early clinical outcomes, i.e. engraftment, donor chimerism, acute graft-versus-host disease, treatment-related mortality, and overall survival."

"It seems that a relationship between treosulfan exposure and early regimen-related toxicity and clinical outcome is still unresolved."

"As far as the future introduction of therapeutic monitoring of treosulfan is considered, it is worth mentioning that 2- or 3-point limited sampling strategies have been developed for determination of the drug AUC in HSCT children."

Conclusions

"This holistic review of the currently available literature indicates that three processes contribute to the Cl_{tot} of treosulfan: glomerular filtration, tubular reabsorption, and nonenzymatic epoxy transformation of the prodrug."

"Blood pH, body temperature, and intravenous fluid delivery should not be neglected as covariates of the Cl_{tot} of treosulfan in dose optimization efforts in HSCT patients, particularly infants."

"Organ disposition of treosulfan and S,S-EBDM in rats provides support for a lack of graft exposure to the compounds after at least a 2-day washout period preceding HSCT, the low organ toxicity of treosulfan-based conditioning compared with busulfan-based treatment, and the higher odds of neurological adverse effects in infants compared with older children."

"In terms of future therapeutic drug monitoring, larger studies are needed to verify the association of early and long-term toxicity and clinical outcomes with systemic exposure of not only treosulfan but also its active epoxy-transformers, at least S,S-EBDM."

Development of a particle swarm optimization-backpropagation artificial neural network model and effects of age and gender on pharmacokinetics study of omeprazole enteric-coated tablets in Chinese population [94] This is a machine-generated summary of:

Xu, Yichao; Chen, Jinliang; Yang, Dandan; Hu, Yin; Jiang, Bo; Ruan, Zourong; Lou, Honggang: Development of a particle swarm optimization-backpropagation artificial neural network model and effects of age and gender on pharmacokinetics study of omeprazole enteric-coated tablets in Chinese population [94]

Published in: BMC Pharmacology and Toxicology (2022)

Link to original: https://doi.org/10.1186/s40360-022-00594-2

If you want to cite the papers, please refer to the original.

For technical reasons we could not place the page where the original quote is coming from.

Abstract-Summary "The effects of age and gender were explored on pharmacokinetics study of omeprazole enteric-coated tablets in Chinese population and a plasma concentration prediction model was developed."

"All the data (demographic characteristics and results of clinical laboratory tests) were collected from healthy Chinese subjects in pharmacokinetics study using 20 mg omeprazole enteric-coated tablets."

"Pharmacokinetic data from the low-age and high-age groups or male and female groups were compared by Student t-test."

"The model was fully validated and used to predict the plasma concentration in Chinese population."

"It was noticed that the C_{max}, $AUC_{0\text{-}t}$, $AUC_{0\text{-}\infty}$ and $t_{1/2}$ values have significant differences when omeprazole was administered by low-age groups or high-age groups while there were slight or no significant differences of pharmacokinetic data were found between male and female subjects."

"It is necessary to pay attention to the age and gender effects on omeprazole and PSO-BPANN model could be used to predict omeprazole concentration in Chinese subjects to minimize the associated morbidity and mortality with peptic ulcer."

Introduction

"In China, omeprazole enteric-coated capsules (AstraZeneca Pharmaceutical Co. Ltd.) were approved for marketing in 2013 by the National Medical Products Administration."

"No import or product registration application for the original AstraZeneca omeprazole enteric-coated capsules has been submitted yet."

"There is no systematic study on the pharmacokinetics of omeprazole enteric-coated tablets in a Chinese population."

"A pharmacokinetics study of omeprazole enteric-coated tablets was conducted and the effects of age and gender were explored in Chinese population."

Methods

"We enrolled male and female volunteers aged from 18 to 45 years and with a body mass index between 19 and 26 kg•m^{-2} The inclusion criteria were as follows: (1) no clinically relevant abnormalities identified by subjects' medical history, physical examination, clinical laboratory tests, vital signs, chest radiography, and 12-lead ECG; (2) no tobacco, drug, or alcohol abuse; (3) no breastfeeding, pregnancy or childbearing potential of female subjects during the study."

"The main calculation procedures of PCA were as follows: (1) the data collection from the subjects was conducted on standardized processing; (2) the characteristic value and feature vector of the correlation coefficient matrix R were calculated to define new indicator variables; (3) the principal components were chosen and the information contribution rate and accumulated contribution rate were calculated;

(4) when the accumulated contribution was close to 1, we chose the principal components to replace the original variables and thereby obtain the key factors."

"Through the global search ability of the PSO algorithm, the initial weights and biases of the BPANN were obtained and the true global optimization and performance improvement were found."

Results

"Of PCA, the final established BPANN model consisted of one input layer with 7 neurons, 1 hidden layer with 13 nodes, and 1 output layer with one node (plasma concentration of omeprazole)."

"After the model was trained, the performance of the network was evaluated by the following four metrics: the MSE, magnitude of the gradient, number of validation checks, and correlation coefficient."

"The constructed PSO-BPANN model was used to predict the plasma concentration of omeprazole in the test group."

"The calculation process of MIVs was as follows: (1) new training samples were formed by increasing and decreasing 10% of each input variable values; (2) new training samples were incorporated into the PSO-BPANNs, and the two results were subtracted and divided by the number of observations; (3) the sign of the MIV results represented the positive and negative correlation between the variable and the result, and the absolute values of the MIV results represent the importance of the variables."

Discussion

"We analyzed the effects of age and gender on the plasma concentration of omeprazole in a Chinese population."

"This is the first time that the effects of age and gender and the pharmacokinetics parameters of omeprazole enteric-coated tablets were explored in a Chinese population."

"The MIV of gender and age was -127.56 and -150.14, respectively, which indicated strong negative effects on the plasma concentration of omeprazole."

"Although the low number of subjects included in the study did not allow us to explore the effect of more variables on the pharmacokinetics of omeprazole enteric-coated tablets, such as the CYP genetic polymorphisms [95], the model we established still had a good predictive effect on the plasma concentration of omeprazole in the Chinese population."

2.4 Biodistribution of Novel Drugs and Nanoparticles in Solid Tumors

Introduction by the Editor

Biodistribution is a critical aspect for understanding the interaction of novel drugs and nanoparticles within solid tumors. In cancer therapy, achieving effective drug delivery to tumor sites while minimising off-target effects is a paramount. Solid tumors possess a unique microenvironment characterized by abnormal blood vessel structures, impaired lymphatic drainage, and increased interstitial pressure, which collectively influence the distribution of therapeutic agents. Novel drugs and nanoparticles designed for cancer treatment aim to exploit these tumor-specific features. Nanoparticles, in particular, offer opportunities to enhance drug accumulation in tumors. Their small size allows them to exploit the enhanced permeability and retention (EPR) effect, where leaky blood vessels in tumors allow nanoparticles to extravasate and accumulate preferentially within the tumor tissue. Surface modifications with ligands further enhance tumor targeting by promoting interaction with specific receptors overexpressed on tumor cells. However, the biodistribution of novel drugs and nanoparticles involves a complex interaction of many factors. Drug physicochemical properties, nanoparticle size, surface charge, and composition influence their circulation time, tissue penetration, and clearance. Additionally, the tumor type and location within the body affect the drug distribution pattern. Imaging techniques like positron emission tomography (PET), magnetic resonance imaging (MRI), and fluorescence imaging play a crucial role in visualising the biodistribution of these agents. They allow researchers to track drug or nanoparticle movement, accumulation, and retention, aiding in the optimisation of drug delivery strategies. In conclusion, understanding the biodistribution of novel drugs and nanoparticles in solid tumors is pivotal for developing effective cancer therapies. By capitalising on the unique attributes of tumor microenvironment and leveraging advanced imaging techniques, researchers strive to enhance drug accumulation within tumors, ultimately improving treatment outcomes and reducing side effects in cancer patients.

Machine generated summaries

Disclaimer: The summaries in this chapter were generated from Springer Nature publications using extractive AI auto-summarization: An extraction-based summarizer aims to identify the most important sentences of a text using an algorithm and uses those original sentences to create the auto-summary (unlike generative AI). As the constituted sentences are machine selected, they may not fully reflect the body of the work, so we strongly advise that the original content is read and cited. The auto generated summaries were curated by the editor to meet Springer Nature publication standards.

To cite this content, please refer to the original papers.

Machine generated keywords: tumor, solid tumor, patient advanced, advanced, kinase, safety, growth factor, growth, phase study, survival, patient, normal, safety efficacy, patient receive, colorectal

A Phase I dose-escalation, pharmacokinetics and food-effect study of oral donafenib in patients with advanced solid tumours [96] This is a machine-generated summary of:

Li, Xiaoyu; Qiu, Meng; Wang, ShengJun; Zhu, Hong; Feng, Bi; Zheng, Li: A Phase I dose-escalation, pharmacokinetics and food-effect study of oral donafenib in patients with advanced solid tumours [96]

Published in: Cancer Chemotherapy and Pharmacology (2020)
Link to original: https://doi.org/10.1007/s00280-020-04031-1

If you want to cite the papers, please refer to the original.

For technical reasons we could not place the page where the original quote is coming from.

Abstract-Summary "This Phase I study evaluated the safety, tolerability, food effects, pharmacodynamics, and pharmacokinetics of donafenib in patients with advanced solid tumours."

"Eligible patients received a single dose of donafenib (50 mg, 100 mg, 200 mg, 300 mg, or 400 mg) and were then observed over a 7-day period; thereafter, each patient received the corresponding dose of donafenib twice daily for at least 4 weeks."

"Safety assessment and pharmacokinetic sampling were performed for all patients at the given time points; preliminary tumour response was also assessed."

"The maximum tolerated dose (MTD) was 300 mg bid."

"The dose-limiting toxicities (DLTs) were grade 3 diarrhoea and fatigue at 300 mg bid and grade 3 skin toxicity at 400 mg bid."

"Oral donafenib was generally well tolerated and appeared to provide some clinical benefits; adverse events were manageable."

"Of this study, oral donafenib at 200 mg ~ 300 mg twice daily is recommended for further studies."

Introduction

"Sorafenib, the first oral multi-kinase inhibitor, was approved by the Food and Drug Administration (FDA) for the first-line treatment of advanced HCC [97, 98]."

"Although sorafenib has been used in the clinic for more than a decade, it is still one of the best options in advanced HCC treatment [99]."

"In patients with advanced HCC, the disease control rate of sorafenib group was 43%, and the median progression-free survival time was nearly 3 months longer than that of the placebo group [97]."

"The cost of sorafenib has limited its clinical accessibility in developing countries [100, 101]."

"Donafenib is a patented novel small molecule drug developed by Zelgen Biopharmaceuticals by creatively substituting a trideuteriomethyl group for a methyl on a sorafenib molecule."

"The purpose of this study was to evaluate the tolerability, dose-limiting toxicity (DLT) and maximum tolerated dose (MTD), pharmacodynamics and pharmacokinetics of donafenib in patients with advanced solid tumours."

Materials and methods

"The study used a 3 + 3 method for both single ascending dose (SAD) and multiple ascending dose (MAD) escalations except in the pilot, where three patients received only a single dose of 50 mg."

"In each dose cohort, patients first received a single dose and were observed for one week; if no DLT was observed, they entered the MAD phase and received the same dose twice daily for 28 consecutive days (defined as one treatment cycle)."

"For the MAD period, patients started with donafenib twice a day for 4 weeks at doses of 100 mg, 200 mg, 300 mg, or 400 mg, corresponding to their original single doses."

"On Day 1, seven patients were given a single dose of donafenib orally under fasting conditions, washed out and observed for 7 days."

"On Day 8, the patients administered a single dose of donafenib within 5 min after receiving a high-fat and high-calorie diet."

Results

"In the 300 mg group, one out of seven patients developed grade 3 anorexia 3 days after initiating the drug."

"In the 400 mg group, three out of six patients developed DLT, including grade 3 rash (1 case) and hand-foot syndrome reaction (2 cases); all three patients were discontinued the study."

"Patients in the 300 mg group received a single 300 mg dose while fasting, and the drug concentration of donafenib was the highest in plasma."

"The plasma exposure of M4 was the lowest, with the majority of patients having a concentration lower than the quantitative limit of 2.00 ng/mL. A total of eight metabolites were detected after oral administration, and the metabolic pathways of donafenib include pyridine N-oxidation, hydroxylation, N-deuterium methylation, amide hydrolysis, and glucuronic acid-binding."

Discussion

"The results of this Phase I trial showed that 100 mg ~ 300 mg of donafenib administered twice a day was generally well tolerated in patients with advanced solid tumours."

"In the dose range of 200–400 mg, the half-lives of donafenib and sorafenib are similar."

"The results of this study show that the $AUC_{0-12\,h}$ of donafenib was slightly higher than that of sorafenib at the same dose after multiple doses."

"We may consider the $AUC_{0-12\,h}$ values of 200–300 mg for donafenib could be similar to the $AUC_{0-12\,h}$ of sorafenib at a dose of 400 mg."

"A previous phase I study of sorafenib reported that bid dosing seems to overcome absorption saturation, thereby leading to higher exposure than once-daily dosing [102]."

"Another phase I study used a continuous dosing schedule of 200 mg of sorafenib, and compared with once-daily dosing, the $AUC_{0-\infty}$ values were twofold greater following bid dosing [103]."

Phase 1 study of safety, pharmacokinetics, and pharmacodynamics of tivantinib in combination with bevacizumab in adult patients with advanced solid tumors [104] This is a machine-generated summary of:

Maguire, William F.; Schmitz, John C.; Scemama, Jonas; Czambel, Ken; Lin, Yan; Green, Anthony G.; Wu, Shaoyu; Lin, Huang; Puhalla, Shannon; Rhee, John; Stoller, Ronald; Tawbi, Hussein; Lee, James J.; Wright, John J.; Beumer, Jan H.; Chu, Edward; Appleman, Leonard J.: Phase 1 study of safety, pharmacokinetics, and pharmacodynamics of tivantinib in combination with bevacizumab in adult patients with advanced solid tumors [104]

Published in: Cancer Chemotherapy and Pharmacology (2021)

Link to original: https://doi.org/10.1007/s00280-021-04317-y

If you want to cite the papers, please refer to the original.

For technical reasons we could not place the page where the original quote is coming from.

Abstract-Summary "We investigated the combination of tivantinib, a c-MET tyrosine kinase inhibitor (TKI), and bevacizumab, an anti-VEGF-A antibody."

"Patients with advanced solid tumors received bevacizumab (10 mg/kg intravenously every 2 weeks) and escalating doses of tivantinib (120–360 mg orally twice daily)."

"No exposure-toxicity relationship was observed for tivantinib or metabolites."

"Tivantinib reduced levels of both phospho-MET (7/11 patients) and tubulin (4/11 patients) in skin."

"The combination of tivantinib and bevacizumab produced toxicities that were largely consistent with the safety profiles of the individual drugs."

"Tivantinib reversed the upregulation of bFGF caused by bevacizumab, which has been considered a potential mechanism of resistance to therapies targeting the VEGF pathway."

"The findings from this study suggest that the mechanism of action of tivantinib in humans may involve inhibition of both c-MET and tubulin expression."

Introduction

"The c-MET receptor tyrosine kinase (RTK) pathway has attracted considerable interest as a potential target for cancer therapy given its involvement in numerous hallmark features of cancer development and progression [105]."

"Preclinical studies have suggested that the c-MET pathway may serve as an alternative angiogenic pathway that is upregulated in tumors treated with VEGF-directed therapies [106], and preclinical models have shown synergistic effects of targeting both signaling pathways simultaneously [107]."

"To further explore the hypothesis that combined inhibition of c-MET and VEGF pathways may represent a potential therapeutic strategy, we conducted a phase 1 trial of tivantinib in combination with anti-VEGF-A antibody bevacizumab in patients with advanced solid malignancies."

"During the conduct of this trial, studies were published that questioned the mechanism of action of tivantinib and suggested that its preclinical activity may result from effects on tubulin causing microtubule destabilization as opposed to inhibition of c-MET signaling [108, 109]."

Materials and methods

"Secondary objectives were to determine the dose-limiting toxicity (DLT) and other toxicities associated with the combination as assessed by CTCAE v4.0; to document anti-tumor activity; to determine the pharmacokinetics of tivantinib when given in combination with bevacizumab; to assess the effect of the combination on plasma components of the HGF-MET signaling pathway and VEGF signaling pathway; and to assess tissue (skin) protein biomarkers before and after study treatment including MET and phospho-MET-Tyr1234."

"A lead-in dose of bevacizumab (commercially obtained) was administered on day − 15 of cycle 1 to allow assessment of plasma biomarkers on cycle 1 day 1 prior to administration of tivantinib."

"With respect to the trial-related assessments, tissue samples (plasma, PBMCs, and skin biopsies) were obtained from patients prior to starting treatment (cycle 1, day − 15), 15 days post-bevacizumab administration (pre-dose on cycle 1, day 1), and 15 days post-bevacizumab/daily tivantinib administration (cycle 1, day 15)."

Results

"The study then proceeded to stage 2 and enrolled seven additional patients on dose level 3, of which one patient was taken off study before receiving tivantinib due to adverse events that were possibly related to his dose of bevacizumab on day − 15."

"Of the 11 patients who received the combination of tivantinib and bevacizumab, 6 patients experienced stable disease as their best response, while 5 patients experienced progressive disease."

"One patient with metastatic colorectal cancer who had previously progressed on a regimen containing bevacizumab received eight cycles of treatment with tivantinib and bevacizumab before experiencing progressive disease."

"On day 15 of the treatment cycle following 14 days of tivantinib treatment, concentrations of bFGF returned to baseline in most patients ($p = 0.047$)."

"SVEGFR1 concentrations following tivantinib increased in four patients, including those patients with low basal expression levels at baseline."

Discussion

"As noted above, preclinical studies suggested that tivantinib may actually exert its effects through microtubule destabilization as opposed to inhibition of c-Met signaling."

"BFGF levels were elevated in 8 of 11 patients following the lead-in treatment with bevacizumab and normalized after tivantinib was added."

"HGF levels were elevated in six patients following bevacizumab treatment and subsequently normalized after administration of tivantinib."

"Given the fact that we did see evidence of c-MET inhibition in human skin, it is conceivable that the degree of c-MET inhibition at the dose levels of tivantinib used in our study was simply not sufficient to produce a clinically relevant effect."

"The findings of this study do not support the further clinical development of the combination of tivantinib and bevacizumab in patients with advanced cancer."

Panitumumab: A Review of Clinical Pharmacokinetic and Pharmacology Properties After Over a Decade of Experience in Patients with Solid Tumors [110] This is a machine-generated summary of:

Kast, Johannes; Dutta, Sandeep; Upreti, Vijay V.: Panitumumab: A Review of Clinical Pharmacokinetic and Pharmacology Properties After Over a Decade of Experience in Patients with Solid Tumors [110]

Published in: Advances in Therapy (2021)

Link to original: https://doi.org/10.1007/s12325-021-01809-4

If you want to cite the papers, please refer to the original.

For technical reasons we could not place the page where the original quote is coming from.

Abstract-Summary "Over the last 10 years, the pharmacokinetic and pharmacodynamic profile of panitumumab has been studied to further evaluate its safety, efficacy, and optimal dosing regimen."

"Panitumumab has a nonlinear pharmacokinetic profile and its approved dosing regimen (6 mg/kg every 2 weeks) is based on body weight; dose adjustments are not needed based on sex, age, or renal or hepatic impairment."

"The level of tumor EGFR expression was found to have no effect on panitumumab pharmacokinetics or efficacy."

"The incidence of anti-panitumumab antibodies is low; when anti-panitumumab antibodies are produced, they do not affect the efficacy, safety, or pharmacokinetics of panitumumab."

"The pharmacokinetic and pharmacodynamic profile of panitumumab is well suited for standard dosing, and the approved body weight–based dosing regimen maintains efficacy and safety in the treatment of wild-type RAS metastatic colorectal cancer across a broad range of patients."

Digital Features

"This article is published with digital features, including a summary slide, to facilitate understanding of the article."

"To view digital features for this article go to https://doi.org/10.6084/m9.figshare.14673549."

Introduction

"Panitumumab is an epidermal growth factor receptor (EGFR) antagonist first approved in the USA in 2006 for the treatment of wild-type RAS (defined as wild-type in both KRAS and NRAS) metastatic colorectal cancer (mCRC) [111]."

"In the USA, panitumumab is currently indicated as first-line treatment for mCRC in combination with the cytotoxic doublet 5-fluorouracil with folinic acid plus oxaliplatin (FOLFOX)."

"Panitumumab is also indicated as monotherapy in patients with disease progression after fluoropyrimidine-, oxaliplatin-, and irinotecan-containing chemotherapy."

"Panitumumab is approved in Europe for similar indications, and in combination with 5-fluorouracil with folinic acid plus irinotecan (FOLFIRI) in patients progressing after treatment with fluoropyrimidine [112]."

In Vitro Pharmacology

"Binding of panitumumab to EGFR results in internalization of the receptor, induction of apoptosis, inhibition of cell growth, and decreased interleukin-8 and vascular endothelial growth factor production [111, 113]."

"In vitro assays and in vivo animal studies have shown that panitumumab inhibits the growth and survival of tumor cells lines expressing EGFR [113]."

"In vitro, panitumumab inhibited growth of EGFR-expressing tumor cell lines from breast and epidermal origin in a dose-dependent manner [114]."

"EGF-induced phosphorylation was inhibited by panitumumab in human non-small cell lung carcinoma cells expressing wild-type or mutant EGFR [115] and in epidermoid carcinoma cells [116]."

"Panitumumab inhibited EGFR-expressing tumors of breast (MDA-MB-468), epidermal (A431), renal (SK-RC-29), pancreatic (BxPC-3 and HS766T), prostate (PC-3), and ovarian (IGROVI) origin in xenograft models in the absence of concomitant chemotherapy [114, 117]."

"The antitumor effects of panitumumab on mutant EGFR-expressing non-small cell lung carcinoma xenografts were increased with chemotherapy and/or targeted therapeutic agents compared with chemotherapy or targeted agents alone [115]."

Clinical Pharmacokinetics

"A body weight–based dosing regimen results in less variability in panitumumab exposure than a fixed dose [118]."

"The pharmacokinetics of panitumumab in patients with severe renal or hepatic impairment have not been specifically studied."

"The approved panitumumab dosing regimen is based on body weight and is 6 mg/kg Q2W [111, 112]."

"When dosed according to actual body weight, exposures of panitumumab were greater in heavier patients compared with lighter patients [118, 119]."

"In a phase 2 study of panitumumab administered 2.5 mg/kg once weekly in patients with mCRC, no differences in pharmacokinetics were observed between high (at least 10%) and low (less than 10%) EGFR expression groups [120]."

"In phase 2 and 3 studies of patients with mCRC receiving panitumumab 6 mg/kg Q2W, EGFR expression level had no impact on the serum concentrations of panitumumab [121]."

Exposure–Response Relationship

"In an in vitro study, expression levels of EGFR in human tumor xenograft did not predict response to panitumumab [122]."

"After 8 weeks of treatment with panitumumab 2.5 mg/kg once weekly, response rates between EGFR expression groups were not significantly different."

"Another phase 2 study of patients with mCRC examined response rates in patients with low (1–9%) or no (less than 1%) EGFR expression (determined by IHC) after 16 weeks of treatment with panitumumab 6 mg/kg Q2W [123]."

"In phase 2 and phase 3 studies of patients with mCRC receiving panitumumab 6 mg/kg Q2W, EGFR expression levels of 1% or higher had no effect on response rate [124, 125]."

"Further studies are required to establish whether there is any association between EGFR expression level in tumors at more granular cutoff levels and the clinical response to panitumumab."

Other Important Factors

"Acid dissociation bridging enzyme-linked immunosorbent assay (ELISA) and Biacore® (surface plasmon resonance assay) have been used to screen for anti-panitumumab antibodies (Abs) on the basis of their ability to bind high- and low-affinity Abs, respectively [126]."

"The reported incidence of non-transient binding anti-panitumumab Abs is less than 1% by ELISA and ranges from 3.8% to 5.3% by Biacore assay [111, 112, 126]."

"The incidence of anti-panitumumab Abs determined by ELISA or Biacore assay in a population analysis of panitumumab was slightly higher in patients receiving the dose regimen 9 mg/kg Q3W compared with 6 mg/kg Q2W (2/26 [7.7%] vs 16/498 [3.2%], respectively); however, the overall incidence of 18/530 patients (3.4%) was comparable to previous reports [121]."

"When combined with oxaliplatin- or irinotecan-based chemotherapy, the incidence of anti-panitumumab Abs and neutralizing Abs remained low (1.8% and 0.2%, respectively) [127]."

Conclusions

"Over the last 10 years, the pharmacokinetic profile and dose regimen of panitumumab have been further explored, including in combination with chemotherapy drugs and in patients with hepatic and renal impairment."

"The approved dosing regimen of 6 mg/kg Q2W maintains a C_{min} to achieve clinical efficacy, reduces inter-individual variability, has a manageable safety profile, and is administered on a schedule that is compatible with the chemotherapy regimens typically used with it in mCRC."

"Panitumumab has a favorable pharmacokinetic/pharmacodynamic profile which makes it well suited for standard dosing for the treatment of wild-type RAS mCRC across a broad range of patients."

Pharmacokinetics and safety of rucaparib in patients with advanced solid tumors and hepatic impairment [128] This is a machine-generated summary of:

Grechko, Nikolay; Skarbova, Viera; Tomaszewska-Kiecana, Monika; Ramlau, Rodryg; Centkowski, Piotr; Drew, Yvette; Dziadziuszko, Rafal; Zemanova, Milada; Beltman, Jeri; Nash, Eileen; Habeck, Jenn; Liao, Mingxiang; Xiao, Jim: Pharmacokinetics and safety of rucaparib in patients with advanced solid tumors and hepatic impairment [128]

Published in: Cancer Chemotherapy and Pharmacology (2021)

Link to original: https://doi.org/10.1007/s00280-021-04278-2

If you want to cite the papers, please refer to the original.

For technical reasons we could not place the page where the original quote is coming from.

Abstract-Summary "This study investigated whether hepatic impairment affects the pharmacokinetics, safety, and tolerability of rucaparib in patients with advanced solid tumors."

"Patients with normal hepatic function or moderate hepatic impairment according to the National Cancer Institute Organ Dysfunction Working Group (NCI-ODWG) criteria were enrolled and received a single oral dose of rucaparib 600 mg."

"Rucaparib maximum concentration (C_{max}) was similar, while the area under the concentration–time curve from time 0 to infinity ($AUC_{0\text{–}inf}$) was mildly higher in the moderate hepatic impairment group than in the normal control group (geometric mean ratio, 1.446 [90% CI 0.668–3.131]); similar trends were observed for M324."

"Eight (50%) patients experienced ≥ 1 treatment-emergent adverse event (TEAE); 2 had normal hepatic function and 6 had moderate hepatic impairment."

"Patients with moderate hepatic impairment showed mildly increased $AUC_{0\text{–}inf}$ for rucaparib compared to patients with normal hepatic function."

"Although more patients with moderate hepatic impairment experienced TEAEs, only 2 TEAEs were considered treatment related."

Introduction

"The clinical pharmacokinetics (PK) of rucaparib have been well characterized among patients with advanced solid tumors."

"Across the dose range of 240–840 mg twice daily (BID), rucaparib displays linear PK with time-independent and dose-proportional increases in plasma exposure."

"The apparent steady-state clearance ranges from 15.3 to 79.2 L/h following rucaparib 600 mg BID administration [129–132]."

"There are limited data available from patients with hepatic impairment treated with rucaparib."

"Because cytochrome P450 3A (CYP3A) and CYP1A2 mediate the formation of M324 [133], it is possible that hepatic impairment could affect rucaparib metabolism and M324 PK."

"We report results from Part 1 of a phase 1 trial designed to collect clinical PK and safety data to characterize the effect of hepatic impairment on rucaparib and its metabolite M324 in patients with advanced solid tumors."

Materials and methods

"The primary study objective was to compare the PK parameters of a single dose of rucaparib in patients with advanced solid tumors and normal hepatic function to those in patients with advanced solid tumors and moderate hepatic impairment based on the National Cancer Institute Organ Dysfunction Working Group (NCI-ODWG) criteria for hepatic dysfunction (total bilirubin > 1.5× and ≤ 3× upper limit of normal with any level of aspartate aminotransferase) [134]."

"Key exclusion criteria included anticancer treatment (chemotherapy, radiation, or other targeted agents) within 14 days or five times the half-life of the drug administered; unresolved grade ≥ 2 adverse events from prior therapies (except conditions associated with underlying liver disease in patients in the moderate hepatic impairment group); prior treatment with a PARP inhibitor (unless the PARP inhibitor was not the latest treatment, and it was discontinued > 3 months prior to the first dose on the study); pre-existing duodenal stent and/or any gastrointestinal disorder or defect that could interfere with absorption of rucaparib; corrected QT interval using Fridericia's formula (QTcF) ≥ 480 ms; clinically significant arrythmias or electrocardiogram (ECG) abnormalities; and arterial or venous thrombosis, myocardial infarction, unstable angina, cardiac angioplasty, stenting, or uncontrolled hypertension within 3 months."

Results

"The geometric mean (GM) C_{max} values for rucaparib in patients with moderate hepatic impairment and normal hepatic function were 583 ng/mL and 642 ng/mL, respectively."

"PK parameters for M324 showed similar GM C_{max} values in patients with moderately impaired hepatic function and normal hepatic function, 76.7 ng/mL and 72.7 ng/mL, respectively."

"Despite comparable baseline estimated glomerular filtration rates between the two groups, CL_R for rucaparib was lower in patients with moderate hepatic impairment (GM, 36.9 mL/min) than in patients with normal hepatic function (GM, 52.5 mL/min)."

"CL_R for M324 was also lower in patients with moderate hepatic impairment (GM, 47.7 mL/min) than in patients with normal hepatic function (GM, 118 mL/min)."

"A total of eight patients (50%) experienced at least 1 TEAE, including two patients (25.0%) in the normal hepatic function group and six patients (75.0%) in the moderate hepatic impairment group."

Discussion

"The results from Part 1 of this trial suggest that moderate hepatic impairment, as defined by the NCI-ODWG criteria, has no apparent effect on the oral absorption of rucaparib, based on similar C_{max} and t_{max} values observed in both hepatic function groups."

"Although patients with moderate hepatic impairment based on NCI-ODWG criteria showed mildly increased AUC_{0-inf} as compared to patients with normal hepatic function, overall, PK variability was moderate, and no statistically significant differences were observed for C_{max}, AUC, or t_{max} for rucaparib and M324 between the groups."

"The results of this study also imply that the effects of moderate hepatic impairment on rucaparib PK is not be considered clinically significant, suggesting that no starting dose adjustment is necessary for patients with moderate hepatic impairment; however, patients with moderate hepatic impairment should be carefully monitored for hepatic function and adverse reactions."

An open-label, crossover study to compare different formulations and evaluate effect of food on pharmacokinetics of pimitespib in patients with advanced solid tumors [135] This is a machine-generated summary of:

Komatsu, Yoshito; Shimokawa, Tsuneo; Akiyoshi, Kohei; Karayama, Masato; Shimomura, Akihiko; Kawamoto, Yasuyuki; Yuki, Satoshi; Tambo, Yuichi; Kasahara, Kazuo: An open-label, crossover study to compare different formulations and evaluate effect of food on pharmacokinetics of pimitespib in patients with advanced solid tumors [135]

Published in: Investigational New Drugs (2022)

Link to original: https://doi.org/10.1007/s10637-022-01285-9

If you want to cite the papers, please refer to the original.

For technical reasons we could not place the page where the original quote is coming from.

Abstract-Summary "This study compared the bioavailability of two pimitespib formulations (Formulations A and B), evaluated the food effect on Formulation A, and evaluated the safety and efficacy of multiple pimitespib doses in patients with solid tumors."

"A single dose of Formulation A or B was administered in a crossover design to compare the pharmacokinetics in Cohort 1."

"In Cohort 2, the effects of fed vs fasting conditions were evaluated among those receiving Formulation A. Subsequently, multiple Formulation A doses were administered to all patients for safety and efficacy assessments."

"In Cohorts 1 and 2, 12 and 16 patients, respectively, were analyzed for pharmacokinetics."

"Maximum concentration (C_{max}), area under the curve $(AUC)_{last}$, and AUC_{inf} geometric mean ratios for Formulations A and B (90% confidence interval [CI]) were 0.8078 (0.6569–0.9933), 0.7973 (0.6672–0.9529), and 0.8094 (0.6697–0.9782), respectively; 90% CIs were not within the bioequivalence range (0.80–1.25)."

"In Cohort 2, mean C_{max}, AUC_{last}, and AUC_{inf} were higher in fed vs fasting conditions."

"Systemic exposure of Formulation A was approximately 20% less than Formulation B. A high-fat/calorie meal increased the relative pharmacokinetics and bioavailability of a single 160-mg dose."

Introduction

"A phase II study of pimitespib was conducted in patients with advanced GIST who had failed or were intolerant to imatinib, sunitinib, and regorafenib treatments."

"The phase III CHAPTER-GIST-301 trial [136] found that pimitespib significantly increased the median PFS of patients with advanced GIST refractory or intolerant to treatment with imatinib, sunitinib, and regorafenib."

"It is necessary to compare the pharmacokinetics (PK) profiles of pimitespib from Formulation A with Formulation B. The new 40 mg strength Formulation A was developed to allow for more convenient drug administration because the administration of pimitespib starts at 160 mg/patient."

"The primary objective of this study was to compare the PK parameters between pimitespib Formulations A and B and compare the PK parameters of Formulation A under fasting and fed conditions in patients with advanced malignant tumors, including malignant soft tissue tumors or stromal tumors, refractory to conventional therapy or without standard therapy available."

Materials and methods

"In Cohort 1, during the PK evaluation period, pimitespib was administered under fasting conditions as a single administration of Formulation A (40 mg × 4) followed by a single dose of Formulation B (10 mg × 1 and 50 mg × 3), or a single dose of Formulation B followed by a single dose of Formulation A. Patient enrollment in Cohort 2 was initiated after enrollment of Cohort 1 was completed."

"The primary PK outcome included the following parameters for Formulations A and B administered under fasting conditions in Cohort 1, and Formulation A administered under fasting and fed conditions in Cohort 2 during the PK evaluation period: maximum observed plasma concentration (C_{max}), area under the plasma concentration–time curve from time 0 to the time of the last measurable plasma concentration (AUC_{last}), and area under the plasma concentration–time curve from time 0 to infinity (AUC_{inf}) in the PK evaluable population."

Results

"In Cohort 2, 17 patients received Formulation A under fasting and fed conditions."

"In Cohort 1, the median (range) age was 64.0 (38–74) years, 61.5% of patients were male, and 61.5% had an ECOG PS of 0."

"In Cohort 2, the mean C_{max}, AUC_{last}, and AUC_{inf} were 3046 ng/mL, 45479 ng·h/mL, and 49345 ng·h/mL, respectively, under fed conditions compared with 1625 ng/mL, 29922 ng·h/mL, and 29384 ng·h/mL, respectively, under fasting conditions."

"The t_{max} (90% CIs) for Formulation A under fasting conditions (n = 16) was 2.03 (0.93–7.53) and that for fed conditions (n = 16) was 4.02 (1.07–10.03), p = 0.0490."

"In the consecutive administration period, 83.3% (25/30) of patients experienced TRAEs, and 33.3% (10/30) had Grade 3 or higher TRAEs."

Discussion

"The primary objectives of this study were to compare the bioavailability of pimitespib Formulations A and B and to evaluate the effect of food on the bioavailability of Formulation A. Secondarily, we also assessed the safety and anti-tumor efficacy of multiple dosing of pimitespib 160 mg/day orally in patients with malignant tumors, including malignant soft tissue tumors or stromal tumors, that were refractory to conventional therapy."

"During the PK evaluation period, patients in Cohort 1 receiving Formulation B had a higher mean C_{max} (1519 and 1237 ng/mL), AUC_{last} (29538 and 23511 ng·h/mL), and AUC_{inf} (31933 and 25192 ng·h/mL) compared with those receiving Formulation A; thus, the results indicate that pimitespib Formulations A and B did not fulfill the bioequivalence criteria."

"In a phase III (patients with GIST) study using Formulation A, pimitespib significantly increased the median PFS [136]."

A phase I delayed-start, randomized and pharmacodynamic study of metformin and chemotherapy in patients with solid tumors [137] This is a machine-generated summary of:

Saif, Mohammad Wasif; Rajagopal, Shrikar; Caplain, Jennifer; Grimm, Elizabeth; Serebrennikova, Oksana; Das, Madhumita; Tsichlis, Philip N.; Martell,

Robert: A phase I delayed-start, randomized and pharmacodynamic study of metformin and chemotherapy in patients with solid tumors [137]

Published in: Cancer Chemotherapy and Pharmacology (2019)
Link to original: https://doi.org/10.1007/s00280-019-03967-3

If you want to cite the papers, please refer to the original.

For technical reasons we could not place the page where the original quote is coming from.

Abstract-Summary "We conducted a prospective phase I study to assess the safety of metformin in combination with chemotherapy in patients with solid tumors."

"We conducted a delayed-start randomized trial of non-diabetic patients in two stages."

"In Stage 1, we randomized patients to two arms: concurrent arm (metformin with chemo) vs. delayed arm (chemo alone)."

"In Stage 2, patients in delayed arm were crossed over to receive metformin."

"Patients received metformin 500 mg twice daily with chemotherapy to define dose-limiting toxicities (DLTs) in both stages."

"A total of 100 patients were enrolled (51 in delayed arm vs. 49 concurrent arm)."

"Rate of DLTs in patients receiving metformin with chemotherapy was 6.1% vs. 7.8% in patients receiving chemotherapy alone."

"DLTs seen with addition of metformin included those associated with established chemo adverse events."

"AMPK phosphorylation showed a four- to sixfold increase in AMPK phosphorylation after metformin."

"Post-metformin increase in AMPK phosphorylation may potentially explain lack of disease progression in nearly half of our patients."

Introduction

"Metformin, a biguanide is a relatively safe oral anti-diabetic medication that decreases hepatic glucose production and intestinal glucose absorption, increases insulin sensitivity and sulfonylureas act to stimulate pancreatic islet beta cell insulin release [138]."

"A study examining all cancer-related mortality among diabetic users of insulin, metformin, and sulfonylureas found that insulin users had a mortality risk almost double that of non-insulin users, and sulfonylurea use was associated with an increased mortality as compared to metformin users [139]."

"Retrospective case–control and cohort studies have also indicated an association between metformin use and decreased nonspecific cancer mortality, higher pathologic complete response rate in breast cancer, and a decreased risk of prostate cancer in diabetic patients [140–142]."

"Metformin is a relatively safe drug for use in diabetic populations over many years and compared to other diabetic drugs, metformin is less likely to causes hypoglycemia by itself [143]."

Materials and methods

"Patients deemed eligible during the run-in stage were randomized with equal allocation to one of the following two treatment arms: (1) metformin arm: patients received the next cycle of chemotherapy with concomitant administration of metformin or (2) delayed metformin arm: patients continued with chemotherapy alone for one cycle and then received metformin on the subsequent cycle."

"Following the run-in stage, patients were randomly assigned to receive their next cycle of chemotherapy (Stage 1 of the study) with either (1) concomitant metformin or (2) delayed metformin: continuing with chemotherapy alone for one cycle and then receiving metformin on the subsequent cycle."

"Subjects randomized to the chemotherapy alone arm who had no further DLT during the first cycle of treatment following randomization were then crossed over to receive metformin concomitantly with chemotherapy for the subsequent cycle (Stage 2 of the study)."

Results

"For individual subjects, participation in the trial lasted until metformin is discontinued."

"No lactic acidosis, a known AE associated with metformin, occurred in any patient."

"One patient on metformin with gemcitabine suffered an unrelated fatality, an intracranial hemorrhage from a cerebral metastasis."

"Restaging showed stable disease in 46% at cessation of metformin."

Discussion

"This is the largest prospective phase I study that enrolled 100 non-diabetic patients with different cancers and showed that metformin can be given safely with chemotherapy."

"Many diabetic patients have received metformin during administration of chemotherapy, with no specific safety contraindication for concomitant administration [144, 145]."

"Clinical safety of combining metformin with chemotherapy was not previously tested in a prospective study in either diabetic or non-diabetic patients."

"Our study design also had the advantage of establishing tolerance of the chemotherapy regimen prior to administering concomitant metformin [146]."

"The lower mean glucose levels documented in Stage 2 upon the addition of metformin to chemotherapy were not associated with hypoglycemia in these non-diabetic cancer patients."

"Metformin is a long-approved drug with excellent safety record, and our study suggests the feasibility of combining with both cytotoxic and targeted agents in patients with solid tumors."

Extrapolation of pharmacokinetics and pharmacodynamics of sunitinib in children with gastrointestinal stromal tumors [147] This is a machine-generated summary of:

Khosravan, Reza; DuBois, Steven G.; Janeway, Katherine; Wang, Erjian: Extrapolation of pharmacokinetics and pharmacodynamics of sunitinib in children with gastrointestinal stromal tumors [147]

Published in: Cancer Chemotherapy and Pharmacology (2021)

Link to original: https://doi.org/10.1007/s00280-020-04221-x

If you want to cite the papers, please refer to the original.

For technical reasons we could not place the page where the original quote is coming from.

Abstract-Summary "The starting dose of sunitinib in children with gastrointestinal stromal tumors (GIST) was extrapolated based on data in adults with GIST or solid tumors and children with solid tumors."

"Integrated population pharmacokinetics (PK), PK/pharmacodynamics (PD), and exposure–response analyses using nonlinear mixed-effects modeling approaches were performed to extrapolate PK and PD of sunitinib in children with GIST at projected dose(s) with plasma drug exposures comparable to 50-mg/day in adults with GIST."

"The effect of age on sunitinib apparent clearance (CL/F) and body surface area on SU012662 CL/F was statistically significant ($P \leq 0.001$): children who were younger or of smaller body size had lower CL/F; however, age and body size did not appear to negatively affect safety or efficacy response to plasma drug exposure."

"Based on PK, safety, and efficacy trial simulations, a sunitinib starting dose of ~ 25 mg/m^2/day was predicted to provide comparable plasma drug exposures in children with GIST as in adults with GIST treated with 50 mg/day."

"In the absence of a tumor type effect of sunitinib on CL/F in children, the projected equivalent dose for this population would be ~ 20 mg/m^2/day."

Introduction

"Inhibition of KIT and PDGFRA signaling pathways by therapeutic agents (i.e., imatinib and sunitinib) has been demonstrated to have survival benefit in adult patients with metastatic GIST [148, 149]."

"These models have also been used to examine covariates that potentially contribute to variability in sunitinib and SU012662 exposure, and to predict efficacy and safety [150–153]."

"The objectives of the current study were to develop a Pop-PK model for sunitinib and SU012662 using pooled PK data from all available studies in adults and children with GIST or solid tumors and to develop sequential PK/pharmacodynamics (PD) modeling/analysis with respect to key safety and efficacy endpoints, using PK-model post hoc predictions."

"Further, we aimed to extrapolate PK, safety, and efficacy of sunitinib for children with GIST using the developed models and to identify covariates that may account for the inter-individual variability in sunitinib PK/PD."

Methods

"Data from safety and efficacy assessments were used for the PK/PD modeling portions."

"Two previous Pop-PK analyses of sunitinib and SU01266 suggested sunitinib concentration–time data were well described using a population approach with a two-compartment PK model with first-order absorption and elimination (see Online resource 2)."

"The base and final models were validated using visual predictive check techniques containing 1000 simulations, and the median and upper and lower bounds of the 95% prediction interval (PI) for PK were compared against the observed median and confidence intervals (CIs)."

"Based on the final Pop-PK and PK/PD models, trial simulations were carried out to provide predictions with respect to PK, safety, and efficacy of sunitinib in children with GIST ages 6–11 (n = 210) and 12–17 years (n = 210) in comparison to adults (n = 210) with GIST."

Results

"Based on the analyses of the SLD PK/PD trial simulations, the predicted medians (95% CI) for TTP and ORR, respectively, were 20.9 (15.8–32.4) weeks and 4.9% (0.0–21.5%) in children aged 6–11 years and 23.8 (14.9–31.2) weeks and 8.2% (0.0–15.2%) in children aged 12–17 years, indicating slightly lower TTP and ORR in children dosed with sunitinib 15 mg/m^2/day as compared with adults dosed at 50 mg/day."

"Based on the trial simulations, the predicted medians (95% CI) for TTP and ORR, respectively, at the maximum sunitinib dose were similar to those at the starting dose: 25.1 (10.5–42.6) weeks and 9.0% (0.0–36.0%) for patients in the Janeway and Agaram studies [154, 155] versus 24.7 (12.7–42.6) weeks and 9.0% (0.0–27.0%) in adults."

Discussion

"In these previous studies, the predicted measured changes from baseline in safety in children dosed with sunitinib 15 mg/m^2 were fewer than those reported in adults who received 50 mg/day, which indicated doses of sunitinib higher than 15 mg/m^2 could potentially be well-tolerated in children with GIST."

"The observed ORR within the Janeway and Agaram studies was ~ 18% [154, 155], which is higher than the median predictions, but still within the 95% CI

(0.0–36.0%) for the median, confirming the predictiveness of the SLD PK/PD model in children with GIST."

"The trial simulation results for PK, safety, and efficacy indicated that sunitinib starting doses of ~ 25 mg/m^2 would be more appropriate in children with GIST and provide PK, safety, and efficacy results comparable to adults with GIST treated with 50 mg."

Conclusion

"Key safety and efficacy endpoints of sunitinib in children with GIST were successfully described using mechanism-based and semi-mechanistic PK/PD models."

"Of the PK, safety, and efficacy trial simulations, a sunitinib starting dose of ~ 15 mg/m^2/day appears to be inadequate in the treatment of children with GIST."

"The simulations in the current study predict that a sunitinib starting dose of ~ 25 mg/m^2/day, in the presence of tumor type effect (as in patients with GIST vs. patients with solid tumors) on CL/F, equivalent to ~ 20 mg/m^2/day in the absence of tumor type effect on CL/F, to be more appropriate in children with GIST, as this dose is predicted to provide comparable plasma drug exposures and subsequently, safety and efficacy to 50 mg/day on Schedule 4/2 in adult patients with GIST."

Bibliography

1. Zhang A, Meng K, Liu Y, Pan Y, Qu W, Chen D, Xie S (2020) Absorption, distribution, metabolism, and excretion of nanocarriers in vivo and their influences. Adv Colloid Interf Sci 284:102261
2. Ernsting MJ, Murakami M, Roy A, Li SD (2013) Factors controlling the pharmacokinetics, biodistribution and intratumoral penetration of nanoparticles. J Control Release 172(3):782–794
3. Alexis F, Pridgen E, Molnar LK, Farokhzad OC (2008) Factors affecting the clearance and biodistribution of polymeric nanoparticles. Mol Pharm 5(4):505–515
4. Xia Q, Li H, Xiao K (2016) Factors affecting the pharmacokinetics, biodistribution and toxicity of gold nanoparticles in drug delivery. Curr Drug Metab 17(9):849–861
5. Cho WS, Kang BC, Lee JK, Jeong J, Che JH, Seok SH (2013) Comparative absorption, distribution, and excretion of titanium dioxide and zinc oxide nanoparticles after repeated oral administration. Particle Fibre Toxicol 10(1):1–9
6. Swetledge S, Jung JP, Carter R, Sabliov C (2021) Distribution of polymeric nanoparticles in the eye: implications in ocular disease therapy. J Nanobiotechnol 19(1):1–9
7. Yan Y, Chen B, Yin Q, Wang Z, Yang Y, Wan F, Wang Y, Tang M, Xia H, Chen M, Liu J (2022) Dissecting extracellular and intracellular distribution of nanoparticles and their contribution to therapeutic response by monochromatic ratiometric imaging. Nat Commun 13(1):2004
8. Christensen J, Litherland K, Faller T, van de Kerkhof E, Natt F, Hunziker J, Boos J, Beuvink I, Bowman K, Baryza J, Beverly M (2014) Biodistribution and metabolism studies of lipid nanoparticle–formulated internally [3H]-labeled siRNA in mice. Drug Metab Dispos 42(3):431–440
9. Zhang YN, Poon W, Tavares AJ, McGilvray ID, Chan WC (2016) Nanoparticle–liver interactions: cellular uptake and hepatobiliary elimination. J Control Release 240:332–348
10. Zielińska A, Carreiró F, Oliveira AM, Neves A, Pires B, Venkatesh DN, Durazzo A, Lucarini M, Eder P, Silva AM, Santini A (2020) Polymeric nanoparticles: production, characterization, toxicology and ecotoxicology. Molecules 25(16):3731

11. Longmire M, Choyke PL, Kobayashi H (2008) Clearance properties of nano-sized particles and molecules as imaging agents: considerations and caveats. Nanomedicine (Lond) 3(5):703–717
12. Seo HJ, Nam SH, Im HJ, Park JY, Lee JY, Yoo B, Lee YS, Jeong JM, Hyeon T, Who Kim J, Lee JS (2015) Rapid hepatobiliary excretion of micelle-encapsulated/radiolabeled upconverting nanoparticles as an integrated form. Sci Rep 5(1):15685
13. Poon W, Zhang YN, Ouyang B, Kingston BR, Wu JL, Wilhelm S, Chan WC (2019) Elimination pathways of nanoparticles. ACS Nano 13(5):5785–5798
14. Yu M, Zheng J (2015) Clearance pathways and tumor targeting of imaging nanoparticles. ACS Nano 9(7):6655–6674
15. Rajput A, Sevalkar G, Pardeshi K, Pingale P (2023) COMPUTATIONAL NANOSCIENCE AND TECHNOLOGY. OpenNano:100147
16. Abla N, Naik A, Guy RH, Kalia YN (2005) Effect of charge and molecular weight on transdermal peptide delivery by iontophoresis. Pharm Res. https://doi.org/10.1007/s11095-005-8110-2
17. Charro D, Guy M (1998) Iontophoresis of peptides. In B. Berner and S. M. Dinh (eds.), Electronically Controlled Drug Delivery, CRC Press, Boca Raton 129–157
18. Kalia YN, Naik A, Garrison J, Guy RH (2004) Iontophoretic drug delivery. Advanced drug delivery reviews 56(5):619–58. https://doi.org/10.1016/j.addr.2003.10.026
19. R Haak, SK Gupta (1993) Pulsatile drug delivery from electrotransport therapeutic systems. In R. Gurny, H. E. Junginger, and N. A. Peppas R. Gurny H. E. Junginger N. A. Peppas (eds.), Pulsatile Drug Delivery Current Applications and Future Trends, Wiss. Verl.-Ges., Stuttgart 99–112
20. Green PG (1996) Iontophoretic delivery of peptide drugs. Journal of controlled release 41(1–2):33–48. https://doi.org/10.1016/0168-3659(96)01354-5
21. Zhao G, Long L, Zhang L, Peng M, Cui T, Wen X, Zhou X, Sun L, Che L (2017) Smart pH-sensitive nanoassemblies with cleavable PEGylation for tumor targeted drug delivery. Sci Rep. https://doi.org/10.1038/s41598-017-03111-2
22. Tang M et al (2016) Disulfide-bridged cleavable PEGylation of poly-L-lysine for SiRNA delivery. Methods Mol Biol 1364:49–61. https://doi.org/10.1007/978-1-4939-3112-5_5
23. Dong H et al (2015) Disulfide-bridged cleavable PEGylation in polymeric nanomedicine for controlled therapeutic delivery. Nanomedicine (Lond) 10:1941–1958. https://doi.org/10.2217/nnm.15.38
24. Zhu H et al (2014) Cleavable PEGylation and hydrophobic histidylation of polylysine for siRNA delivery and tumor gene therapy. ACS Appl Mater Interfaces 6:10393–10407. https://doi.org/10.1021/am501928p
25. Zalipsky S, Mullah N, Engbers C, Hutchins MU, Kiwan R (2007) Thiolytically cleavable dithiobenzyl urethane-linked polymer-protein conjugates as macromolecular prodrugs: reversible PEGylation of proteins. Bioconjug Chem 18:1869–1878. https://doi.org/10.1021/bc7001902
26. Choi SH, Lee H, Park TG (2003) PEGylation of G-CSF using cleavable oligo-lactic acid linkage. J Control Release 89:271–284. https://doi.org/10.1016/S0168-3659(03)00100-7
27. Lee ES et al (2008) Super pH-sensitive multifunctional polymeric micelle for tumor pH(e) specific TAT exposure and multidrug resistance. J Control Release 129:228–236. https://doi.org/10.1016/j.jconrel.2008.04.024
28. Lee ES, Na K, Bae YH (2005) Super pH-sensitive multifunctional polymeric micelle. Nano Lett 5:325–329. https://doi.org/10.1021/nl0479987
29. Knorr V, Allmendinger L, Walker GF, Paintner FF, Wagner E (2007) An acetal-based PEGylation reagent for pH-sensitive shielding of DNA polyplexes. Bioconjug Chem 18:1218–1225. https://doi.org/10.1021/bc060327a
30. Zhu Y et al (2011) Highly efficient nanomedicines assembled via polymer-drug multiple interactions: tissue-selective delivery carriers. J Control Release 152:317–324. https://doi.org/10.1016/j.jconrel.2011.03.013

31. Zhou X et al (2014) Facile route to versatile nanoplatforms for drug delivery by one-pot self-assembly. Acta Biomater 10:2630–2642. https://doi.org/10.1016/j.actbio.2014.01.024
32. Zhang JX, Ma PX (2009) Polymeric core-shell assemblies mediated by host-guest interactions: versatile nanocarriers for drug delivery. Angew Chem Int Ed 48:964–968. https://doi.org/10.1002/anie.v48:5
33. Che L et al (2012) Assembled nanomedicines as efficient and safe therapeutics for articular inflammation. Int J Pharm 439:307–316. https://doi.org/10.1016/j.ijpharm.2012.09.017
34. Liu B, Gao W, Wu H, Liu H, Pan H (2021) New PTX-HS15/T80 mixed micelles: cytotoxicity, pharmacokinetics and tissue distribution. AAPS PharmSciTech. https://doi.org/10.1208/s12249-021-01929-8
35. Taxol (paclitaxel) injection [package insert] (2017) Bristol-Myers Squibb S.R.L.
36. Cagelac M, Tesanb FC, Bernabeu E et al (2017) Polymeric mixed micelles as nanomedicines: achievements and perspectives. Eur J Pharm Biopharm 113:211–228
37. Su QJ, Mo T, Liu L et al (2015) Paclitaxel-loaded Kolliphor (R) HS15/polysorbate 80-mixed nanomicelles: formulation, in vitro characterization and safety evaluation. Lat Am J Pharm 34(4):702–711
38. Marupudi NI, Han JE, Li KW, Renard VM, Tyler BM, Brem H (2007) Paclitaxel: a review of adverse toxicities and novel delivery strategies. Expert Opin Drug Saf 6:609–621
39. Albarki MA, Donovan MD (2020) Bigger or smaller? Size and loading effects on nanoparticle uptake efficiency in the nasal mucosa. AAPS PharmSciTech. https://doi.org/10.1208/s12249-020-01837-3
40. de Oliveira ER Jr, Santos LCR, Salomão MA, Nascimento TL, de Almeida Ribeiro Oliveira G, Lião LM, Lima EM (2020) Nose-to-brain drug delivery mediated by polymeric nanoparticles: influence of PEG surface coating. Drug Deliv Transl Res. https://doi.org/10.1007/s13346-020-00816-2
41. Feng Y, He H, Li F, Lu Y, Qi J, Wu W (2018) An update on the role of nanovehicles in nose-to-brain drug delivery. Drug Discov Today 23(5):1079–1088. https://doi.org/10.1016/j.drudis.2018.01.005
42. Islam SU, Shehzad A, Ahmed MB, Lee YS (2020) Intranasal delivery of nanoformulations: a potential way of treatment for neurological disorders. Molecules 25(8). https://doi.org/10.3390/molecules25081929
43. de Oliveira Junior ER, Nascimento TL, Salomao MA, da Silva ACG, Valadares MC, Lima EM (2019) Increased nose-to-brain delivery of melatonin mediated by polycaprolactone nanoparticles for the treatment of glioblastoma. Pharm Res 36(9):131. https://doi.org/10.1007/s11095-019-2662-z
44. de Oliveira Junior ER, Truzzi E, Ferraro L, Fogagnolo M, Pavan B, Beggiato S et al (2020) Nasal administration of nanoencapsulated geraniol/ursodeoxycholic acid conjugate: towards a new approach for the management of Parkinson's disease. J Control Release 321:540–552. https://doi.org/10.1016/j.jconrel.2020.02.033
45. Bhattamisra SK, Shak AT, Xi LW, Safian NH, Choudhury H, Lim WM et al (2020) Nose to brain delivery of rotigotine loaded chitosan nanoparticles in human SH-SY5Y neuroblastoma cells and animal model of Parkinson's disease. Int J Pharm 579:119148. https://doi.org/10.1016/j.ijpharm.2020.119148
46. Sonvico F, Clementino A, Buttini F, Colombo G, Pescina S, Staniscuaski Guterres S et al (2018) Surface-modified nanocarriers for nose-to-brain delivery: from bioadhesion to targeting. Pharmaceutics 10(1). https://doi.org/10.3390/pharmaceutics10010034
47. Nigam K, Kaur A, Tyagi A, Nematullah M, Khan F, Gabrani R et al (2019) Nose-to-brain delivery of lamotrigine-loaded PLGA nanoparticles. Drug Deliv Transl Res 9(5):879–890. https://doi.org/10.1007/s13346-019-00622-5
48. Hao RB, Sun BX, Yang LH, Ma C, Li SL (2020) RVG29-modified microRNA-loaded nanoparticles improve ischemic brain injury by nasal delivery. Drug Deliv 27(1):772–781. https://doi.org/10.1080/10717544.2020.1760960

49. Ullah I, Chung K, Bae S, Li Y, Kim C, Choi B et al (2020) Nose-to-brain delivery of cancer-targeting paclitaxel-loaded nanoparticles potentiates antitumor effects in malignant glioblastoma. Mol Pharm 17(4):1193–1204. https://doi.org/10.1021/acs.molpharmaceut.9b01215
50. Brandl F, Bertrand N, Lima EM, Langer R (2015) Nanoparticles with photoinduced precipitation for the extraction of pollutants from water and soil. Nat Commun 6:7765. https://doi.org/10.1038/Ncomms8765
51. Gao N, Chen Z, Xiao X, Ruan C, Mei L, Liu Z, Zeng X (2015) Surface modification of paclitaxel-loaded tri-block copolymer PLGA-b-PEG-b-PLGA nanoparticles with protamine for liver cancer therapy. J Nanopart Res. https://doi.org/10.1007/s11051-015-3121-3
52. Chen Z, Wang Z, Chen X, Xu H, Liu J (2013) Chitosan-capped gold nanoparticles for selective and colorimetric sensing of heparin. J Nanopart Res 15:1930
53. Geever LM, Cooney CC, Lyons JG, Kennedy JE, Nugent MJD, Devery S, Higginbotham CL (2008) Characterisation and controlled drug release from novel drug-loaded hydrogels. Eur J Pharm Biopharm 69(3):1147–1159
54. Yanagihara K, Takigahira M, Kubo T, Ochiya T, Hamaguchi T, Matsumura Y (2014) Marked antitumor effect of NK012, a SN-38-incorporating micelle formulation, in a newly developed mouse model of liver metastasis resulting from gastric cancer. Ther Deliv 5(2):129–138
55. Zeng XW, Tao W, Wang ZY, Zhang XD, Zhu HJ, Wu YP, Gao YF, Liu KW, Jiang YY, Huang LQ, Mei L, Feng SS (2015) Docetaxel-loaded nanoparticles of dendritic amphiphilic block copolymer H40-PLA-b-TPGS for cancer treatment. Part Part Syst Charact 32(1):112–122
56. Gao YH, Yang CH, Liu X, Ma RJ, Kong DL, Shi LQ (2012) A multifunctional nanocarrier based on nanogated mesoporous silica for enhanced tumor-specific uptake and intracellular delivery. Macromol Biosci 12(2):251–259
57. Romberg B, Hennink WE, Storm G (2008) Sheddable coatings for long-circulating nanoparticles. Pharm Res 25(1):55–71
58. Gomez JMM, Csaba N, Fischer S, Sichelstiel A, Kundig TM, Gander B, Johansen P (2008) Surface coating of PLGA microparticles with protamine enhances their immunological performance through facilitated phagocytosis. J Control Release 130(2):161–167
59. Zeng XW, Tao W, Mei L, Huang LG, Tan CY, Feng SS (2013) Cholic acid-functionalized nanoparticles of star-shaped PLGA-vitamin E TPGS copolymer for docetaxel delivery to cervical cancer. Biomaterials 34(25):6058–6067
60. Alayoubi A, Alqahtani S, Kaddoumi A, Nazzal S (2013) Effect of PEG surface conformation on anticancer activity and blood circulation of nanoemulsions loaded with tocotrienol-rich fraction of palm oil. AAPS J. https://doi.org/10.1208/s12248-013-9525-z
61. Le UM, Cui Z (2006) Long-circulating gadolinium-encapsulated liposomes for potential application in tumor neutron capture therapy. Int J Pharm 312(1–2):105–112. https://doi.org/10.1016/j.ijpharm.2006.01.002
62. Maruyama K, Yuda T, Okamoto A, Kojima S, Suginaka A, Iwatsuru M (1992) Prolonged circulation time in vivo of large unilamellar liposomes composed of distearoyl phosphatidylcholine and cholesterol containing amphipathic poly(ethylene glycol). Biochim Biophys Acta 1128(1):44–49
63. Moghimi SM (1997) Prolonging the circulation time and modifying the body distribution of intravenously injected polystyrene nanospheres by prior intravenous administration of poloxamine-908. A ‘hepatic-blockade’ event or manipulation of nanosphere surface in vivo? Biochim Biophys Acta Gen Subj 1336(1):1–6
64. Hammad RW, Sanad RAB, Abdelmalk NS, Aziz RL, Torad FA (2018) Intranasal surface-modified mosapride citrate-loaded nanostructured lipid carriers (MOS-SMNLCs) for treatment of reflux diseases: in vitro optimization, pharmacodynamics, and pharmacokinetic studies. AAPS PharmSciTech. https://doi.org/10.1208/s12249-018-1142-9
65. Tack J, Camilleri M, Chang L, Chey WD, Galligan JJ, Lacy BE et al (2012) Systematic review: cardiovascular safety profile of 5-HT 4 agonists developed for gastrointestinal disorders. Aliment Pharmacol Ther 35(7):745–767

66. Gaba B, Fazil M, Khan S, Ali A, Baboota S, Ali J (2015) Nanostructured lipid carrier system for topical delivery of terbinafine hydrochloride. Bull Fac Pharm Cairo Univ [Internet] 53(2):147–159. https://doi.org/10.1016/j.bfopcu.2015.10.001
67. Li B, Ge Z-Q (2012) Nanostructured lipid carriers improve skin permeation and chemical stability of idebenone. AAPS PharmSciTech [Internet] 13(1):276–283. http://www.ncbi.nlm.nih.gov/pubmed/22234598. Accessed 16 Mar 2018
68. Ugwoke MI, Agu RU, Verbeke N, Kinget R (2005) Nasal mucoadhesive drug delivery: background, applications, trends and future perspectives. Adv Drug Deliv Rev 57(11):1640–1665
69. Gartziandia O, Herran E, Pedraz JL, Carro E, Igartua M, Hernandez RM (2015) Chitosan coated nanostructured lipid carriers for brain delivery of proteins by intranasal administration. Colloids Surfaces B Biointerfaces [Internet] 134:304–313. https://doi.org/10.1016/j.colsurfb.2015.06.054
70. Cui F, Qian F, Yin C (2006) Preparation and characterization of mucoadhesive polymer-coated nanoparticles. Int J Pharm 316(1–2):154–161
71. Abdelbary G (2011) Ocular ciprofloxacin hydrochloride mucoadhesive chitosan-coated liposomes. Pharm Dev Technol 16(1):44–56
72. Sharma M, Gupta N, Gupta S (2016) Implications of designing clarithromycin loaded solid lipid nanoparticles on their pharmacokinetics, antibacterial activity and safety. RSC Adv [Internet] 6:76621–76631. https://doi.org/10.1039/C6RA12841F
73. Shafiei M, Galettis P, Beale P, Reuter SE, Martin JH, McLachlan AJ, Blinman P (2023) Influence of age on pharmacokinetics of capecitabine and its metabolites in older adults with cancer: a pilot study. Cancer Chemother Pharmacol. https://doi.org/10.1007/s00280-023-04552-5
74. Vasudev A, Chaudhari S, Sethi R, Fu R, Sandieson RM, Forester BP (2018) A review of the pharmacological and clinical profile of newer atypical antipsychotics as treatments for bipolar disorder: considerations for use in older patients. Drugs Aging. https://doi.org/10.1007/s40266-018-0579-6
75. Yatham LN, Kennedy SH, Parikh SV, Schaffer A, Bond DJ, Frey BN et al (2018) Canadian Network for Mood and Anxiety Treatments (CANMAT) and International Society for Bipolar Disorders (ISBD) 2018 guidelines for the management of patients with bipolar disorder. Bipolar Disord 20(2):97–170
76. Young RC, Mulsant BH, Sajatovic M, Gildengers AG, Gyulai L, Al Jurdi RK et al (2017) GERI-bd: a randomized double-blind controlled trial of lithium and divalproex in the treatment of mania in older patients with bipolar disorder. Am J Psychiatry 174(11):1086–1093
77. Durgam S, Starace A, Li D, Migliore R, Ruth A, Németh G et al (2015) The efficacy and tolerability of cariprazine in acute mania associated with bipolar I disorder: a phase II trial. Bipolar Disord 17(1):63–75
78. Janssen (2003) Product monograph: asenapine. Janssen, New York, pp 1–52
79. Berwaerts J, Melkote R, Nuamah I, Lim P (2012) A randomized, placebo- and active-controlled study of paliperidone extended-release as maintenance treatment in patients with bipolar i disorder after an acute manic or mixed episode. J Affect Disord 138(3):247–258
80. Romański M, Wachowiak J, Główka FK (2018) Treosulfan pharmacokinetics and its variability in pediatric and adult patients undergoing conditioning prior to hematopoietic stem cell transplantation: current state of the art, in-depth analysis, and perspectives. Clin Pharmacokinet. https://doi.org/10.1007/s40262-018-0647-4
81. Romański M, Urbaniak B, Kokot Z, Główka FK (2015a) Activation of prodrug treosulfan at pH 7.4 and 37°C accompanied by hydrolysis of its active epoxides: kinetic studies with clinical relevance. J Pharm Sci 104:4433–4442
82. Romański M, Ratajczak W, Główka F (2017a) Kinetic and mechanistic study of the pH-dependent activation (epoxidation) of prodrug treosulfan including the reaction inhibition in a borate buffer. J Pharm Sci 106:1917–1922

83. Romański M, Mikołajewski J, Główka FK (2017b) Effect of temperature on the kinetics of the activation of treosulfan and hydrolytic decomposition of its active epoxy derivatives. J Pharm Sci 106:3156–3160
84. Główka F, Kasprzyk A, Romański M, Wróbel T, Wachowiak J, Szpecht D et al (2015) Pharmacokinetics of treosulfan and its active monoepoxide in pediatric patients after intravenous infusion of high-dose treosulfan prior to HSCT. Eur J Pharm Sci 68:87–93
85. Romański M, Kasprzyk A, Karbownik A, Szałek E, Główka FK (2016) Formation rate-limited pharmacokinetics of biologically active epoxy transformers of prodrug treosulfan. J Pharm Sci 105:1790–1797
86. Romański M, Baumgart J, Böhm S, Główka FK (2015b) Penetration of treosulfan and its active monoepoxide transformation product into central nervous system of juvenile and young adult rats. Drug Metab Dispos 43:1946–1954
87. Romański M, Kasprzyk A, Walczak M, Ziółkowska A, Główka F (2017c) Disposition of treosulfan and its active monoepoxide in a bone marrow, liver, lungs, brain, and muscle: studies in a rat model with clinical relevance. Eur J Pharm Sci 109:616–623
88. Ten Brink MH, Ackaert O, Zwaveling J, Bredius RG, Smiers FJ, den Hartigh J et al (2014) Pharmacokinetics of treosulfan in pediatric patients undergoing hematopoietic stem cell transplantation. Ther Drug Monit 36:465–472
89. Chiesa R, Winter R, Nademi Z, Standing J, Amrolia P, Veys P et al (2014) Pharmacokinetics of high dose intravenous treosulfan in children prior to allogeneic HCT [abstract no. PH-P584]. Bone Marrow Transplant 49(Suppl 1):S380–S381
90. Van den Berg PJ, Ruppert M, Sykora K-W, Beier R, Beelen DW, Hilger RA et al (2014) A preliminary population pharmacokinetic model for dose selection of treosulfan used in conditioning treatment prior to haematopoietic stem cell transplantation (HSCT) in children [abstract PH-P543]. Bone Marrow Transplant 49(Suppl 1):S360–S361
91. Danielak D, Twardosz J, Kasprzyk A, Wachowiak J, Kałwak K, Główka F (2017) Population pharmacokinetics of treosulfan and development of a limited sampling strategy in children prior to hematopoietic stem cell transplantation. Eur J Clin Pharmacol 74:79–89
92. Van der Stoep MYEC, Bertaina A, ten Brink MH, Bredius RG, Smiers FJ, Wanders DCM et al (2017) High interpatient variability of treosulfan exposure is associated with early toxicity in paediatric HSCT: a prospective multicentre study. Br J Haematol 179:772–780
93. Mohanan E, Panetta JC, Lakshmi KM, Edison ES, Korula A, Na F et al (2017) Pharmacokinetics and pharmacodynamics of treosulfan in patients with thalassemia major undergoing allogeneic hematopoietic stem cell transplantation. Clin Pharmacol Ther. https://doi.org/10.1002/cpt.988
94. Xu Y, Chen J, Yang D, Hu Y, Jiang B, Ruan Z, Lou H (2022) Development of a particle swarm optimization-backpropagation artificial neural network model and effects of age and gender on pharmacokinetics study of omeprazole enteric-coated tablets in Chinese population. BMC Pharmacol Toxicol. https://doi.org/10.1186/s40360-022-00594-2
95. Koukoula M, Dotsikas Y, Molou E, Schulpis KH, Thodi G, Chatzidaki M, Triantafylli O, Loukas YL (2017) Study of the effect of CYP2C19 polymorphisms on omeprazole pharmacokinetics by utilizing validated LC-MS/MS and Real Time-PCR methods. J Chromatogr B Anal Technol Biomed Life Sci 1047:173–179
96. Li X, Qiu M, Wang SJ, Zhu H, Feng B, Zheng L (2020) A Phase I dose-escalation, pharmacokinetics and food-effect study of oral donafenib in patients with advanced solid tumours. Cancer Chemother Pharmacol. https://doi.org/10.1007/s00280-020-04031-1
97. Llovet JM, Ricci S, Mazzaferro V, Hilgard P, Gane E, Blanc JF, de Oliveira AC, Santoro A, Raoul JL, Forner A, Schwartz M, Porta C, Zeuzem S, Bolondi L, Greten TF, Galle PR, Seitz JF, Borbath I, Haussinger D, Giannaris T, Shan M, Moscovici M, Voliotis D, Bruix J, Group SIS (2008) Sorafenib in advanced hepatocellular carcinoma. N Engl J Med 359(4):378–390. https://doi.org/10.1056/NEJMoa0708857
98. Escudier B, Eisen T, Stadler WM, Szczylik C, Oudard S, Siebels M, Negrier S, Chevreau C, Solska E, Desai AA, Rolland F, Demkow T, Hutson TE, Gore M, Freeman S, Schwartz B,

Shan M, Simantov R, Bukowski RM, Group TS (2007) Sorafenib in advanced clear-cell renal-cell carcinoma. N Engl J Med 356(2):125–134. https://doi.org/10.1056/NEJMoa060655
99. Cheng AL, Kang YK, Chen Z, Tsao CJ, Qin S, Kim JS, Luo R, Feng J, Ye S, Yang TS, Xu J, Sun Y, Liang H, Liu J, Wang J, Tak WY, Pan H, Burock K, Zou J, Voliotis D, Guan Z (2009) Efficacy and safety of sorafenib in patients in the Asia-Pacific region with advanced hepatocellular carcinoma: a phase III randomised, double-blind, placebo-controlled trial. Lancet Oncol 10(1):25–34. https://doi.org/10.1016/S1470-2045(08)70285-7
100. Parikh ND, Marshall VD, Singal AG, Nathan H, Lok AS, Balkrishnan R, Shahinian V (2017) Survival and cost-effectiveness of sorafenib therapy in advanced hepatocellular carcinoma: an analysis of the SEER-Medicare database. Hepatology 65(1):122–133. https://doi.org/10.1002/hep.28881
101. Leung HW, Liu CF, Chan AL (2016) Cost-effectiveness of sorafenib versus SBRT for unresectable advanced hepatocellular carcinoma. Radiat Oncol 11:69. https://doi.org/10.1186/s13014-016-0644-4
102. Clark JW, Eder JP, Ryan D, Lathia C, Lenz HJ (2005) Safety and pharmacokinetics of the dual action Raf kinase and vascular endothelial growth factor receptor inhibitor, BAY 43–9006, in patients with advanced, refractory solid tumors. Clin Cancer Res 11(15):5472–5480. https://doi.org/10.1158/1078-0432.CCR-04-2658
103. Strumberg D, Awada A, Piccart P et al (2003) Final report of the multiple single agent phase I clinical trials of the novel Raf kinase inhibitor BAY 43-9006 in patients with refractory solid tumors [Abstract]. Proc Am Soc Clin Oncol 22(203)
104. Maguire WF, Schmitz JC, Scemama J, Czambel K, Lin Y, Green AG, Wu S, Lin H, Puhalla S, Rhee J, Stoller R, Tawbi H, Lee JJ, Wright JJ, Beumer JH, Chu E, Appleman LJ (2021) Phase 1 study of safety, pharmacokinetics, and pharmacodynamics of tivantinib in combination with bevacizumab in adult patients with advanced solid tumors. Cancer Chemother Pharmacol. https://doi.org/10.1007/s00280-021-04317-y
105. Corso S, Giordano S (2013) Cell-autonomous and non-cell-autonomous mechanisms of HGF/MET-driven resistance to targeted therapies: from basic research to a clinical perspective. Cancer Discov 3:978–992. https://doi.org/10.1158/2159-8290.CD-13-0040
106. Shojaei F, Lee JH, Simmons BH et al (2010) HGF/c-Met acts as an alternative angiogenic pathway in sunitinib-resistant tumors. Cancer Res 70:10090–10100. https://doi.org/10.1158/0008-5472.CAN-10-0489
107. You W-K, Sennino B, Williamson CW et al (2011) VEGF and c-Met blockade amplify angiogenesis inhibition in pancreatic islet cancer. Cancer Res 71:4758–4768. https://doi.org/10.1158/0008-5472.CAN-10-2527
108. Katayama R, Aoyama A, Yamori T et al (2013) Cytotoxic activity of tivantinib (ARQ 197) is not due solely to c-MET inhibition. Cancer Res 73:3087–3096. https://doi.org/10.1158/0008-5472.CAN-12-3256
109. Basilico C, Pennacchietti S, Vigna E et al (2013) Tivantinib (ARQ197) displays cytotoxic activity that is independent of its ability to bind MET. Clin Cancer Res 19:2381–2392. https://doi.org/10.1158/1078-0432.CCR-12-3459
110. Kast J, Dutta S, Upreti VV (2021) Panitumumab: a review of clinical pharmacokinetic and pharmacology properties after over a decade of experience in patients with solid tumors. Adv Ther. https://doi.org/10.1007/s12325-021-01809-4
111. Vectibix® (panitumumab) (2017) Full prescribing information. Amgen Inc., Thousand Oaks, CA
112. Vectibix® (panitumumab) (2018) Summary of product characteristics. Amgen Europe B.V., Breda, Netherlands
113. Center for Drug Evaluation and Research (2006) Vectibix® (Panitumumab), application number 125147/0 (Clinical Pharmacology and Biopharmaceutics Review). https://www.accessdata.fda.gov/drugsatfda_docs/nda/2006/125147s0000_ClinPharmR.pdf. Accessed 1 Nov 2020

114. Yang XD, Jia XC, Corvalan JR et al (1999) Eradication of established tumors by a fully human monoclonal antibody to the epidermal growth factor receptor without concomitant chemotherapy. Cancer Res 59(6):1236–1243
115. Freeman DJ, Bush T, Ogbagabriel S et al (2009) Activity of panitumumab alone or with chemotherapy in non-small cell lung carcinoma cell lines expressing mutant epidermal growth factor receptor. Mol Cancer Ther 8(6):1536–1546
116. Freeman DJ, McDorman K, Ogbagabriel S et al (2012) Tumor penetration and epidermal growth factor receptor saturation by panitumumab correlate with antitumor activity in a preclinical model of human cancer. Mol Cancer 11:47
117. Yang XD, Jia XC, Corvalan JR et al (2001) Development of ABX-EGF, a fully human anti-EGF receptor monoclonal antibody, for cancer therapy. Crit Rev Oncol Hematol 38(1):17–23
118. Liao MZ, Berkhout M, Prenen H et al (2020a) Dose regimen rationale for panitumumab in cancer patients: to be based on body weight or not. Clin Pharmacol 12:109–114
119. Yang BB, Lum P, Chen A et al (2010) Pharmacokinetic and pharmacodynamic perspectives on the clinical drug development of panitumumab. Clin Pharmacokinet 49(11):729–740
120. Hecht JR, Patnaik A, Berlin J et al (2007) Panitumumab monotherapy in patients with previously treated metastatic colorectal cancer. Cancer 110(5):980–988
121. Ma P, Yang BB, Wang YM et al (2009) Population pharmacokinetic analysis of panitumumab in patients with advanced solid tumors. J Clin Pharmacol 49(10):1142–1156
122. European Medicines Evaluation Agency (2007) Vectibix: scientific discussion. European Medicines Agency, Amsterdam, the Netherlands. Accessed 21 Dec 2020
123. Mitchell EP, Hecht JR, Baranda J et al (2007) Panitumumab activity in metastatic colorectal cancer (mCRC) patients (pts) with low or negative tumor epidermal growth factor receptor (EGFr) levels: an updated analysis. J Clin Oncol 25(18_Suppl):4082
124. Muro K, Yoshino T, Doi T et al (2009) A phase 2 clinical trial of panitumumab monotherapy in Japanese patients with metastatic colorectal cancer. Jpn J Clin Oncol 39(5):321–326
125. Van Cutsem E, Peeters M, Siena S et al (2007) Open-label phase III trial of panitumumab plus best supportive care compared with best supportive care alone in patients with chemotherapy-refractory metastatic colorectal cancer. J Clin Oncol 25(13):1658–1664
126. Lofgren JA, Dhandapani S, Pennucci JJ et al (2007) Comparing ELISA and surface plasmon resonance for assessing clinical immunogenicity of panitumumab. J Immunol 178(11):7467–7472
127. Weeraratne D, Chen A, Pennucci JJ et al (2011) Immunogenicity of panitumumab in combination chemotherapy clinical trials. BMC Clin Pharmacol 11:17
128. Grechko N, Skarbova V, Tomaszewska-Kiecana M, Ramlau R, Centkowski P, Drew Y, Dziadziuszko R, Zemanova M, Beltman J, Nash E, Habeck J, Liao M, Xiao J (2021) Pharmacokinetics and safety of rucaparib in patients with advanced solid tumors and hepatic impairment. Cancer Chemother Pharmacol. https://doi.org/10.1007/s00280-021-04278-2
129. Rubraca (rucaparib) tablets [prescribing information] (2020) Clovis Oncology, Inc., Boulder. https://clovisoncology.com/pdfs/RubracaUSPI.pdf . Accessed 27 Oct 2020
130. Shapiro GI, Kristeleit R, Burris HA, LoRusso P, Patel MR, Drew Y, Giordano H, Maloney L, Watkins S, Goble S, Jaw-Tsai S, Xiao JJ (2018) Pharmacokinetic study of rucaparib in patients with advanced solid tumors. Clin Pharmacol Drug Dev 8(1):107–118. https://doi.org/10.1002/cpdd.575
131. Liao M, Watkins S, Nash E et al (2020b) Evaluation of absorption, distribution, metabolism, and excretion of [14C]-rucaparib, a poly(ADP-ribose) polymerase inhibitor, in patients with advanced solid tumors. Investig New Drugs 38(3):765–775. https://doi.org/10.1007/s10637-019-00815-2
132. Kristeleit R, Shapiro GI, Burris HA et al (2017) A phase I–II study of the oral PARP inhibitor rucaparib in patients with germline BRCA1/2-mutated ovarian carcinoma or other solid tumors. Clin Cancer Res 23(15):4095–4106. https://doi.org/10.1158/1078-0432.CCR-16-2796

133. Liao M, Jaw-Tsai S, Beltman J, Simmons AD, Harding T, Xiao JJ (2020c) Evaluation of in vitro absorption, distribution, metabolism and excretion and assessment of drug–drug interaction of rucaparib, an orally potent poly(ADP-ribose) polymerase inhibitor. Xenobiotica 50(9):1032–1042. https://doi.org/10.1080/00498254.2020.1737759
134. National Cancer Institute (2015) Cancer Therapy Evaluation Program (CTEP) protocol template for organ dysfunction studies. https://ctep.cancer.gov/protocolDevelopment/docs/CTEP_Organ_Dysfunction_Protocol_Template.docx . Accessed 25 Sept 2020
135. Komatsu Y, Shimokawa T, Akiyoshi K, Karayama M, Shimomura A, Kawamoto Y, Yuki S, Tambo Y, Kasahara K (2022) An open-label, crossover study to compare different formulations and evaluate effect of food on pharmacokinetics of pimitespib in patients with advanced solid tumors. Investig New Drugs. https://doi.org/10.1007/s10637-022-01285-9
136. Honma Y, Kurokawa Y, Sawaki A, Naito Y, Iwagami S, Baba H, Komatsu Y, Nishida T, Doi T (2021) Randomized, double-blind, placebo (PL)-controlled, phase III trial of pimitespib (TAS-116), an oral inhibitor of heat shock protein 90 (HSP90), in patients (pts) with advanced gastrointestinal stromal tumor (GIST) refractory to imatinib (IM), sunitinib (SU) and regorafenib (REG). J Clin Oncol 39(15_Suppl):11524. https://doi.org/10.1200/JCO.2021.39.15_suppl.11524
137. Saif MW, Rajagopal S, Caplain J, Grimm E, Serebrennikova O, Das M, Tsichlis PN, Martell R (2019) A phase I delayed-start, randomized and pharmacodynamic study of metformin and chemotherapy in patients with solid tumors. Cancer Chemother Pharmacol. https://doi.org/10.1007/s00280-019-03967-3
138. DeFronzo RA, Goodman AM (1995) Efficacy of metformin in patients with non-insulin-dependent diabetes mellitus. The Multicenter Metformin Study Group. N Engl J Med 333(9):541–549
139. Soranna D, Scotti L, Zambon A et al (2012) Cancer risk associated with use of metformin and sulfonylurea in type 2 diabetes; a meta-analysis. Oncologist 17(6):813–822
140. Jiralerspong S, Palla SL, Giordano SH et al (2009) Metformin and pathologic complete responses to neoadjuvant chemotherapy in diabetic patients with breast cancer. J Clin Oncol 27(20):3297–3302
141. Zhang Y, Storr SJ, Johnson K et al (2014) Involvement of metformin and AMPK in the radioresponse and prognosis of luminal versus basil-like breast cancer treated with radiotherapy. Oncotarget 5(24):12936–12949
142. Spratt DE, Zhang C, Zumsteg ZS, Pei X, Zhang Z, Zelefsky MJ (2013) Metformin and prostate cancer: reduced development of castration-resistant disease and prostate cancer mortality. Eur Urol 63(4):709–716
143. Mehenni H, Gehrig C, Nezu J et al (1998) Loss of LKB1 kinase activity in Peutz–Jeghers syndrome, and evidence for allelic and locus heterogeneity. Am J Hum Genet 63:1641–1650
144. Home P, Kahn S, Jones N, Noronha D, Beck-Nielsen H, Viberti G (2010) Experience of malignancies with oral glucose-lowering drugs in the randomised controlled ADOPT (A Diabetes Outcome Progression Trial) and RECORD (Rosiglitazone Evaluated for Cardiovascular Outcomes and Regulation of Glycaemia in Diabetes) clinical trials. Diabetologia 53(9):1838–1845
145. Lee AJ (1996) Metformin in noninsulin-dependent diabetes mellitus. Pharmacotherapy 16(3):327–351
146. D'Agostino RB Sr (2009) The delayed-start study design. N Engl J Med 361:1304–1306
147. Khosravan R, DuBois SG, Janeway K, Wang E (2021) Extrapolation of pharmacokinetics and pharmacodynamics of sunitinib in children with gastrointestinal stromal tumors. Cancer Chemother Pharmacol. https://doi.org/10.1007/s00280-020-04221-x
148. Demetri GD, van Oosterom AT, Garrett CR, Blackstein ME, Shah MH, Verweij J, McArthur G, Judson IR, Heinrich MC, Morgan JA et al (2006) Efficacy and safety of sunitinib in patients with advanced gastrointestinal stromal tumour after failure of imatinib: a randomised controlled trial. Lancet 368:1329–1338. https://doi.org/10.1016/S0140-6736(06)69446-4

149. Demetri GD, von Mehren M, Blanke CD, Van den Abbeele AD, Eisenberg B, Roberts PJ, Heinrich MC, Tuveson DA, Singer S, Janicek M et al (2002) Efficacy and safety of imatinib mesylate in advanced gastrointestinal stromal tumors. N Engl J Med 347:472–480. https://doi.org/10.1056/NEJMoa020461
150. Houk BE, Bello CL, Kang D, Amantea M (2009) A population pharmacokinetic meta-analysis of sunitinib malate (SU11248) and its primary metabolite (SU12662) in healthy volunteers and oncology patients. Clin Cancer Res 15:2497–2506. https://doi.org/10.1158/1078-0432.ccr-08-1893
151. Houk BE, Bello CL, Poland B, Rosen LS, Demetri GD, Motzer RJ (2010) Relationship between exposure to sunitinib and efficacy and tolerability endpoints in patients with cancer: results of a pharmacokinetic/pharmacodynamic meta-analysis. Cancer Chemother Pharmacol 66:357–371. https://doi.org/10.1007/s00280-009-1170-y
152. Lindauer A, Di Gion P, Kanefendt F, Tomalik-Scharte D, Kinzig M, Rodamer M, Dodos F, Sorgel F, Fuhr U, Jaehde U (2010) Pharmacokinetic/pharmacodynamic modeling of biomarker response to sunitinib in healthy volunteers. Clin Pharmacol Ther 87:601–608. https://doi.org/10.1038/clpt.2010.20
153. Khosravan R, Motzer RJ, Fumagalli E, Rini BI (2016) Population pharmacokinetic/pharmacodynamic modeling of sunitinib by dosing schedule in patients with advanced renal cell carcinoma or gastrointestinal stromal tumor. Clin Pharmacokinet 55:1251–1269. https://doi.org/10.1007/s40262-016-0404-5
154. Agaram NP, Laquaglia MP, Ustun B, Guo T, Wong GC, Socci ND, Maki RG, DeMatteo RP, Besmer P, Antonescu CR (2008) Molecular characterization of pediatric gastrointestinal stromal tumors. Clin Cancer Res 14:3204–3215. https://doi.org/10.1158/1078-0432.ccr-07-1984
155. Janeway KA, Albritton KH, Van Den Abbeele AD, D'Amato GZ, Pedrazzoli P, Siena S, Picus J, Butrynski JE, Schlemmer M, Heinrich MC et al (2009) Sunitinib treatment in pediatric patients with advanced GIST following failure of imatinib. Pediatr Blood Cancer 52:767–771. https://doi.org/10.1002/pbc.21909
156. Wiernik PH, Schwartz EL, Strauman JJ, Dutcher JP, Lipton RB, Paietta E (1987) Phase I clinical and pharmacokinetic study of Taxol. Cancer Res 47:2486–2493
157. Verschuur AC, Bajciova V, Mascarenhas L, Khosravan R, Lin X, Ingrosso A, Janeway KA (2019) Sunitinib in pediatric patients with advanced gastrointestinal stromal tumor: results from a phase I/II trial. Cancer Chemother Pharmacol 84:41–50. https://doi.org/10.1007/s00280-019-03814-5

Chapter 3
Pharmacokinetic, Pharmacodynamic, Preclinical and Clinical Models for Evaluation of Nanoparticles

Sankalp A. Gharat, Munira M. Momin, and Tabassum Khan

Introduction by the Editor

Pharmacokinetic (PK) and pharmacodynamic (PD) models are essential tools for the evaluation of nanoparticles in both preclinical and clinical settings [1]. PK models focus on ADME of nanoparticles within the body, providing insights into their biodistribution, clearance rates, and potential accumulation in specific tissues [2]. These models help determine optimal dosing regimens, predict drug concentrations at target sites, and estimate the systemic exposure of nanoparticles. Compartmental modeling involves describing the body as a series of interconnected compartments, each representing a specific physiological space where the drug or nanoparticle distributes. Compartmental models assume that the distribution and elimination processes can be approximated using exponential equations. This approach is suitable for understanding the overall behavior of nanoparticles in different body compartments. In the context of nanoparticles, compartments represent different tissues, organs, or physiological spaces where nanoparticles accumulate. Parameters in the model include transfer rates between compartments, volumes of distribution, and elimination rates [3, 4]. However, compartmental modeling may oversimplify the complex behavior of nanoparticles, especially when considering their unique interactions with biological components. Non-compartmental modeling involves direct analysis of concentration-time profiles without assuming a specific compartmental

S. A. Gharat · M. M. Momin (✉)
Department of Pharmaceutics, SVKM's Dr. Bhanuben Nanavati College of Pharmacy, Mumbai, Maharashtra, India
e-mail: sankalp.gharat@bncp.ac.in; munira.momin@bncp.ac.in

T. Khan
Department of Pharmaceutical Chemistry and Quality Assurance, SVKM's Dr. Bhanuben Nanavati College of Pharmacy, Mumbai, Maharashtra, India
e-mail: tabassum.khan@bncp.ac.in

S. A. Gharat et al. (eds.), *Pharmacokinetics and Pharmacodynamics of Novel Drug Delivery Systems: From Basic Concepts to Applications*,
https://doi.org/10.1007/978-981-99-7858-8_3

structure. This approach is often used when the pharmacokinetics of nanoparticles do not fit well into compartmental models due to their complex behaviour [3]. In non-compartmental modeling, various pharmacokinetic parameters are calculated directly from the observed concentration-time data. These parameters include area under the concentration-time curve (AUC), maximum concentration (Cmax), time to reach Cmax (Tmax), and terminal half-life. Non-compartmental modeling is useful for nanoparticles when the distribution and elimination processes are not well-characterized by a simple compartmental model. It provides a more flexible way to understand nanoparticle behavior in vivo [5]. The choice between compartmental and non-compartmental modeling depends on the specific behavior of the nanoparticles being studied and the availability of the in-vitro data [6].

PD models, on the other hand, investigate the relationship between the concentration of nanoparticles and their pharmacological effects, shedding light on their efficacy, potency, and safety profiles. These models aim to quantify how nanoparticle concentration influences the desired therapeutic response or other relevant biological outcomes [7]. Direct and indirect effect pharmacodynamic models differentiate between the direct effects of the drug on the target and the indirect effects that might result from secondary processes or feedback mechanisms. Both types of models can be adapted to nanoparticles, considering their unique properties and interactions. In a direct effect model, the nanoparticle itself directly interacts with a target molecule or receptor, triggering a biological response. This interaction might involve binding to a receptor, enzyme inhibition, or activation of signalling pathways. The relationship between nanoparticle concentration and the observed effect is described directly through the drug-receptor interaction [8]. Direct effect models can be described using equations like the Emax (maximum effect) model or other mechanistic equations. For nanoparticles, a direct effect model would involve understanding how the concentration of nanoparticles directly influences the target receptors or molecules and leads to the desired pharmacological response. This could be used for targeted drug delivery, where nanoparticles are designed to specifically interact with certain cells or tissues. In an indirect effect model, the nanoparticle triggers a cascade of events or secondary processes that ultimately lead to the observed pharmacological effect. These processes involve feedback loops, multiple steps of signalling, or modulation of various components within the biological system. Indirect effect models capture the complexity of how nanoparticles can influence various elements in the biological pathway. An indirect effect model considers the influence of nanoparticles on various cellular processes, such as immune response, cytokine release, or tissue remodeling. The nanoparticles themselves may not directly interact with the target, but their presence can set off a series of events that result in the desired therapeutic outcome. Developing PD models for nanoparticles can be complex due to the unique properties of nanoparticles and their interactions with biological systems [9].

Further, preclinical models are essential for evaluating the safety, efficacy, pharmacokinetics, and pharmacodynamics of nano formulations before they can be translated into clinical trials. Preclinical models, including in-vitro cell cultures and animal models, are employed to assess the behaviour and effects of nanoparticles. They provide information on nanoparticle behaviour, toxicity, and potential therapeutic benefits. In-vitro cell culture models allow researchers to study the interactions between nano formulations and specific cell types. These models provide insights into cellular uptake, internalization pathways, release of cargo, and potential cytotoxic effects [10, 11]. Tissue explant models involve maintaining sections of living tissues in culture [12]. 3D cell culture systems, such as spheroids or organoids, provide a more realistic representation of tissue architecture and cellular interactions. These models provide understanding of the behaviour of nanoparticles in the three-dimensional environment and their potential interactions with various cell types in tissues [13]. Mouse and rat models are commonly used for assessing the in vivo behavior of nano formulations [14]. Zebrafish model is another such cost-effective model that provides insights into whole-body effects [15]. These models provide information about biodistribution, accumulation in specific organs, metabolism, and overall toxicity. Tumor-bearing mice have also been widely explored for evaluating the targeting efficiency of nano formulations in cancer therapy [16]. Non-human primate models are closer to humans in terms of physiological and anatomical similarities. While they are more expensive and ethically complex, they provide valuable data on nanoparticle behavior, safety, and potential translation to humans. These models allow researchers to study the interactions between nano formulations and complex tissue environments. It can be particularly useful for assessing the penetration, distribution, and therapeutic effects of the nanoparticles [17].

Clinical models, involving human subjects, evaluate the safety, efficacy, and pharmacokinetics of nanoparticles, helping to establish their clinical utility and guide regulatory decisions. These trials follow a well-defined process with different phases, each focusing on specific aspects of behavior of the nanoparticles in human subjects. Clinical trials for nano formulations require a well-coordinated effort to ensure the safety and efficacy of these innovative drug delivery systems in human patients [18].

Dermatopharmacokinetics (DPK) involves the study of ADME of drugs or nanoparticles applied to the skin. Developing pharmacokinetic models for nanoparticles applied to the skin is complex due to the unique properties of the skin barrier and the potential interactions of nanoparticles with skin components. Various techniques like tape stripping, vasoconstrictor assay, microdialysis, open flow perfusion, spectroscopic and microscopic methods are employed for DPK studies. These models can provide insights into nanoparticle behavior on the skin surface, their penetration potential, and their potential for systemic absorption or local effects [19]. The integration of PK, PD, DPK preclinical, and clinical models is crucial for a comprehensive evaluation of nanoparticles, facilitating their translation from the laboratory to clinical applications. This chapter covers various PK-PD models required for understanding the fate of nanoparticles in the body.

3.1 Pharmacokinetic (PK) Model

Introduction by the Editor

The field of pharmacokinetics has expanded to comprehend the complex world of nanoparticles, revolutionising drug delivery and therapeutic strategies. A pharmacokinetic (PK) model of nanoparticles is a mathematical framework used to describe the time-based changes in the concentration of nanoparticles in the body after administration. This model accounts for the unique characteristics of nanoparticles, including their size, surface properties, and interaction with biological systems. It encompasses parameters related to the fate of nanoparticles, allowing for predictions of their biodistribution, target localization, and elimination kinetics. PK models of nanoparticles give critical insights for optimising drug delivery systems and assessing the potential therapeutic benefit of these advanced drug carriers by providing information on the ADME of nanoparticles in the body.

3.1.1 Compartmental Modeling

Compartmental modeling for nanoparticles represents biological systems as interconnected compartments, each describing a distinct body region or tissue where nanoparticles can accumulate. These compartments are connected by transfer rates that simulate nanoparticle movement. The model quantifies nanoparticle distribution, clearance, and interactions with physiological processes, enabling prediction of biodistribution, targeting efficiency, and overall systemic behaviour.

3.1.2 Noncompartmental Modeling

Noncompartmental modeling of nanoparticles is crucial due to the complex and diverse nature of nanoparticle behaviour within biological systems. It offers a data-driven approach to assessing pharmacokinetic parameters directly from concentration-time profiles, avoiding the need for assumptions about compartmental structures. This method provides insight into nanoparticle biodistribution, elimination kinetics, and overall systemic behaviour, enabling more accurate predictions of therapeutic efficacy and potential toxicity. Noncompartmental modeling accommodates the unique characteristics of nanoparticles and enhances the understanding of nanoparticles pharmacokinetics, ultimately aiding in the rational design and optimisation of nanoparticle-based drug delivery systems.

Machine generated summaries

Disclaimer: The summaries in this chapter were generated from Springer Nature publications using extractive AI auto-summarization: An extraction-based summarizer aims to identify the most important sentences of a text using an algorithm and uses those original sentences to create the auto-summary (unlike generative AI). As the constituted sentences are machine selected, they may not fully reflect the body of the work, so we strongly advise that the original content is read and cited. The auto generated summaries were curated by the editor to meet Springer Nature publication standards.

To cite this content, please refer to the original papers.

Machine generated keywords: pbpk, skin, model, topical, nonlinear, product, bioequivalence, dermal, drug product, microdialysis, pbpk model, probe, elimination, linear, drug concentration, pbpk, pbpk model, elimination, change, pharmacokinetic model, model, plasma concentration, linear, base pharmacokinetic, plasma, hepatic, nonlinear, physiologically base, two compartment, concentration time

New Perspectives in Clinical Pharmacokinetics-1: the Importance of Updating the Teaching in Pharmacokinetics that both Clearance and Elimination Rate Constant Approaches Are Mathematically Proven Equally Valid [20] This is a machine-generated summary of:

Jelliffe, Roger; Bayard, David: New Perspectives in Clinical Pharmacokinetics-1: the Importance of Updating the Teaching in Pharmacokinetics that both Clearance and Elimination Rate Constant Approaches Are Mathematically Proven Equally Valid [20]

Published in: The AAPS Journal (2018)

Link to original: https://doi.org/10.1208/s12248-018-0185-x

If you want to cite the papers, please refer to the original.

For technical reasons we could not place the page where the original quote is coming from.

Abstract-Summary "The healing professions have only about four main therapeutic tools at their disposal—surgery, drugs, physical therapy, and psychotherapy."

"For the general profession of internal medicine, drug therapy is its primary tool."

"Providing an understanding of the state-of-the-art in therapeutic methods, grounded in solid scientific and mathematical rigor, is therefore of the utmost clinical importance for both physicians and clinical pharmacists."

"Relatively little attention has been paid to training clinical pharmacokineticists and physicians to manage drug therapy optimally for patients under their care in their everyday practice."

"The conclusion of this analysis is that both approaches are rigorously proven mathematically to be equally valid."

"We also discuss some implications of these equally valid approaches from the framework of mechanistic and non-compartmental models."

INTRODUCTION: THE "BASIC EQUATION" OF PHARMACOKINETICS (PK)

"Proost has stated that "The actual rate of removal of a drug, expressed in amount/time, is dependent on the plasma concentration of the drug, since the plasma concentration passing the eliminating organs, primarily liver and kidneys, is the driving force of elimination."

“It relates the actual rate of removal or elimination rate, on the left, to the two factors determining it, i.e., clearance (CL) and plasma concentration.”

“Many use the word clearance when they really mean elimination.”

“We suggest that the word “clearance” might best describe the overall capacity for elimination, or a volume from which a substance has been eliminated.”

INDEPENDENCE OF VOLUME AND CLEARANCE, AND DEPENDENCE OF KE?

“Example of displacement of drug from tissue-binding sites, it is quite possible that an amount of drug might be displaced from peripheral tissue-binding sites into an actual central or vascular volume which might be either decreased, unchanged, or perhaps even greater than before, depending on the many other physiological processes which may also be taking place at the same time in a patient, such as alterations in V from hemorrhage, changes in permeability of cells, sepsis, or over-hydration, for example.”

“Rather than speculate that clearance probably remains constant while the volume of distribution is decreased and Ke is increased in this situation, if this happens to a real patient, along with the many other traumatic physiological changes associated with losing both legs, for example, one needs to get actual data from each patient and to quantify the actual clinical situation present at each relevant point in time.”

A NEW EXAMINATION OF THE CONTROVERSY USING THE MATHEMATICAL PRINCIPLE OF INVARIANCE

“In order to apply this principle to our current situation, let us define an “elimination parameterization” as the paired parameters $\theta_1 = (Ke, V)$, and a “clearance parameterization” as the paired parameters $\theta_2 = (Cl, V)$, where it is assumed that both Ke and V are > 0.”

“The Principle of Invariance basically states that a maximum likelihood estimate (MLE) of a function of a parameter is equal to that function of the original MLE of the parameter itself.”

“When both are given the same data set to analyze, using maximum likelihood estimation, Joe gets estimates V_{Joe} and Ke_{Joe}, while Alice gets estimates V_{Alice} and Cl_{Alice}.”

“The fact is that because of the principle of invariance, both Alice and Joe get identical parameter estimates: $V_{Alice} = V_{Joe}$ and $Cl_{Alice} = V_{Joe}Ke_{Joe}$.”

THE SCIENTIFIC CONTROVERSY IS OVER: ALL THREE PARAMETERIZATIONS ARE EQUALLY VALID

“Because of the principle of invariance, the longstanding debate concerning these two parameterizations has now been resolved.”

“The utility (usefulness, worth) of any particular parameterization is in the eye of the beholder (researcher or clinician).”

“Further, for real-world patient care, answers are not to be found in abstract speculation about directional changes that may or may not take place in one parameter (V, Ke, or Cl) if another one changes in isolation by itself, or by questions as to what happens to V, Ke, and Cl if a patient should lose both legs without suffering any other associated clinical trauma.”

TESTING THE PRINCIPLE

"Further, the calculation of the likelihood must be done by an exact, not an approximate, method."

"The great majority of software packages for PK analysis make various approximations in the course of pharmacokinetic modeling which corrupt the results obtained to various degrees."

"Further, it is absolutely essential that the likelihood be calculated exactly, without any approximation."

CONSEQUENCES OF THIS PROOF

"While the conclusions of this paper apply mathematically, the PK analysis usually involves estimation of V and CL (based upon the totality of patient input information), with Ke being defined by its relationships to CL and V. When using bottom-up approaches, predictions of CL and V are generated from patient data using the existing bottom-up model (already derived from many different sources of information) coupled with patient-specific input on physiological status (e.g., serum protein binding, hematocrit, and renal and liver function)."

"This contrasts with a top-down clinical approach using population models, therapeutic drug monitoring (TDM), Bayesian analysis, model individualization, and maximally precise individualized dosage, where physiological considerations are often treated as covariates."

"Unless one can assume that drug is strictly contained within the central compartment, the volume of distribution (typically expressed as steady state volume of distribution, Vss) is estimated by use of statistical moment theory."

A CLINICAL PERSPECTIVE ON PRACTICAL BEDSIDE PHARMACOKINETICS—PROVIDING OUR FUTURE CLINICIANS WITH PROPER EDUCATION IN OPTIMAL THERAPEUTIC METHODS

"To parameterizing models for research purposes, we should also consider the most useful way to parameterize a model for clinical practice—for a patient in an intensive care unit, for example."

"We, and other clinicians, often find the V and Ke parameterization more useful for practical patient care."

"Many clinicians prefer to parameterize PK models for patients using V and Ke, to obtain the most direct information about these two separate and important therapeutic issues."

"Further, since the clearance and Ke parameterizations are now shown to be equivalent, it might well be helpful for the pharmaceutical community, which generally finds the V and clearance approach more useful for its research environment, to consider the needs of practicing clinicians as a drug comes to market, by converting their population PK models to the V and Ke format, and to make their population models, even fairly large physiologically based PK models, available for patient care using appropriate clinical software."

CONCLUSION

"Because of the mathematical principle of invariance, the scientific controversy over whether or not the V and Cl are somehow "better" than the V and Ke approach is now over and has been settled."

"Both approaches are now, for the first time to our knowledge, rigorously proven to be equally valid."

"Each community can now feel quite free scientifically to use the approach it finds most useful."

"Neither parameterization is any "better" or "worse" than the other."

Fundamentals of Pharmacokinetics to Assess the Correlation Between Plasma Drug Concentrations and Different Blood Sampling Methods [21] This is a machine-generated summary of:

Chen, Wei-Ching; Huang, Pei-Wei; Yang, Wan-Ling; Chen, Yen-Lun; Shih, Ying-Ning; Wang, Hong-Jaan: Fundamentals of Pharmacokinetics to Assess the Correlation Between Plasma Drug Concentrations and Different Blood Sampling Methods [21]

Published in: Pharmaceutical Research (2019)

Link to original: https://doi.org/10.1007/s11095-018-2550-y

If you want to cite the papers, please refer to the original.

For technical reasons we could not place the page where the original quote is coming from.

Abstract-Summary "Various blood collection methods were developed and used in the pharmacokinetic evaluation of drugs."

"The influence of different blood sampling methods on plasma drug concentrations has not been clarified."

"Blood samples were collected from different sites at the same individual, and pharmacokinetic properties of the drugs were then evaluated."

"Study results showed that the maximum plasma concentration or area under the curve of three study drugs was significantly higher in rats when blood was sampled from the carotid artery than when it was sampled from the caudal vein or by tail snip."

"Pharmacokinetics of certain drugs may differ based on the blood sampling site."

"The acid-base properties of drugs may influence pharmacokinetic evaluation."

Introduction

"Although, a few studies have explored that the plasma drug concentrations are dependent on the sampling site used [22, 23], there is no clear mechanism currently to identify which factors will cause such variation."

"Since 1980, the effect of different blood sampling methods on pharmacokinetic evaluations has been debated."

"Large differences in drug concentrations in plasma obtained from different sampling sites, will lead to errors in the assessment of pharmacokinetic parameters and alterations in pharmacokinetic/pharmacodynamics evaluations [24]."

"In such studies, the lack of sufficient testing drugs and the inconsistent methods of blood collection used limit the overall conclusions that can be drawn [22, 23]; therefore, no definitive theory has been raised based on those research results."

"We compared the pharmacokinetic properties of eight clinical drugs in rats to determine whether the plasma concentrations of these drugs were influenced by the sampling method."

Materials and Methods

"A standard stock solution of rifabutin, atorvastatin, simvastatin, nifedipine, pravastatin, atenolol, rosuvastatin, and ketorolac were prepared together by dissolving the compounds in methanol, to give a final concentration of 1 mg/mL. Subsequently, the working solution containing a mixture of all eight analytes were serially diluted (10–10,000 ng/mL) with 50% (v/v) aqueous methanol solution."

"The calibration standards and quality control (QC) samples were prepared by adding 10 μL of known working solution to 90 μL of drug-free rat plasma, the final concentrations for each analyte were prepared to be 1, 5, 10, 20, 50, 100, 250, 500 and 1000 ng/mL, except for ketorolac which lacked 1 ng/mL. The QC samples were also prepared in the same way to obtain the final concentrations of 1 ng/mL (lower limit of quantification; LLOQ), 2.5 ng/mL (low concentration of quality control; LQC), 400 ng/mL (medium concentration of quality control; MQC) and 750 ng/mL (high concentration of quality control; HQC) for all tested drugs, but 5 ng/mL of LLOQ and 15 ng/mL of LQC for ketorolac."

Results

"Both in the cardiac and oral administration groups, rifabutin presented statistically significant differences in C_{max} in blood obtained by the three different sampling methods."

"There were statistically significant differences in AUC_t between carotid artery and caudal vein cannulation sampling in both administration groups."

"In the oral administration group, the V_D in blood obtained by caudal vein cannulation was 1.7-times that in blood obtained by carotid artery cannulation, and the $t_{1/2}$ was 1.3-times greater; these differences were all statistically significant."

"After cardiac administration of nifedipine, the C_{max} in blood sampled from the carotid artery was 1.4-times significantly higher than that in blood sampled from the caudal vein."

"In the cardiac and oral administration groups, the C_{max} values of atenolol in blood obtained by carotid artery cannulation were 1.5- and 1.3-times higher, respectively, than that in blood obtained by caudal vein cannulation, and the increases were significantly different."

Discussion

"After utilizing different sampling methods, variations in the plasma concentrations of the BDDCS class II drugs, rifabutin and nifedipine, and the class III drug atenolol resulted in variations in their pharmacokinetic parameters."

"Statistically significant differences in plasma concentrations due to the sampling site selection were not observed for all the study drugs."

"Rifabutin, nifedipine, and atenolol are non-acidic drugs, and their C_{max} values obtained from different blood sources were significantly different, whereas the AUC, $t_{1/2}$, and CL of only rifabutin in plasma obtained via the three sampling methods varied substantially."

"As found in this experiment, the acid-base properties of the drug itself and its V_D should be considered; significant differences in the plasma concentration and pharmacokinetic estimates are common for alkaline drugs with high V_D (e.g."

"Simvastatin, an acidic drug with an extremely high V_D [25], had non-significant different pharmacokinetic parameters across the three different blood sampling sites."

Conclusion

"A traditionally held belief is that blood is a homogeneous central compartment system, and the collection of blood from different sites should not affect the analysis of drug concentration and estimates of pharmacokinetics."

"The experimental results described herein show that this is not necessarily true."

"A preliminary inference that we suggest based on this study is that non-acidic drugs are prone to measurement errors when pharmacokinetic parameters are determined using different blood sampling methods, and that the level of influence is determined by the V_D."

"This work was supported by grants (MOST106-2320-B-016-001) from the Ministry of Science and Technology, Taipei, Taiwan, and a grant (MAB-106-078) from the National Defense Medical Center, Taipei, Taiwan."

3.1.1 Compartmental Modeling

Machine generated keywords: pbpk, pbpk model, pharmacokinetic model, model, elimination, base pharmacokinetic, change, hepatic, nonlinear, linear, physiologically base, twocompartment, integrate, physiological, compartment

Impact of saturable distribution in compartmental PK models: dynamics and practical use [26] This is a machine-generated summary of:

Peletier, Lambertus A.; de Winter, Willem: Impact of saturable distribution in compartmental PK models: dynamics and practical use [26]

Published in: Journal of Pharmacokinetics and Pharmacodynamics (2017)

Link to original: https://doi.org/10.1007/s10928-016-9500-2

If you want to cite the papers, please refer to the original.

For technical reasons we could not place the page where the original quote is coming from.

Abstract-Summary "We explore the impact of saturable distribution over the central and the peripheral compartment in pharmacokinetic models, whilst assuming that back flow into the central compartiment is linear."

"Using simulations and analytical methods we demonstrate characteristic telltale differences in plasma concentration profiles of saturable versus linear distribution models, which can serve as a guide to their practical applicability."

"For two extreme cases, relating to (i) the size of the peripheral compartment with respect to the central compartment and (ii) the magnitude of the back flow as related to direct elimination from the central compartment, we derive explicit approximations which make it possible to give quantitative estimates of parameters."

Introduction

"Starting from a conventional three compartment PK model, transformation of one of the two peripheral compartments to a low capacity, high affinity compartment with saturable distribution resulted in a highly significant improvement of the model fit."

"The objectives of this paper are (i) To identify characteristic properties of the time courses in the central compartment, and identify differences between linear and saturable models which may serve as handles to determine which class of models should be used to fit a given set of data. (ii) To study the dynamics of the nonlinear model incorporating saturation with a view to understand the impact of the relative capacities and the rate constants of the system and identify the characteristic time-scales. (iii) To identify the impact of saturable distribution in practical applications, such as the exposure resulting from SAD and MAD regimens."

Methods

"In order to study the impact of saturation we compare the dynamics of two distribution models, one with linear and one with nonlinear distribution that involves saturation."

"The transfer from the central compartment to the peripheral compartment is saturable, whilst that from the peripheral back to the central compartment is linear."

"This model has five parameters whereas the linear model has four."

"In the large capacity case, the infusion rate q is assumed to be constant, and initially the system is assumed to be empty, i.e., the amounts in the compartments are all assumed to be zero:In the small capacity case, the infusion rate q is assumed to be zero, and the initial conditions after an iv dose D are given by"

Steady state

"We assume, The capacity of the peripheral compartment is large compared to that of the central compartment."

"The drug flows back from the peripheral compartment into the central compartment at a much smaller rate than it is eliminated from the central compartment."

"In terms of the rate constants we assume that: We assume, The capacity of the peripheral compartment is small compared to that of the central compartment."

Simulations

"We do this separately for the large and the small capacity peripheral compartment."

"We select a series of different values of the infusion rate q in order to demonstrate the differences between the linear and the nonlinear model."

"The dynamics of the system is effectively determined by the interaction between the central and the peripheral compartment."

Results

"During the long second phase, with its slow dynamics, the influence of leakage from the peripheral compartment may well be relevant."

"To fully appreciate the effect of a large capacity of the peripheral compartment combined with a slow exchange between the two compartments, we conclude with a brief discussion of the dynamics of the nonlinear model for the converse situation: small capacity of the peripheral compartment combined with a fast exchange between the two compartments."

"We here assume thatSince in this case the peripheral compartment has small capacity and direct elimination is relatively small, one expects that an iv bolus administration will lead to a large peak in concentration in the central compartment."

"We find that for small capacity and rapid exchange between central and peripheral compartment the dynamics has a brief initial phase followed by a long terminal phase, with an appropriately defined plateau value in between."

Discussion

"We have compared the dynamics of two types of models for the distribution of a compound over a central and a peripheral compartment."

"In one type the elimination of compound from the central compartment into the peripheral compartment is linear, and the other it is saturable and hence nonlinear."

"Distribution over two compartments in which the peripheral compartment has a limited capacity, has much in common with tissue-binding."

"Dimensionless parameters are often a numerical measure of the relative importance of different processes involved in the model, such as direct elimination from the central compartment and distributional transfer between the compartments."

"We have demonstrated a number of interesting dynamic properties of saturable distribution models which can be of value in practical modelling applications."

"This is achieved by relaxing the assumption of linear distribution in the standard model at the cost of only one extra parameter per peripheral compartment."

An Analytical Approach of One-Compartmental Pharmacokinetic Models with Sigmoidal Hill Elimination [27] This is a machine-generated summary of:

Wu, Xiaotian; Zhang, Hao; Li, Jun: An Analytical Approach of One-Compartmental Pharmacokinetic Models with Sigmoidal Hill Elimination [27]

Published in: Bulletin of Mathematical Biology (2022)

Link to original: https://doi.org/10.1007/s11538-022-01078-4

If you want to cite the papers, please refer to the original.

For technical reasons we could not place the page where the original quote is coming from.

Abstract-Summary "We aim to develop the analytical solutions of one-compartment pharmacokinetic models with sigmoidal Hill elimination and quantitatively revisit some widely used pharmacokinetic indexes."

"In the case of a single dose, we have obtained the explicit formulas for several pharmacokinetic surrogates, such as the clearance, elimination half-life and partial/total drug exposure."

"The present findings elucidate the intrinsic quantitative structural properties of pharmacokinetic models with Hill elimination and provide new knowledge for non-linear pharmacokinetics and guidance for rational drug designs."

Introduction

"These examples indicate that the sigmoidal behavior of drug elimination represented by the Hill equation is a non-negligible issue in pharmacokinetic modeling."

"For more complex compartmental pharmacokinetic models with simultaneous first-order and Michaelis–Menten elimination, Wu and coauthors have recently proposed a new way for the analytical solutions through introducing well-defined transcendent X and Y functions for the cases of IV bolus administration and constant IV infusion (Wu et al. [28–31])."

"For the one-compartment pharmacokinetic model with a single Michaelis–Menten elimination pathway, it was discovered that the closed-form solution can be expressed using the transcendent Lambert W function (Beal [32]; Goliňik [33]; Goličnik [34]; Tang and Xiao [35]; Wilkinson et al. [36])."

"By introducing a transcendent function H, we will propose the closed-form solution of one-compartment PK model with a single sigmoidal Hill elimination for the case of IV bolus administration."

Closed-Form Solution of Single-Dose Regimen

"To have a meaningful pharmacokinetics, all model parameter values are assumed to be positive."

"A complete investigation of H function is complex and out of scope for the current work."

"Comparing the closed-form solutions for Hill and Michaelis–Menten eliminations, we have"

Pharmacokinetic Indexes for Hill Elimination

"The clearance at a specific time can be defined as the ratio of the drug elimination rate in amount to the drug concentration at the same time."

"For the case of linear elimination, the clearance is a constant at any time and independent to the variation of drug concentration."

"Following this definition, the clearance of nonlinear elimination is not a constant with respect to time but varies with drug concentrations."

"If we consider the drug elimination rate as a function of drug concentration, we can also ask how the tendency of the elimination rate changes with regard to the change in concentration."

Closed-Form Solution of Multiple-Dose Regimens

"Drugs are most commonly prescribed for multiple repeated administrations with fixed doses."

"We assume the first dose starts at time zero."

"The proof of the Lemma is provided in "Appendix C"."

"The existence of periodic solution at steady state."

Drug Examples

"We illustrate with two drug examples the explicit formulas in terms of drug concentration-time course, elimination half-life and drug exposure."

"These succinct expressions allow the readers an intuitive characterization of the pharmacokinetics and a direct calculation."

Discussion

"We have proposed and developed the analytical solutions for the one-compartment pharmacokinetic model with a single sigmoidal Hill elimination pathway in the case of a single or multiple bolus administrations."

"Motivated by the former works (Beal [32]; Goliňik [33]; Goličnik [34]; Tang and Xiao [35]) on that of the pharmacokinetic models with Michaelis–Menten elimination kinetics, for which the transcendent Lambert W function is proved a necessary part for expressing the closed-form solutions, we are guided to propose a new transcendent H function to fulfill our goal."

"During our review of published models, we found that kinetic models of Hill elimination are largely reported, and they are mainly for in vitro experiences or pharmacokinetics in animals, or those liposoluble drug substrates involving enzyme kinetics in drug metabolism (Shou et al. [37]; Yadav et al. [38])."

"The Hill elimination in humans is usually modeled with an additional linear elimination pathway (Konsil et al. [39])."

Steady-state volume of distribution of two-compartment models with simultaneous linear and saturated elimination [40] This is a machine-generated summary of:

Wu, Xiaotian; Nekka, Fahima; Li, Jun: Steady-state volume of distribution of two-compartment models with simultaneous linear and saturated elimination [40]

Published in: Journal of Pharmacokinetics and Pharmacodynamics (2016)

Link to original: https://doi.org/10.1007/s10928-016-9483-z

If you want to cite the papers, please refer to the original.

For technical reasons we could not place the page where the original quote is coming from.

Abstract-Summary "In the current paper, two-compartment models with simultaneous first-order and Michaelis–Menten elimination are considered."

Introduction

"Many compounds, such as monoclonal antibodies and growth factors, often exhibit complex mechanisms where, for example, the elimination can involve both linear and nonlinear pathways [41, 42]."

"Compartment models with simultaneous linear and nonlinear saturated elimination mechanism have been proposed for these compounds [43–49]."

"Two-compartment structures are the most common in population PK modeling, representing more than 83 % of published models, 8 % of them include the nonlinear elimination mechanism (according to an internal survey by the Advanced Quantitative Sciences group [44])."

"For compounds exhibiting simultaneous linear and nonlinear elimination mechanism, the Michaelis–Menten elimination via intracellular metabolism or internalization by their pharmacological targets is suggested to occur from the central compartment [45–48]."

"Two-compartment models with intermediate elimination, where, additional to the central Michaelis–Menten elimination, the linear elimination occurs from central and/or peripheral compartments can be envisaged."

Indistinguishable models with simultaneous first-order and Michaelis–Menten elimination

"Conditions have been established for the indistinguishability of two-compartment pharmacokinetic models when linear elimination is exclusively from either central or peripheral [49]."

"Models are indistinguishable in the sense that their input-output behaviors are identical, which means that they share exactly the same concentration-time curve for the same administered dose."

"Indistinguishability can also hold for two-compartment models exhibiting simultaneous first-order and nonlinear Michaelis–Menten elimination that we study here."

[Section 4]

"The conditions for their indistinguishability were also given."

"We here provide a simple derivation for these conditions."

[Section 7]

"This will allow the understanding of the role of non-linearity and/or peripheral elimination and the contribution of each to the steady-state volume of distribution."

"Clearance is one of the most important pharmacokinetic parameters, defined as the volume of plasma or blood that is totally cleared from its content of drug per unit time."

Theoretical considerations

"When no model structure is taken into account, this is the only known formula relying on the drug concentration data for the estimation of volume of distribution at steady state."

"Following the law of conservation of mass, the total eliminated dose amount is equal to the administered dose."

Application

"The models discussed in the current paper have been proposed to characterize the pharmacokinetics of some therapeutic agents such as macromolecular drugs [41, 47, 48, 50]."

"To illustrate the utility of our proposed formulas for the steady-state volume of distribution, we consider the case of recombinant human erythropoietin (rHuEPO), a glycoprotein hormone drug used to stimulate the red blood cells production during cancer chemotherapy or other diseases."

"As a macromolecular protein drug, we cannot exclude that rHuEPO can be eliminated, in a large proportion, from tissues not in rapid equilibrium with the central compartment [41]."

"These two models cannot be distinguished solely based on drug concentration data."

Discussion

"We have shown that, under certain conditions on model parameters, these models are indistinguishable in the sense that they produce the same PK profile for a same dose, while having different steady-state volumes of distribution."

"Their expressions explicitly delineate the intrinsic implication of the linear, non-linear and peripheral elimination."

"The nonlinear Michaelis–Menten amplifies the NCA-based steady-state volume of distribution, while peripheral elimination acts as an additive factor."

Statistical Analysis of Two-Compartment Pharmacokinetic Models with Drug Non-adherence [51] This is a machine-generated summary of:

Yan, Dingding; Wu, Xiaotian; Li, Jun; Tang, Sanyi: Statistical Analysis of Two-Compartment Pharmacokinetic Models with Drug Non-adherence [51]

Published in: Bulletin of Mathematical Biology (2023)

Link to original: https://doi.org/10.1007/s11538-023-01173-0

If you want to cite the papers, please refer to the original.

For technical reasons we could not place the page where the original quote is coming from.

Abstract-Summary "In the current paper, we aim to investigate the impact of dose omission on the plasma concentrations of two-compartment pharmacokinetic models with two typical routes of drug administration, namely the intravenous bolus and extravascular first-order absorption."

"We numerically simulate the impact of drug non-adherence to different extents on the variability and regularity of drug concentration and compare the drug pharmacokinetic preference between one and two compartment pharmacokinetic models."

"The results of sensitivity analysis also suggest the drug non-adherence as one of the most sensitive model parameters to the expectation of limit concentration."

"Our modelling and analytical approach can be integrated into the chronic disease models to estimate or quantitatively predict the therapy efficacy with drug pharmacokinetics presumably affected by random dose omissions."

Introduction

"A number of PK model-based studies exist on the impact of non-adherence on the variability and regularity of drug concentration."

"In a seminal work of Li and others, they provided a stochastic formalism of one-compartment PK models while integrating poor drug adherence in terms of general distribution of random dosing times, and analytically studied the mean and variance of drug concentration at any time (Li and Nekka [52])."

"Under this reality, it has become imperative the quantitative investigation of the impact of poor adherence on the variability of plasma drug concentration in the context of 2-compartment PK models."

"We will adopt an analytical approach to measure the impact of a patient's poor drug adherence on the 2-compartment PK models, by assuming the patient's drug intake follows the prescribed dosing schedule (i.e., fixed dosing intervals, equal doses), whereas some doses could be missed by forgiveness."

A Uniform Formalism of a Stochastic Pharmacokinetic Model

"To model the non-adherence in a patient' drug intake, we suppose that a dose is missed on a prescribed dosing time with a certain probability, and this probability only depends on the precedent information of dose taking."

"Q is the limit probability of the medication taken at every dosing time."

"We consider the situation that a patient either takes or omits a dose D, thus the problem of drug adherence only occurs at the exact dosing times."

"It requires a deposit compartment to reflect the process of dose absorption."

[Section 3]

"The truthiness of this condition can be verified."

"The drug concentration after multiple random doses presents the regular behaviour, whereas the extent of drug non adherence will affect the statistical properties of the limit distribution."

"Once the limit distribution is reached, the drug concentration is maintained at the steady-state level and translational with time, no matter what the initial drug distribution is."

Simulations

"We will simulate the classical two-compartment PK models by taking into account the patient's poor drug adherence and analyse the variability in drug concentrations."

"The drug concentrations significantly deviate from the expected concentrations of perfect adherence."

"Although the explicit probability density functions of limit drug concentration distribution cannot be found as aforementioned, it is interesting to know how the regularity of the limit drug concentrations in central compartment is changed if the extent of drug non-adherence pattern changes."

"It is of interest, in the scenario of drug non-adherence, to investigate the difference of drug concentrations in the central compartment between one and two-compartment PK models, and how the drug transfer rates between these two compartments impact the PK behaviour in the central compartment."

"We consider the drug concentration in the central compartment as a comparable criteria since drug data in central compartment is measurable."

Discussion and Conclusion

"We considered the random dose omissions raised in a patient's drug intake and explicitly formalized this behaviour model using a probabilistic approach."

"Based on these stochastic PK models, we were able to analytically assess the impact of various patterns of dose omissions on drug PK for blood/plasma concentrations (central compartment) as well as drug amounts in tissues (peripheral compartment)."

"The former is the exact solution of the two-compartment deterministic PK models with perfect adherence and the latter reflects the deviation caused by a patient's dose missing behaviour."

"Since a large part of nowadays published PK models are of two compartment nature, the current work on two-compartment stochastic PK models provide an easy quantitative transition of non adherence studies for a lot of medications."

Development of a Physiologically Based Pharmacokinetic Model for Sinogliatin, a First-in-Class Glucokinase Activator, by Integrating Allometric Scaling, In Vitro to In Vivo Exploration and Steady-State Concentration–Mean Residence Time Methods: Mechanistic Understanding of its Pharmacokinetics [53] This is a machine-generated summary of:

Song, Ling; Zhang, Yi; Jiang, Ji; Ren, Shuang; Chen, Li; Liu, Dongyang; Chen, Xijing; Hu, Pei: Development of a Physiologically Based Pharmacokinetic Model for Sinogliatin, a First-in-Class Glucokinase Activator, by Integrating Allometric Scaling, In Vitro to In Vivo Exploration and Steady-State Concentration–Mean Residence Time Methods: Mechanistic Understanding of its Pharmacokinetics [53]

Published in: Clinical Pharmacokinetics (2018)

Link to original: https://doi.org/10.1007/s40262-018-0631-z

If you want to cite the papers, please refer to the original.

For technical reasons we could not place the page where the original quote is coming from.

Abstract-Summary "The objective of this study was to develop a physiologically based pharmacokinetic (PBPK) model for sinogliatin (HMS-5552, dorzagliatin) by integrating allometric scaling (AS), in vitro to in vivo exploration (IVIVE), and steady-state concentration–mean residence time (C_{ss}-MRT) methods and to provide mechanistic insight into its pharmacokinetic properties in humans."

"Human major pharmacokinetic parameters were analyzed using AS, IVIVE, and C_{ss} MRT methods with available preclinical in vitro and in vivo data to understand sinogliatin drug metabolism and pharmacokinetic (DMPK) characteristics and underlying mechanisms."

"An initial mechanistic PBPK model of sinogliatin was developed."

"The final model was validated by simulating sinogliatin pharmacokinetics under a fed condition."

"For the PBPK approach, the 90% confidence intervals (CIs) of the simulated maximum concentration (C_{max}), CL, and area under the plasma concentration–time curve (AUC) of sinogliatin were within those observed and the 90% CI of simulated time to C_{max} (t_{max}) was closed to that observed for a dose range of 5–50 mg in the SAD study."

"The final PBPK model was validated by simulating sinogliatin pharmacokinetics with food."

"The 90% CIs of the simulated C_{max}, CL, and AUC values for sinogliatin were within those observed and the 90% CI of the simulated t_{max} was partially within that observed for the dose range of 25–200 mg in the multiple ascending dose (MAD) study."

"Sinogliatin pharmacokinetic properties were mechanistically understood by integrating all four methods and a mechanistic PBPK model was successfully developed and validated using clinical data."

"This PBPK model was applied to support the development of sinogliatin."

Introduction

"Although several classes of therapies for T2DM are available for clinical use, the need for novel therapies still remains to be met in order to improve the effectiveness of glycemic control; so far, current therapies can control glucose levels well in only 60% of T2DM patients [54]."

"In order to guide the study design and dose selection of clinical pharmacology studies and to fully understand the potential pharmacokinetic and safety profiles of sinogliatin in the broader population during its late-stage development, a mechanistic physiologically based pharmacokinetic (PBPK) model is required to accurately predict sinogliatin pharmacokinetics in different populations under various conditions."

"On top of this, a mechanistic PBPK model was developed, optimized with human pharmacokinetic data from a single ascending dose (SAD) study in healthy subjects, and validated with sinogliatin pharmacokinetic data under fed conditions from a multiple ascending dose (MAD) study in T2DM patients."

Methods

"The model was optimized with various methods based on human pharmacokinetic data from the SAD study in healthy subjects, and was validated by simulating human pharmacokinetic data under fed conditions from the MAD study in T2DM patients."

"In the elimination part, three different models were used to predict human clearance: (i) in vivo CL_{iv} calculated from the two-species scaling method using rat and dog data (TS-$_{rat,dog}$) (Model A); (ii) in vitro CYP3A4-mediated CL_{int} values (Model B); and (iii) in vitro CYP3A4-mediated maximum rate of the metabolite formation (V_{max}) and Michaelis-Menten constant (K_m) (Model C)."

"The simulated pharmacokinetic parameters and profiles were compared with data observed in a clinical study to assess the acceptance of the model estimation."

"The geometric means values and 90% CIs of key pharmacokinetic parameters from ten simulated trials were compared with the observed clinical data to validate the final PBPK model."

Results

"The estimated CL/F and V_{ss} values were 0.98- and 0.99-fold of the observed data, respectively."

"For doses from 5 to 50 mg, 90% CIs of simulated CL, C_{max}, and AUC values were within those observed in the healthy population, indicating the final model could accurately simulate the metabolism and absorption extent."

"The simulation results demonstrate that the sinogliatin PBPK model could capture the observed pharmacokinetic characteristics well in healthy populations."

"The 90% CIs of simulated CL, C_{max}, and AUC values were within that observed and the 90% CI of the simulated t_{max} was partially within that observed, indicating that the final model performed a good simulation of food effect."

"The simulated AUC values in three Sim-Cirrhosis-CP populations were higher than that in the Sim-Healthy (simulation in healthy patients) population."

"The simulated AUC in Sim-Cirrhosis-C was 2-fold and 1.5-fold higher than that in Sim-Cirrhosis-A and in Sim-Cirrhosis-B, respectively."

Discussion

"The final PBPK model could not only provide a knowledge base to learn and confirm input data and understanding but also evaluate the effect of extrinsic (DDI) and intrinsic (hepatic cirrhosis, genetic) factors on drug exposure."

"AS, IVIVE, and C_{ss}-MRT provided mechanistic understanding to support sinogliatin PBPK model development; furthermore, according to the learning exercises of this case study, we proposed an effective strategy for PBPK model construction."

"The in vitro V_{max} and K_m of rhCYP3A4 were input into the final PBPK model to capture the metabolism properties of sinogliatin based on the IVIVE concept."

"The PBPK model of sinogliatin provided the following positive impacts on drug development: Provided a knowledge base to learn and confirm the data and understanding PBPK simulated results provided judgement of input data quality to help decide whether supplementary experiments were needed."

Conclusion

"We proposed an effective PBPK development strategy based on understanding of mechanistic pharmacokinetics."

"The final model was validated using food effect data from the MAD study."

"The validated PBPK model provided positive impacts on the drug development of sinogliatin: it was use for the selection of the final clinical DDI study design and evaluated the effects of intrinsic (hepatic cirrhosis, genetic) factors on drug exposure."

"Our study has three main implications: (i) provides an effective strategy for PBPK development based on mechanistic understandings provided by AS, IVIVE, and C_{ss}-MRT; (ii) a PBPK model was developed to simulate the effects of extrinsic and intrinsic factors on drug exposure to support clinical study design; and (iii) provides a methodology of learning and confirms preclinical and clinical data by integrating four methods for the FIH research."

Virtual Clinical Studies to Examine the Probability Distribution of the AUC at Target Tissues Using Physiologically-Based Pharmacokinetic Modeling: Application to Analyses of the Effect of Genetic Polymorphism of Enzymes and Transporters on Irinotecan Induced Side Effects [55] This is a machine-generated summary of:

Toshimoto, Kota; Tomaru, Atsuko; Hosokawa, Masakiyo; Sugiyama, Yuichi: Virtual Clinical Studies to Examine the Probability Distribution of the AUC at Target Tissues Using Physiologically-Based Pharmacokinetic Modeling: Application to Analyses of the Effect of Genetic Polymorphism of Enzymes and Transporters on Irinotecan Induced Side Effects [55]

Published in: Pharmaceutical Research (2017)

Link to original: https://doi.org/10.1007/s11095-017-2153-z

If you want to cite the papers, please refer to the original.

For technical reasons we could not place the page where the original quote is coming from.

Abstract-Summary "To establish a physiologically-based pharmacokinetic (PBPK) model for analyzing the factors associated with side effects of irinotecan by using a computer-based virtual clinical study (VCS) because many controversial associations between various genetic polymorphisms and side effects of irinotecan have been reported."

"VCS also indicated that the frequency of significant association of biliary index with diarrhea was higher than that of UGT1A1 *28 polymorphism."

"The VCS confirmed the importance of genetic polymorphisms of UGT1A1 *28 and SLCO1B1 c.521T>C in the irinotecan induced side effects."

"The VCS also indicated that biliary index is a better biomarker of diarrhea than UGT1A1 *28 polymorphism."

Introduction

"The potential of virtual clinical study (VCS), which uses computational simulation with PBPK, pharmacodynamic, and toxicodynamic models—with virtual patients generated by the variability of physiological and pharmacokinetic parameters based on genetic polymorphism, ethnic differences, and inter- and intra-individual variability—is an advantage of PBPK model analysis [56]."

"The parameters for the PBPK model, which could reproduce the average blood concentration–time profile of irinotecan and its metabolites, were estimated as the first step of our analyses."

"46 unknown parameters in the PBPK model of the metabolism of irinotecan and its metabolites were optimized by CNM."

"Obtained sets of parameters were evaluated regarding whether the PBPK model with optimized parameters could reproduce the outcomes of clinical studies, focusing on the effect of UGT1A1 *28 and SLCO1B1 c.521T>C polymorphism, as reported by Teft and others [57] and in some other clinical studies, and the reproduction of clinical outcomes was attempted using a VCS approach."

Materials and Methods

"The clinical study reported by van der Bol and others [58] was used for observed AUC and blood concentration–time profiles of irinotecan, SN-38, SN-38G, NPC, and APC."

"Using a VCS approach, 30 sets of parameters optimized by CNM so that the blood concentration–time data of irinotecan and its metabolites were well described [59], were evaluated whether some sets of parameters can describe well other clinical findings [44 and References 60, 61 in Supplementary Text]."

"The clinical findings showing the similarities to the prediction by VCSs adopted in the present analyses consist of the following 4 categories: (I) the association between SN-38 plasma concentration at the end of infusion (90 min) and UGT1A1 *28 polymorphism or SLCO1B1 c.521T>C polymorphism; (II) the association between neutropenia and each genetic polymorphism of enzymes and transporters; (III) the association between diarrhea and each genetic polymorphism of enzymes and transporters; and (IV) the association between diarrhea and "biliary index (= $AUC_{(irinotecan)} \times AUC_{(SN\text{-}38)} / AUC_{(SN\text{-}38G)}$)," which has been reported as a good biomarker of diarrhea [62]."

Results

"In these VCSs, only the outcomes for the set of parameters ID 1, 7, and 9 could reproduce the results of the target study with regard to the effects of polymorphism of both UGT1A1 and SLCO1B1 on the SN-38 plasma concentrations."

"To examine whether the 100 times of VCSs using each set of parameters reproduce the outcomes of clinical studies, the following 7 criteria were used: (I) the effect of UGT1A1 *28 heterozygous on plasma concentration of SN-38 was reproduced over 75 times; (II) the effect of UGT1A1 *28 homozygous on plasma concentration of SN-38 was reproduced over 75 times; (III) the effect of UGT1A1 *28 polymorphism on neutropenia using either dominant model (UGT1A1 *1/*1 vs. *1/*28 and *28/*28) or recessive model (UGT1A1 *1/*1 and *1/*28 vs. *28/*28) was reproduced over 75 times; (IV) the effect of UGT1A1 *28 polymorphism on diarrhea using either dominant model or recessive model was reproduced over 25 times; (V) the effect of SLCO1B1 c.521T>C on plasma concentration of SN-38 was reproduced over 25 times; (VI) the effect of SLCO1B1 c.521T>C on neutropenia using either dominant model (SLCO1B1 521 T/T vs. T/C and C/C) or recessive model (SLCO1B1 521 T/T and T/C vs. C/C) was reproduced over 25 times; and (VII) the association between biliary index and diarrhea was reproduced over 25 times."

Discussion

"The sets of parameters obtained were also evaluated in terms of whether they could reproduce the results of reported clinical outcomes of irinotecan-induced side effects such as neutropenia and delayed diarrhea using a VCS approach."

"The VCS results indicated three important findings: (I) inter-individual variability of patients has the potential to cause a different clinical outcome with a small number of patients; $n \geq 192$ is needed to obtain a significant result with the effect of SLCO1B1 c.521T>C on the plasma concentration of SN-38, (II) specific parameters were strongly correlated with the frequency of reproducing clinical observations, and (III) the biliary index is considered to be a better biomarker than UGT1A1 *28 polymorphism regarding susceptibility to diarrhea."

"It was thus estimated that $n \geq 192$ is needed to obtain the significant result of the effect of SLCO1B1 c.521T>C. Generally, it is difficult to perform a clinical study using such large number of patients."

Conclusion

"This work successfully obtained multiple sets of PBPK model parameters that could reproduce the effects of genetic polymorphisms of UGT1A1 *28 and SLCO1B1 c.521T>C on the plasma concentration of SN-38 and side effects such as neutropenia and diarrhea using a VCS approach."

"The current VCS confirmed the importance of the biliary index as a better biomarker of irinotecan-induced diarrhea compared with only UGT1A1 *28 polymorphism."

"VCS suggested that the importance of polymorphism of other factors (CES1, ABCC2, ABCB1, and ABCG2) should be interpreted carefully."

Can Population Modelling Principles be Used to Identify Key PBPK Parameters for Paediatric Clearance Predictions? An Innovative Application of Optimal Design Theory [63] This is a machine-generated summary of:

Calvier, Elisa A. M.; Nguyen, Thu Thuy; Johnson, Trevor N.; Rostami-Hodjegan, Amin; Tibboel, Dick; Krekels, Elke H. J.; Knibbe, Catherijne A. J.: Can Population Modelling Principles be Used to Identify Key PBPK Parameters for Paediatric Clearance Predictions? An Innovative Application of Optimal Design Theory [63]

Published in: Pharmaceutical Research (2018)

Link to original: https://doi.org/10.1007/s11095-018-2487-1

If you want to cite the papers, please refer to the original.

For technical reasons we could not place the page where the original quote is coming from.

Abstract-Summary "We developed a methodology to investigate the feasibility and requirements for precise and accurate estimation of PBPK parameters using population modelling of clinical data and illustrate this for two key PBPK parameters for hepatic metabolic clearance, namely whole liver unbound intrinsic clearance ($CLint_{,u,WL}$) and hepatic blood flow (Qh) in children."

"Requirements for the trial components to yield precise estimation of the PBPK parameters and their inter-individual variability were established using a novel application of population optimal design theory."

"Precise estimation of $CLint_{,u,WL}$ and Qh and their inter-individual variability was found to require a trial with two drugs, of which one has an extraction ratio (ER) ≤ 0.27 and the other has an ER ≥ 0.93."

"The proposed clinical trial design was found to lead to precise and accurate parameter estimates and was robust to parameter uncertainty."

Introduction

"System-specific parameter values for PBPK models can be either obtained experimentally by direct measurements or they can be derived from clinical PK data through model parameter estimation [64–66]."

"Once global structural identifiability of model parameters is achieved, population modelling of longitudinal data can be used for estimation of the PBPK parameters and their inter-individual variability."

"For drugs undergoing hepatic metabolism, the part of a paediatric PBPK model describing this clearance contains two key parameters that cannot be directly measured and that cannot be simultaneously estimated from the PK data of one drug, due to identifiability issues, namely whole liver unbound intrinsic clearance ($CLint_{,u,WL}$) and hepatic blood flow (Qh)."

"The aim of this paper was to develop an analysis framework to investigate whether population modelling approach can be used to estimate PBPK model parameters from clinical PK data and establish the required criteria for such estimations."

Materials and Methods

"In order to obtain precise estimates of Θ and ω^2 for $CLint_{,u,WL}$ and Qh, a global sensitivity analysis of the uncertainty of these parameters with respect to the drugs ERs was undertaken using PFIM 4.0 (see under uncertainty section in methods)."

"In order to assess the impact of the drug's ER on the performance of the design without any confounding impact of sampling times, drug dosing, and number of patients and sampling times, these parameters were adapted to the drugs' properties."

"The clinical design parameter to optimize for precise estimation of these parameters was the ER of each of the two drugs in the design."

"Since the clinical trial design depends on the half-life and CLp of the two drugs for establishing sampling times and infusion rate respectively these parameters were used for its implementation."

Results

"To facilitate the identification of model drugs for which clinical data could be obtained to precisely estimate population values and variance of $CLint_{u,WL}$ and Qh at different ages, maturation patterns of different isoenzymes were used to identify isoenzymes and ages for which drugs with the lowest and highest ER required for such estimation are likely to exist."

"We found that drugs with an ER ≤ 0.27 are likely to exist for all investigated isoenzymes at all ages (results not shown)."

"Drugs metabolized by slowly maturing isoenzymes such as CYP2E1 and UGT2B7, are unlikely to have a high ER at young ages."

"Drugs metabolized by fast maturing isoenzymes, such as CYP1A2 or UGT1A4, with a very high ER over a wide range of paediatric ages are likely to exist and could be used as model drugs."

Discussion

"The use of optimal design not only allowed to optimize the characteristics of drugs to include in the clinical trial design in order to solve numerical identifiability issues, but also acts as a safeguard ensuring global identifiability of the model parameters."

"These results do highlight the importance of investigating the clinical trial requirements a priori, as otherwise the chances of successfully estimating PBPK model parameters from clinical PK data using population approach will be very low."

"This step allowed for the assessment of whether further investigation to optimize the clinical trial design would be needed to ensure unbiased parameter estimates."

"The clinical design parameter optimized was the ER of the drugs included in the clinical trial."

"This work presents an analysis framework that allows for the a priori identification of clinical trial requirements that would allow for the estimation of PBPK model parameters from clinical data using population modelling."

ACKNOWLEDGMENTS AND DISCLOSURES

"We would like to thank Sinziana Cristea for reviewing all model codes used in this study."

"Trevor Johnson is a paid employee of Simcyp Limited."

"Professor Amin Rostami-Hodjegan holds shares in Certara, a company focusing on Model-Informed Drug Development."

"This study was supported by the Innovational Research Incentives Scheme (Vidi grant, June 2013) of the Netherlands Organization for Scientific Research (NWO) to Catherijne A. J. Knibbe (2013)."

Semi-physiological Enriched Population Pharmacokinetic Modelling to Predict the Effects of Pregnancy on the Pharmacokinetics of Cytotoxic Drugs [67] This is a machine-generated summary of:

Janssen, J. M.; Damoiseaux, D.; van Hasselt, J. G. C.; Amant, F. C. H.; van Calsteren, K.; Beijnen, J. H.; Huitema, A. D. R.; Dorlo, T. P. C.: Semi-physiological Enriched Population Pharmacokinetic Modelling to Predict the Effects of Pregnancy on the Pharmacokinetics of Cytotoxic Drugs [67]
Published in: Clinical Pharmacokinetics (2023)
Link to original: https://doi.org/10.1007/s40262-023-01263-1

If you want to cite the papers, please refer to the original.

For technical reasons we could not place the page where the original quote is coming from.

Abstract-Summary "Of changes in physiology during pregnancy, the pharmacokinetics (PK) of drugs can be altered."

"We aimed to develop and evaluate a semi-physiological enriched model that incorporates physiological changes during pregnancy into available population PK models developed from non-pregnant patient data."

"Gestational changes in plasma protein levels, renal function, hepatic function, plasma volume, extracellular water and total body water were implemented in existing empirical PK models for docetaxel, paclitaxel, epirubicin and doxorubicin."

"These models were used to predict PK profiles for pregnant patients, which were compared with observed data obtained from pregnant patients."

"For paclitaxel, epirubicin and doxorubicin, the semi-physiological enriched model performed better in predicting PK in pregnant patients compared with a model that was not adjusted for pregnancy-induced changes."

"By incorporating gestational changes into existing population pharmacokinetic models, it is possible to adequately predict plasma concentrations of drugs in pregnant patients which may inform dose adjustments in this population."

Introduction

"During pregnancy, pharmacokinetics (PK) of drugs can be altered as a result of various changes in several physiological processes."

"These changes may result in increased or decreased drug concentrations compared with non-pregnant women, and this may change over the course of pregnancy [68]."

"Given the complexity of the physiological changes during pregnancy, the magnitude and relevance of these alterations on the PK of anticancer drugs is not straightforward."

"With this work, we aimed to develop a methodology in which the advantages of physiologically based PK models are combined with relevant existing knowledge of the PK in non-pregnant patients, enabling the prediction of individual PK profiles of a range of cytotoxic drugs in pregnant patients."

"We implemented a semi-physiological enriched pregnancy model including changes over the gestational time that allows the prediction of the PK of cytotoxic drugs in pregnant women using only available empirical compartmental models based on non-pregnant patient PK data."

Methods

"To describe the typical change in PK parameters during pregnancy, a selection of relevant empirical equations for physiological changes from Abduljalil and others were implemented in our semi-physiological predictions [68]."

"Limited and contradictory data are reported on the change in $Q_{H,blood}$ during pregnancy, we therefore assumed that the hepatic blood flow remained unchanged over pregnancy and was thus fixed to a typical non-pregnant value of 109 L/h [69]."

"These models were extended with the above described semi-physiological gestational changes to provide PK predictions for pregnant individuals."

"Covariates were excluded from the predictions to predict the typical change in PK parameters during pregnancy."

"The fit of the semi-physiological model during pregnancy was compared to the fit of the model parameters for the non-pregnant state (EGA = 0)."

Results

"Comparison of the model fit for the individual predictions based on the semi-physiological pregnant parameter estimates versus non-pregnant parameter estimates resulted in an increase in OFV of 70.1 points."

"The model fit for the individual predictions based on the semi-physiological pregnant parameter estimates showed decrease of the OFV of 279.3 compared with non-pregnant parameter estimates."

"The model fit for the individual predictions based on the semi-physiological pregnant parameter estimates showed a decrease of the OFV of 74.0 compared with non-pregnant parameter estimates."

"The semi-physiological enriched model adequately predicted the epirubicin concentration-time curves that were observed during pregnancy."

"Comparison of the model fit for the individual predictions based on the non-pregnant versus the semi-physiological pregnant parameter estimates showed a significantly improved fit for the latter, with a decrease in OFV of 21.2 points observed for epirubicin."

Discussion

"With this work, we have demonstrated the feasibility and relevance of a semi-physiological prediction approach in which prior knowledge of both the human population PK of a cytotoxic drug and physiological changes during pregnancy are combined to predict changes in PK in pregnant patients."

"For docetaxel, the semi-physiological approach did not perform better than the non-pregnant PK model parameters and only modest changes in typical parameters were observed."

"This shows that pregnancy has a limited effect on the PK of docetaxel and predictions using the parameters based on non-pregnant data perform well for the pregnant population."

"The semi-physiological approach performed significantly better than the model-based predictions based on non-pregnant parameter estimates."

"The semi-physiological model resulted in a significant better fit of the PK data from pregnant patients than the literature-based models for paclitaxel, epirubicin and doxorubicin."

Conclusions

"The semi-physiological enriched model provided an adequate prediction of the PK for four cytotoxic agents of two distinct drug classes in women over varying stages of gestation."

"It can be concluded that this proof of principle for a semi-physiological enriched model is applicable to the four cytotoxic drugs in our manuscript and can be extended to drugs with different pharmacological characteristics by the addition of relevant metabolic properties."

"This method may therefore be used for extrapolation purposes to adjust dosing regimens in pregnant women for drugs for which PK data from pregnant women are unavailable."

Physiologically Based Pharmacokinetic Modelling to Identify Physiological and Drug Parameters Driving Pharmacokinetics in Obese Individuals [70] This is a machine-generated summary of:

Berton, Mattia; Bettonte, Sara; Stader, Felix; Battegay, Manuel; Marzolini, Catia: Physiologically Based Pharmacokinetic Modelling to Identify Physiological and Drug Parameters Driving Pharmacokinetics in Obese Individuals [70]

Published in: Clinical Pharmacokinetics (2022)

Link to original: https://doi.org/10.1007/s40262-022-01194-3

If you want to cite the papers, please refer to the original.

For technical reasons we could not place the page where the original quote is coming from.

Abstract-Summary “This study aimed to investigate which physiological parameters and drug properties determine drug disposition changes in obese using our physiologically based pharmacokinetic (PBPK) framework, informed with obese population characteristics.”

“Simulations were performed for ten drugs with clinical data in obese (i.e., midazolam, triazolam, caffeine, chlorzoxazone, acetaminophen, lorazepam, propranolol, amikacin, tobramycin, and glimepiride).”

“PBPK drug models were developed and verified first against clinical data in non-obese (body mass index (BMI) $\leq$ 30 kg/m^2) and subsequently in obese (BMI $\geq$ 30 kg/m^2) without changing any drug parameters.”

“The PBPK model was used to study the effect of obesity on the pharmacokinetic parameters by simulating drug disposition across BMI, starting from 20 up to 60 kg/m^2.”

“Clearance increased by 1.6% per BMI unit up to 64% for a BMI of 60 kg/m^2, which was explained by the increased hepatic and renal blood flows.”

“Volume of distribution increased for all drugs up to threefold for a BMI of 60 kg/m^2; this change was driven by pK_a for ionized drugs and logP for neutral and unionized drugs.”

“Both physiological changes and drug properties impact drug pharmacokinetics in obese subjects.”

Introduction

“The logP represents the lipophilicity of the compound in the unionized form, the pK_a or acid dissociation constant determines the ionisation state of the drug, and the logD or distribution coefficient is a parameter combining logP and pK_a and expressing the lipophilicity of a compound at a given pH. Physiologically based pharmacokinetic (PBPK) modelling allows simulation of virtual clinical trials by using prior knowledge on human physiology and drug properties, thereby enabling exploration of drug disposition in virtual subjects with different degrees of obesity.”

“The first aim of this study was to verify the capability of our PBPK framework informed by our previously developed obese population repository to predict drug pharmacokinetics in obese individuals [71].”

“The second aim was to determine the physiological and drug physicochemical properties driving drug disposition changes in obese individuals.”

Methods

"We carried out a thorough literature search to identify drugs with available pharmacokinetic data in obese subjects in order to verify the PBPK drug models predictions against clinically observed data."

"Once all the observed data were collected, we developed and verified the PBPK model by predicting the pharmacokinetics in non-obese subjects first after intravenous administration (if available in the literature) then after single and multiple oral administration (more detailed workflow for midazolam is reported in the OSM)."

"The verified drug model was then used to predict the pharmacokinetics in virtual obese individuals without changing any drug parameter."

"Comparison with the observed clinical data was also carried out for the simulations in obese both visually and numerically; the latter was retained successful if the pharmacokinetic parameters were predicted within a twofold error margin."

Results

"PBPK models for the ten selected drugs were developed and parametrized to predict both the observed concentration-time profiles and pharmacokinetic parameters in non-obese subjects."

"The predicted:observed ratios in non-obese subjects were in good agreement for AUC_{inf} (0.92) but showed a slight underprediction for C_{max} and overprediction of $t_{1/2}$ (0.65 and 1.29, respectively), while for obese all three ratios were within 1.25-fold."

"The increase in volume of distribution was very variable, being the lowest for drugs such as amikacin and tobramycin with a 3% increase per BMI unit and the highest for midazolam with a 9% increase per BMI unit."

"This is supported by the fact that the volume of distribution of the lipophilic drug glimepiride does not increase as much as other drugs with similar logP values (i.e., midazolam)."

Discussion

"The pharmacokinetics of ten drugs eliminated through different pathways were successfully predicted a priori in obese subjects using our PBPK framework informed with our obese population repository [72], thereby demonstrating the predictive power of this approach to predict drug disposition in special populations."

"The effect of obesity on the pharmacokinetics was different across drugs."

"We demonstrated that for compounds mostly ionized at physiological pH, logP is not representative of the drug behaviour in obese subjects."

"Drugs metabolized by enzymes whose abundance is changed in obese subjects (i.e., UGT and CYP3A) had a different pattern [72–74]."

"Since metabolism is increased in obese subjects, the AUCs of drugs tend to decrease, and due to the higher elimination half-life, AUC_{inf} was shown to better describe differences in drug exposure."

Conclusions

"Drug exposure changes in obese subjects are due to increases in clearance and volume of distribution."

"The volume of distribution was higher in obese subjects for all drugs; however, different trends could not be explained by the logP, instead $logD_{7.4}$ was a better descriptor of lipophilicity in a physiological system since it considers the ionization of the compound."

"The study also highlights that the physiological changes and physicochemical parameters of the drug both play a role in the drug disposition in obese subjects, therefore making it difficult to have a fixed rule for scaling doses from non-obese to obese."

"PBPK modelling takes into consideration both the physiology and physicochemical properties, and can be used to simulate virtual clinical trials for those drugs for which clinical data are still not available in the obese, thereby filling the pharmacokinetic knowledge gap still present in the literature."

Hepatic Impairment Physiologically Based Pharmacokinetic Model Development: Current Challenges [75] This is a machine-generated summary of:

Han, Agnes Nuo; Han, Beatrice Rae; Zhang, Tao; Heimbach, Tycho: Hepatic Impairment Physiologically Based Pharmacokinetic Model Development: Current Challenges [75]

Published in: Current Pharmacology Reports (2021)

Link to original: https://doi.org/10.1007/s40495-021-00266-5

If you want to cite the papers, please refer to the original.

For technical reasons we could not place the page where the original quote is coming from.

Abstract-Summary "This review summarizes the development processes of hepatic impairment (HI) PBPK models, examines current challenges, and proposes potential solutions."

"Because hepatic impairment can significantly alter a patient's physiology, HI PBPK models must consider complex in vivo processes leading to potential changes in PK parameters."

"Despite recent progress, HI PBPK models face multiple challenges and may overpredict drug exposure with increasing severity of liver dysfunction."

"Foremost among these challenges is the use of the Child–Pugh scoring system in designing HI PBPK models."

"Most HI PBPK models do not account for changes in certain drug parameters, potentially skewing resulting predictions."

"Recent advancements have enhanced the predictive power of HI PBPK models, enabling accurate reflections of clinical trials."

Introduction

"As one of the top 10 causes of death in 25–74-year-olds, hepatic impairment (HI) includes multiple distinct diseases, including liver cirrhosis, hepatitis, and non-alcoholic steatohepatitis (NASH)."

"Despite the drastic physiological and PK-related changes caused by HI, there are few published physiologically based pharmacokinetic (PBPK) predictions for HI patients evaluating diverse drugs and specific populations [76••]."

"In a recent FDA workshop, presenters noted that current PBPK models tend to overpredict the impact of HI on drug exposure [77••]."

"These findings were confirmed by the first large-scale study of HI PBPK models across a range of compounds and disease stages, which found the degree of model overprediction sometimes increased with disease severity [76••]."

"Because these flaws in existing HI PBPK models make it difficult to determine accurate dose adjustments, few FDA-approved labels meet established criteria for evaluation of dosing in hepatic disease [78]."

Drug Regulatory Guidelines (EMA and FDA)

"The FDA had released the "Pharmacokinetics in Patients with Impaired Hepatic Function" guidance in 2003, providing recommendations for study design, data analysis, dosing, and labeling [79]."

"It recommended that studies on hepatic impairment be conducted if (a) either hepatic metabolism or excretion accounts for over 20% of the absorbed parent drug or active metabolite's elimination, (b) the drug has a narrow therapeutic range, or (c) the drug's metabolism is unknown."

"Labels should indicate caution for severe HI if the drug has significant hepatic clearance."

"The EMA recommends pharmacokinetic studies if (a) the drug is likely to be used in patients with hepatic impairment or (b) hepatic impairment will likely significantly alter the PK of the drug/its metabolites [80]."

Limitations with Existing Data

"HI PBPK models for compounds are typically constructed via data from clinical studies which often have difficulty recruiting participants, resulting in small sample sizes and skewed results."

"Of the existing published studies, few contain a systematic evaluation of the predictive performance of PBPK in different stages of HI across multiple compounds."

"The dataset did not include compounds with significant transporter involvement in the elimination [76••] Since most published studies only focus on small numbers of compounds, drawing conclusions on the accuracy of HI PBPK models remains difficult and more clinical data should be compiled across the pharmaceutical industry."

Child–Pugh Classification System

"These results were slightly challenged in a study evaluating the impacts of shunting on lidocaine (a high extraction ratio drug) PK for mild and severely hepatically impaired patients."

"The study concluded that reduction in hepatic blood flow resulting from portacaval shunting minimally affects drug disposition in patients with severe liver disease but significantly reduces the clearance of high extraction drugs in mildly cirrhotic patients."

"Despite the potential importance of hepatic impairment-induced absorption changes for lipophilic drugs, the impact of liver cirrhosis on drug absorption has not been widely accounted for in hepatic impairment PBPK modeling, possibly because measured bile salts concentrations in liver cirrhosis have not yet been reported [76••]."

"The study demonstrated that while there was no net difference in total clearance between cirrhotic and healthy patients was observed, metabolism is significantly altered, and the decrease in hepatic metabolic capacity was offset by the increase in the fraction of unbound drug."

Case Example

"Although both observed and predicted data found a negligible effect from a mild hepatic impairment, the PBPK model had overestimated the effect of moderate and severe hepatic impairment on drug exposure."

"Observed data found a 1.6-fold increase in AUC in moderate hepatic impairment compared to a PBPK model predicted ratio of a 2.25-fold."

"The most likely explanation for the variability between the predicted and observed data is the failure of the PBPK model to account for changes to the rate (ka) and extent (fa) of absorption with hepatic impairment."

"For moderate hepatic impairment, it is possible that the PBPK model overpredicted AUC compared with observed data from the clinical trial conducted under fed conditions."

"Future PBPK predictions should employ a complete mechanistic absorption model that accounts for changes in absorption with hepatic impairment."

Conclusion

"While the accuracy of HI PBPK models has improved considerably, scientific challenges to predict ADME remain, especially for severe impairment, with complex comorbidities."

"Future models should include alternatives to the Child–Pugh classification system, incorporate changes in bile salt concentrations in absorption processes, factor in the impact of ascites on drug distribution, resolve contradictions in the literature on changes in CYP and UGT enzyme levels, and establish a consensus on transporter expression level changes associated with hepatic impairment."

A Preterm Physiologically Based Pharmacokinetic Model. Part I: Physiological Parameters and Model Building [81] This is a machine-generated summary of:

Abduljalil, Khaled; Pan, Xian; Pansari, Amita; Jamei, Masoud; Johnson, Trevor N.: A Preterm Physiologically Based Pharmacokinetic Model. Part I: Physiological Parameters and Model Building [81]

Published in: Clinical Pharmacokinetics (2019)

Link to original: https://doi.org/10.1007/s40262-019-00825-6

If you want to cite the papers, please refer to the original.

For technical reasons we could not place the page where the original quote is coming from.

Abstract-Summary "Developmental physiology can alter pharmacotherapy in preterm populations."

"Because of ethical and clinical constraints in studying this vulnerable age group, physiologically based pharmacokinetic models offer a viable alternative approach to predicting drug pharmacokinetics and pharmacodynamics in this population."

"Such models require comprehensive information on the changes of anatomical, physiological and biochemical variables, where such data are not available in a single source."

"The objective of this study was to integrate the relevant physiological parameters required to build a physiologically based pharmacokinetic model for the preterm population."

"Data on organ size show different growth patterns that were quantified as functions of bodyweight to retain physiological variability and correlation."

"Despite the limitations identified in the availability of some tissue composition values, the data presented in this article provide an integrated resource of system parameters needed for building a preterm physiologically based pharmacokinetic model."

Introduction

"While the preterm PBPK model is not intended to replace clinical studies, it can facilitate making decisions regarding first-time dosing and study design in this population."

"The availability of the PBPK modelling approach can facilitate this study design process and help to explore different scenarios that can affect drug kinetics, such as a change in physiological parameters in a specific clinical condition (e.g. ductus arteriosus), presence of variants in metabolising enzymes or transporters, impact of co-medications, optimal sampling and dosing suggestions."

"The objectives of this work are: (1) to collect data for physiological parameters that are required for building a preterm PBPK model; and (2) to incorporate system parameters as a function of individual age or/and weight within the Simcyp® Simulator to create a PBPK model that is capable of modelling altered pharmacokinetics in a preterm population."

Materials and Methods

"Data inclusion criteria were as follows: (1) individuals between 28 and 44 weeks of postmenstrual age (PMA) but within their first month of postnatal age (PNA); PMA is calculated as the sum of GA and PNA, where PNA is the age after birth, at birth (PNA = 0) PMA equals GA. (2) Where possible, data from extreme premature neonates (GA < 28 weeks) were not included. (3) In the case of mixed-population studies, the extreme preterm individuals should not comprise more than 30% of the overall preterm sample size under study. (4) When preterm data were not available, foetal data were included. (5) Caucasian data; where Caucasian data for a specific parameter were not available, data from non-Caucasian populations were used."

"Meta-analyses of organ volume data and gross composition have been published earlier as a function of an individual's body weight, where longitudinal changes in systems parameters were tested using a number of commonly used growth equations, the optimal function was selected based on Akaike information criteria [82, 83]."

Results

"'The following equations were derived to describe the changes in total, intracellular and extracellular water during growth for both male and female subjects:where WT is the neonatal body weight in kilograms and PMA is the postmenstrual age in years."

"The propagated variability from total body weight seems sufficient to recover the observed inter-individual variability for total and extracellular water in preterm neonates."

"The following equations determine the tissue volumes based on the total body weight (WT), in kilograms, for both male and female preterm subjects: The propagated variability from total body weight seems sufficient to recover the observed inter-individual variability for each tissue (see figures in the ESM)."

"The preterm GFR function within the Simcyp Simulator to describe drug filtration renal clearance was a modification of Rhodin and others's equation [84]:where PMA is the postmenstrual age in years, and WT is the total body weight in kilograms for a preterm subject."

Discussion

"Physiological parameters required to build preterm PBPK population models have been collated and equations derived to describe their change (mean and the associated variability) during development."

"Claassen and others published a preterm PBPK model without providing information on growth functions for all organs and their inter-individual variability, nor how changes to binding proteins and haematocrit were accounted for."

"To Claassen and others's model, Yang and others's model provided detailed methodology including model equations for physiological parameters and the associated variability; however, performance of the model in the prediction of preterm pharmacokinetics was lacking A major drawback of the available data is the lack of longitudinal investigation of system parameters and the high variability in the available cross-sectional studies in this special population."

"The preterm population model is flexible and allows the changing of physiological parameters to mimic different disease conditions."

Conclusions

"Integrated system parameters for building a stochastic preterm PBPK model were compiled and incorporated within the Simcyp Simulator population library."

"These population data aim to mechanistically predict pharmacokinetics in the preterm population and to account for inter-individual variability at different PMA, GA or PNA when coupled with the in vitro-in vivo extrapolation approach."

"This approach will facilitate the investigation of drug exposure and any associated toxicity in different body organs and age-varying drug–drug interactions, and to inform the design of clinical studies in the preterm patient population."

"Future developments should expand the model to the very preterm populations."

3.1.2 Noncompartmental Modeling

Machine generated keywords: bioequivalence, curve, maximum concentration, auc, concentrationtime, nca, determination, partial, renal, area concentrationtime, clinical practice, indication, adult, regression, ratio

A Systematic Review of Multiple Linear Regression-Based Limited Sampling Strategies for Mycophenolic Acid Area Under the Concentration–Time Curve Estimation [85] This is a machine-generated summary of:

Sobiak, Joanna; Resztak, Matylda: A Systematic Review of Multiple Linear Regression-Based Limited Sampling Strategies for Mycophenolic Acid Area Under the Concentration–Time Curve Estimation [85]

Published in: European Journal of Drug Metabolism and Pharmacokinetics (2021)

Link to original: https://doi.org/10.1007/s13318-021-00713-0

If you want to cite the papers, please refer to the original.

For technical reasons we could not place the page where the original quote is coming from.

Abstract-Summary "The aim of this systematic review was to review the MPA LSSs and define the most frequent time points for MPA determination in patients with different indications for mycophenolate mofetil (MMF) administration."

"A total of 27, 17, and 11 studies were found for groups 1, 2, and 3, respectively, and 126 MLR-based LSS formulae (n = 120 for MPA, n = 6 for fMPA) were included in the review."

"Four MPA LSSs: 2.8401 + 5.7435 × C0 + 0.2655 × C0.5 + 1.1546 × C1 + 2.8971 × C4 for adult renal transplant recipients, 1.783 + 1.248 × C1 + 0.888 × C2 + 8.027 × C4 for adults after islet transplantation, 0.10 + 11.15 × C0 + 0.42 × C1 + 2.80 × C2 for adults after heart transplantation, and 8.217 + 3.163 × C0 + 0.994 × C1 + 1.334 × C2 + 4.183 × C4 for pediatric renal transplant recipients, plus one fMPA LSS, 34.2 + 1.12 × C1 + 1.29 × C2 + 2.28 × C4 + 3.95 × C6 for adult liver transplant recipients, seemed to be the most promising and should be validated in independent patient groups before introduction into clinical practice."

"The review includes updated MPA LSSs, e.g., for different MPA formulations (suspension, dispersible tablets), generic form, and intravenous administration for adult and pediatric patients, and emphasizes the need of individual therapeutic approaches according to MMF indication."

"Five MLR-based MPA LSSs might be implemented into clinical practice after evaluation in independent groups of patients."

Introduction

"Due to the pharmacokinetic variability, therapeutic drug monitoring (TDM) in the case of MPA is recommended in clinical practice [86, 87]."

"The most accurate approach to TDM is the determination of the full pharmacokinetic profile of the drug and calculation of the area under the concentration–time curve from 0 to 12 h (AUC_{0-12}), as the concentration determined before the next dose (C_{trough}) does not reflect the overall exposure to MPA [88]."

"LSS may be calculated using a Bayesian approach or multiple linear regression (MLR) analysis, which uses an equation derived from stepwise regression analysis based on concentrations measured at pre-defined times after dosing [87, 89]."

"As it does not depend on the pharmacokinetic model of the drug and can be calculated with simple software or manually [90], MLR is easier to use in clinical practice than Bayesian analysis; however, the MLR approach has some limitations."

Methods

"Original articles determining LSS based on MLR calculations for MPA and fMPA were included."

"There are some studies establishing MPA LSSs with a Bayesian estimator, and although this approach has some advantages (e.g., better accuracy and precision, the lack of strict adherence to sampling times, and number of samples [87]), we decided to include only MLR-based MPA LSSs due to the excessive amount of data and the difficulty in analyzing MLR-based LSSs and Bayesian-approach LSSs."

"The data were analyzed according to the most frequently used time points in three groups of patients treated with MMF: (1) adult renal transplant recipients, (2) adults receiving MMF due to other indication than renal transplantation, and (3) pediatric patients."

"Whenever possible, the predictive performance results of the LSSs (bias, precision, validation group) were included in the review, as was the information whether the validation was internal or external."

Results

"The LSSs established in the early post-transplant period (n = 22) most frequently included C2 (25% of all time points, 73% of the equations), and C4 (22% of all time points, 64% of the equations)."

"For LSSs established in the stable post-transplant period (n = 16), the concentrations most often used included C1 (20% of all time points, 63% of the equations) and C3 (14% of all time points, 44% of the equations)."

"For LSSs established for patients treated with MMF less than 1 month, the most frequently included time points were C2 (27% of all time points, 87% of the equations) and C6 (23% of all time points, 73% of the equations)."

"Among 14 LSSs established for children after renal transplantation, the most frequently included time points were the same as for all the pediatric studies."

Discussion

"The authors [91] observed that the application of LSS established for lung transplant recipients to predict MPA AUC in patients after heart transplant yielded satisfactory prediction results (bias and precision within ± 15%); however, they concluded that the LSSs seem to be center-specific."

"Comparing LSSs between patients treated concomitantly with CsA or Tac, the time points beyond 6 h were more frequently included in LSS when Tac was co-administered."

"For MMF and CsA administration, in MPA LSSs, the most frequently included time points were C2, C0, and C0.5."

"Constantly improving renal function after transplantation affects MPA pharmacokinetics [92]; therefore, some differences in time points included in LSSs which were established for patients treated with MMF less than 1 month after renal transplantation and longer than 3 months after renal transplantation were expected."

"In MPA LSSs developed for patients with other than renal transplantation indication for MMF treatment, different sets of time points were used more frequently."

Conclusions

"We found five MLR-based MPA LSSs which might be considered as useful in clinical practice; however, they require further evaluation in independent groups of patients."

"For adult patients, MPA LSSs most frequently included C2 and C4, while, for pediatric patients, C0 and C2 were the most often used."

"The fact that the time points of MPA concentrations most frequently included in LSSs were different for adult renal transplant recipients, adults after other than renal transplantation, and in children treated with MMF, emphasizes the need of individual therapeutic approaches for each group of MMF-treated patients."

"Whereas the methodology of developing MPA LSS is rather a simple method enabling TDM, establishing the most accurate MPA LSSs require numerous factors to be considered, such as the drugs co-administered with MMF (particularly calcineurin inhibitors), the time elapsed from the transplantation or the duration of treatment with MMF, and the indication for MMF treatment."

Comparison of non-compartmental and mixed effect modelling methods for establishing bioequivalence for the case of two compartment kinetics and censored concentrations [93] This is a machine-generated summary of:

Hughes, Jim H.; Upton, Richard N.; Foster, David J. R.: Comparison of non-compartmental and mixed effect modelling methods for establishing bioequivalence for the case of two compartment kinetics and censored concentrations [93]

Published in: Journal of Pharmacokinetics and Pharmacodynamics (2017)

Link to original: https://doi.org/10.1007/s10928-017-9511-7

If you want to cite the papers, please refer to the original.
For technical reasons we could not place the page where the original quote is coming from.

Abstract-Summary "Non-compartmental analysis (NCA) is regarded as the standard for establishing bioequivalence, despite its limitations and the existence of alternative methods such as non-linear mixed effects modelling (NLMEM)."

"Comparisons of NCA and NLMEM in bioequivalence testing have been limited to drugs with one-compartment kinetics and have included a large number of different approaches."

"A simulation tool was developed with the ability to rapidly compare NCA and NLMEM methods in determining bioequivalence using both R and NONMEM and applied to a drug with two-compartment pharmacokinetics."

"The simulations showed NLMEM having a consistent 20–40% higher accuracy and sensitivity in identifying bioequivalent studies when compared to NCA, while NCA was found to have a 1–10% higher specificity than NLMEM."

"Increasing data censoring by increasing the LLOQ resulted in decreases of ~10% to the accuracy and sensitivity of NCA, with minimal effects on NLMEM."

"The tool provides a platform for comparing NCA and NLMEM methods and its use can be extended beyond the scenarios reported here."

Introduction

"Non-linear mixed effects modelling (NLMEM) is an alternate method that can minimise the effect of censored data and RUV."

"NCA and NLMEM have been compared in the literature to examine their ability to determine bioequivalence in drugs with one-compartment kinetics [94–96]."

"The decrease in specificity, combined with the extra effort required to undergo NLMEM, has left NCA as the preferred method for determining bioequivalence."

"A comparison of these NCA and NLMEM in drugs with two-compartment kinetics has not been tried in the literature."

"The aims of this work were to: (1) Create a flexible simulation tool that facilitated the comparison of the two methods, where the accuracy and precision of bioequivalence results were the main interest. (2) Use this tool to assess the relative merits of NCA and NLMEM for the case of two compartment kinetics and censored data."

Method

"The tool was written so that data were simulated and NLMEM could also be initiated directly from the R script, with the population pharmacokinetic modelling process being undertaken by non-linear mixed effect modelling (NONMEM®) Version 7.2 [97] with the Wings for NONMEM (Version 720) interface and the Intel complier."

"The raw data provided by the simulation then had RUV added and its sampling times truncated to replicate observed data representative of 'real world' study data that could subsequently be used in NCA and NLMEM."

"Once the data were simulated, the rich dataset was analysed to determine which studies demonstrated bioequivalence, as these are the 'true' concentrations without random error for each individual."

"The accuracy, sensitivity and specificity for each method were calculated using the following: Whether a bioequivalent study was a true positive (TP) or false positive (FP) was determined by comparing it to the result from the original rich simulated data, as this represented "the truth"."

Results

"The scenarios compared the bioequivalence results of each study for NCA and the four NLMEM methods to the bioequivalence results received from the simulated or 'true' data."

"The accuracy of the NLMEM 'F estimate' methods in determining relative bioavailability were consistently higher than all other methods for all scenarios, however when using the NLMEM 'post hoc' method achieved similar reduced accuracy to those achieved by NCA."

"Simulation Run 2 and 4 (which were characterised by increased LLOQ and fewer blood samples, respectively) resulted in decreased accuracy and sensitivity for NCA while having a minimal effect on the NLMEM 'F estimate' method."

"Simulation Run 3 had all methods relatively unchanged by its increase in proportional RUV, however when this was combined with a higher LLOQ in Scenario 10 NCA showed a large decrease in accuracy in similar scale to that seen in Simulation Run 2."

Discussion

"Due to this reason the increased accuracy seen by using NLMEM over NCA could be considered a direct result of the drug in question having two-compartmental kinetics, however previous studies in this have not published accuracy of methods, with a bigger emphasis being placed on Type I error, a determinant in the calculation of specificity."

"The accuracy and sensitivity of the method are likely to decrease as more data is censored, due to poor characterisation of the terminal phase giving a non-bioequivalent result to more studies, truly bioequivalent or not."

"The increase in specificity by ~5% seen between the scenarios mentioned earlier are likely explained by both the poor characterisation of the terminal phase in simulations in Scenario 1 and the reduced number of usable concentration points in Scenario 10."

"A 'real-world' result could be achieved in similar simulation studies, with the use of a tool which creates study-specific models as opposed to using a single template for all studies."

Conclusion

"This study has shown that NLMEM is more accurate and has a higher sensitivity than NCA when used on two-compartmental drugs."

"NCA has a higher specificity than NLMEM and is less likely to falsely assign bioequivalence, however the number of false positives that NLMEM produced was still relatively low."

"Some may approve of NLMEM, as NCA has been shown here to reject drugs despite being bioequivalent."

A Simulation Study of the Comparative Performance of Partial Area under the Curve (pAUC) and Partial Area under the Effect Curve (pAUEC) Metrics in Crossover Versus Replicated Crossover Bioequivalence Studies for Concerta and Ritalin LA [98] This is a machine-generated summary of:

Jackson, Andre J.; Foehl, Henry C.: A Simulation Study of the Comparative Performance of Partial Area under the Curve (pAUC) and Partial Area under the Effect Curve (pAUEC) Metrics in Crossover Versus Replicated Crossover Bioequivalence Studies for Concerta and Ritalin LA [98]

Published in: The AAPS Journal (2022)

Link to original: https://doi.org/10.1208/s12248-022-00726-w

If you want to cite the papers, please refer to the original.

For technical reasons we could not place the page where the original quote is coming from.

Abstract-Summary "The current purpose was to compare via simulation, using literature MPH models, the performance of the pAUC metrics in establishing BE via the standard crossover design versus a replicated design, and the relationship of the pAUC metrics to PD (pharmacodynamics) metrics, e.g., SKAMP (Swanson, Kotkin, Agler, M-Flynn, and Pelham rating scale) composite scores pAUEC (partial area under the effect curve)."

"An indirect response model described the SKAMP composite scores corrected for placebo."

"Performance of the pAUC metrics was demonstrated by the calculation of 90% confidence intervals (CIs) for each k0fast (fast absorption rate constant) and kaslow (slow absorption rate constant) test/reference (T/R) ratio."

"The 90% CIs resulting from changes in the k0fast and kaslow ratios, e.g., T/R, showed greater sensitivity to changes in the ratios at quotients below 0.8 than above for both Concerta and Ritalin LA."

"Ritalin LA pAUC values were insensitive to increases in either ratio once the ratio exceeded 1.0 and the study design."

"Correlations between least squares means (LSM) for pAUC and the SKAMP pAUEC for the composite scores were near 90%."

INTRODUCTION

"For MPH products which are multiphasic modified release (MR) products, 90% CI limits (i.e., upper and lower limits) are applied throughout the time span corresponding to a typical school day of 8 h. Several authors (Chen and others [99]) have proposed an extension of the use of the pAUC metric to multiphasic modified release formulations, which was a logical extension of the prior recommended use."

"The purpose of the simulation study described herein was to compare how changes in k0fast (fast absorption rate constant) and kaslow (slow absorption rate constant) T/R ratios affect the performance of the pAUC BE metrics pAUC0–3 h, pAUC3–7 h, and pAUC7–12 h for the MPH MR products Concerta and Ritalin LA when using a two-formulation, two-sequence, two-period crossover design (referred to herein as the "crossover" design) vs. a two-formulation, two-sequence, four-period, replicated crossover design (referred to herein as the "replicated" design)."

MATERIALS AND METHODS

"Simulated PK data from the previously described scenarios were coupled with the above PD model to give 1000 PD studies for Concerta replicated designed study at T/R ratios for k0fast and kaslow between 0.25 and 1.50 for Concerta and 1000 PD studies for Concerta two-way crossover by using only the first two periods of the replicated study."

"The PROC POWER procedure in SAS [100] was used to estimate the sample size for the simulated crossover design and replicated design studies for Concerta and Ritalin LA."

"The residual variance for the crossover design studies and subject-by-formulation interaction variance for the replicated design studies were extracted from the ANOVA analysis for the respective study designs to determine these values for the different pAUC metrics at the k0fast (test)/k0fast (reference) and kaslow (test)/kaslow (reference) ratios of 0.30, 1.00, and 1.50."

RESULTS

"The pAUEC3–7 h values decreased as kaslow was increased for both statistical designs but was non-responsive to changes in k0fast T/R values."

"The pAUEC7–12 h values were not responsive to changes in either the k0fast or kaslow T/R ratios for either statistical design."

"As also observed for Concerta, the pAUEC0–3 h metric was responsive only to k0fast T/R value increases for both study designs but was not responsive to changes in the kaslow T/R ratios."

"The pAUEC3–7 h values were influenced by both k0fast test and kaslow test values up to a T/R ratio of 1.00; above that, neither statistical design exhibited any response to pAUEC values."

"Since the pAUC7–12 h metric was not responsive to increases in the kaslow T/R ratio for Concerta and Ritalin LA, alternative pAUC values of 7–9 h and 7–10 h were investigated."

DISCUSSION

"The pAUC0–3 h metric is a dynamic metric that responds to changes in the k0fast T/R ratios, especially at low values."

"The pAUC3–7 h metric had a static response to changes in the k0fast T/R ratio with pAUC7–12 h correlating negatively with k0fast T/R ratios."

"The pAUC3–7 h metric rose above the 1.25 UCL limit when the kaslow T/R ratio was 1.50, whereas pAUC7–12 h did not respond."

"The pAUC3–7 h and pAUC7–12 h values correlated negatively with decreases in the k0fast T/R ratio but the LCL did not exceed the 1.25 UCL as seen with Concerta."

"The pAUC0–3 h metric did not respond to changes in the kaslow T/R ratio."

"The pAUC7–12 h metric was correlated negatively with kaslow because as the T/R ratio decreased, the pAUC7–12 h UCL exceeded 1.25."

"Changes in the kaslow T/R ratios had no measurable effect on the pAUEC7–12 h values for Concerta or Ritalin LA."

CONCLUSION

"Not actively responding to kaslow test/reference changes after 7 h, BE determination seems to have worked since the implementation of the new FDA guidance for methylphenidates [101]."

"In the future, it may be prudent to check via simulation to ensure that each proposed BE pAUC study metric has the desired dynamic performance."

3.2 Pharmacodynamic (PD) Model

Introduction by the Editor

Pharmacodynamic (PD) modeling of nanoparticles is vital for deciphering their therapeutic impact within biological systems. It quantifies the relationship between nanoparticle concentrations and their pharmacological effect, guiding optimal dosing strategies. By accounting for nanoparticle-specific properties, PD modeling enhances our understanding of target engagement, cellular responses, and therapeutic outcomes, aiding the design of effective drug delivery systems. It enables the prediction of dose-response relationship, assists in modifying nanoparticles for desired effects, and contributes to the development of safer and more potent nanoparticle-based therapies.

Machine generated summaries

Disclaimer: The summaries in this chapter were generated from Springer Nature publications using extractive AI auto-summarization: An extraction-based summarizer aims to identify the most important sentences of a text using an algorithm and uses those original sentences to create the auto-summary (unlike generative AI). As the constituted sentences are machine selected, they may not fully reflect the body of the work, so we strongly advise that the original content is read and cited. The auto generated summaries were curated by the editor to meet Springer Nature publication standards.

To cite this content, please refer to the original papers.

Machine generated keywords: response, relationship, child, cumulative, concentration drug, measurement, empirical, interindividual variability, iontophoresis, detail, interindividual, model parameter, continuous, accurately, probability

Population Pharmacodynamic Modeling Using the Sigmoid E_{max} Model: Influence of Inter-individual Variability on the Steepness of the Concentration–Effect Relationship. a Simulation Study [102] This is a machine-generated summary of:

Proost, Johannes H.; Eleveld, Douglas J.; Struys, Michel M. R. F.: Population Pharmacodynamic Modeling Using the Sigmoid E_{max} Model: Influence of Inter-individual Variability on the Steepness of the Concentration–Effect Relationship. a Simulation Study [102]

Published in: The AAPS Journal (2020)

Link to original: https://doi.org/10.1208/s12248-020-00549-7

If you want to cite the papers, please refer to the original.

For technical reasons we could not place the page where the original quote is coming from.

Abstract-Summary "We investigated the relationship between the steepness of the concentration–effect relationship and inter-individual variability (IIV) of the parameters of the sigmoid E_{max} model, using the similarity between the sigmoid E_{max} model and the cumulative log-normal distribution."

"It is investigated whether IIV in the model parameters can be estimated accurately by population modeling."

"Multiple data sets, consisting of 40 individuals with 4 binary observations in each individual, were simulated with varying values for the model parameters and their IIV."

"The steepness of the population-predicted concentration–effect relationship (γ*) is less than that of the individuals (γ)."

"Using γ^*, the population-predicted drug effect represents the drug effect, for binary data the probability of drug effect, at a given concentration for an arbitrary individual."

INTRODUCTION

"This implies that it can be used in cases where the concentration–effect is likely to follow a cumulative log-normal distribution, e.g., in the case of binary responses, where the probability of response is modeled as a function of drug concentration."

"A simultaneous analysis of the data of these three studies revealed that the steepness of the concentration–effect relationship is, among other factors, dependent on the inclusion or exclusion of IIV in model parameters [103]."

"It is the aim of this paper (procedure 1) to describe quantitatively the relationship between the steepness of the concentration–effect relationship and IIV in the model parameters, and (procedure 2) to investigate whether IIV in the model parameters can be estimated by population analysis with data obtained from study designs as used in reported clinical research studies."

METHODS

"If we assume that the drug effect in an individual can be predicted from the sigmoid E_{max} model, the drug effect P is defined: where C is the effect-site concentration, C50 is the effect-site concentration resulting in $P = 0.5$, and γ is a dimensionless model parameter, reflecting the steepness of the concentration–effect relationship."

"The steepness of the population-predicted P may be expressed by the sigmoid E_{max} model (using symbol γ^*, to discriminate from the model parameter γ) or cumulative log-normal distribution (using symbol σ^*, to discriminate from the model parameter σ) and is a function of γ (or σ) and the IIV in C50 and γ (or σ)."

"Using the similarity between the probability of drug effect according to the sigmoid E_{max} model and the cumulative log-normal distribution, an equation for the relationship between γ and IIV in C50 and γ on the population-predicted γ^* was derived in a Monte Carlo simulation study."

RESULTS

"To investigate whether the parameters C50 and γ as well as their IIV (ω_{C50} and ω_γ) could be estimated from clinical study data with acceptable precision, a series of simulations was performed using NONMEM."

"For $\gamma = 30$ and IIV in C50 during simulation, γ became very high, resulting in a "near boundary" message in NONMEM."

"When IIV was absent during simulation, the estimate of the corresponding variance became very low, resulting in a "near boundary" message in NONMEM, indicating the absence of variance in the corresponding parameter."

"Where IIV was present during simulation, the corresponding variance could be estimated, but the estimated values were far from the variance used in the simulations."

"A similar series of simulations were performed assuming that IIV in γ was absent during simulation and estimation."

"Ω_{C50} was fixed to 0.1 and $\omega_\gamma = 0$ during simulation and IIV was assumed to be absent during estimation (naive pooling)."

DISCUSSION

"From our simulations (procedure 2) it can be concluded that the tested study design (40 individuals with 4 binary observations in each individual) is not suited to estimate IIV in both C50 and γ with reasonable precision."

"Our simulations show that it does not seem easy to design and perform a study to estimate C50 and γ as well as their IIV."

"Besides, it may be noted that the value of γ cannot be determined precisely in all cases, since its value becomes very high during the population analysis with IIV, as shown in clinical studies [103, 104] as well as in the simulations in the current paper."

"The tested study design (40 individuals with 4 binary observations in each individual) is not suited to estimate the IIV in C50 and γ with reasonable precision."

Population pharmacokinetic/pharmacodynamic modeling of histamine response measured by histamine iontophoresis laser Doppler [105]

This is a machine-generated summary of:

Liu, Xiaoxi; Jones, Bridgette L.; Roberts, Jessica K.; Sherwin, Catherine M.: Population pharmacokinetic/pharmacodynamic modeling of histamine response measured by histamine iontophoresis laser Doppler [105]

Published in: Journal of Pharmacokinetics and Pharmacodynamics (2016)

Link to original: https://doi.org/10.1007/s10928-016-9478-9

If you want to cite the papers, please refer to the original.

For technical reasons we could not place the page where the original quote is coming from.

Abstract-Summary "The EH method is limited in providing an objective and qualitative assessment of histamine pharmacodynamic response."

"The histamine iontophoresis with laser Doppler (HILD) monitoring method, an alternative method, allows a fixed dose of histamine to be delivered and provides an objective, continuous, and dynamic measurement of histamine epicutaneous response in children and adults."

"The reduced data was further analyzed and a population linked effect pharmacokinetic/pharmacodynamic (PK/PD) model was developed to describe the local histamine response."

"The model consisted of a one-compartment PK model and a direct-response fractional maximum effect (Emax) model."

Introduction

"The epicutaneous histamine (EH) test is the current gold standard method for the clinical evaluation of allergic conditions (e.g. allergic rhinitis) [106]."

"EH testing assesses the microvasculature response to histamine through the observation and measurement of the 'wheal and flare' response which results from vasodilatation and vascular leakage after activation of histamine receptors."

"The EH method is limited in providing an objective and quantitative assessment of histamine pharmacodynamic response."

"We have shown that the Histamine iontophoresis with laser Doppler (HILD) monitoring method allows a fixed dose of histamine to be delivered and provides an objective, continuous, and dynamic measurement of histamine epicutaneous response in children and adults [107, 108]."

"Among children with allergic rhinitis, we have identified distinct histamine response phenotypes via the HILD method [107]."

"One of the advantages of HILD over EH method is that the former allows continuous histamine measurements over time."

Methods

"A PK/PD linked effect model was developed to describe the histamine response over time monitored by HILD."

"The model is expressed mathematically as below: where A0 is the amount of histamine remaining to be absorbed; A1 is the amount of histamine in the local depot; A2 is the amount of histamine in the central plasma compartment; V1 is the local volume of distribution; V2 is the central plasma volume of distribution; CLpar is the partitioning clearance from local depot to systemic circulation; CL is the central clearance; ka is the absorption rate constant; F is the fraction of histamine absorbed; E is the local response observed; E0, is the baseline effect; Emax is the maximum effect; EC50 is the histamine concentration at half of the maximum effect."

"Between-subject variability (BSV) was initially investigated for all PK and PD parameters assuming an exponential error model:where P_i is the individual parameter value for ith subject; θ is the population mean of the parameter; η_i is the individual variability from the population mean in the ith subject."

"The variability was best described with an additive error model:where $Obs_{i,j}$ denotes the jth HILD measurement in the ith subject."

Results

"The original data for each subject contained up to 3600 observations (measured at 40 Hz, sampled at 0.5 Hz) for up to 120 min."

"To reduce the data size to a suitable and manageable level for pharmacometric modeling, an averaging algorithm was applied to reduce the size to 15–20 observations per subject."

"The final data set contained a total of 2494 observations from 156 subjects."

"The model is composed of a one-compartment PK model with first-order absorption, and a direct-response fractional Emax PD model."

"Different error models (additive, proportional, combined, etc) were tested and additive error model was found to best describe the data."

Discussion

"A population PK/PD linked effect model was developed and described the histamine response over time data with high confidence."

"Although previous investigators have utilized the PD model to describe antihistamine effect [106, 109–111], this study is the first to explore the feasibility of modeling histamine response measured by HILD method."

"This study is the first attempt to access local histamine response over time through a population modeling approach."

"In recognition of the Occam's razor principle [112] and due to the lack of systemic or local (at measurement site) histamine exposure data, the simplest 1-compartmental model structure was assumed and applied."

"This was likely due to the fact that local histamine concentrations were not available in the current study design, and therefore the more complex indirect model structure could not be sufficiently supported by the data structure."

Conclusion

"The richly sampled HILD data was reduced to a suitable size for pharmacometric analysis using an averaging algorithm."

"The reduced data set was best described by a population PK/PD linked effect model."

"The model consists of a one-compartment PK model with first-order absorption, first-order elimination and an absorption lag time, and a direct-response fractional Emax model."

"This study has demonstrated the feasibility of modeling HILD data for pharmacodynamic purpose."

3.3 Nonlinear Mixed-Effects (NLME) Model

Introduction by the Editor

A Nonlinear Mixed-Effects (NLME) model for nanoparticles is a sophisticated statistical approach used to analyse pharmacokinetic and pharmacodynamic data from populations of individuals. This modeling technique accounts for both inter-individual variability and the complex nonlinear behaviour often exhibited by nanoparticles within biological systems. NLME models incorporate fixed and random effects, enabling the characterization of population-level trends while considering individual variations. This method allows for the exploration of how nanoparticle properties, dosing regimen, and physiological factors collectively influence drug behaviour and response. NLME modeling is particularly valuable in optimizing nanoparticle-based therapies by capturing the complex interaction between nanoparticles and the body across diverse patient population.

Machine generated summaries

Disclaimer: The summaries in this chapter were generated from Springer Nature publications using extractive AI auto-summarization: An extraction-based summarizer aims to identify the most important sentences of a text using an algorithm and uses those original sentences to create the auto-summary (unlike generative AI). As the constituted sentences are machine selected, they may not fully reflect the body of the work, so we strongly advise that the original content is read and cited. The auto generated summaries were curated by the editor to meet Springer Nature publication standards.

To cite this content, please refer to the original papers.

Machine generated keywords: residual, prediction, nonlinear, nca, line, covariate, error, effect model, noncompartmental, marker, noncompartmental analysis, bayesian, model, variability, guideline

Metabolic Profiling of Human Long-Term Liver Models and Hepatic Clearance Predictions from In Vitro Data Using Nonlinear Mixed-Effects Modeling [113] This is a machine-generated summary of:

Kratochwil, Nicole A.; Meille, Christophe; Fowler, Stephen; Klammers, Florian; Ekiciler, Aynur; Molitor, Birgit; Simon, Sandrine; Walter, Isabelle; McGinnis, Claudia; Walther, Johanna; Leonard, Brian; Triyatni, Miriam; Javanbakht, Hassan; Funk, Christoph; Schuler, Franz; Lavé, Thierry; Parrott, Neil J.: Metabolic Profiling of Human Long-Term Liver Models and Hepatic Clearance Predictions from In Vitro Data Using Nonlinear Mixed-Effects Modeling [113]

Published in: The AAPS Journal (2017)

Link to original: https://doi.org/10.1208/s12248-016-0019-7

If you want to cite the papers, please refer to the original.

For technical reasons we could not place the page where the original quote is coming from.

Abstract-Summary "Long-term in vitro models have been developed which enable sophisticated hepatic drug disposition studies and improved clearance predictions."

"The cell line HepG2, iPSC-derived hepatocytes (iCell®), the hepatic stem cell line HepaRG™, and human hepatocyte co-cultures (HμREL™ and HepatoPac®) were compared to primary hepatocyte suspension cultures with respect to their key metabolic activities."

"Similar metabolic activities were found for the long-term models HepaRG™, HμREL™, and HepatoPac® and the short-term suspension cultures when averaged across all 11 enzyme markers, although differences were seen in the activities of CYP2D6 and non-CYP enzymes."

"The micropatterned HepatoPac® model was further evaluated with respect to clearance prediction."

"The determination of intrinsic clearance by nonlinear mixed-effects modeling in a long-term model significantly increased the confidence in the parameter estimation and extended the sensitive range towards 3% of liver blood flow, i.e., >10-fold lower as compared to suspension cultures."

"Further research is needed to understand whether transporter activity and drug metabolism by non-CYP enzymes, such as UGTs, SULTs, AO, and FMO, is comparable to the in vivo situation in these long-term culture models."

INTRODUCTION

"We report the metabolic activity across a diverse set of phase I and phase II enzyme markers in different in vitro liver models as this is an important prerequisite for successful clearance determination across structural classes."

"For the first time, we applied pharmacokinetic (PK) modeling using a nonlinear mixed-effects approach for low in vitro clearance estimations with a long-term liver model for reference and in-house Roche compounds."

"By applying the nonlinear mixed-effects approach to in vitro data, where the statistical unit is the well, the confidence in the clearance estimate is increased as the pharmacokinetic parameters and residual variability can be derived."

"We compare clearance predictions in man for reference compounds in the long-term in vitro liver model, HepatoPac®, and in suspension cultures using the same hepatocyte donor."

MATERIALS AND METHODS

"After the addition of test compounds at different concentrations to the wells (100 μL), the 96-well hepatocyte suspension cultures plates were shaken (900 rpm) in a 5% CO_2 atmosphere at 37°C and sampling was done at different time points up to 2 h. HepaRG™ cells were cultured at 37°C in a humidified atmosphere with 5% CO_2 in complete HepaRG™ growth medium consisting of William's E medium, growth medium supplement, and 1% (v/v) GlutaMAX-I. To initiate differentiation, 0.9% (v/v) DMSO was added to the growth medium on confluent cells."

"At defined time points, either an aliquot of the cell incubation medium was taken or the whole well (for suspension cultures) was quenched with acetonitrile (volume ratio, 1:2) containing 0.1 mM chlorpromazine as internal standard."

"The iPSC-derived hepatocyte-like cells (iCell®) had a protein content of 1.4 mg/mL. The micropatterned HepatoPac® co-culture plates contained 3200 (in a 96-well plate) hepatocytes per well, as specified in the HepatoPac® donor specification (3121A) sheet provided by the supplier."

RESULTS

"Comparing the single activities for the different enzyme markers in HepaRG™, HμREL™, and HepatoPac® versus primary pooled cryopreserved hepatocytes in suspension, several differences were observed."

"The CYP2D6 activity differed between the three in vitro liver models, e.g., 56-fold lower in HepaRG™, 14-fold lower in HμREL™, and 2.5-fold lower in HepatoPac® as compared to suspension cultures."

"The UGT1A1 activity as determined by SN-38 glucuronide formation was two to threefold lower in HepaRG™, HμREL™, and HepatoPac® as compared to the suspension cultures."

"For the SULT activity using 7-hydroxycoumarin as the enzyme marker, an eightfold higher formation of 7-hydroxycoumarin sulfate was observed in HepatoPac® as compared to the suspension cultures, whereas only a twofold increase or similar activity was seen in HμREL™ and HepaRG™, respectively."

"HepaRG™, HμREL™, and HepatoPac® demonstrated metabolic activity similar to primary suspension cultures and could qualify for use in intrinsic clearance determinations of compounds."

DISCUSSION

"The better performance of direct scaling for the in vivo clearance prediction of Roche compounds might be due to the involvement of non-CYP enzymes and transporters, which gives rise to larger underprediction compared to CYP-metabolized reference compounds [114, 115]."

"For drugs highly bound in plasma and drugs bound to albumin, Poulin and others [116] could demonstrate that their scaling approach performed better in terms of human clearance predictions from suspension cultures, giving rise to an average fold error (afe) of 1.1, followed by the regression method, direct scaling, and conventional approach with afe values of 1.3, 2.2, and 0.52, respectively, for a data set of 38 reference compounds."

"The addition of plasma proteins may also benefit clearance prediction from long-term in vitro liver models, as demonstrated for suspension and plated hepatocyte cultures."

CONCLUSION

"The human liver cancer cell line HepG2, induced pluripotent stem cell-derived hepatocyte-like cells (iCell®), the hepatic stem cell line HepaRG™, and hepatocyte co-cultures (HμREL™ and HepatoPac®) were compared to primary hepatocyte suspension cultures with respect to their key metabolic activities."

"The long-term in vitro liver models, HepaRG™, HμREL™, and HepatoPac®, showed similar mean metabolic activities across 11 metabolic enzyme markers as compared to primary hepatocyte suspension cultures using different hepatocyte donors and qualify for use in intrinsic clearance determinations of compounds."

"Clearance predictions for low-clearance clinical drug candidates, which differ in their physicochemical properties and clearance routes from literature compounds, could be reliably improved by applying in vitro long-term models in combination with pharmacokinetic modeling using a nonlinear mixed-effects approach."

"Further research is needed to understand whether transporter activity and drug metabolism by non-CYP enzymes, such as UGTs, SULTs, AO, and FMO, are comparable to the in vivo situation in these long-term culture models."

Use of population approach non-linear mixed effects models in the evaluation of biosimilarity of monoclonal antibodies [117] This is a machine-generated summary of:

Reijers, Joannes A. A.; van Donge, T.; Schepers, F. M. L.; Burggraaf, J.; Stevens, J.: Use of population approach non-linear mixed effects models in the evaluation of biosimilarity of monoclonal antibodies [117]

Published in: European Journal of Clinical Pharmacology (2016)

Link to original: https://doi.org/10.1007/s00228-016-2101-6

If you want to cite the papers, please refer to the original.

For technical reasons we could not place the page where the original quote is coming from.

Abstract-Summary "We investigated whether PPK could also be useful in biosimilarity testing for monoclonal antibodies (MAbs)."

"Data from a biosimilarity trial with two trastuzumab products were used to build population pharmacokinetic models."

"A combined model was developed and similarity between test and reference product was evaluated by performing a covariate analysis with trastuzumab drug product (test or reference) on all model parameters."

"Two separate models were developed, one for each drug product."

"The model structure and parameters were compared and evaluated for differences."

"Drug product could not be identified as statistically significant covariate on any parameter in the combined model, and the addition of drug product as covariate did not improve the model fit."

"A similar structural model described both the test and reference data best."

Introduction

"Mentré's group has extensively studied the use of population pharmacokinetic techniques in bioequivalence testing and found that it yielded similar results, with the modelling approach leading to a better understanding of the underlying biological system and the NCA being a relatively easy approach that does not require modelling and whose results can be used in a statistical analysis."

"We investigated whether a population pharmacokinetic analysis (PPK) could also be useful in bioequivalence testing for monoclonal antibodies (MAbs), which display complex elimination mechanisms, including non-linear routes, and have a plasma half-life of one to multiple weeks."

"We developed a combined model built on all available data for both the test and reference product and tested whether adding product (test/reference) as a covariate would improve the model, indicating non-similarity."

Methods

"Based on the trastuzumab content of the used test and reference product vials, the actual dose levels were determined to be 0.49, 1.48, 2.96 and 5.96 mg/kg for T and 6.44 mg/kg for R. Trastuzumab was quantitated in serum samples collected pre-dose and at 45 min, 1.5 h, 2 h, 3 h, 4 h, 5 h, 6 h, 8 h and 24 h, and at 2, 4, 8, 14, 21, 28, 35, 42, 49 and 63 days after start of administration."

"A general model for trastuzumab, hereafter referred to as 'combined model', was developed based on all available PK data for both test and reference product, including dose levels of the dose escalation part (0.49, 1.48 and 2.96 mg/kg)."

"Trastuzumab concentration at the start of administration was assumed to be 0 μg/mL. For comparison to a standard NCA, AUCs were derived using model simulated (predicted) individual concentrations at the original sampling times."

Results

"After identification of the structural model, individual estimates of random effects for between-subject variability were identified for the parameters V1, K_M and k_e, with final coefficient of variation values of 14.8, 35.9 and 17.2 %, respectively."

"For the separate models, individual estimates of random effects for the between-subject variability were identified for the parameters V1, V_{max} and k_e, with final coefficient of variation values in model T of 16.5, 12.8 and 19 %, respectively."

"To the combined model, the best model fit with the greatest reduction in OFV for both separate models was obtained by incorporating LBW as linear covariate on V1 and BMI on k_e."

"The shrinkage observed for the parameters for which between-subject variability was identified (V1, V_{max}, k_e) is not significant (<17.80 % for model T, <15.50 % for model R)."

Discussion

"Testing for (statistically) significant differences between drug product can be done for all the model parameters via covariate analysis."

"Adding trastuzumab drug product as covariate to the model could not explain any residual variability, which not only strongly supports the biosimilarity claim but also indicates that the difference in AUCs must be attributed to population characteristics."

"With a PK model, multiple scenarios can be simulated within these extremes, which can be used to build the case that the test product achieves therapeutic drug concentrations, similar to the reference product, when administered according to a certain dosing regimen."

"Questions that need to be addressed before a PPK can fully substitute the NCA in demonstrating biosimilarity relate to selection of the most meaningful PK or pharmacodynamic parameter from the model, and the minimal population size to detect with sufficient statistical power relevant (model) differences."

Comparison of Nonlinear Mixed Effects Models and Noncompartmental Approaches in Detecting Pharmacogenetic Covariates [118] This is a machine-generated summary of:

Tessier, Adrien; Bertrand, Julie; Chenel, Marylore; Comets, Emmanuelle: Comparison of Nonlinear Mixed Effects Models and Noncompartmental Approaches in Detecting Pharmacogenetic Covariates [118]

Published in: The AAPS Journal (2015)

Link to original: https://doi.org/10.1208/s12248-015-9726-8

If you want to cite the papers, please refer to the original.

For technical reasons we could not place the page where the original quote is coming from.

Abstract-Summary "There is no consensus on methods to test the association between pharmacokinetics and genetic covariates."

"We performed a simulation study inspired by real clinical trials, using the pharmacokinetics (PK) of a compound under development having a nonlinear bioavailability along with genotypes for 176 single nucleotide polymorphisms (SNPs)."

"We compared 16 combinations of four association tests, a stepwise procedure and three penalised regressions (ridge regression, Lasso, HyperLasso), applied to four pharmacokinetic phenotypes, two observed concentrations, area under the curve estimated by noncompartmental analysis and model-based clearance."

"The different combinations were compared in terms of true and false positives and probability to detect the genetic effects."

"In presence of nonlinearity and/or variability in bioavailability, model-based phenotype allowed a higher probability to detect the SNPs than other phenotypes."

"In a realistic setting with a limited number of subjects, all methods showed a low ability to detect genetic effects."

"Ridge regression had the best probability to detect SNPs, but also a higher number of false positives."

INTRODUCTION

"Information about the phenotypes (obtained by NCA or modelling), the design of studies, the genetic data and the statistical analysis have been extracted from each publication, and summarised through descriptive statistics."

"During the period 2010–2012, on 85 pharmacogenetic studies using PK parameters as phenotypes, 69% used NCA and 31% used modelling-based phenotypes for association analysis."

"Model-based phenotypes were explored using mostly stepwise regression (78%), univariate methods applied on individual parameter estimates (7%) or descriptive methods."

"Although health authorities strongly recommend studying the pharmacogenetics of new chemical entities in development [119, 120], there is no consensus on analysis methods to explore a large number of polymorphisms in association with PK phenotypes."

"We propose to compare those three penalised regression methods with a stepwise approach through a simulation study, to assess their ability to detect the influence of genetic variables on the PK."

MATERIALS AND METHODS

"The regularisation parameter ξ is calculated to achieve a target FWER, using the following expression [121]:where α is the type I error per SNP, Φ^{-1} is the inverse normal distribution function, N the number of subjects and σ the standard error of the phenotype considered."

"Each method, ridge regression, Lasso, HyperLasso and the stepwise procedure, was applied to each PK phenotype: observed C24h and C192h, AUC estimated by NCA and model parameters CL, V and Q estimated by NLMEM."

"Under H_1, we recorded for each method and phenotype the number of true positives (TP, corresponding to the selection of a SNP which was indeed associated to CL in the simulation, its maximum over the 200 simulations being 1200) and false positives (FP, corresponding to a SNP selected in the model but not present in the simulation)."

RESULTS

"Ridge regression yielded a higher TPR more often than the other methods ($p < 0.003$ for all pairwise comparison), while Lasso, HyperLasso and the stepwise procedure were comparable."

"Concerning the comparison between phenotypes, we found similar results as for S_{real}: all methods were more powerful when applied to CL compared to the other phenotypes ($p < 0.001$ for all pairwise comparison)."

"The TPR for ridge regression was higher than Lasso and HyperLasso ($p < 0.001$ with methods applied to CL)."

"Departure in methods was observed on the power to detect at least three and more variants, with the ridge regression and stepwise procedure showing higher power."

"In both scenarios, regardless of the association method, the power to detect a gene effect was higher using PK parameter obtained by NLMEM than AUC estimated by NCA or observations."

DISCUSSION

"All methods showed a higher number of TP when used on individual clearances CL from NLMEM, compared to the other phenotypes (AUC, C24h and C192h)."

"The power to detect a gene effect was higher for the model-based approach than for the phenotypes estimated by NCA or observed due to the specific features in the PK model, the nonlinearity in the absorption model and the variability in the bioavailability."

"In the scenarios we simulated, the number of sampling points per subject was large, so that all model parameters, included the phenotype, were well estimated."

"The present work complements a previous study by Bertrand and Balding [122], who compared four association methods (ridge regression, Lasso, HyperLasso and a stepwise procedure) on only one kind of PK phenotype, CL estimated by NLMEM."

Developing Tools to Evaluate Non-linear Mixed Effect Models: 20 Years on the npde Adventure [123] This is a machine-generated summary of:

Comets, Emmanuelle; Mentré, France: Developing Tools to Evaluate Non-linear Mixed Effect Models: 20 Years on the npde Adventure [123]

Published in: The AAPS Journal (2021)

Link to original: https://doi.org/10.1208/s12248-021-00597-7

If you want to cite the papers, please refer to the original.

For technical reasons we could not place the page where the original quote is coming from.

Abstract-Summary "This article revisits 20 years of our work in developing evaluation tools adapted to non-linear mixed effect models."

INTRODUCTION

"Their usage has expanded, to help design clinical trials by optimising the number and arrangement of samples to maximise information, to therapeutic drug monitoring by individualising drug regimens, to managing long-term therapy by incorporating disease models."

"This last question of model evaluation was also addressed very early by Lewis Sheiner and Stuart Beal in 1981 [124] and is crucial to the use of NLMEM."

"Model evaluation was integrated into the guidelines on population pharmacokinetic analyses, first in the European guidance [125] and more recently into the revised version of the guideline by the Food and Drug Administration [126]."

"We will first recall the history of model evaluation and how the npde have taken their place amongst the other evaluation tools."

HISTORY

"In attempting to evaluate a population PK analysis of mizolastine in sparse data using the non-parametric maximum likelihood approach [127], we realised that the discrete nature of non-parametric distributions allowed an explicit computation of the quantile of an observation in the predictive distribution, a quantity we termed prediction discrepancy (pd)."

"Using Monte-Carlo simulation to derive population weighted residuals (PWRES) helped alleviate the bias with WRES, and using a first-order conditional approximation (FOCE) to adjust WRES to the individual predictions of mean and variance led to residuals called conditional weighted residuals (CWRES) which provide similar evaluation graphs as npde, although their distribution is strictly known only if the structural model is linear [128, 129]."

"The evaluation stresses the robustness of simulation-based tools such as npd, npde and VPC, although they may require more computational effort to implement the CWRES being the only prediction-based residual which could compare in terms of versatility to detect model misspecifications."

CONCEPTS

"In red, we represent the construction when y_{ij} is below the LOQ: the probability of being LOQ for an observation at time t_{ij} is computed from the predictive distribution as $p(y_{ij} < LOQ)$."

"To correct this, Brendel and others proposed to take into account the longitudinal nature of the data, using estimated individual variance-covariance matrix obtained from the data simulated under the model to decorrelate both simulated and observed data before computing the quantiles [130], and called the resulting variable prediction distribution errors (pde)."

"Distribution graphs include QQ-plots and histograms, while scatterplots versus the independent variable and versus model predictions can be used to detect when the model deviates from the observed data, suggesting adjustments in the structural or variability models [130]."

"Evaluation graphs for simple TTE models are less informative, as outcome and time are confounded, and evaluation relies on distribution plots such as QQ-plots to assess the shape of the distribution."

THE npde PACKAGE

"This simulated data was based on the phase II clinical trial COPHAR 3-ANRS 134 trial [131], where viral loads after the initiation of an antiviral combination in HIV patients were modelled using a bi-exponential decrease."

"We show the graphs for the complete dataset (without censoring), when the same model is used to compute npde as the one used to simulate the original data."

"The two plots on the bottom are scatterplots of npd versus the independent variable (left) and the population predictions (right) and help assess whether there are any trends in the model."

"In these graphs, prediction intervals are produced for the median (in pink), to show deviations for the typical profile, and for the 2.5th and 97.5th percentile of the npd, to a 95% prediction interval (in light blue), to assess whether the interindividual variability is taken into account properly in the model."

THE npde FOLLOWING

"A measure of the popularity in the community is the number of citations generated by an article, and we looked at the citations for the 3 seminal methodological articles [130, 132, 133], the article presenting the npde package [134] and the tutorial on evaluation metrics [135], which were the most cited on the topic in our bibliography (F. Mentré and E. Comets)."

"The graph shows that our foundation articles have been cited worldwide, with the largest number in France (141 citations), followed by the Netherlands (83), the USA (79) and China (50), and including 2 articles in South America and one in Africa."

"Forty methodological articles not authored by one of us cited one of the foundation articles and could be classified in various categories: articles dealing with VPC or their derivatives (n = 10), new model evaluation approaches (n = 8), estimation methods or algorithms (n = 6), modelling approaches (n = 10) and tools or guidelines (n = 6)."

CONCLUSION

"While npde have been shown to perform well in simulation studies, in practice we find the statistical tests to be sensitive to outliers which may not be relevant for the purpose of modelling in pharmacometrics, with a sensitivity that increases with the number of observations as the prediction intervals become increasingly narrow."

"Evaluation graphs are therefore the recommended approach to guide model building and communicate on the performances of alternative models in order to select the best model for a given purpose."

"npd and npde can be used both in internal evaluation, using the data used to build the model, and in external evaluation on a separate dataset, which is considered a more robust assessment of the model's predictive ability."

"We performed a survey on PK/PD analyses reported in the two preceding years and found that only a fourth of articles included any kind of model evaluation [136, 137]."

A Bayesian hierarchical nonlinear mixture model in the presence of artifactual outliers in a population pharmacokinetic study [138] This is a machine-generated summary of:

Choi, Leena; Caffo, Brian S.; Kohli, Utkarsh; Pandharipande, Pratik; Kurnik, Daniel; Ely, E. Wesley; Stein, C. Michael: A Bayesian hierarchical nonlinear mixture model in the presence of artifactual outliers in a population pharmacokinetic study [138]

Published in: Journal of Pharmacokinetics and Pharmacodynamics (2011)
Link to original: https://doi.org/10.1007/s10928-011-9211-7

If you want to cite the papers, please refer to the original.

For technical reasons we could not place the page where the original quote is coming from.

Abstract-Summary "The purpose of this study is to develop a statistical methodology to handle a large proportion of artifactual outliers in a population pharmacokinetic (PK) modeling."

"The motivating PK data were obtained from a population PK study to examine associations between PK parameters such as clearance of dexmedetomidine (DEX) and cytochrome P450 2A6 phenotypes."

"Conventional population PK analysis of these data revealed several challenges and intricacies."

"We propose a novel population PK model, a Bayesian hierarchical nonlinear mixture model, to accommodate the artifactual outliers using a finite mixture as the residual error model."

"Our results showed that the proposed model handles the outliers well."

"These simulation results showed that the proposed model can accommodate the outliers well so that the estimated PK parameters are less biased."

Introduction

"We analyzed the pharmacokinetic (PK) data of a sedative, dexmedetomidine (DEX), using a population PK modeling to estimate drug exposure and examine source of variability in drug kinetics in ICU patients."

"Some outlier values may not be artifactual but rather represent individuals with truly high drug concentrations due to biological variability—a problem relevant to the ICU where the clinical status of patients can change rapidly, potentially affecting drug elimination."

"In the presence of these artifactual outliers, the standard assumptions on residual errors in conventional population PK models may not be valid."

"We consider a mixture model as a contaminant-generating model, since the artifactual outliers in our PK data can be reasonably assumed to come from a population different from that of the main body of data; i.e. our data consist of a mixture of two different populations: valid and invalid concentrations."

Data and conventional population PK analysis

"The motivating PK data consist of DEX plasma concentrations (ng/ml) with 1–15 measurements per patient obtained over a time interval that ranges from 13 h to 5 days, with a total of 247 measurements from 43 patients in the ICU at Vanderbilt University Medical Center, Nashville, Tennessee, and Washington Hospital Center, Washington, DC [139]."

"Blood samples were collected at three predefined time points each day during continuous DEX infusion: at 05:00 ± 2 h by the bedside nurse, at 10:00 ± 2 h by the investigators during the morning sedation level assessment, and at 16:00 ± 2 h with the evening sedation level evaluation."

"The model has two PK parameters, clearance (Cl), the key parameter of interest, and volume of distribution (V)."

"This leads us to conclude that the commonly used residual error models are not appropriate to fit the DEX data."

Evidence of artifactual outliers

"The maximum tolerated plasma concentration of DEX in humans described in the literature is 16 ng/ml during continuous infusion at stepwise increasing dose rates [140]."

"An earlier study reported peak plasma concentration was about 10 ng/ml at the end of a 5-min infusion of DEX (2 μg/kg) [141]."

"The average plasma concentrations for healthy Caucasians receiving low doses of DEX in a study conducted by the authors [142, 143] was 0.14 ng/ml after three 10-min infusions of DEX (at 0.1, 0.15, and 0.15 μg/kg; cumulative dose, 0.4 μg/kg) over 90 min."

"The individual plots, along with the blood sampling process, provide strong evidence that some of the blood samples were contaminated with high concentrations of the infused drug and thus the estimated PK parameters would be seriously biased, if the outliers are not accommodated."

A Bayesian hierarchical nonlinear mixture model

"The second component is the convolution of a $N(0, \tau_2)$ and $Ga(a, b)$, which accommodates large values of contaminants close to the boundary of the valid concentrations as well as more obviously large outliers (e.g. the density represented by red dashed line)."

"As a note, a normal proportional error model would not be a good choice for the valid population, since it would compete with the Gamma component of the model to model outliers, making the MCMC sampling for posterior distributions difficult."

"Even if the valid population follows a normal proportional error model, the use of a normal additive error model would result in little bias in the estimates for PK parameters, since some skewed valid concentrations on the boundary of the invalid would be categorized as the valid in many MCMC iterations (so still high probability of being valid, although not 1), allowing them to contribute in estimating the PK parameters most of times."

Model checking and inferences

"Each data point was weighted by its corresponding the posterior probability of w_1 (probability of being valid) in estimating PK parameters."

"The proportion of MCMC iterations for which it is classified to the second component is the estimate of w_2; in this example, $w_2 \approx 1$, hence this data point will be almost entirely predicted by the normal-gamma convolution, in no relationship with other data points and the PK parameters."

"They will be sometimes included in the estimation at $100w_1\%$ of MCMC iterations, during which their prediction is in connection with other data points, and hence the posterior means would not be perfectly matched with the observed ones."

"The large reduction in the posterior mean deviance and the smaller D(k) along with the goodness-of-fit plots and the posterior predictive checking, all suggest that the proposed model greatly improves the fit and is preferable."

Simulation studies

"Both normal additive and proportional residual error models were considered for the main population (valid concentrations) for each outlier-generating model."

"The simulated data from the combination of the outliers and the main population models were fit to evaluate a sensitivity of modeling assumption for our proposed model."

"From each outlier-generating model, a proportion of the large outliers was simulated in a range of 10–30%, which in our experience are reasonably supported values for clinical data."

"Although the model misspecification for the main population results in more bias in general than the misspecification of the outlier-generating mechanism, the maximum bias in the estimates of Cl and V was at most 5% across all models in this simulation study."

"Our simulation results support that our proposed model can well accommodate the outliers even generated from incorrect models in a reasonable range of proportions and variability of data."

Summary and discussion

"Model checking revealed that the commonly used residual error models in conventional population PK analysis cannot handle this large proportion of outliers and lead to biased estimates."

"We proposed a novel population PK model to accommodate the artifactual outliers using a mixture distribution as the residual error model within a Bayesian framework."

"Our model checking supported the ability of our model to accommodate the large proportion of artifactual outliers well compared to conventional population PK analysis."

"The results showed that our proposed model can accommodate the outliers well, with small to large proportion of outliers, findings which are relevant to most observational PK studies."

3.4 Preclinical Model

Introduction by the Editor
Preclinical models form an indispensable bridge between the lab-scale preparation of nanoparticles and their clinical trials, serving as crucial testing ground for these novel medical interventions. These models, typically involving animals or in vitro systems, allow researchers to rigorously assess the safety, efficacy, and mechanism of action of these nanosystems before they are administered to humans. Preclinical studies offer valuable insight into the potential benefits and risks of nanosystems, enabling researchers to refine their approaches, optimise dosing, and identify potential pitfalls. Through meticulous observation and controlled experimentation, preclinical models pave the way for more informed and ethical decisions during subsequent clinical trials, ultimately shaping the trajectory of medical progress and patient well-being.

Machine generated summaries
Disclaimer: The summaries in this chapter were generated from Springer Nature publications using extractive AI auto-summarization: An extraction-based summarizer aims to identify the most important sentences of a text using an algorithm and uses those original sentences to create the auto-summary (unlike generative AI). As the constituted sentences are machine selected, they may not fully reflect the body of the work, so we strongly advise that the original content is read and cited. The auto generated summaries were curated by the editor to meet Springer Nature publication standards.

To cite this content, please refer to the original papers.

Machine generated keywords: gel, vitro release, film, loading, soluble, release, load, potent, poorly, vitro, unlike, entry, guide, formulate, reduction

A Dose Ranging Pharmacokinetic Evaluation of IQP-0528 Released from Intravaginal Rings in Non-Human Primates [144] This is a machine-generated summary of:

Su, Jonathan T.; Teller, Ryan S.; Srinivasan, Priya; Zhang, Jining; Martin, Amy; Sung, Samuel; Smith, James M.; Kiser, Patrick F.: A Dose Ranging Pharmacokinetic Evaluation of IQP-0528 Released from Intravaginal Rings in Non-Human Primates [144]

Published in: Pharmaceutical Research (2017)

Link to original: https://doi.org/10.1007/s11095-017-2224-1

If you want to cite the papers, please refer to the original.

For technical reasons we could not place the page where the original quote is coming from.

Abstract-Summary "Design of intravaginal rings (IVRs) for delivery of antiretrovirals is often guided by in vitro release under sink conditions, based on the assumption that in vivo release will follow a similar release profile."

"We conducted a dose-ranging study in the female reproductive tract of pigtail macaques using matrix IVRs containing IQP-0528, a poorly soluble but highly potent antiretroviral drug with an IC_{90} of 146 ng/mL. These IVRs consisted of drug-loaded segments, 15.6% IQP-0528 in Tecoflex 85A, comprising either all, half, or a quarter of the entire ring."

"While drug concentration in vaginal fluid is well in excess of IQP-0528's EC_{90}, we find no statistical difference between the different ring loadings in either swab drug levels or drug released from our rings."

INTRODUCTION

"IQP-0528 has been formulated in a number of ways for topical vaginal drug delivery, including vaginal films, IVRs, and gel formulations."

"Srinivasan and others [145] evaluated vaginal films loaded with 1.5 w/w% IQP-0528 in a pigtail macaque model and measured vaginal fluid concentrations one to five logs higher than the in vitro IC_{90}."

"Pereira and others [146] conducted a pharmacokinetic and pharmacodynamics evaluation of a 1% IQP-0528 vaginal gel in a rhesus macaque model and found that they were able to achieve drug levels 5 to 7 logs higher than the in vitro EC_{50}."

"They were able to demonstrate use of the IVR to achieve micro-molar IQP-0528 concentrations in the pigtail macaque vaginal tract over 28 days, far in excess of the EC_{50} [147, 148]."

"In a follow-up to these studies, we have conducted a dose ranging topical vaginal pharmacokinetic study in pigtailed macaques using matrix rings containing IQP-0528."

MATERIALS AND METHODS

"For matrix-controlled systems the drug release is directly proportional to the surface area of the drug-loaded segment of the ring, varying length of drug-loaded segment allows for precise control of the relative drug release."

"The quantitative levels of IQP-0528 remaining within the IVRs were determined using a previously determined polyurethane (PU) dissolution method similar to that used to analyze the pellets (above). [147] The drug-containing portion of the IVR was excised from the ring and dissolved in a 10 mL volumetric flask of a 50:50 mixture of dimethylacetamide and dichloromethane."

"A 1.2 mg/mL spike solution was prepared by weighing out 30 mg of drug into a 25 mL volumetric of methanol."

"The Wilcoxon rank sum test was used to evaluate differences in vaginal fluid and tissue drug levels between the three ring loadings."

RESULTS

"As is typical with these experiments, drug levels are highly variable: for days 3 to 28, with the ring present, the drug levels in the proximal vagina ranged from 6×10^2 to 3.2×10^5 ng/mL for the full ring (median 1.9×10^5 ng/mL); to 3×10^2 to 7.4×10^5 ng/mL for the half ring (median 3.8×10^4 ng/mL); and 1.4×10^2 to 1.0×10^5 ng/mL for the quarter ring (median 5×10^3 ng/mL)."

"Drug levels in the distal vagina ranged from 2 × 10^2 to 1.5 × 10^5 ng/mL for the full ring (median 1.6 × 10^3 ng/mL); to 2.5 to 3.0 × 10^5 ng/mL for the half ring (median 1.9 × 10^3 g/mL); and 1.1 × 10^2 to 1.9 × 10^5 ng/mL for the quarter ring (median 1.2 × 10^3 ng/mL)."

STATISTICAL ANALYSIS

"For the half ring, at 14 days, vaginal tissue levels ranged from 5.4 × 10^2–6.2 × 10^3 (median 1.7 × 10^3) ng/g tissue in the proximal vagina; 1.4 × 10^2–1.9 × 10^3 (median 9.0 × 10^2) ng/g tissue in the medial vagina; and 84–8.4 × 10^2 (median 3.7 × 10^2) ng/g tissue in the distal vagina."

"For the quarter ring at 14 days, tissue levels ranged from 42–4.1 × 10^3 (median 1.7 × 10^3) ng/g tissue in the proximal vagina to 2.5–2.2 × 10^3 (median 61) ng/g tissue in the medial vagina, and 2.5–7.8 × 10^3 (median 1.1 × 10^3) ng/g tissue in the distal vagina."

"The Wilcoxon rank sum test ($P = 0.05$) showed no statistically significant difference between AUMC based on location (proximal, medial, or distal vagina) for each of the full, half, or quarter rings, indicating that drug is distributed relatively evenly throughout the vaginal tissue for all rings."

DISCUSSION

"Despite our clearly dose-proportional in vitro release, the amount of drug released from these different rings in vivo is dependent on neither the total amount of drug in our IVR nor the surface area available for release."

"In vivo release of IQP-0528 from the ring may be better modeled as partition-controlled: because in vivo diffusion of drug from the IVR is dominated by the solubility limit of drug in the surrounding vaginal fluid, release can be thought of as governed by this solubility limit [149]."

"200 mg and 25 mg reservoir dapivirine rings were evaluated in vivo and found to achieve similar vaginal fluid concentrations; as expected, given that release rate from reservoir devices is governed largely by the rate controlling membrane [150, 151]."

"Nel and others evaluated a silicone elastomer matrix ring loaded with 25 mg of DPV and observed higher initial in vivo release and 42% release of loaded drug over 28 days of use [152]."

CONCLUSION

"By using full, half, and quarter segment macaque rings, the design of our study allowed us to measure in vivo and in vitro release from rings with an overall loading ranging from 15.6 to 3.9 wt% but with the same mechanism of release, varying only the surface area available for release."

"Despite the dose-proportional release found in vitro, no statistically significant difference in average daily release was found between the differently loaded IVRs in vivo for tissue samples or for vaginal fluid samples."

"This indicates that in vitro release may be not be a good guide for in vivo release, particularly for poorly soluble drugs."

"Despite a range of in vitro releases, all in vivo rings demonstrated the ability to achieve vaginal fluid levels well in excess of the IC_{90}, indicating protective levels of IQP-0528 may be achieved even with very low loading."

3.5 Dermatopharmacokinetic Model

Introduction by the Editor

The skin, being the largest organ in the body, presents unique challenges and opportunities for drug delivery. A dermatopharmacokinetic model is an essential tool that allows researchers to interpret the complex interaction between formulations, skin physiology, and drug disposition. The emerging field of dermatopharmacokinetics for nanoparticles offers a compelling perspective on the interaction between advanced drug delivery systems and the skin. This intricate model examines how nanoparticles pass through the complex layers of the skin, elucidating their absorption, distribution, and potential systemic effects. By analysing the mechanism of nanocarrier penetration into the skin, their permeation through various layers, and subsequent systemic absorption, this model provides a roadmap to understanding their unique behaviour within the skin's microenvironment. A dermatopharmacokinetics study reveals the complexity of skin as a route of drug delivery, influencing the development of successful dermatological therapies.

Machine generated summaries

Disclaimer: The summaries in this chapter were generated from Springer Nature publications using extractive AI auto-summarization: An extraction-based summarizer aims to identify the most important sentences of a text using an algorithm and uses those original sentences to create the auto-summary (unlike generative AI). As the constituted sentences are machine selected, they may not fully reflect the body of the work, so we strongly advise that the original content is read and cited. The auto generated summaries were curated by the editor to meet Springer Nature publication standards.

To cite this content, please refer to the original papers.

Machine generated keywords: skin, dermal, topical, microdialysis, probe, technique, product, drug product, human skin, topical drug, bioequivalence, open, assay, fluid, deep

Variability of Skin Pharmacokinetic Data: Insights from a Topical Bioequivalence Study Using Dermal Open Flow Microperfusion [153] This is a machine-generated summary of:

Bodenlenz, Manfred; Augustin, Thomas; Birngruber, Thomas; Tiffner, Katrin I.; Boulgaropoulos, Beate; Schwingenschuh, Simon; Raney, Sam G.; Rantou, Elena; Sinner, Frank: Variability of Skin Pharmacokinetic Data: Insights from a Topical Bioequivalence Study Using Dermal Open Flow Microperfusion [153]

Published in: Pharmaceutical Research (2020)

Link to original: https://doi.org/10.1007/s11095-020-02920-x

If you want to cite the papers, please refer to the original.

For technical reasons we could not place the page where the original quote is coming from.

Abstract-Summary "The overall variability of logAUC values (CV: 39% for R and 45% for T) was dominated by inter-subject variability (R: 82%, T: 91%) which correlated best with the subject's skin conductance."

"Intra-subject variability was 18% (R) and 9% (T) of the overall variability; skin treatment sites or methodological factors did not significantly contribute to that variability."

"Inter-subject variability was the major component of overall variability for acyclovir, and treatment site location did not significantly influence intra-subject variability."

"These results support a dOFM BE study design with T and R products assessed simultaneously on the same subject, where T and R treatment sites do not necessarily need to be next to each other."

Introduction

"Although such methods for topical in vivo permeation studies are promising for BE assessments, the resulting data are highly variable - like most of topical PK data."

"Characterization of data variability has been performed previously with clinical dMD data in studies on topically applied drugs [154–156]."

"As part of a U.S. Food and Drug Administration (FDA) funded collaborative research effort to evaluate PK-based methods for topical BE assessment, we have recently performed a clinical dOFM study to assess the BE of commercially available acyclovir cream, 5% products [157]."

"The resulting data set included data on the dermal PK and on multivariate biological-methodological parameters that might potentially have been associated with the observed variability, thus representing an ideal data set with which to investigate skin PK data variability after topical drug application, and with which to evaluate the sources of that variability."

Material and Methods

"Dermal interstitial fluid was continuously sampled with a flow rate of 1 μL/min using sterile perfusate (physiological saline containing 1% albumin and 600 mg/dL glucose) from 1 h pre-dose to 36 h post-dose, with post-dose sampling intervals of 4 h. Before dosing, the skin temperature of each of the 6 treatment sites was measured (Infrared thermometer TDT8806, Thomson Health Care, France) and transepidermal water loss (TEWL) was measured in duplicate on each thigh (TEWL; Aquaflux AF200; Biox Ltd., London, UK)."

"Its main components, the inter- and intra-subject variability, were determined by performing analyses of variance (ANOVAs) with the fixed factors subject, treatment site and probe."

"The following variables were analyzed: sex, age, body mass index (BMI), transepidermal water loss (TEWL), skin conductance, skin temperature (pre-dose), dOFM probe depths after sampling, dOFM sample volumes, and exchange rates of glucose and lactate."

Results and Discussion

"The low intra-subject dOFM variability found in our study indicates a relatively minor influence of any localized variations in skin permeation on overall variability as well as a relatively low contribution of methodological factors to overall variability, which we attribute to the extensive optimization and standardization of the dOFM materials and study procedures."

"The correlation of lactate concentrations to the logAUC values was statistically significant, but rather low, which indicates a small contribution of the factor relative recovery to the inter-subject variability."

"We hypothesized that when using a hydrophilic drug with low permeation, like acyclovir, a significant portion of the observed intra-subject variability might be caused by local skin-related factors, which may influence the skin barrier function, and the drug delivery into the region of skin immediately above the probe."

Conclusion

"This comprehensive analysis of what is, to date, the largest dermal acyclovir data set obtained by continuous dOFM sampling, has characterized the main sources of variability in a topical dOFM BE study and provided information on the relevance of these sources of variability for topical BE assessments."

"None of the methodological factors accounted for this intra-subject variability, and the characteristics of our data support the hypothesis that a significant proportion of the observed intra-subject variability in topical studies might be caused by local skin-related biological factors, e.g. hair follicles, when using a hydrophilic drug with low permeation, like acyclovir."

"Additional BE studies would help to further characterize the variability of BE data for topical drugs with different physicochemical properties and/or with greater amounts permeating into the skin."

"The low intra-subject variability supports the concept of a head-to-head BE analysis of topical products (T vs R) within a relatively small number of subjects that can adequately power statistical conclusions."

Evaluating Dermal Pharmacokinetics and Pharmacodymanic Effect of Soft Topical PDE4 Inhibitors: Open Flow Microperfusion and Skin Biopsies [158] This is a machine-generated summary of:

Eirefelt, Stefan; Hummer, Joanna; Basse, Line Hollesen; Bertelsen, Malene; Johansson, Fredrik; Birngruber, Thomas; Sinner, Frank; Larsen, Jens; Nielsen, Simon Feldbæk; Lambert, Maja: Evaluating Dermal Pharmacokinetics and

Pharmacodymanic Effect of Soft Topical PDE4 Inhibitors: Open Flow Microperfusion and Skin Biopsies [158]

Published in: Pharmaceutical Research (2020)

Link to original: https://doi.org/10.1007/s11095-020-02962-1

If you want to cite the papers, please refer to the original.

For technical reasons we could not place the page where the original quote is coming from.

Abstract-Summary "To investigate the difference in clinical efficacy in AD patients between two topical PDE4 inhibitors using dermal open flow microperfusion and cAMP as a pharmacodynamic read-out in fresh human skin explants."

"Clinical formulations were applied to intact or barrier disrupted human skin explants and both skin biopsy samples and dermal interstitial fluid was sampled for measuring drug concentration."

"CAMP levels were determined in the skin biopsies as a measure of target engagement."

"Low unbound drug concentration in dISF in combination with minimal target engagement of LEO 39652 in barrier impaired human skin explants supports that lack of clinical efficacy of LEO 39652 in AD patients is likely due to insufficient drug availability at the target."

Introduction

"In these clinical studies both compounds showed high dermal exposure as measured by drug concentrations in skin punch biopsies (6–10 μM, unpublished results, [159, 160]) and based on this and the in vitro potencies both compounds were prior to the clinical studies expected to show similar clinical efficacy."

"It was evident from the skin drug levels measured in punch biopsies in the clinical studies (10 μM for LEO 29102 and 6 μM for LEO 39652) that they did not reflect the free concentration at the target site of these equipotent compounds."

"As one possible reason for the difference in target engagement and clinical efficacy between the two PDE4 inhibitors could be related to fast clearance due to instability in skin, we also investigate the extent of metabolism and the concentration of metabolites in all PK experiments in fresh human skin explants."

Materials and Methods

"The unbound ISF concentration was calculated using the following formula Skin concentrations of LEO 29102 and LEO 39652 were measured in traditional punch biopsies taken from parallel sites on the human skin explants treated similar to the dOFM sites."

"Skin biopsy analysis LEO 29102, LEO 39652 and 3 potential metabolites was performed by mixing 500 μL homogenization buffer (100 mM citrate buffer, pH 4.0:acetonitrile, 1:1) and 100 μL of internal standard solution (containing deuterated internal standards of all analytes) with the pre-weighed skin samples in 2 mL in a BeadRuptor-vial containing ceramic beads."

"Sample preparation was performed by addition of 15 μL of internal standard solution (containing deuterated internal standards of all analytes) to 30 μL of sample, followed by protein precipitation with five volumes of acetonitrile, centrifugation and dilution of the supernatant with 100 μL of water prior to analysis by LC-MS/MS in MRM positive ion ESI."

Results

"In human skin explants with intact stratum corneum total dISF concentrations could be detected by dOFM from 2 to 4 h up to 24 h after application (end of experiment), for all 3 evaluated doses of LEO 29102."

"LEO 39652 was applied topically to human skin explants with the stratum corneum removed by tape-stripping at the dose corresponding to the middle dose tested for LEO 29102 (4.25 $\mu g/cm^2$, 2.5 mg/g)."

"In previous sections LEO 29102 was tested in multiple doses in intact and tape-stripped human explants to measure the effect on cAMP and in OFM studies to measure dISF concentrations."

"Using the clinical doses of LEO 29102 only the 2.5 mg/g doses with and without tape stripping resulted in unbound drug concentrations in the interstitial fluid above the EC_{50} of the compound (i.e. a $dISF/EC_{50,u}$ ratio > 1)."

Discussion

"The purpose of this work has been to investigate the reason behind the difference in clinical efficacy between two topical PDE4 inhibitors [159, 160], LEO 29102 and LEO 39652, with similar in vitro potencies and similar skin exposure in the clinical studies as measured by traditional skin biopsies."

"As both compounds had shown similar and very high (6–10 μM) skin concentrations based on traditional skin punch biopsy measurements in the clinical studies it is reasonable to assume that the concentrations measured by skin punch biopsies did not reflect the free and available concentrations at the target site in the skin."

"These data suggest that high concentrations of LEO 29102 can be reached even in intact skin after topical administration of the clinical formulation and overall it supports the results obtained in the studies using cAMP as a pharmacodynamic read-out."

Conclusion

"LEO 29102 showed a significant and clinically meaningful effect in a phase 2a study in AD patients and this was accompanied by a high skin concentration in sampled skin punch biopsies."

"From the skin PK data obtained using dOFM to measure skin concentrations of the two PDE4 inhibitors we conclude that the lack of clinical efficacy observed with LEO 39652 in AD patients is likely to be related to too low concentrations of LEO 39652 at the site of action in the skin."

"The study shows the importance of having a marker for target engagement and illustrates very well that dOFM has the potential to measure relevant skin concentrations and that the method is suitable to establish PK/PD relations and to be used in a preclinical setting to predict the clinical efficacy of topical candidate drugs."

A Microfluidic Perfusion Platform for In Vitro Analysis of Drug Pharmacokinetic-Pharmacodynamic (PK-PD) Relationships [161] This is a machine-generated summary of:

Guerrero, Yadir A., Desai, Diti; Sullivan, Connor; Kindt, Erick; Spilker, Mary E.; Maurer, Tristan S.; Solomon, Deepak E.; Bartlett, Derek W.: A Microfluidic Perfusion Platform for In Vitro Analysis of Drug Pharmacokinetic-Pharmacodynamic (PK-PD) Relationships [161]

Published in: The AAPS Journal (2020)

Link to original: https://doi.org/10.1208/s12248-020-0430-y

If you want to cite the papers, please refer to the original.

For technical reasons we could not place the page where the original quote is coming from.

Abstract-Summary "Static in vitro cell culture studies cannot capture the dynamic concentration profiles of drugs, nutrients, and other factors that cells experience in physiological systems."

"Proof-of-concept studies using doxorubicin and gemcitabine demonstrated the ability of the microfluidic PK-PD device to examine dose- and time-dependent effects of doxorubicin as well as schedule-dependent effects of doxorubicin and gemcitabine combination therapy on cell viability using both step-wise drug concentration profiles and species-specific (i.e., mouse, human) drug PK profiles."

"The results demonstrate the importance of including physiologically relevant dynamic drug exposure profiles during in vitro drug testing to more accurately mimic in vivo drug effects, thereby improving translatability across nonclinical studies and reducing the reliance on animal models during drug development."

INTRODUCTION

"Modeling and simulation approaches have improved the ability to extrapolate PK-PD relationships from preclinical species to humans, and differences in dynamic drug exposures must be carefully addressed in the design of nonclinical studies [162–164]."

"Significant effort is being placed on the development of more physiologically relevant systems for in vitro testing such as the MPS devices, but there has been less emphasis on addressing the inability to faithfully recapitulate physiologically relevant drug exposure profiles in these in vitro systems."

"Several macroscale perfusion devices, such as hollow fiber systems, have been developed to enable evaluation of PD responses to dynamic drug exposures, and these have been primarily applied in the context of antibiotic development [165, 166]."

"Integration of programmable, dynamic perfusion systems with advanced MPS devices may hopefully enable elimination of the majority of animal testing required during preclinical drug development."

MATERIALS AND METHODS

"The gas-tight syringe assemblies with removable needles (Hamilton 1000 series, Hamilton Company, Reno, NV, USA) were filled with cell media or drug solution in cell media (DOX or GEM) and equilibrated for 30 min in a cell incubator at 37 °C and 5% CO_2."

"MCF-7 human breast cancer cells (ATCC, Manassas, VA) were cultured in MEM media (Gibco, Dublin, Ireland) with 10% fetal bovine serum (FBS) (Thermo Fisher Scientific)."

"MDA-MB-231 human breast cancer cells (ATCC) were cultured in RPMI-1640 media (Gibco) supplemented with 10% FBS (Thermo Fisher Scientific)."

"For static experiments, cell media (MEM + 10% FBS or RPMI-1640 + 10% FBS) was manually replenished every 24 h, whereas flow conditions under perfusion studies (constant and dynamic perfusion) were enough to replenish the cells."

"Cells were cultured in the microwell of the microfluidic platform and exposed to drug under static or perfusion conditions."

RESULTS

"Drug PK profiles are converted to flow rates over time for a media or drug stream using modeling software (MATLAB, Natick, MA, USA), enabling proper dilution of the input drug concentration to achieve the desired temporal concentration profile for perfusion of the cell culture region."

"The modified physiologically inspired PK exposure profiles based on previous work infusing drug for 0.5 h, 4 h, and 16 h [167] illustrate how the microfluidic system can be used to simulate species-specific drug PK profiles in an in vitro system."

"Schedule-dependent synergism was first confirmed in the MDA-MB-231 cells under static conditions by treating with either DOX or GEM for 24 h followed by exposure to the other drug for the remaining 48 h in an analogous manner to the studies performed in vitro by Vogus and others [168]."

DISCUSSION

"We have developed a microfluidic PK-PD platform that enables exposure of cells in 2D or 3D to any desired dynamic drug exposure profile, including combination drug treatments."

"The proof-of-concept studies described here demonstrate how the microfluidic PK-PD platform can be employed to examine exposure- and time-dependent drug effects on cellular systems."

"Being able to perform such dose fractionation studies in a controlled and repeatable manner with any desired exposure profile using the microfluidic PK-PD platform has significant advantages over existing approaches."

"When these studies were repeated in the microfluidic PK-PD platform using the in vivo-relevant drug exposure profiles instead, the results were consistent with the observed in vivo effects."

"To enabling analysis using limited biopsy material, the microfluidic PK-PD platform could provide dosing regimen optimization using patient-specific drug PK profiles."

CONCLUSIONS

"A novel microfluidic PK-PD platform has been developed for the delivery of user-defined, dynamic drug exposure profiles to existing 2D or 3D in vitro cell systems, helping to fill the current translational gap between static in vitro studies and nonclinical in vivo or clinical studies."

"This microfluidic PK-PD platform is envisioned to find application across the spectrum of drug development, from early compound design and target evaluation to clinical trial design and patient-specific dosing."

"The introduction of microfluidic PK-PD platforms, including those leveraging advancements in microphysiological systems, into the drug development paradigm will improve translatability of nonclinical studies, reduce the reliance on animal models, and increase the efficiency with which new medicines can be brought to patients."

Assessment of Topical Bioequivalence Using Dermal Microdialysis and Tape Stripping Methods [169] This is a machine-generated summary of:

Incecayir, Tuba; Agabeyoglu, Ilbeyi; Derici, Ulver; Sindel, Sukru: Assessment of Topical Bioequivalence Using Dermal Microdialysis and Tape Stripping Methods [169]

Published in: Pharmaceutical Research (2011)

Link to original: https://doi.org/10.1007/s11095-011-0444-3

If you want to cite the papers, please refer to the original.

For technical reasons we could not place the page where the original quote is coming from.

Abstract-Summary "To assess the bioequivalence of two commercial topical formulations of oxytetracycline HCl by tape stripping and microdialysis in healthy volunteers."

"Tape stripping study was conducted on 12 healthy volunteers."

"After a 30-minute application of the formulations, adhesive tapes were used to sample stratum corneum at 0.25, 0.5, 1, 1.5, 2, 3, 4 hr."

"Pharmacokinetic evaluation by microdialysis yielded that the test could not be said to be bioequivalent to the reference at 90% CI."

"The test was found to be bioequivalent to reference according to the dermatopharmacokinetic evaluation by tape stripping."

"No significant correlations were found between microdialysis and tape stripping methods as regarding the topical bioequivalence of oxytetracycline HCl formulations."

INTRODUCTION

"Possible methods for the determination of BE of multisource topical drug products include clinical trials, pharmacodynamics, tape stripping (TS), dermal microdialysis (DMD), and other techniques."

"The TS method, which was also mentioned as the dermatopharmacokinetics (DPK) method, is attracting increasing attention as a method with which to assess the rate and extent of topical drug bioavailability (BA) in the rate-limiting barrier of skin, the stratum corneum (SC)."

"Microdialysis (MD) is an in vivo sampling method for measuring endogenous and exogenous compounds in extracellular spaces of tissues."

"This method needs standardization and more BA and BE studies on the field of topical drug administration [170]."

"The aim of the present study was to assess the BE of two commercial topical formulations of a wide spectrum antibiotic, oxytetracycline HCl (OTC), by TS and DMD in healthy volunteers and to evaluate the correlation between these methods."

MATERIALS AND METHODS

"In vitro RR studies for OTC across the MD membrane were performed before in vivo MD studies to ensure reproducible and concentration-independent sampling of OTC during the experiment."

"The probe was placed into constantly stirred OTC solutions with concentrations of 0.250–50.0 μg/ml at 35°C and perfused with saline (0.9% NaCl sol.) at a flow rate of 2 μl/min up to 5 hr."

"The probes were perfused at a flow rate of 2 μl/min for 4 hr with sterile and isotonic saline containing 5 μg/ml OTC."

"OTC concentrations in TS samples were determined by HPLC."

"Area under the OTC content-time profile (AUC) values were calculated over the 0 to t_{last}-hour time interval using the log-trapezoidal rule method [171]."

"For MD PK data, mean ± standard deviation (SD) dermis concentrations of OTC were plotted versus time."

RESULTS

"The AUC ratio for test/reference was 2.25 ± 2.16 (±SD)."

"The BE of the test versus the reference was 89–135% for AUC and 82–161% for C_{max} with a 90% Cl, which is outside of 80–125% BE criteria."

"The test product could not be said to be bioequivalent to the reference as regarding log transformed AUC and C_{max} at 90% CI."

"The AUC ratio for test/reference was found to be 0.838 ± 0.460 (±SD)."

"The BE of the test versus the reference was 88–100% for AUC and 82–97% for C_{max} with a 90% CI."

"The test product was found to be bioequivalent to the reference as regarding log transformed AUC and C_{max} at 90% CI."

DISCUSSION

"DMD and TS methods were used to continously sample the unbound concentration of OTC in dermis and to sample the SC levels, respectively."

"DPK of OTC in the skin has not previously been studied, and two commercial OTC topical products existing on the market would enable a comparison of DMD and TS methods for assessing the topical BE of OTC in the same healthy volunteers."

"MD data demonstrated that the measurable amounts of OTC in dialysates were determined from the first sampling points after the topical administration of the test and the reference formulations in the present study."

"A BE study has been conducted to evaluate the topical BE of three marketed topical metronidazole formulations by simultaneous DMD and TS methods, showing no correlation between them either [172]."

CONCLUSION

"Although it is accepted that the TS results may be indicative of dermis concentrations, SC concentration was found not to be predictive of OTC concentration in the dermis in the present study."

"DMD as a method to be routinely used in topical BE studies has to be explored extensively."

"It seems that more studies are needed for a wide range of topical drugs to understand whether a correlation between DMD and TS methods exists and to investigate the effect of different formulations on this relationship between the methods."

Dermal Microdialysis Technique to Evaluate the Trafficking of Surface-Modified Lipid Nanoparticles upon Topical Application [173] This is a machine-generated summary of:

Desai, Pinaki R.; Shah, Punit P.; Patlolla, Ram R.; Singh, Mandip: Dermal Microdialysis Technique to Evaluate the Trafficking of Surface-Modified Lipid Nanoparticles upon Topical Application [173]

Published in: Pharmaceutical Research (2012)

Link to original: https://doi.org/10.1007/s11095-012-0789-2

If you want to cite the papers, please refer to the original.

For technical reasons we could not place the page where the original quote is coming from.

Abstract-Summary "To evaluate the skin pharmacokinetics and tissue distribution of cell penetrating peptides (CPP) modified nano-structured lipid carrier (NLC) using an in vivo dermal microdialysis (MD) technique."

"Celecoxib (Cxb) encapsulated NLCs (CXBN), CPP modified CXBN (CXBN-CPP) and Cxb-Solution (CXBS) formulations were prepared and tested for in vitro skin distribution."

"The effect of pre-treatment with Cxb formulations was evaluated for expression of prostaglandin-E2 (PGE_2) and Interleukin-6 (IL-6) after exposure of xylene using MD."

"The cumulative permeation of Cxb in MD dialysate after 24 h for CXBN-CPP was significantly higher ($p< 0.001$) than CXBN and CXBS."

"Further, pre-treatment with CXBN-CPP significantly inhibited PGE_2 and IL-6 expression compared to CXBS and CXBN ($p< 0.001$)."

INTRODUCTION

"To overcome these limitations, dermal microdialysis (MD) can be used as an in vivo technique for investigation of pharmacokinetic and therapeutic response of a drug in the skin [174, 175]."

"To study the effect of CPP modified NLCs on the Cxb delivery into the skin, specific aims of the current study were to: 1) prepare and characterize CPP modified NLCs; 2) study the effects of arginine chain length on the in vitro skin permeation of Cxb using hairless rat skin as a model; 3) evaluate the Cxb concentration at the epidermal-dermal junction for NLCs and NLC-CPP using MD; 4) assess the effect of NLCs and NLC-CPP pre-treatment followed by exposure to xylene on the expression of PGE_2 and IL-6 biomarkers by dermal MD and 5) investigate the therapeutic response of NLC-CPP in an in vivo inflammatory mouse model of allergic contact dermatitis (ACD)."

MATERIALS AND METHODS

"Cxb formulation exposed skin was collected after pre-determined time points of 2, 4, 6, 12 and 24 h. Cryotome was used to separate all three layers of the skin SC, epidermis and dermis."

"Cxb standard stock in 0.5% (w/v) volpo-20 in Krebs-Ringer solution was passed through the probe using an infusion pump at flow rate of 2 µl/ml and dialysate samples were collected every 60 min for 480 min."

"For the recovery of IL-6, equilibrated CMA 20 probe was placed in 2000 pg/ml IL-6 stock solution and perfused with 0.1% w/v BSA in Krebs-Ringer solution at flow rates of 0.5, 1 and 2 µl/min."

"In order to study the effect of modified and unmodified CXBN on the expression of PGE_2 and IL-6, the exposure site on the back of rat was pre-treated with 200 µl of CXBS, CXBN, CXBN-R_{11} and CXBN-YKA dispersion All the Cxb formulations were applied unocclusively 16 h prior to the probe insertion."

RESULTS

"Cxb permeation from CPPs modified NLCs was significantly higher ($p < 0.001$) at different collection points (2, 6, 12 and 24 h) compared to CXBS, CXBN and CXBN-YKA."

"The concentration vs. time profile showed that the levels of permeated Cxb were high for the NLCs and NLCs modified with CPPs at the end of 24 h. These results indicate that the permeation of Cxb from CXBN-R_{11} was significantly higher than CXBN, CXBN-YKA and CXBS ($p < 0.001$) after 24 h. The AUC_{0-24h} for Cxb was 0.541 ± 0.063, 1.345 ± 0.094, 2.714 ± 0.135 and 1.549 ± 0.075 µg × hr/ml from CXBS, CXBN, CXBN-R_{11} and CXBN-YKA, respectively."

"Following application of xylene, IL-6 release was increased gradually and reached the maximum levels after removal of the xylene within 3 h. The expression levels for IL-6 were significantly less ($p < 0.001$) for CXBN-R_{11} compared to CXBS, CXBN, CXBN-YKA."

DISCUSSION

"Based on in vitro skin permeation studies, CXBN-R_{11} was selected for further in vivo investigation of pharmacokinetic parameters of Cxb in normal skin and expression of biomarkers in the damaged skin along with therapeutic response in ACD model."

"Our in vitro skin permeation results suggested that the Cxb retention in the skin for unmodified and CPP modified NLCs was 4.4 times higher at 24 h than 12 h. Calibration methods in MD are important when quantitative measurement of the extracellular fluid concentrations of the drug is desired."

"These results are in agreement with the in vitro skin permeation and in vivo pharmacokinetic studies and verify that the surface modification of CXBN with R_{11} is responsible for the increase in the skin permeation of Cxb along with reduced PGE_2 and IL-6 levels in the skin."

CONCLUSION

"Our studies demonstrate that the NLCs modified with R_{11} significantly improve the retention of Cxb in the skin layers as a function of time."

"The pre-treatment of CXBN-R_{11} was responsible for the increase in the skin permeation of Cxb and thus decrease in the PGE_2 and IL-6 levels during the skin inflammation induced by topical application of xylene, an irritant chemical."

"This further resulted in reduction of ear thickness and inflammation associated with an in vivo ACD model, suggesting the potential of R_{11} modified NLCs to treat various skin diseases and disorders like contact dermatitis, psoriasis and skin cancer where delivery of drug is required to be delivered in the deep skin layers."

Comparison of Open-Flow Microperfusion and Microdialysis Methodologies When Sampling Topically Applied Fentanyl and Benzoic Acid in Human Dermis Ex Vivo [176] This is a machine-generated summary of:

Holmgaard, R.; Benfeldt, E.; Nielsen, J. B.; Gatschelhofer, C.; Sorensen, J. A.; Höfferer, C.; Bodenlenz, M.; Pieber, T. R.; Sinner, F.: Comparison of Open-Flow Microperfusion and Microdialysis Methodologies When Sampling Topically Applied Fentanyl and Benzoic Acid in Human Dermis Ex Vivo [176]

Published in: Pharmaceutical Research (2012)

Link to original: https://doi.org/10.1007/s11095-012-0705-9

If you want to cite the papers, please refer to the original.

For technical reasons we could not place the page where the original quote is coming from.

Abstract-Summary "The purpose of this study is to compare two sampling methods—dermal Open-Flow Microperfusion (dOFM) and dermal Microdialysis (dMD) in an international joint experiment in a single-laboratory setting."

"The dOFM and dMD techniques are compared in equal set-ups using three probe-types (one dOFM probe and two dMD probe-types) in donor skin (n = 9) - 27 probes of each type sampling each penetrant in solutions applied in penetration-chambers glued to the skin surface over a time range of 20 h. Pharmacokinetic results demonstrated concordance between dOFM and dMD sampling technique under the given experimental conditions."

"When planning a study of cutaneous penetration the advantages and limitations of each probe-type have to be considered in relation to the scientific question posed, the physico-chemical characteristics of the substance of interest, the choice of experimental setting e.g. ex vivo/in vivo and the analytical skills available."

INTRODUCTION

"Sampling of large, highly lipophilic and/or protein-bound substances has, however, always been a challenge in MD when using the traditional MD probes and perfusates."

"Lipophilic molecules have a low affinity for the traditional aqueous perfusate compared to that for the tissue or the probe material, which can hamper MD sampling [177]."

"The exchange/diffusion of a penetrant between the tissue and the perfusate (recovery or delivery) is determined by the probe membrane — pore area and MW cut-off value — as well as by the physico-chemical properties of the substance — size, charge, and solubility."

"The in vitro recovery studies are used to determine the basic efficacy of each probe type for sampling of the penetrant of interest and in order to evaluate the degree of non-specific adsorption."

"We compare the two sampling methods used in skin penetration studies: dOFM and dMD in human skin ex vivo."

MATERIALS AND METHODS

"The dialysates from the CMA66 and 2 kDa probe outlets were collected in plastic vials (0.2 mL Domed Cap Maximum Recovery PCR Tubes®, Axygen, USA) Fentanyl: 3 mL fentanyl was added to each penetration-chamber."

"The commercially available aqueous fentanyl solution (Janssen-Cilag, Belgium) was modified by ethanol (40 µg Fentanyl/mL in 20% ethanol) in order to assure skin penetration and to assure the recovery by the dMD probes."

"All probes were inserted using 21-gauge (0.8 mm) guide cannulas, three under each outlined penetration-chamber area, nine probes in each half-flap and a total of 18 probes in each experiment/donor."

"The 2 kDa probe was inserted through entry and exit points marked on the skin ~20 mm apart."

"Sample fluids from dOFM were collected in capillaries connected between the probe outlets on one side and the pump on the other for 'pull' functionality."

RESULTS

"There were no significant differences in AUC between the probe-types sampling fentanyl and benzoic acid, respectively (P-values all > 0.05)."

"Another donor (donor 7) displayed a significantly increased penetration for fentanyl, most pronounced in the samples from the 2 kDa probe where the lag-time was as short as 5 min and the AUC 240 ng*hr/mL. These results are > 3SD away from the mean, and since this unusual penetration profile is seen for both 2 kDa probes, this could indicate a damaged skin barrier."

"For the fentanyl studies, where the perfusate contained albumin, the following numbers of probes were malfunctioning: dOFM 2, CMA66 3 and 2 kDa 8."

"For the benzoic acid studies, where the perfusate did not contain albumin, the following numbers of probes were malfunctioning and therefore excluded: dOFM 1, CMA66 1 and 2 kDa 3."

DISCUSSION

"All three probes sample benzoic acid more effectively than they sample fentanyl in these ex vivo studies."

"In the ex vivo study the 2 kDa probe has no outlet tubing, since the hemodialysis fiber continues from the outlet opening in the skin directly into the sampling vial, thus limiting the potential adsorption area."

"The experimental variability for sampling varied between penetrants as well as between probes."

"Variability in probe depth has been demonstrated to affect the pharmacokinetic parameters of topically applied penetrants sampled by dMD ex vivo [178]."

"All probe-types sample an almost negligible amount of fentanyl in this donor, but the absorption of benzoic acid in donor 8 is close to average."

"Prior to the present study we investigated the feasibility of fentanyl sampling by the dOFM probe (ex vivo) and the 2 kDa probe (in vitro)."

CONCLUSION

"The three probe-types show concordance in AUC and Cmax for both benzoic acid and fentanyl."

"This should be considered in relation to study type, penetrant/drug and analytical techniques available as well as experience."

"Our study needs to be repeated with more lipophilic drugs as well as conducted under in vivo conditions in order to confirm the present experimental observations."

Role of Microdialysis in Pharmacokinetics and Pharmacodynamics: Current Status and Future Directions [179] This is a machine-generated summary of:

Azeredo, Francine Johansson; Dalla Costa, Teresa; Derendorf, Hartmut: Role of Microdialysis in Pharmacokinetics and Pharmacodynamics: Current Status and Future Directions [179]

Published in: Clinical Pharmacokinetics (2014)

Link to original: https://doi.org/10.1007/s40262-014-0131-8

If you want to cite the papers, please refer to the original.

For technical reasons we could not place the page where the original quote is coming from.

Abstract-Summary "Microdialysis is a semi-invasive technique that is able to measure concentrations of the free, active drug or endogenous compounds in almost all human tissues and organs."

Introduction

"This technique is well-established and used in in vitro and in vivo experiments for sampling drugs and their metabolites as well as endogenous substances from body fluids or the interstitial cell fluid of selected tissues in animal and human studies [180]."

"In order to determine free drug concentrations in interstitial spaces, in vivo microdialysis has found important applications in the field of pharmacokinetics, specifically in the investigation of metabolic processes and drug distribution [181]."

"Since the microdialysis probe is placed in the interstitial space of the tissue of interest, this technique is attractive for clinical distribution studies of drugs that have their pharmacological effect in peripheral tissues, such as anti-infectives and chemotherapeutics [182], and is considered by regulatory authorities to be a promising technique for the bioequivalence investigation of topically applied medicines [183]."

"Besides the investigation of brain disorders, clinical microdialysis has also successfully been used to assess biomarkers in other body regions."

"We review the clinical data published on microdialysis in determining pharmacokinetics and pharmacodynamics in different fields of application during the past 5 years, providing future directions."

Anti-Infective Drugs

"In the field of anti-infectives, the pharmacokinetic/pharmacodynamic indices are used to evaluate the antibiotic pharmacological effect by employing the minimum inhibitory concentration (MIC) as the pharmacodynamic parameter and classifying the drug activity as time dependent, where the index that correlates with efficacy is the percentage of time that the antimicrobial concentration remains above the MIC of the microorganism (T>MIC), or concentration dependent, when the ratio between antimicrobial peak plasma concentration (C_{max}) and MIC (C_{max}/MIC) or between the area under the plasma concentration–time curve (AUC) and MIC (AUC/MIC) better forecast microbial eradication."

"The free tissue concentrations determined by clinical microdialysis investigations have been used to correlate with anti-infective pharmacological effect using pharmacokinetic/pharmacodynamic indices instead of the more descriptive and predictive approach offered by pharmacokinetic/pharmacodynamic mathematical modeling."

"Pharmacokinetic/pharmacodynamic modeling allows establishment of rational dosing regimens when free tissue concentrations of anti-infectives are investigated against bacteria in vitro."

Anti-Cancer Drugs

"Since accurate knowledge of the intratumoral pharmacokinetics of anti-cancer agents is of extreme importance, especially during drug development, microdialysis is often used to assess drug exposure in solid tumors to show that the compounds of interest reach their target site in sufficiently high concentrations [184]."

"The results showed a large difference in the concentration values in both locations (tenfold higher when the microdialysis probe was inserted close to the tumor), showing that glioma alters BBB permeability at the site of the tumor and emphasizing that microdialysis probe location must be well-established to properly assess the significance of the microdialysis data obtained in brain tumor investigation [185]."

"The authors showed that imaging techniques such as contrast-enhanced MRI are fundamental to ensure the correct location of microdialysis probes in brain tumor investigation and assure the reliability of the anti-cancer drug concentrations obtained."

Topical Treatment

"The application of dermal microdialysis as a valuable tool to assess the bioavailability/bioequivalence of topically applied drug formulations was successfully demonstrated by Tettey-Amlalo and others [186] in a study using a ketoprofen gel formulation applied to the skin of healthy subjects."

"Another study that corroborates the feasibility of using dermal microdialysis in analyzing bioequivalence of topical drugs was conducted by Brunner and others [187]."

"They investigated the penetration of five different diclofenac formulations into the skin which were comparatively analyzed by dermal microdialysis following topical application to human arms."

"Two recent dermal microdialysis studies performed by the same research group found increased dermal penetration in patients with atopic dermatitis [188] and irritant dermatitis [189] after topical application of a metronidazole cream compared with unmodified skin, underlying the importance of obtaining data regarding dermal penetration in diseased skin in future studies."

Endogenous Compounds

"In order to verify the effect of prostacyclin in severe traumatic brain injury (sTBI), a microdialysis catheter was inserted in the bilateral intracerebral area of sTBI patients and the levels of lactate/pyruvate and brain glucose were evaluated."

"An important study regarding the evaluation of Aβ amyloids in patients with severe brain injuries using microdialysis technique reported by Brody and others [190] showed increased brain levels of Aβ amyloids during several hours, up to days, in most patients after the injury."

"A recent microdialysis study of the neurotransmitter alterations was conducted in conscious patients with Parkinson's disease to verify the catecholamine levels as well as the pharmacokinetics of levodopa after oral dosing, during deep brain stimulation (DBS)."

"The use of microdialysis to monitor blood glucose is still controversial since intravenous microdialysis is technically feasible but not accurate, especially in critical ill patients [191]."

Conclusions

"Clinical microdialysis has become one of the most important and promising sampling techniques for the determination of tissue concentrations of endogenous biomolecules and drugs."

"Since microdialysis offers the advantage of measuring the unbound target-site concentrations of drugs, it plays an important role in drug discovery and development and is able to satisfy regulatory requirements for pharmacokinetic distribution and bioequivalence studies."

"Microdialysis is a very useful technique for better understanding drug efficacy and disease progression in the clinical setting, and one can expect its increased importance as a tool to help clinicians in the decision-making process for patients' treatment."

Assessing Topical Bioavailability and Bioequivalence: A Comparison of the In vitro Permeation Test and the Vasoconstrictor Assay [192] This is a machine-generated summary of:

Lehman, Paul A.; Franz, Thomas J.: Assessing Topical Bioavailability and Bioequivalence: A Comparison of the In vitro Permeation Test and the Vasoconstrictor Assay [192]

Published in: Pharmaceutical Research (2014)
Link to original: https://doi.org/10.1007/s11095-014-1439-7

If you want to cite the papers, please refer to the original.

For technical reasons we could not place the page where the original quote is coming from.

Abstract-Summary "To compare the sensitivity of a pharmacokinetic assay, the in vitro permeation test (IVPT), with that of a pharmacodynamic assay, the human skin blanching or vasoconstrictor (VC) assay, in assessing the relative bioavailability of topical clobetasol propionate products."

"IVPT found total clobetasol absorption varying ten-fold from highest to lowest product, whereas the VC assay found this same difference was less than two-fold."

"Due to its greater variability as well as saturation of the pharmacodynamic response at higher flux levels, the VC assay found all products except the solution to be equipotent."

"IVPT was found to be substantially more sensitive and less variable than the VC assay for assessing clobetasol bioavailability."

Introduction

"The long history of use of excised human skin as an accepted in vitro model for the study of percutaneous absorption has logically led to its consideration as a surrogate for clinical trials or human pharmacokinetic studies in determining the bioavailability (BA) and bioequivalence (BE) of topical drug products [193]."

"In vitro assessment of the seventh test product, the glucocorticoid mometasone furoate, found a test/reference ratio of only 0.63; yet, by vasoconstrictor assay, it too was found to be BE to the reference product (test/reference = 1.11) and subsequently approved."

"Because of the aforementioned observations regarding the glucocorticoids mometasone furoate and betamethasone valerate (BMV), suggesting that the in vitro permeation test (IVPT) might be more sensitive than the vasoconstrictor (VC) assay, this study was undertaken to evaluate the use of IVPT a possible surrogate for the determination of glucocorticoid BA/BE."

Materials and Methods

"Following the integrity test the receptor solution was changed several times to remove all traces of radioactivity and then replaced with a 1:10 dilution of PBS. (This was done because all receptor samples required concentrating to quantify CP and a high salt content interfered with the analytical procedure.) Subsequently, the chimney portion of the chamber was removed to facilitate access to the skin surface and the four semi-solid products applied using the rounded end of a thin glass rod at a target dose of five milligrams."

"The six test products, five CP and the BMV control, were randomly assigned to one of eight test sites. (Two additional research formulations were included to make a total of eight products, but their data are not reported here.) A 10 μL quantity of product was applied and evenly spread over the entire site using the rounded end of a small glass rod previously shown to leave >98% of the dispensed drug on the skin."

Results

"Statistical analysis of relative BA showed that CP absorption from the ointment was significantly different from all other products and that absorption from the gel and cream were significantly different from the ointment, solution and emollient cream products."

"In one respect the results tended to parallel those of the permeation study in that the CP products with a high rate of absorption (ointment, cream, gel) elicited a numerically greater blanching response than those with a low rate of absorption (emollient cream, solution), though the rank order for the emollient cream and solution was reversed from that seen with IVPT."

"The parallel between the rate of absorption and blanching profiles may at first seem puzzling in that the CP dose is removed after 3 hours in the VC assay but not in the permeation test."

Discussion

"IVPT was used to determine the relative BA of 0.05% CP from five marketed products and the results compared to those obtained using the VC assay."

"The results make it quite clear that IVPT is a substantially more sensitive and less variable test than the VC assay at detecting differences in BA between these five products."

"In a series of blinded, paired comparison studies involving application of two different products to symmetrical lesions on the trunk or extremities of psoriatic subjects, they found the efficacy results in 20 of 23 comparisons to be in line with the results obtained by VC assay."

"The lack of clinical data in which there was a direct comparison of any two of the five CP products tested in this study makes it difficult to evaluate the clinical implications of the large differences in absorption that were observed here."

Conclusions

"A comparison of two surrogate tests has revealed that the use of the IVPT to quantify differences in relative BA between five marketed clobetasol propionate products provides a much greater level of sensitivity than that afforded by the VC assay."

"The permeation test found total clobetasol absorption from the five products to vary over a ten-fold range whereas the vasoconstrictor assay found this same difference was less than two-fold."

"In agreement with earlier studies, these data continue to support use of IVPT for the determination of relative BA and "in vitro" BE of many, if not all, topical products."

Follow-up of drug permeation through excised human skin with confocal Raman microspectroscopy [194] This is a machine-generated summary of:

Tfayli, Ali; Piot, Olivier; Pitre, Franck; Manfait, Michel: Follow-up of drug permeation through excised human skin with confocal Raman microspectroscopy [194]

Published in: European Biophysics Journal (2007)
Link to original: https://doi.org/10.1007/s00249-007-0191-x

If you want to cite the papers, please refer to the original.

For technical reasons we could not place the page where the original quote is coming from.

Abstract-Summary "The efficacy of several topically applicated drugs is directly related to their penetration through the skin barrier."

"The use of nondestructive and structurally informative techniques permits a real breakthrough in the investigations on skin penetration at a microscopic scale."

"Confocal Raman microspectroscopy is a nondestructive and rapid technique which allows information to be obtained from deep layers under the skin surface, giving the possibility of a real-time tracking of the drug in the skin layers."

"We try to follow the penetration of Metronidazole, a drug produced by Galderma as a therapeutic agent for Rosacea treatment, through the skin."

"Micro-axial profiles were conducted to follow the penetration of the drug in the superficial layers, on excised human skin specimens."

"The collected spectra permit us to follow the structural modifications, induced by the Metronidazole on the skin, by studying the changes in the spectral signature of the skin constituents."

Introduction

"Several techniques are used for evaluating drug delivery and diffusion into the skin and skin structures."

"This method lacks accuracy and cannot be used for drugs whose penetration is limited to the outermost layers (stratum corneum) of the skin."

"Biophysical techniques like Electron Microscopy (EM) (Hofland and others [195]), Small Angle X-ray Scattering technique (SAXS) (Bouwstra and others [196]), Confocal Laser Scanning Microscopy (CLSM) (Grams and others [197, 198]; Veiro and Cummins [199]), and Two Photon Microscopy (TPFM) (Yu and others [200]), have provided invaluable information about the speed and depth of penetration and the interaction of exogenous molecules and substances with the skin."

"Vibrational spectroscopy presents the advantage of providing molecular information directly from the skin."

"Several works have used confocal Raman spectroscopy to study the skin and skin hydration by assessing water concentrations and the effects of moisturizing factors (Caspers and others [201, 202]; Chrit and others [203]; Sieg and others [204])."

"In a final step, we investigated the effect of the Metronidazole on the skin molecular constituents by studying the spectral changes at the level of certain vibrations, specific for the lipid or protein content."

Materials and methods

"For confocal Raman analysis, a 1.5 × 1.5 cm^2 of skin were cut."

"Once the in-depth confocal measurement was performed, the skin was cryo-fixed and 20 μm transverse sections were cut by cryo-microtome, and deposited on CaF_2 slides."

"Raman spectra were collected at different focus points, from −10 μm (above the skin surface) to 40 μm (under the skin surface) with a 4 μm step."

"Raman spectral acquisitions were performed with a Labram microspectrometer (Horiba Jobin Yvon, Lille France)."

"The spectral range was from 620 to 1,810 cm^{-1}, the acquisition time of each spectrum was 2 times 10 s and the confocal pinhole was adjusted to 150 μm."

"To reduce the influence of the instrument response in the spectra, firstly, we corrected the collected signal by subtracting the black current signal (the signal collected on the CCD detector with the laser switched off)."

Results

"Spectral features arising from lipid conformation in the skin clearly appear at 1,063 cm^{-1} (hydrocarbon chain, trans conformation), and at 1,085 and 1,128 cm^{-1} (hydrocarbon chain, gauche conformation)."

"Other features of current interest are the sharp phenylalanine-ring breathing vibration around 1,003 cm^{-1}, suitable for studying the protein content of the skin."

"The relative intensity of these bands, with respect of the skin protein bands intensities, indicates the relative concentration of Metronidazole, and the variation in their intensities reveals the deposition of the drug at different depths."

"These bands mask the signal of the tyrosine Fermi doublet of the endogenous skin protein at 830–850 cm^{-1}; and the tryptophan vibration at 883 cm^{-1}."

"The deeper we progress into the skin, then the less intense is the 1,211 cm^{-1} feature and the higher is the intensity of the drug feature, especially between 16 and 40 μm (14–34 μm on the measuring scale)."

Discussion

"Since the drug and the solvent, both, present vibrations in the 600–1,800 cm^{-1} region, it was very important to determine the pertinent spectral features that could be used as signature for the Metronidazole solution."

"These features enables us to detect the drug in the skin despite the very small concentrations employed for topical application."

"One hour after the application of the Metronidazole on the skin surface, it is detected at a depth of 23–24 μm and 1 h later it has been observed between 16 and 40 μm."

"The changes in the skin spectral features on these lines are due to the effect of the drug solution on the skin components."

"Several changes were detected due to the application of the Metronidazole–Transcutol solution on the skin, especially for tyrosine and tryptophan vibrations."

"It gives direct structural information, and enables the changes in the skin to be monitored following the topical application of drug."

3.6 Clinical Model

Introduction by the Editor

The integration of nanoparticles into clinical practice represents a significant advancement in modern medicine. Clinical models for nanoparticles involve a comprehensive approach to evaluating the safety, efficacy, and translational potential of these innovative delivery systems in human subjects. These models involve thorough study designs, ethical considerations, and robust methodologies that allow researchers and clinicians to explore the fate of nanoparticles and their therapeutic outcomes. Clinical models provide invaluable insights into the feasibility of nanoparticle-based therapies, aiding in dose optimization, patient stratification, and the identification of potential challenges or adverse effects. This intersection of cutting-edge technology and patient care heralds a new era of personalized medicine, where nanoparticles hold the promise of revolutionizing treatment approaches and improving patient outcomes.

Machine generated summaries

Disclaimer: The summaries in this chapter were generated from Springer Nature publications using extractive AI auto-summarization: An extraction-based summarizer aims to identify the most important sentences of a text using an algorithm and uses those original sentences to create the auto-summary (unlike generative AI). As the constituted sentences are machine selected, they may not fully reflect the body of the work, so we strongly advise that the original content is read and cited. The auto generated summaries were curated by the editor to meet Springer Nature publication standards.

To cite this content, please refer to the original papers.

Machine generated keywords: rate constant, rat, plasma, bind, drug disposition, nonlinear, constant, pharmacological, disposition, exhibit, peptide, mouse, inhibitor, protein bind, tissue

Incorporating Pharmacological Target-Mediated Drug Disposition (TMDD) in a Whole-Body Physiologically Based Pharmacokinetic (PBPK) Model of Linagliptin in Rat and Scale-up to Human [205] This is a machine-generated summary of:

Wu, Nan; An, Guohua: Incorporating Pharmacological Target-Mediated Drug Disposition (TMDD) in a Whole-Body Physiologically Based Pharmacokinetic (PBPK) Model of Linagliptin in Rat and Scale-up to Human [205]

Published in: The AAPS Journal (2020)

Link to original: https://doi.org/10.1208/s12248-020-00481-w

If you want to cite the papers, please refer to the original.

For technical reasons we could not place the page where the original quote is coming from.

Abstract-Summary "We established a novel whole-body physiologically-based pharmacokinetic (PBPK)-TMDD model for linagliptin."

"Our final model adequately captured the concentration-time profiles of linagliptin in both plasma and various tissues in both wildtype rats and DPP4-deficient rats following different doses."

"The binding affinity of linagliptin to DPP-4 (Kd) was predicted to be higher in plasma (0.0740 nM) than that in tissue (1.29 nM)."

"When scaled up to a human, this model captured the substantial and complex nonlinear pharmacokinetic behavior of linagliptin in human adults that is characterized by less-than dose-proportional increase in plasma exposure, dose-dependent clearance and volume of distribution, as well as long terminal half-life with minimal accumulation after repeated doses."

INTRODUCTION

"Coupled with its high-affinity and long-lasting binding to the DPP-4 enzyme, linagliptin exhibits a unique nonlinear pharmacokinetic profile that is different from the linear pharmacokinetics observed in other DPP-4 inhibitors [206–210]."

"Consistent with the target-mediated nonlinear pharmacokinetics observed in vitro and in animals, linagliptin exhibits substantial and complex nonlinear pharmacokinetics in human that is characterized by less-than dose-proportional increase in plasma exposure, concentration-dependent plasma protein binding, as well as long terminal half-life (110–130 h) with minimal accumulation after repeated doses [206]."

"In order to predict linagliptin's pharmacokinetics and tissue distribution more accurately, a mechanistic PBPK model that incorporates DPP-4 binding affinity and binding kinetics (i.e., TMDD process) is needed."

"Reported rat data, a whole-body PBPK model incorporating TMDD components was developed to quantitatively characterize the time-course of linagliptin in plasma and various tissues in both wildtype and DPP-4 deficient rats simultaneously."

METHODS

"Part I determined linagliptin concentrations in various tissues of wildtype and DPP-4 deficient rats at different time points (3, 24, and 168 h) after an iv dose of 2 mg/kg."

"2 reported linagliptin plasma concentration-time profiles following single iv doses of 0.01, 0.1, 0.3, 1, 3, 10, and 50 mg/kg of [^{14}C] linagliptin in both wild-type and DPP-4 deficient rats."

"3 evaluated linagliptin plasma concentrations at various time points (13 time points ranging from 0.25 to 72 h) following a single oral dose of 1 or 15 mg/kg linagliptin in wildtype rats."

"The total amount of DPP-4 targets in human tissues and organs ($R_{max,human\ tissue}$) are fixed in the simulation based on the following equation: Here, $C_{DPP\text{-}4,total}$ represents DPP-4 targets concentration levels in each tissue, which were estimated from the rat PBPK-TMDD model."

RESULTS

"A comprehensive whole-body PBPK-TMDD model comprising plasma and 14 tissue compartments has been successfully established to characterize linagliptin pharmacokinetics and tissue distribution in wildtype rats and DPP-4 deficient rats."

"The model predicted linagliptin concentrations were in close agreement with the experimental observations in most tissues."

"To 2 mg/kg, the robustness of our model in estimating the tissue distribution of linagliptin in various other doses, ranging from 0.01 to 50 mg/kg, was also evaluated."

"Our model predicted that linagliptin plasma concentrations in wildtype rats were higher than that in DPP-4 deficient rats when low doses were given, while there was no pharmacokinetics difference between these two types of rats when linagliptin was given at doses higher than 10 mg/kg."

"As it shows, the model estimated linagliptin pharmacokinetics was in good agreement with the observed one in wildtype rats following oral doses."

DISCUSSION

"Reported data, a novel and comprehensive whole-body PBPK-TMDD model was successfully developed, which adequately captured the concentration-time profiles of linagliptin in both plasma and various tissues in both wildtype rats and DPP4-deficient rats."

"Based on the model estimated DPP-4 amount (i.e., R_{max}) in those 8 organs and the literature reported organ volume, we calculated the tissue concentration of DPP-4, and then compared them with the literature reported gene expression level of DPP-4 in those organs; these literature data was obtained using transcriptomics experiment in the rat [211]."

"It is worth emphasizing that linagliptin human pharmacokinetics profiles were simulated based entirely on human physiological parameters (i.e., human organ flow rate and human organ weight), parameters obtained from linagliptin rat PBPK-TMDD model, as well as a few parameters obtained from in vitro experiments (such as DPP-4 amount in human plasma)."

"This model adequately captured the concentration-time profiles of linagliptin in both plasma and various tissues in both wildtype rats and DPP4-deficient rats."

Bibliography

1. Liu E, Zhang M, Huang Y (2016) Pharmacokinetics and pharmacodynamics (PK/PD) of bionanomaterials. Biomed Nanomater:1–60
2. Abdifetah O, Na-Bangchang K (2019) Pharmacokinetic studies of nanoparticles as a delivery system for conventional drugs and herb-derived compounds for cancer therapy: a systematic review. Int J Nanomed:5659–5677
3. Lebreton V, Legeay S, Saulnier P, Lagarce F (2021) Specificity of pharmacokinetic modeling of nanomedicines. Drug Discov Today 26(10):2259–2268
4. Prajapati BG, Paliwal H, Patel JK (2022) Nanoparticle pharmacokinetic profiling in vivo using magnetic resonance imaging. In: Pharmacokinetics and pharmacodynamics of nanoparticulate drug delivery systems. Springer International Publishing, Cham, pp 399–416

5. Feitosa RC, Geraldes DC, Beraldo-de-Araújo VL, Costa JS, Oliveira-Nascimento L (2019) Pharmacokinetic aspects of nanoparticle-in-matrix drug delivery systems for oral/buccal delivery. Front Pharmacol 10:1057
6. Osipova N, Budko A, Maksimenko O, Shipulo E, Vanchugova L, Chen W, Gelperina S, Wacker MG (2023) Comparison of compartmental and non-compartmental analysis to detect biopharmaceutical similarity of intravenous nanomaterial-based rifabutin formulations. Pharmaceutics 15(4):1258
7. Miller HA, Frieboes HB (2019) Pharmacokinetic/pharmacodynamics modeling of drug-loaded PLGA nanoparticles targeting heterogeneously vascularized tumor tissue. Pharm Res 36:1–5
8. Zou H, Banerjee P, Leung SS, Yan X (2020) Application of pharmacokinetic-pharmacodynamic modeling in drug delivery: development and challenges. Front Pharmacol 11:997
9. Joshi M, Dora CP, Kaushik L, Patel J, Raza K (2022) Models used in pharmacodynamic evaluation of nanoparticulate drug delivery systems (NPDDS). In: Pharmacokinetics and pharmacodynamics of nanoparticulate drug delivery systems. Springer International Publishing, Cham, pp 69–77
10. Fröhlich E (2018) Comparison of conventional and advanced in vitro models in the toxicity testing of nanoparticles. Artif Cells Nanomed Biotechnol 46(Suppl 2):1091–1107
11. Moore TL, Urban DA, Rodriguez-Lorenzo L, Milosevic A, Crippa F, Spuch-Calvar M, Balog S, Rothen-Rutishauser B, Lattuada M, Petri-Fink A (2019) Nanoparticle administration method in cell culture alters particle-cell interaction. Sci Rep 9(1):900
12. Gokulan K, Williams K, Orr S, Khare S (2020) Human intestinal tissue explant exposure to silver nanoparticles reveals sex dependent alterations in inflammatory responses and epithelial cell permeability. Int J Mol Sci 22(1):9
13. Tofani LB, Luiz MT, Paes Dutra JA, Abriata JP, Chorilli M (2023) Three-dimensional culture models: emerging platforms for screening the antitumoral efficacy of nanomedicines. Nanomedicine 18(7):633–647
14. Dong C, Ma A, Shang L (2021) Animal models used in the research of nanoparticles for cardiovascular diseases. J Nanopart Res 23(8):172
15. Chakraborty C, Sharma AR, Sharma G, Lee SS (2016) Zebrafish: a complete animal model to enumerate the nanoparticle toxicity. J Nanobiotechnol 14(1):1–3
16. Liu Y, Rohrs J, Wang P (2014) Advances and challenges in the use of nanoparticles to optimize PK/PD interactions of combined anti-cancer therapies. Curr Drug Metab 15(8):818–828
17. Chiarelli PA, Revia RA, Stephen ZR, Wang K, Jeon M, Nelson V, Kievit FM, Sham J, Ellenbogen RG, Kiem HP, Zhang M (2017) Nanoparticle biokinetics in mice and nonhuman primates. ACS Nano 11(9):9514–9524
18. Liu Y, Cheng W, Xin H, Liu R, Wang Q, Cai W, Peng X, Yang F, Xin H (2023) Nanoparticles advanced from preclinical studies to clinical trials for lung cancer therapy. Cancer Nanotechnol 14(1):1–25
19. Palmer BC, DeLouise LA (2016) Nanoparticle-enabled transdermal drug delivery systems for enhanced dose control and tissue targeting. Molecules 21(12):1719
20. Jelliffe R, Bayard D (2018) New perspectives in clinical pharmacokinetics-1: the importance of updating the teaching in pharmacokinetics that both clearance and elimination rate constant approaches are mathematically proven equally valid. AAPS J. https://doi.org/10.1208/s12248-018-0185-x
21. Chen W-C, Huang P-W, Yang W-L, Chen Y-L, Shih Y-N, Wang H-J (2019) Fundamentals of pharmacokinetics to assess the correlation between plasma drug concentrations and different blood sampling methods. Pharm Res. https://doi.org/10.1007/s11095-018-2550-y
22. Hui YH, Huang NH, Ebbert L, Bina H, Chiang A, Maples C et al (2007) Pharmacokinetic comparisons of tail-bleeding with cannula- or retro-orbital bleeding techniques in rats using six marketed drugs. J Pharmacol Toxicol Methods 56(2):256–264
23. Illum L, Hinchcliffe M, Davis SS (2003) The effect of blood sampling site and physicochemical characteristics of drugs on bioavailability after nasal administration in the sheep model. Pharm Res 20(9):1474–1484

24. Chiou WL (1989) The phenomenon and rationale of marked dependence of drug concentration on blood sampling site. Implications in pharmacokinetics, pharmacodynamics, toxicology and therapeutics (part I). Clin Pharmacokinet 17(3):175–199
25. Ahmad M, Qamar-uz-Zaman M, Madni MA, Minhas M, Atif M, Naveed A et al (2011) Pharmacokinetic and bioavailability studies of commercially available simvastatin tablets in healthy and moderately hyperlipidemic human subjects. J Chem Soc Pak 33(1):49–54
26. Peletier LA, de Winter W (2017) Impact of saturable distribution in compartmental PK models: dynamics and practical use. J Pharmacokinet Pharmacodyn. https://doi.org/10.1007/s10928-016-9500-2
27. Wu X, Zhang H, Li J (2022) An analytical approach of one-compartmental pharmacokinetic models with sigmoidal hill elimination. Bull Math Biol. https://doi.org/10.1007/s11538-022-01078-4
28. Wu X, Li J, Nekka F (2015) Closed form solutions and dominant elimination pathways of simultaneous first-order and Michaelis–Menten kinetics. J Pharmacokinet Pharmacodyn 42:151–161
29. Wu X, Nekka F, Li J (2018) Mathematical analysis and drug exposure evaluation of pharmacokinetic models with endogenous production and simultaneous first-order and Michaelis–Menten elimination: the case of single dose. J Pharmacokinet Pharmacodyn 45(5):693–705
30. Wu X, Nekka F, Li J (2019) Analytical solution and exposure analysis of a pharmacokinetic model with simultaneous elimination pathways and endogenous production: the case of multiple dosing administration. Bull Math Biol 81:3436–3459
31. Wu X, Chen M, Li J (2021) Constant infusion case of one compartment pharmacokinetic model with simultaneous first-order and Michaelis–Menten elimination: analytical solution and drug exposure formula. J Pharmacokinet Pharmacodyn 48(4):495–508
32. Beal SL (1982) On the solution to the Michaelis–Menten equation. J Pharmacokinet Biopharm 10:109–119
33. Goličnik M (2011) Explicit reformulations of the Lambert W-omega function for calculations of the solutions to one-compartment pharmacokinetic models with Michaelis-Menten elimination kinetics. Eur J Drug Metab Pharmacokinet 36:121
34. Goličnik M (2012) On the Lambert W function and its utility in biochemical kinetics. Biochem Eng J 63:116–123
35. Tang S, Xiao Y (2007) One-compartment model with Michaelis–Menten elimination kinetics and therapeutic window: an analytical approach. J Pharmacokinet Pharmacodyn 34:807–827
36. Wilkinson PK, Sedman AJ, Sakmar E, Kay DR, Wagner JG (1977) Pharmacokinetics of ethanol after oral administration in the fasting state. J Pharmacokinet Biopharm 5(3):207–224
37. Shou M, Mei Q, Ettore MW Jr, Dai R, Baillie TA, Rushmore TH (1999) Sigmoidal kinetic model for two co-operative substrate-binding sites in a cytochrome P450 3A4 active site: an example of the metabolism of diazepam and its derivatives. Biochem J 340:845–853
38. Yadav J, Korzekwa K, Nagar S (2021) Numerical methods for modeling enzyme kinetics. Methods Mol Biol 2342:147–168
39. Konsil J, Dechasathian S, Mason DH Jr, Stevens RE (2002) Reanalysis of carbamazepine and carbamazepine-epoxide pharmacokinetics after multiple dosing of extended release formulation. J Pharm Pharm Sci 5(2):169–175
40. Wu X, Nekka F, Li J (2016) Steady-state volume of distribution of two-compartment models with simultaneous linear and saturated elimination. J Pharmacokinet Pharmacodyn. https://doi.org/10.1007/s10928-016-9483-z
41. Wang W, Wang EQ, Balthasar JP (2008) Monoclonal antibody pharmacokinetics and pharmacodynamics. Clin Pharmacol Ther 84(5):548–558
42. Kozawa S, Yukawa N, Liu J, Shimamoto A, Kakizaki E, Fujimiya T (2007) Effect of chronic ethanol administration on disposition of ethanol and its metabolites in rat. Alcohol 41(2):87–93
43. Craig M, Humphries AR, Nekka F, Bélair J, Li J, Mackey MC (2015) Neutrophil dynamics during concurrent chemotherapy and G-CSF administration: mathematical modelling guides dose optimisation to minimise neutropenia. J Theor Biol 385:77–89

44. Schmidt H, Radivojevic A (2014) Enhancing population pharmacokinetic modeling efficiency and quality using an integrated workflow. J Pharmacokinet Pharmacodyn 41(4):319–334
45. Scholz M, Engel C, Apt D, Sankar SL, Goldstein E, Loeffler M (2009) Pharmacokinetic and pharmacodynamic modelling of the novel human G-CSF derivative Maxy-G34 and pegfilgrastim in the rat. Cell Prolif 42:823–837
46. Scholz M, Schirm S, Wetzler M, Engel C, Loeffler M (2012) Pharmacokinetic and -dynamic modelling of G-CSF derivatives in humans. Theor Biol Med Model 9:32
47. Woo S, Krzyzanski W, Jusko WJ (2006) Pharmacokinetic and pharmacodynamic modeling of recombinant human erythropoietin after intravenous and subcutaneous administration in rats. J Pharmacol Exp Ther 319(3):1297–1306
48. Wang B, Ludden TM, Cheung EN, Schwab GG, Roskos LK (2001) Population pharmacokinetic–pharmacodynamic modeling of filgrastim (r-metHuG-CSF) in healthy volunteers. J Pharmacokinet Pharmacodyn 28(4):321–342
49. Yates JW, Arundel PA (2008) On the volume of distribution at steady state and its relationship with two-compartmental models. J Pharm Sci 97(1):111–122
50. Shen HW, Jiang XL, Winter JC, Yu AM (2010) Psychedelic 5-methoxy-N, N-dimethyltryptamine: metabolism, pharmacokinetics, drug interactions, and pharmacological actions. Curr Drug Metab 11(8):659–666
51. Yan D, Wu X, Li J, Tang S (2023) Statistical analysis of two-compartment pharmacokinetic models with drug non-adherence. Bull Math Biol. https://doi.org/10.1007/s11538-023-01173-0
52. Li J, Nekka F (2007) A pharmacokinetic formalism explicitly integrating the patient drug compliance. J Pharmacokinet Pharmacodyn 34(1):115–139
53. Song L, Zhang Y, Jiang J, Ren S, Chen L, Liu D, Chen X, Hu P (2018) Development of a physiologically based pharmacokinetic model for Sinogliatin, a first-in-class glucokinase activator, by integrating allometric scaling, in vitro to in vivo exploration and steady-state concentration–mean residence time methods: mechanistic understanding of its pharmacokinetics. Clin Pharmacokinet. https://doi.org/10.1007/s40262-018-0631-z
54. Inzucchi SE, Bergenstal RM, Buse JB, Diamant M, Ferrannini E, Nauck M et al (2015) Management of hyperglycemia in type 2 diabetes, 2015: a patient-centered approach: update to a position statement of the American Diabetes Association and the European Association for the Study of Diabetes. Diabetes Care 38(1):140–149. https://doi.org/10.2337/dc14-2441
55. Toshimoto K, Tomaru A, Hosokawa M, Sugiyama Y (2017) Virtual clinical studies to examine the probability distribution of the AUC at target tissues using physiologically-based pharmacokinetic modeling: application to analyses of the effect of genetic polymorphism of enzymes and transporters on irinotecan induced side effects. Pharm Res. https://doi.org/10.1007/s11095-017-2153-z
56. Yamada A, Maeda K, Kiyotani K, Mushiroda T, Nakamura Y, Sugiyama Y (2014) Kinetic interpretation of the importance of OATP1B3 and MRP2 in docetaxel-induced hematopoietic toxicity. CPT Pharmacometrics Syst Pharmacol 3:e126
57. Teft WA, Welch S, Lenehan J, Parfitt J, Choi YH, Winquist E, Kim RB (2015) OATP1B1 and tumour OATP1B3 modulate exposure, toxicity, and survival after irinotecan-based chemotherapy. Br J Cancer 112(5):857–865
58. van der Bol JM, Loos WJ, de Jong FA, van Meerten E, Konings IR, Lam MH, de Bruijn P, Wiemer EA, Verweij J, Mathijssen RH (2011) Effect of omeprazole on the pharmacokinetics and toxicities of irinotecan in cancer patients: a prospective cross-over drug-drug interaction study. Eur J Cancer 47(6):831–838
59. Sharkey I, Boddy AV, Wallace H, Mycroft J, Hollis R, Picton S (2001) Body surface area estimation in children using weight alone: application in paediatric oncology. Br J Cancer 85:23–28
60. Marsh S, Hoskins JM (2010) Irinotecan pharmacogenomics. Pharmacogenomics 11(7):1003–1010
61. Fujita K, Masuo Y, Okumura H, Watanabe Y, Suzuki H, Sunakawa Y, Shimada K, Kawara K, Akiyama Y, Kitamura M, Kunishima M, Sasaki Y, Kato Y (2016) Increased plasma concen-

trations of unbound SN-38, the active metabolite of irinotecan, in cancer patients with severe renal failure. Pharm Res 33(2):269–282
62. Gupta E, Lestingi TM, Mick R, Ramirez J, Vokes EE, Ratain MJ (1994) Metabolic fate of irinotecan in humans: correlation of glucuronidation with diarrhea. Cancer Res 54(14):3723–3725
63. Calvier EAM, Nguyen TT, Johnson TN, Rostami-Hodjegan A, Tibboel D, Krekels EHJ, Knibbe CAJ (2018) Can population modelling principles be used to identify key PBPK parameters for paediatric clearance predictions? An innovative application of optimal design theory. Pharm Res. https://doi.org/10.1007/s11095-018-2487-1
64. Salem F, Johnson TN, Abduljalil K, Tucker GT, Rostami-Hodjegan A (2014) A re-evaluation and validation of ontogeny functions for cytochrome P450 1A2 and 3A4 based on in vivo data. Clin Pharmacokinet 53(7):625–636
65. Zhao W, Leroux S, Biran V, Jacqz-Aigrain E (2018) Developmental pharmacogenetics of CYP2C19 in neonates and young infants: omeprazole as a probe drug. Br J Clin Pharmacol 28
66. Upreti VV, Wahlstrom JL (2016) Meta-analysis of hepatic cytochrome P450 ontogeny to underwrite the prediction of pediatric pharmacokinetics using physiologically based pharmacokinetic modeling. J Clin Pharmacol 56(3):266–283
67. Janssen JM, Damoiseaux D, van Hasselt JGC, Amant FCH, van Calsteren K, Beijnen JH, Huitema ADR, Dorlo TPC (2023) Semi-physiological enriched population pharmacokinetic modelling to predict the effects of pregnancy on the pharmacokinetics of cytotoxic drugs. Clin Pharmacokinet. https://doi.org/10.1007/s40262-023-01263-1
68. Abduljalil K, Furness P, Johnson TN, Rostami-Hodjegan A, Soltani H (2012) Anatomical, physiological and metabolic changes with gestational age during normal pregnancy a database for parameters required in physiologically based pharmacokinetic modelling. Clin Pharmacokinet 51:365–396
69. Nakai A, Sekiya I, Oya A, Koshino T, Araki T (2002) Assessment of the hepatic arterial and portal venous blood flows during pregnancy with Doppler ultrasonography. Arch Gynecol Obstet 266:25–29
70. Berton M, Bettonte S, Stader F, Battegay M, Marzolini C (2022) Physiologically based pharmacokinetic modelling to identify physiological and drug parameters driving pharmacokinetics in obese individuals. Clin Pharmacokinet. https://doi.org/10.1007/s40262-022-01194-3
71. Berton M, Bettonte S, Stader F, Battegay M, Marzolini C (2022) Repository describing the anatomical, physiological, and biological changes in an obese population to inform physiologically based pharmacokinetic models. Clin Pharmacokinet 61(9):1251–1270
72. Berton M, Bettonte S, Stader F, Battegay M, Marzolini C (2022) Repository describing the anatomical, physiological, and biological changes in an obese population to inform physiologically based pharmacokinetic models. Clin Pharmacokinet. (in press)
73. Ulvestad M, Skottheim IB, Jakobsen GS, Bremer S, Molden E, Asberg A et al (2013) Impact of OATP1B1, MDR1, and CYP3A4 expression in liver and intestine on interpatient pharmacokinetic variability of atorvastatin in obese subjects. Clin Pharmacol Ther 93(3):275–282
74. Krogstad V, Peric A, Robertsen I, Kringen MK, Vistnes M, Hjelmesaeth J et al (2021) Correlation of body weight and composition with hepatic activities of cytochrome P450 enzymes. J Pharm Sci 110(1):432–437
75. Han AN, Han BR, Zhang T, Heimbach T (2021) Hepatic impairment physiologically based pharmacokinetic model development: current challenges. Curr Pharmacol Rep. https://doi.org/10.1007/s40495-021-00266-5
76. •• Heimbach T et al (2020) Physiologically-based pharmacokinetic modeling in renal and hepatic impairment populations: a pharmaceutical industry perspective. Clin Pharmacol Ther 110:297–310. (This paper evaluated the predictive accuracy of current HI PBPK models across a range of compounds. It suggests that for moderate to severe cases of HI, existing models tend to overpredict exposure because they do not account for changes in absorption induced by HI.)
77. •• Assessing Changes in Pharmacokinetics of Drugs in Liver Disease, October 8, 2020. This workshop evaluated the impact of hepatic impairment on drug pharmacokinetics. It

reviewed the shortcomings of existing hepatic impairment PBPK models. https://www.fda.gov/drugs/news-events-human-drugs/assessing-changes-pharmacokinetics-drugs-liver--disease-10082020-10082020. Accessed 11 Jul 2021
78. Chang Y et al (2013) Evaluation of hepatic impairment dosing recommendations in FDA-approved product labels. J Clin Pharmacol 53(9):962–966
79. Food and drug administration, pharmacokinetics in patients with impaired hepatic function: study design, data analysis, and impact on dosing and labeling (2003). https://www.fda.gov/regulatory-information/search-fda-guidance-documents/pharmacokinetics-patients-impaired-hepatic-function-study-design-data-analysis-and-impact-dosing-and. Accessed 11 Jul 2021
80. Guideline on the evaluation of the pharmacokinetics of medicinal products in patients with impaired hepatic function (2005). https://www.emaeuropaeu/en/documents/scientific--guideline/guideline-evaluation-pharmacokinetics-medicinal-products-patients-impaired-hepatic-function_enpdf. Accessed 11 Jul 2021
81. Abduljalil K, Pan X, Pansari A, Jamei M, Johnson TN (2019) A preterm physiologically based pharmacokinetic model. Part I: Physiological parameters and model building. Clin Pharmacokinet. https://doi.org/10.1007/s40262-019-00825-6
82. Abduljalil K, Jamei M, Johnson TN (2019) Fetal Physiologically based pharmacokinetic models: systems information on the growth and composition of fetal organs. Clin Pharmacokinet 58:235–262
83. Abduljalil K, Johnson TN, Rostami-Hodjegan A (2018) Fetal physiologically-based pharmacokinetic models: systems information on fetal biometry and gross composition. Clin Pharmacokinet 57:1149–1171
84. Rhodin MM, Anderson BJ, Peters AM, Coulthard MG, Wilkins B, Cole M et al (2009) Human renal function maturation: a quantitative description using weight and postmenstrual age. Pediatr Nephrol 24:67–76
85. Sobiak J, Resztak M (2021) A systematic review of multiple linear regression-based limited sampling strategies for mycophenolic acid area under the concentration–time curve estimation. Eur J Drug Metab Pharmacokinet. https://doi.org/10.1007/s13318-021-00713-0
86. Staatz CE, Tett SE (2014) Pharmacology and toxicology of mycophenolate in organ transplant recipients: an update. Arch Toxicol 88:1351–1389
87. Abd Rahman AN, Tett SE, Staatz CE (2014) How accurate and precise are limited sampling strategies in estimating exposure to mycophenolic acid in people with autoimmune disease? Clin Pharmacokinet 53:227–245
88. Filler G, Alvarez-Elías AC, McIntyre C, Medeiros M (2017) The compelling case for therapeutic drug monitoring of mycophenolate mofetil therapy. Pediatr Nephrol 32:21–29
89. Saint-Marcoux F, Guigonis V, Decramer S, Gandia P, Ranchin B, Parant F et al (2011) Development of a Bayesian estimator for the therapeutic drug monitoring of mycophenolate mofetil in children with idiopathic nephrotic syndrome. Pharmacol Res 63:423–431
90. Bruchet NK, Ensom MH (2009) Limited sampling strategies for mycophenolic acid in solid organ transplantation: a systematic review. Expert Opin Drug Metab Toxicol 5:1079–1097
91. Ting LSL, Partovi N, Levy RD, Ignaszewski AP, Ensom MHH (2008) Performance of limited sampling strategies for predicting mycophenolic acid area under the curve in thoracic transplant recipients. J Heart Lung Transplant 27:325–328
92. Staatz CE, Tett SE (2007) Clinical pharmacokinetics and pharmacodynamics of mycophenolate in solid organ transplant recipients. Clin Pharmacokinet 46:13–58
93. Hughes JH, Upton RN, Foster DJR (2017) Comparison of non-compartmental and mixed effect modelling methods for establishing bioequivalence for the case of two compartment kinetics and censored concentrations. J Pharmacokinet Pharmacodyn. https://doi.org/10.1007/s10928-017-9511-7
94. Dubois A, Gsteiger S, Pigeolet E, Mentre F (2010) Bioequivalence tests based on individual estimates using non-compartmental or model-based analyses: evaluation of estimates of

sample means and type I error for different designs. Pharm Res 27(1):92–104. https://doi.org/10.1007/s11095-009-9980-5
95. Panhard X, Mentré F (2005) Evaluation by simulation of tests based on non-linear mixed-effects models in pharmacokinetic interaction and bioequivalence cross-over trials. Stat Med 24(10):1509–1524. https://doi.org/10.1002/sim.2047
96. Pentikis HS, Henderson JD, Tran NL, Ludden TM (1996) Bioequivalence: individual and population compartmental modeling compared to the noncompartmental approach. Pharm Res 13(7):1116–1121. https://doi.org/10.1023/A:1016083429903
97. Beal S, Sheiner LB, Boeckmann A, Bauer RJ (2009) NONMEM user's guides (1989–2009). Icon Development Solutions, Ellicott City
98. Jackson AJ, Foehl HC (2022) A simulation study of the comparative performance of partial area under the curve (pAUC) and partial area under the effect curve (pAUEC) metrics in crossover versus replicated crossover bioequivalence studies for concerta and ritalin la. AAPS J. https://doi.org/10.1208/s12248-022-00726-w
99. Chen ML, Lesko L, Williams RL (2001) Measures of exposure versus measures of rate and extent of absorption. Clin Pharmacokinet 40:565–572
100. SAS Help Center: PROC POWER Statement
101. Methylphenidate hydrochloride (fda.gov)
102. Proost JH, Eleveld DJ, Struys MMRF (2020) Population pharmacodynamic modeling using the sigmoid E_{max} model: influence of inter-individual variability on the steepness of the concentration–effect relationship. A simulation study. AAPS J. https://doi.org/10.1208/s12248-020-00549-7
103. Hannivoort LN, Vereecke HEM, Proost JH, Heyse BE, Eleveld DJ, Bouillon TW et al (2016) Probability to tolerate laryngoscopy and noxious stimulation response index as general indicators of the anaesthetic potency of sevoflurane, propofol, and remifentanil. Br J Anaesth 116(5):624–631. https://doi.org/10.1093/bja/aew060
104. Bouillon TW, Bruhn J, Radulescu L, Andresen C, Shafer TJ, Cohane C et al (2004) Pharmacodynamic interaction between propofol and remifentanil regarding hypnosis, tolerance of laryngoscopy, bispectral index, and electroencephalographic approximate entropy. Anesthesiology 100:1353–1372
105. Liu X, Jones BL, Roberts JK, Sherwin CM (2016) Population pharmacokinetic/pharmacodynamic modeling of histamine response measured by histamine iontophoresis laser Doppler. J Pharmacokinet Pharmacodyn. https://doi.org/10.1007/s10928-016-9478-9
106. Simons FE, Simons KJ (2005) Levocetirizine: pharmacokinetics and pharmacodynamics in children age 6 to 11 years. J Allergy Clin Immunol 116(2):355–361. https://doi.org/10.1016/j.jaci.2005.04.010
107. Jones BL, Kearns G, Neville KA, Sherwin CM, Spigarelli MM, Leeder JS (2013) Variability of histamine pharmacodynamic response in children with allergic rhinitis. J Clin Pharmacol 53(7):731–737. https://doi.org/10.1002/jcph.93
108. Jones BL, Abdel-Rahman SM, Simon SD, Kearns GL, Neville KA (2009) Assessment of histamine pharmacodynamics by microvasculature response of histamine using histamine iontophoresis laser Doppler flowmetry. J Clin Pharmacol 49(5):600–605. https://doi.org/10.1177/0091270009332247
109. Deschamps C, Dubruc C, Mentre F, Rosenzweig P (2000) Pharmacokinetic and pharmacodynamic modeling of mizolastine in healthy volunteers with an indirect response model. Clin Pharmacol Ther 68(6):647–657. https://doi.org/10.1067/mcp.2000.112341
110. Kong AN, Ludwig EA, Slaughter RL, DiStefano PM, DeMasi J, Middleton E Jr, Jusko WJ (1989) Pharmacokinetics and pharmacodynamic modeling of direct suppression effects of methylprednisolone on serum cortisol and blood histamine in human subjects. Clin Pharmacol Ther 46(6):616–628
111. Abelo A, Holstein B, Eriksson UG, Gabrielsson J, Karlsson MO (2002) Gastric acid secretion in the dog: a mechanism-based pharmacodynamic model for histamine stimulation and irreversible inhibition by omeprazole. J Pharmacokinet Pharmacodyn 29(4):365–382

112. Casti JL (1994) Simple and complex models in science. SFI working paper (1996-06-034)
113. Kratochwil NA, Meille C, Fowler S, Klammers F, Ekiciler A, Molitor B, Simon S, Walter I, McGinnis C, Walther J, Leonard B, Triyatni M, Javanbakht H, Funk C, Schuler F, Lavé T, Parrott NJ (2017) Metabolic profiling of human long-term liver models and hepatic clearance predictions from in vitro data using nonlinear mixed effects modeling. AAPS J. https://doi.org/10.1208/s12248-016-0019-7
114. Di L (2014) The role of drug metabolizing enzymes in clearance. Expert Opin Drug Metab Toxicol 10:379–393
115. Argikar UA, Potter PM, Hutzler JM, Marathe PH (2016) Challenges and opportunities with non-CYP enzymes aldehyde oxidase, carboxylesterase, and UDP-glucuronosyltransferase: focus on reaction phenotyping and prediction of human clearance. AAPS J 18:1391–1405. https://doi.org/10.1208/s12248-016-9962-6
116. Poulin P (2013) Prediction of total hepatic clearance by combining metabolism, transport, and permeability data in the in vitro–in vivo extrapolation methods: emphasis on an apparent fraction unbound in liver for drugs. J Pharm Sci 102:2085–2095
117. Reijers JAA, van Donge T, Schepers FML, Burggraaf J, Stevens J (2016) Use of population approach non-linear mixed effects models in the evaluation of biosimilarity of monoclonal antibodies. Eur J Clin Pharmacol. https://doi.org/10.1007/s00228-016-2101-6
118. Tessier A, Bertrand J, Chenel M, Comets E (2015) Comparison of nonlinear mixed effects models and noncompartmental approaches in detecting pharmacogenetic covariates. AAPS J. https://doi.org/10.1208/s12248-015-9726-8
119. EMA (2012) Guideline on the use of pharmacogenetic methodologies in the pharmacokinetic evaluation of medicinal products. Report No.: EMA/CHMP/37646/2009
120. FDA (2007) Guidance for Industry and FDA staff: pharmacogenetic tests and genetic tests for heritable markers
121. Hoggart CJ, Whittaker JC, De Iorio M, Balding DJ (2008) Simultaneous analysis of all SNPs in genome-wide and re-sequencing association studies. PLoS Genet 4(7):e1000130
122. Bertrand J, Balding DJ (2013) Multiple single nucleotide polymorphism analysis using penalized regression in nonlinear mixed-effect pharmacokinetic models. Pharmacogenet Genomics 23(3):167–174
123. Comets E, Mentré F (2021) Developing tools to evaluate non-linear mixed effect models: 20 years on the npde adventure. AAPS J. https://doi.org/10.1208/s12248-021-00597-7
124. Sheiner LB, Beal SL (1981) Some suggestions for measuring predictive performance. J Pharmacokinet Biopharm 9:503–512
125. European Medicines Agency (2007) Guideline on reporting the results of population pharmacokinetic analysis (CHMP). http://www.ema.europa.eu/docs/en_GB/document_library/Scientific_guideline/2009/09/WC500003067.pdf
126. Food and Drug Administration (2019) Guidance for industry exposure-response relationships– study design, data analysis, and regulatory applications. https://www.fda.gov/media/128793/download
127. Mesnil F, Mentré F, Dubruc C, Thénot JP, Mallet A (1998) Population pharmacokinetic analysis of mizolastine and validation from sparse data on patients using the nonparametric maximum likelihood method. J Pharmacokinet Pharmacodyn 26(2):133–161
128. Hooker AC, Staatz CE, Karlsson MO (2007) Conditional weighted residuals (CWRES): a model diagnostic for the FOCE method. Pharm Res 24(12):2187–2197
129. Nyberg J, Bauer RJ, Hooker AC (2010) Investigations of the weighted residuals in NONMEM 7. PAGE 10, Abstr 1883
130. Brendel K, Comets E, Laffont C, Laveille C, Mentré F (2006) Metrics for external model evaluation with an application to the population pharmacokinetics of gliclazide. Pharm Res 23:2036–2049
131. Savic R, Barrail-Tran A, Duval X, Nembot G, Panhard X, Descamps D et al (2012) Effect of adherence as measured by MEMS, ritonavir boosting, and CYP3A5 genotype on atazanavir pharmacokinetics in treatment-naive HIV-infected patients. Clin Pharmacol Ther 92:575–583

132. Mentré F, Escolano S (2006) Prediction discrepancies for the evaluation of nonlinear mixed-effects models. J Pharmacokinet Pharmacodyn 33:345–367
133. Brendel K, Comets E, Laffont C, Mentré F (2010) Evaluation of different tests based on observations for external model evaluation of population analyses. J Pharmacokinet Pharmacodyn 37:49–65
134. Comets E, Brendel K, Mentré F (2008) Computing normalised prediction distribution errors to evaluate nonlinear mixed-effect models: the npde add-on package for R. Comput Methods Prog Biomed 90:154–166
135. Nguyen TH, Mouksassi MS, Holford N, Al-Huniti N, Freedman I, Hooker AC, John J, Karlsson MO, Mould DR, Perez Ruixo JJ et al (2017) Model evaluation of continuous data pharmacometric models: metrics and graphics. CPT Pharmacometrics Syst Pharmacol 6:87–109. https://doi.org/10.1002/psp4.12161
136. Brendel K, Dartois C, Comets E, Lemmenuel-Diot A, Laveille C, Tranchand B et al (2007) Are population PK and/or PD models adequately evaluated? A 2002 to 2004 literature survey. Clin Pharmacokinet 46:221–234
137. Dartois C, Brendel K, Comets E, Laffont C, Laveille C, Tranchand B et al (2007) Overview of model building strategies in population PK/PD analyses: 2002 to 2004 literature survey. Br J Clin Pharmacol 64:603–612
138. Choi L, Caffo BS, Kohli U, Pandharipande P, Kurnik D, Ely EW, Stein C, Michael A (2011) Bayesian hierarchical nonlinear mixture model in the presence of artifactual outliers in a population pharmacokinetic study. J Pharmacokinet Pharmacodyn. https://doi.org/10.1007/s10928-011-9211-7
139. Pandharipande PP, Pun BT, Herr DL, Maze M, Girard TD, Miller RR, Shintani AK, Thompson JL, Jackson JC, Deppen SA, Stiles RA, Dittus RS, Bernard GR, Ely EW (2007) Effect of sedation with dexmedetomidine vs lorazepam on acute brain dysfunction in mechanically ventilated patients: the mends randomized controlled trial. JAMA J Am Med Assoc 298(22):2644–2653
140. Dutta S, Lal R, Karol MD, Cohen T, Ebert T (2000) Influence of cardiac output on dexmedetomidine pharmacokinetics. J Pharm Sci 89(4):519–527
141. Dyck JB, Maze M, Haack C, Vuorilehto L, Shafer SL (1993) The pharmacokinetics and hemodynamic effects of intravenous and intramuscular dexmedetomidine hydrochloride in adult human volunteers. Anesthesiology 78(5):813–820
142. Kurnik D, Muszkat M, Sofowora GG, Friedman EA, Dupont WD, Scheinin M, Wood AJ, Stein CM (2008) Ethnic and genetic determinants of cardiovascular response to the selective alpha 2-adrenoceptor agonist dexmedetomidine. Hypertension 51(2):406–411
143. Kohli U, Muszkat M, Sofowora GG, Harris PA, Friedman EA, Dupont WD, Scheinin M, Wood AJ, Stein CM, Kurnik D (2010) Effects of variation in the human alpha2a- and alpha2c-adrenoceptor genes on cognitive tasks and pain perception. Eur J Pain 14(2):154–159
144. Su JT, Teller RS, Srinivasan P, Zhang J, Martin A, Sung S, Smith JM, Kiser PF (2017) A dose ranging pharmacokinetic evaluation of IQP-0528 released from intravaginal rings in non-human primates. Pharm Res. https://doi.org/10.1007/s11095-017-2224-1
145. Srinivasan P, Zhang J, Martin A, Kelley K, McNicholl JM, Buckheit RW Jr et al (2016) Safety and pharmacokinetics of quick dissolving polymeric vaginal films delivering the antiretroviral IQP-0528 for pre-exposure prophylaxis. Antimicrob Agents Chemother
146. Pereira LE, Mesquita PM, Ham A, Singletary T, Deyounks F, Martin A et al (2015) Pharmacokinetic and pharmacodynamic evaluation following vaginal application of IQB3002, a dual-chamber microbicide gel containing the nonnucleoside reverse transcriptase inhibitor IQP-0528 in Rhesus Macaques. Antimicrob Agents Chemother 60(3):1393–1400
147. Johnson TJ, Srinivasan P, Albright TH, Watson-Buckheit K, Rabe L, Martin A et al (2012) Safe and sustained vaginal delivery of pyrimidinedione HIV-1 inhibitors from polyurethane intravaginal rings. Antimicrob Agents Chemother 56(3):1291–1299
148. Hartman TL, Yang L, Buckheit RW Jr (2011) Antiviral interactions of combinations of highly potent 2,4(1H,3H)-pyrimidinedione congeners and other anti-HIV agents. Antivir Res 92(3):505–508

149. Chien YW, Lambert HJ (1974) Controlled drug release from polymeric delivery devices II: differentiation between partition-controlled and matrix-controlled drug release mechanisms. J Pharm Sci 63(4):515–519
150. Malcolm KMC, Woolfson D (2008) Correlation between in vitro - in vivo release rates of the antiretroviral candidate, dapivirine, from silicone elastomer vaginal rings, in British Pharmaceutical Conference. Manchester, UK
151. Romano J, Variano B, Coplan P, Van Roey J, Douville K, Rosenberg Z et al (2009) Safety and availability of dapivirine (TMC120) delivered from an intravaginal ring. AIDS Res Hum Retrovir 25(5):483–488
152. Nel A, Bekker LG, Bukusi E, Hellstrm E, Kotze P, Louw C et al (2016) Safety, acceptability and adherence of dapivirine vaginal ring in a microbicide clinical trial conducted in multiple countries in sub-Saharan Africa. PLoS One 11(3):e0147743
153. Bodenlenz M, Augustin T, Birngruber T, Tiffner KI, Boulgaropoulos B, Schwingenschuh S, Raney SG, Rantou E, Sinner F (2020) Variability of skin pharmacokinetic data: insights from a topical bioequivalence study using dermal open flow microperfusion. Pharm Res. https://doi.org/10.1007/s11095-020-02920-x
154. Benfeldt E, Hansen SH, Vølund A, Menné T, Shah VP (2007) Bioequivalence of topical formulations in humans: evaluation by dermal microdialysis sampling and the dermatopharmacokinetic method. J Invest Dermatol 127(1):170–178
155. Kreilgaard M, Kemme MJB, Burggraaf J, Schoemaker RC, Cohen AF (2001) Influence of a microemulsion vehicle on cutaneous bioequivalence of a lipophilic model drug assessed by microdialysis and pharmacodynamics. Pharm Res 18(5):593–599
156. Simonsen L, Jørgensen A, Benfeldt E, Groth L (2004) Differentiated in vivo skin penetration of salicylic compounds in hairless rats measured by cutaneous microdialysis. Eur J Pharm Sci 21(2–3):379–388
157. Bodenlenz M, Tiffner KI, Raml R, Augustin T, Dragatin C, Birngruber T et al (2016) Open flow microperfusion as a dermal pharmacokinetic approach to evaluate topical bioequivalence. Clin Pharmacokinet 56(1):99
158. Eirefelt S, Hummer J, Basse LH, Bertelsen M, Johansson F, Birngruber T, Sinner F, Larsen J, Nielsen SF, Lambert M (2020) Evaluating dermal pharmacokinetics and pharmacodymanic effect of soft topical PDE4 inhibitors: open flow microperfusion and skin biopsies. Pharm Res. https://doi.org/10.1007/s11095-020-02962-1
159. Clinical Study Report Synopsis (2011) LEO 29102 Cream in the treatment of atopic dermatitis. A Phase 2, proof of concept and dose finding study, investigating treatment efficacy of LEO 29102 cream (2.5 mg/g, 1.0 mg/g, 0.3 mg/g, 0.1 mg/g, 0.03 mg/g), LEO 29102 cream vehicle, and Elidel® (pimecrolimus) cream 10 mg/g, after cutaneous administration twice daily for 4 weeks. https://www.clinicaltrialsregister.eu/ctr-search/rest/download/result/attachment/2009-013792-22/1/17100. Accessed 1 May 2020
160. Clinical Study Report (2015) An explorative trial evaluating the effect of LEO 39652 Cream 2.5 mg/g in adults with mild to moderate atopic dermatitis (AD). A phase I, single-centre, prospective, randomised, vehicle-controlled, double-blind, intraindividual comparison trial with twice daily topical administration for 3 weeks. https://www.leoclinicaltrials.com/Frontpage/Our%20clinical%20trials/Trial%20page/?lpNumber=LP0083-1085. Accessed 1 May 2020
161. Guerrero YA, Desai D, Sullivan C, Kindt E, Spilker ME, Maurer TS, Solomon DE, Bartlett DW (2020) A microfluidic perfusion platform for in vitro analysis of drug pharmacokinetic-pharmacodynamic (PK-PD) relationships. AAPS J. https://doi.org/10.1208/s12248-020-0430-y
162. Yamazaki S (2012) Translational pharmacokinetic-pharmacodynamic modeling from nonclinical to clinical development: a case study of anticancer drug, crizotinib. AAPS J 15(2):354–366
163. Spilker ME, Chen X, Visswanathan R, Vage C, Yamazaki S, Li G et al (2017) Found in translation: maximizing the clinical relevance of nonclinical oncology studies. Clin Cancer Res 23(4):1080–1090

164. Yamazaki S, Spilker ME, Vicini P (2016) Translational modeling and simulation approaches for molecularly targeted small molecule anticancer agents from bench to bedside. Expert Opin Drug Metab Toxicol 12(3):253–265
165. Ande A, Vaidya TR, Tran BN, Vicchiarelli M, Brown AN, Ait-Oudhia S (2018) Utility of a novel three-dimensional and dynamic (3DD) cell culture system for PK/PD studies: evaluation of a triple combination therapy at overcoming anti-HER2 treatment resistance in breast cancer. Front Pharmacol 9:403
166. Wu J, Racine F, Wismer MK, Young K, Carr DM, Xiao JC et al (2018) Exploring the pharmacokinetic/pharmacodynamic relationship of relebactam (MK-7655) in combination with imipenem in a hollow-fiber infection model. Antimicrob Agents Chemother 62(5)
167. Ishisaka T, Kishi S, Okura K, Horikoshi M, Yamashita T, Mitsuke Y et al (2006) A precise pharmacodynamic study showing the advantage of a marked reduction in cardiotoxicity in continuous infusion of doxorubicin. Leuk Lymphoma 47(8):1599–1607
168. Vogus DR, Pusuluri A, Chen R, Mitragotri S (2018) Schedule dependent synergy of gemcitabine and doxorubicin: improvement of in vitro efficacy and lack of in vitro-in vivo correlation. Bioeng Transl Med 3(1):49–57
169. Incecayir T, Agabeyoglu I, Derici U, Sindel S (2011) Assessment of topical bioequivalence using dermal microdialysis and tape stripping methods. Pharm Res. https://doi.org/10.1007/s11095-011-0444-3
170. Holmgaard R, Nielsen JB, Benfeldt E (2010) Microdialysis sampling for investigations of bioavailability and bioequivalence of topically administered drugs: current state and future perspectives. Skin Pharmacol Physiol 23(5):225–243
171. Rowland M, Tozer TN (1980) Clinical pharmacokinetics: concept and applications. Lea&Febiger, Philadelphia
172. García Ortiz P, Hansen SH, Shah VP, Sonne J, Benfeldt E (2011) Are marketed topical metronidazole creams bioequivalent? Evaluation by in vivo microdialysis sampling and tape stripping methodology. Skin Pharmacol Physiol 24(1):44–53
173. Desai PR, Shah PP, Patlolla RR, Singh M (2012) Dermal microdialysis technique to evaluate the trafficking of surface-modified lipid nanoparticles upon topical application. Pharm Res. https://doi.org/10.1007/s11095-012-0789-2
174. Mathy FX, Ntivunwa D, Verbeeck RK, Preat V (2005) Fluconazole distribution in rat dermis following intravenous and topical application: a microdialysis study. J Pharm Sci 94:770–780
175. Mathy FX, Denet AR, Vroman B, Clarys P, Barel A, Verbeeck RK et al (2003) In vivo tolerance assessment of skin after insertion of subcutaneous and cutaneous microdialysis probes in the rat. Skin Pharmacol Appl Ski Physiol 16:18–27
176. Holmgaard R, Benfeldt E, Nielsen JB, Gatschelhofer C, Sorensen JA, Höfferer C, Bodenlenz M, Pieber TR, Sinner F (2012) Comparison of open-flow microperfusion and microdialysis methodologies when sampling topically applied fentanyl and benzoic acid in human dermis ex vivo. Pharm Res. https://doi.org/10.1007/s11095-012-0705-9
177. Carneheim C, Stahle L (1991) Microdialysis of lipophilic compounds – a methodological study. Pharmacol Toxicol 69:378–380
178. Holmgaard R, Benfeldt E, Bangsgaard N, Sorensen JA, Brosen K, Nielsen F et al (2012) Probe depth matters in dermal microdialysis sampling of topical penetration. An ex vivo study in human skin. Skin Pharmacol Physiol 25:9–16
179. Azeredo FJ, Dalla Costa T, Derendorf H (2014) Role of microdialysis in pharmacokinetics and pharmacodynamics: current status and future directions. Clin Pharmacokinet. https://doi.org/10.1007/s40262-014-0131-8
180. Plock N, Kloft C (2005) Microdialysis-theoretical background and recent implementation in applied life-sciences. Eur J Pharm Sci 25:1–24
181. Muller M, De La Pena A, Derendorf H (2004) Issues in pharmacokinetics and pharmacodynamics of anti-infective agents: distribution in tissue. Antimicrob Agents Chemother 48:1441–1453
182. Ryan DM (1993) Pharmacokinetics of antibiotics in natural and experimental superficial compartments in animals and humans. Antimicrob Chemother 31 Suppl D:1–16

183. Chaurasia C, Müller M, Bashaw ED et al (2007) AAPS-FDA workshop white paper: microdialysis principles, application, and regulatory perspectives. J Clin Pharmacol 47:589–603
184. Brunner M, Muller M (2002) Microdialysis: an in vivo approach for measuring drug delivery in oncology. Eur J Clin Pharmacol 58(4):227–234
185. Blakeley JO, Olson J, Grossman SA et al (2009) Effect of blood brain barrier permeability in recurrent high grade gliomas on the intratumoral pharmacokinetics of methotrexate: a microdialysis study. J Neuro-Oncol 91:51–58
186. Tettey-Amlalo RNO, Kanfer I, Skinner MF, Benfeldt E, Verbeeck RK (2009) Application of dermal microdialysis for the evaluation of bioequivalence of a ketoprofen topical gel. Eur J Pharm Sci 36(2–3):219–225
187. Brunner M, Davies D, Martin W et al (2011) A new topical formulation enhances relative diclofenac bioavailability in healthy male subjects. Br J Clin Pharmacol 71(6):852–859
188. Garcia Ortiz P, Hansen SH, Shah VP, Menné T, Benfeldt E (2009) Impact of adult atopic dermatitis on topical drug penetration: assessment by cutaneous microdialysis and tape stripping. Acta Derm Venereol 89(1):33–38
189. Ortiz PG, Hansen SH, Shah VP, Menne T, Benfeldt E (2008) The effect of irritant dermatitis on cutaneous bioavailability of a metronidazole formulation, investigated by microdialysis and dermatopharmacokinetic method. Contact Derm 59:23–30
190. Brody DL, Magnoni S, Schwetye KE et al (2008) Amyloid-β dynamics correlate with neurological status in the injured human brain. Science 321(5893):1221–1224
191. Rooyackers O, Blixt C, Mattsson P, Wernerman J (2010) Continuous glucose monitoring by intravenous microdialysis. Acta Anaesthesiol Scand 54(7):841–847
192. Lehman PA, Franz TJ (2014) Assessing topical bioavailability and bioequivalence: a comparison of the in vitro permeation test and the vasoconstrictor assay. Pharm Res. https://doi.org/10.1007/s11095-014-1439-7
193. Franz TJ, Lehman PA, Raney SG (2008) The cadaver skin absorption model and the drug development process. Pharmacopeial Forum 34(5):1349–1356
194. Tfayli A, Piot O, Pitre F, Manfait M (2007) Follow-up of drug permeation through excised human skin with confocal Raman microspectroscopy. Eur Biophys J. https://doi.org/10.1007/s00249-007-0191-x
195. Hofland HE, Bouwstra JA, Bodde HE, Spies F, Junginger HE (1995) Interactions between liposomes and human stratum corneum in vitro: freeze fracture electron microscopical visualization and small angle X-ray scattering studies. Br J Dermatol 132:853–866
196. Bouwstra JA, de Vries MA, Gooris GS, Bras W, Brussee J, Ponec M (1991) Thermodynamic and structural aspects of the skin barrier. J Control Release 15:209–219
197. Grams YY, Whitehead L, Cornwell P, Bouwstra JA (2004) On-line visualization of dye diffusion in fresh unfixed human skin. Pharm Res 21:851–859
198. Grams YY, Whitehead L, Cornwell P, Bouwstra JA (2004) Time and depth resolved visualisation of the diffusion of a lipophilic dye into the hair follicle of fresh unfixed human scalp skin. J Control Release 98:367–378
199. Veiro JA, Cummins PG (1994) Imaging of skin epidermis from various origins using confocal laser scanning microscopy. Clin Lab Invest 189:16–22
200. Yu B, Kim KH, So PT, Blankschtein D, Langer R (2003) Visualization of oleic acid-induced transdermal diffusion pathways using two-photon fluorescence microscopy. J Invest Dermatol 120:448–455
201. Caspers PJ, Lucassen GW, Carter EA, Bruining HA, Puppels GJ (2001) In vivo confocal Raman microspectroscopy of the skin: noninvasive determination of molecular concentration profiles. J Invest Dermatol 116:434–442
202. Caspers PJ, Lucassen GW, Puppels GJ (2003) Combined in vivo confocal Raman spectroscopy and confocal microscopy of human skin. Biophys J 85:572–580
203. Chrit L, Hadjur C, Morel S, Sockalingum G, Lebourdon G, Leroy F, Manfait M (2005) In vivo chemical investigation of human skin using a confocal Raman fiber optic microprobe. J Biomed Opt 10:44007

204. Sieg A, Crowther J, Blenkiron P, Marcott C, Matts PJ (2006) Confocal Raman microspectroscopy—measuring the effects of topical moisturisers on stratum corneum water gradients in vivo. In: Mahadevan-Jansen A, Petrich WH (eds) The international society of optical engineering
205. Wu N, An G (2020) Incorporating pharmacological target-mediated drug disposition (TMDD) in a whole-body physiologically based pharmacokinetic (PBPK) model of linagliptin in rat and scale-up to human. AAPS J. https://doi.org/10.1208/s12248-020-00481-w
206. Horie Y, Kanada S, Watada H, Sarashina A, Taniguchi A, Hayashi N et al (2011) Pharmacokinetic, pharmacodynamic, and tolerability profiles of the dipeptidyl peptidase-4 inhibitor linagliptin: a 4-week multicenter, randomized, double-blind, placebo-controlled phase IIa study in Japanese type 2 diabetes patients. Clin Ther 33(7):973–989
207. Retlich S, Duval V, Ring A, Staab A, Hüttner S, Jungnik A et al (2010) Pharmacokinetics and pharmacodynamics of single rising intravenous doses (0.5 mg–10mg) and determination of absolute bioavailability of the dipeptidyl peptidase-4 inhibitor linagliptin (BI 1356) in healthy male subjects. Clin Pharmacokinet 49(12):829–840
208. Heise T, Graefe-Mody E, Hüttner S, Ring A, Trommeshauser D, Dugi K (2009) Pharmacokinetics, pharmacodynamics and tolerability of multiple oral doses of linagliptin, a dipeptidyl peptidase-4 inhibitor in male type 2 diabetes patients. Diabetes Obes Metab 11(8):786–794
209. Hüttner S, Graefe-Mody E, Withopf B, Ring A, Dugi K (2008) Safety, tolerability, pharmacokinetics, and pharmacodynamics of single oral doses of BI 1356, an inhibitor of dipeptidyl peptidase 4, in healthy male volunteers. J Clin Pharmacol 48(10):1171–1178
210. Retlich S, Withopf B, Greischel A, Staab A, Jaehde U, Fuchs H (2009) Binding to dipeptidyl peptidase-4 determines the disposition of linagliptin (BI 1356)–investigations in DPP-4 deficient and wildtype rats. Biopharm Drug Dispos 30(8):422–436
211. Yu Y, Fuscoe JC, Zhao C, Guo C, Jia M, Qing T et al (2014) A rat RNA-Seq transcriptomic body map across 11 organs and 4 developmental stages. Nat Commun 5(1):1–11
212. HepaRG® (2016). http://www.heparg.com/. Accessed 16 Aug 2016
213. HepatoPac® (Ascendance Corporation) (2016). http://www.hepregen.com/products/human--hepatopac. Accessed 16 Aug 2016
214. Hμrel® Corporation (2016). http://hurelcorp.com/technology/hurel-hepatic-co-cultures/. Accessed 16 Aug 2016
215. Di L, Atkinson K, Orozco CC, Funk C, Zhang H, McDonald TS et al (2013) In vitro-in vivo correlation for low-clearance compounds using hepatocyte relay method. Drug Metab Dispos 41:2018–2023
216. Zamek-Gliszczynski MJ, Ruterbories KJ, Ajamie RT, Wickremsinhe ER, Pothuri L, Rao MVS et al (2011) Validation of 96-well equilibrium dialysis with non-radiolabeled drug for definitive measurement of protein binding and application to clinical development of highly-bound drugs. J Pharm Sci 100:2498–2507
217. Banker MJ, Clark TH, Williams JA (2003) Development and validation of 96-well equilibrium dialysis apparatus for measuring plasma protein binding. J Pharm Sci 92:967–974
218. Liston DR, Davis M (2017) Clinically relevant concentrations of anticancer drugs: a guide for nonclinical studies. Clin Cancer Res 23(14):3489–3498
219. Reichel A, Lienau P (2016) Pharmacokinetics in drug discovery: an exposure-centred approach to optimising and predicting drug efficacy and safety. In: Nielsch U, Fuhrmann U, Jaroch S (eds) New approaches to drug discovery. Springer International Publishing, Cham, pp 235–260
220. Patlolla RR, Mallampati R, Fulzele SV, Babu RJ, Singh M (2009) Dermal microdialysis of inflammatory markers induced by aliphatic hydrocarbons in rats. Toxicol Lett 185:168–174

Chapter 4
Pharmacokinetics and Pharmacodynamics of Nanocarriers and Novel Drug Delivery Systems

Sankalp A. Gharat, Munira M. Momin, and Tabassum Khan

Introduction by the Editor

The pharmacokinetics of nanocarriers and novel drug delivery systems play a pivotal role in optimizing the delivery and therapeutic efficacy of drugs. These systems, such as liposomes, ethosomes, lipomers, transferosomes, cubosomes, phytosomes, polymeric nanoparticles, micelles, solid lipid nanoparticles, nanostructured lipid carriers, nanoemulsions, metallic nanoparticles, implants and other topically administered novel drug delivery systems have unique properties that influence their pharmacokinetic behaviour [1]. Upon administration, nanocarriers affect the absorption, distribution, metabolism, and excretion of encapsulated drugs. They enhance drug stability, protect against enzymatic degradation, and prolong systemic circulation, leading to improved drug bioavailability and target site accumulation [2]. The size, surface charge, and surface modifications of nanocarriers impact their biodistribution and tissue penetration capabilities. Moreover, the choice of materials and formulation strategies can influence drug release kinetics and tissue-specific targeting. Understanding the pharmacokinetics of nanocarriers is crucial to optimize drug delivery systems, ensure proper dosing regimens, maximize therapeutic outcomes and subsequently minimize systemic toxicity.

This chapter will discuss PK-PD of various novel drug delivery systems.

S. A. Gharat · M. M. Momin (✉)
Department of Pharmaceutics, SVKM's Dr. Bhanuben Nanavati College of Pharmacy, Mumbai, Maharashtra, India
e-mail: sankalp.gharat@bncp.ac.in; munira.momin@bncp.ac.in

T. Khan
Department of Pharmaceutical Chemistry and Quality Assurance, SVKM's Dr. Bhanuben Nanavati College of Pharmacy, Mumbai, Maharashtra, India
e-mail: tabassum.khan@bncp.ac.in

S. A. Gharat et al. (eds.), *Pharmacokinetics and Pharmacodynamics of Novel Drug Delivery Systems: From Basic Concepts to Applications*,
https://doi.org/10.1007/978-981-99-7858-8_4

4.1 Pharmacokinetics and Pharmacodynamics of Polymeric Nanoparticles

Introduction by the Editor

Polymeric nanoparticles are engineered nanoscale polymeric constructs, representing a forefront in drug delivery and biomedical innovation. These microscopic carriers, developed from biocompatible and biodegradable polymers, offer a transformative platform to enhance therapeutic outcomes. With the ability to encapsulate drugs, genes and imaging agents, polymeric nanoparticles enable precise control of release kinetics, improve drug stability, and facilitate targeted delivery to specific tissues or cells. A diverse array of polymers finds utility in the construction of polymeric nanoparticles, each designed for specific applications. Biocompatible polymers like poly(lactic-co-glycolic acid) (PLGA) and polyethylene glycol (PEG) are commonly employed due to their safety profiles and controlled release capabilities. Chitosan and gelatin offer biodegradability and mucoadhesive properties, making them valuable for oral and transmucosal drug delivery. Polysaccharides such as alginate and hyaluronic acid provide targeted delivery and compatibility with biological systems. Cationic polymers like polyethylenimine (PEI) facilitate nucleic acid delivery, while thermo-sensitive polymers respond to temperature changes for on-demand drug release. These versatile polymers highlight the customizable nature of polymeric nanoparticles, enabling unique solutions for diverse therapeutic challenges in drug solubility, bio-availability, and site-specificity for improved patient care.

Machine Generated Summaries

Disclaimer: The summaries in this chapter were generated from Springer Nature publications using extractive AI auto-summarization: An extraction-based summarizer aims to identify the most important sentences of a text using an algorithm and uses those original sentences to create the auto-summary (unlike generative AI). As the constituted sentences are machine selected, they may not fully reflect the body of the work, so we strongly advise that the original content is read and cited. The auto generated summaries were curated by the editor to meet Springer Nature publication standards.

To cite this content, please refer to the original papers.

Machine generated keywords: transdermal, skin, microneedle, iontophoresis, mns, nanoparticle, corneum, charge, injection, lipid, transdermal delivery, bioavailability, stratum, stratum corneum, oil, nanoparticle, prepare, uptake, plga, alzheimer, gastric, pump, alzheimer disease, charge, polymer, acid, sustain, dosage form, plga nanoparticle, oral administration

Novel Lansoprazole-Loaded Nanoparticles for the Treatment of Gastric Acid Secretion-Related Ulcers: In Vitro and In Vivo Pharmacokinetic Pharmacodynamic Evaluation [3] This is a machine-generated summary of:

Alai, Milind; Lin, Wen Jen: Novel Lansoprazole-Loaded Nanoparticles for the Treatment of Gastric Acid Secretion-Related Ulcers: In Vitro and In Vivo Pharmacokinetic Pharmacodynamic Evaluation [3]

Published in: The AAPS Journal (2014)
Link to original: https://doi.org/10.1208/s12248-014-9564-0

If you want to cite the papers, please refer to the original.

For technical reasons we could not place the page where the original quote is coming from.

Abstract-Summary "The objective of this study is to combine nanoparticle design and enteric coating technique to sustain the delivery of an acid-labile drug, lansoprazole (LPZ), in the treatment of acid reflux disorders."

"Lansoprazole-loaded Eudragit® RS100 nanoparticles (ERSNP-LPZ) as well as poly(lactic-co-glycolic acid) (PLGA) nanoparticles (PLGANP-LPZ) were prepared using a solvent evaporation/extraction method."

"The effects of nanoparticle charge and permeation enhancers on lansoprazole uptake was assessed in Caco-2 cells."

"The oral administration of enteric-coated capsules filled with nanoparticles sustained and prolonged the LPZ concentration up to 24 h in ulcer-induced Wistar rats, and 92.4% and 89.2% of gastric ulcers healed after a 7-day treatment with either EC-ERSNP1010-Na caprate or EC-PLGANP1005-Na caprate, respectively."

INTRODUCTION

"The once-daily administration of LPZ is popularly used to control gastric acid secretion in the clinic."

"The once-daily oral administration of an extended-release formulation of LPZ (Prevacid capsule, TAP Pharmaceuticals Inc., IL, USA) does not effectively control gastro-esophageal symptoms over the entire day [4–6]."

"PLGA has been approved by the food and drug administration of the USA as a biomedical material and is widely used in controlled release dosage forms [7, 8]."

"The emulsification/solvent evaporation method is popularly applied to prepare micro- and nanoparticles, which are widely applicable to control delivery of drugs [9–11]."

"The objective of the present work was to develop LPZ-loaded sustained release nanoparticles using positively charged Eudragit® RS100 and negatively charged poly(lactic-co-glycolic acid)."

"The pharmacokinetic performance and ulcer healing response were evaluated in ulcer-induced Wistar rats after the oral administration of the enteric coated capsule filled with LPZ-loaded nanoparticles."

MATERIALS AND METHODS

"For the acid resistance study, the enteric coated capsule was suspended in 100 mL of 0.1 N HCl at 37 ± 0.5°C in a shaker bath at 75 rpm for 1 h. The capsule was then washed with deionized water, and the nanoparticles were removed from the capsule."

"ERSNP-C6 and PLGANP-C6 (10 mg) were dispersed in 1 mL HBSS, pH 7.4, in a dialysis bag (MWCO 6,000–8,000 Da) that was immersed in 10 mL HBSS at

37 ± 0.5°C in a mechanical shaker at 50 rpm for 24 h. The sample was collected, and the fluorescence was measured."

"The enteric coated hard gelatin capsules (#9, Torpac Inc., NJ, USA) filled with (1) ERSNP1010 (EC-ERSNP1010), (2) ERSNP1010 and sodium caprate (20 mg/kg) (EC-ERSNP1010-Na caprate), (3) PLGANP1005 (EC-PLGANP1005), or (4) PLGANP1005 and sodium caprate (20 mg/kg) (EC-PLGANP1005-Na caprate) equivalent to 5 mg LPZ/kg were orally administered to ulcer-induced Wistar rats."

RESULTS AND DISCUSSION

"The enteric coated capsules were evaluated for their acid resistance, and 95.1 ± 2.5% and 96.8 ± 1.8% of the LPZ remained stable in the nanoparticle-filled enteric coated capsules, EC-ERSNP1010 and EC-PLGANP1005, respectively, after acid treatment for 1 h. This result confirmed that the HPMCP enteric coated capsule successfully protected LPZ from degradation in an acidic medium [12]."

"Burst releases of 12.1 ± 2.0, 13.7 ± 6.2, 6.7 ± 0.4, and 8.0 ± 0.5% of the available LPZ from ERSNP1010, PLGANP1005, EC-ERSNP1010, and EC-PLGANP1005, respectively, were observed during the first 1 h due to diffusion of the drugs located at or near the surface of the nanoparticles."

"Pharmacokinetic characteristics of LPZ were evaluated in the ulcer-induced Wistar rats following the oral administration of enteric coated capsules filled with either positively charged ERSNP1010 or negatively charged PLGANP1005 nanoparticles in the presence or absence of sodium caprate."

CONCLUSION

"Novel, LPZ-loaded, enteric capsule nanoparticles for the treatment of gastric acid secretion-related ulcers were demonstrated in vitro and in vivo."

"The Eudragit and PLGA nanoparticles sustained and prolonged the LPZ concentration up to 24 h in ulcer-induced Wistar rats."

"The oral administration of the LPZ-loaded, enteric capsule nanoparticles for 7 days resulting in varying abilities to heal the gastric ulcers: EC-ERSNP1010-Na caprate > EC-ERSNP1010 > EC-PLGANP1005-Na caprate > RICH® > EC-PLGANP1005."

"LPZ-loaded, enteric capsule nanoparticles may be an alternative oral dosage form for the treatment of acid-related disorders, especially to control night-time acid disorders."

A Preliminary Pharmacodynamic Study for the Management of Alzheimer's Disease Using Memantine-Loaded PLGA Nanoparticles [13] This is a machine-generated summary of:

Kaur, Atinderpal; Nigam, Kuldeep; Tyagi, Amit; Dang, Shweta: A Preliminary Pharmacodynamic Study for the Management of Alzheimer's Disease Using Memantine-Loaded PLGA Nanoparticles [13]

Published in: AAPS PharmSciTech (2022)

Link to original: https://doi.org/10.1208/s12249-022-02449-9

Copyright of the summarized publication:

The Author(s), under exclusive licence to American Association of Pharmaceutical Scientists 2022

If you want to cite the papers, please refer to the original.

For technical reasons we could not place the page where the original quote is coming from.

Abstract-Summary "To provide an effective treatment option, memantine-encapsulated polymeric nanoparticles were prepared in the study."

"The nanoparticles were prepared by using nanoprecipitation followed by homogenization and ultrasonication methods, characterized on the basis of particle size, polydispersity index, and zeta potential."

"To observe the efficacy of nanoparticles in scopolamine-induced Alzheimer models in vivo studies were also carried out."

"The results showed that nanoparticles were in the nano range with a particle size of 58.04 nm and − 23 mV zeta potential."

"The cytotoxicity results with ~ 98 to 100% cell viability and no morphological changes through Giemsa staining indicated that nanoparticles were not leading to cell toxicity."

"The findings clearly indicated that the developed memantine nanoparticles could act as an alternative approach for the management of Alzheimer's disease."

Introduction

"The gamma images showed higher uptake of the radiolabeled memantine nanoemulsion in the target organ brain of rats administered with the formulation intranasally as compared to the rat treated intravenously and with aqueous drug solution intranasally."

"The results showed that encapsulation of memantine in the nanoparticles and target drug delivery through the intranasal route enhanced the concentration of the drug in the brains of Alzheimer-induced animal models [14]."

"The results clearly indicated that nanoemulsion given by IN route showed a higher concentration of drug within 1.5 h in the brains of the rats as compared to the IV and oral route of administration [15]."

"Shah and others (2021) also formulated clonazepam-loaded PLGA NPs for nose-to-brain drug delivery for the management of epilepsy."

Material and Method

"To assess the efficacy of the developed Mem NPs, a water maze model was used and scopolamine was administered to induce AD in rats."

"Group I: Control group (without treatment) Group II: Scopolamine treatment IP (2 mg/kg) Group III: Mem NPs administration for 9 days IN Group IV: Aqueous drug solution administration for 9 days IN The experimental apparatus consisted of a circular water tank (120 cm in diameter, 45 cm high)."

"Along with the training, the rats were treated with the aqueous drug and Mem NPs for 9 days via intranasal administration of 20 μl of formulation at the dose of 0.1 mg/kg intranasally (with a concentration of 1 mg/ml of drug in Mem NPs) [16]."

Results and Discussion

"The results indicated that the developed Mem NPs showed 89% of entrapment efficiency and 12% of drug loading."

"As these NPs were developed for the intranasal administration, so the particle size less than 100 nm is required for a higher uptake of the drug through the nasal cavity and results in enhanced bioavailability at the target site of action."

"The results clearly indicated that the Mem NPs showed a better release profile in the SNF and ACSF than the PBS, so the formulation could be delivered effectively via the intranasal route to the brain."

"The determination of the cytotoxicity profile of developed NPs is extremely important because drugs showed a slow and sustained pattern of release from the nanoparticles as compared to aqueous drug solutions."

"The results indicated that the developed nanoparticles at Cmax-11 ng/ml showed cell viability of approximately 90%, whereas aqueous drug solution showed cell viability of 78% in the case of the Neuro 2a cell line."

Conclusion

"The Mem NPs were developed using the nanoprecipitation method followed by homogenization and ultrasonication."

"The developed NPs were assessed to investigate their release behavior in different media and showed sustained release pattern in SNF, ACSF, and PBS for 24 h. Further, to explore the intranasal route of administration, nanoparticles were delivered through intranasal, oral, and IV route of administration to evaluate the pharmacodynamic effect of the encapsulated drug."

"The gamma images of the radiolabeled memantine showed that the drug encapsulated in polymeric NPs reached more efficiently at the target site, i.e., the brain through the intranasal route."

"The findings confirmed that the developed Mem NPs could be used for the management of Alzheimer's disease and act as a more effective formulation with less peripheral side effects."

4.2 Pharmacokinetics and Pharmacodynamics of Lipid Based Nanocarriers

Introduction by the Editor

Lipid based nanocarriers like Solid Lipid Nanoparticles (SLN) and Nanostructured Lipid Carriers (NLC) represent pioneering advancements in the realm of lipid-based drug delivery systems. SLN are composed of solid lipids and offer improved stability, enhanced drug encapsulation, and sustained release profiles. Solid lipids like stearic acid, glyceryl monostearate, and cetyl palmitate serve as the foundation for

SLN, providing stability, sustained release, and protection of encapsulated drugs. NLC, on the other hand, combines solid lipids with liquid lipids such as oleic acid, capric/caprylic triglycerides, or monoacyl glycerols, offering a versatile matrix that minimizes drug expulsion and crystalline imperfections. NLCs overcome the limitations of SLNS with respect to drug loading and crystalline imperfections. These lipidic nanoparticles hold the promise to revolutionise pharmaceuticals by enhancing drug solubility, extending therapeutic release, and enabling site-specific targeting. Their adaptability to a wide range of active compounds and their ability to navigate biological barriers make SLN and NLC versatile contenders in modern drug delivery, paving the way for innovative therapeutic solutions.

Machine Generated Summaries

Disclaimer: The summaries in this chapter were generated from Springer Nature publications using extractive AI auto-summarization: An extraction-based summarizer aims to identify the most important sentences of a text using an algorithm and uses those original sentences to create the auto-summary (unlike generative AI). As the constituted sentences are machine selected, they may not fully reflect the body of the work, so we strongly advise that the original content is read and cited. The auto generated summaries were curated by the editor to meet Springer Nature publication standards.

To cite this content, please refer to the original papers.

Machine generated keywords: oil, oral bioavailability, lipid, solvent, formulate, chemical, purchase, oral, lipophilic, bioavailability, solubility, solid lipid, solid, prepared, preparation

Lipid Nanoemulsions of Rebamipide: Formulation, Characterization, and In Vivo Evaluation of Pharmacokinetic and Pharmacodynamic Effects [17] This is a machine-generated summary of:

Narala, Arjun; Guda, Swathi; Veerabrahma, Kishan: Lipid Nanoemulsions of Rebamipide: Formulation, Characterization, and In Vivo Evaluation of Pharmacokinetic and Pharmacodynamic Effects [17]

Published in: AAPS PharmSciTech (2019)

Link to original: https://doi.org/10.1208/s12249-018-1225-7

If you want to cite the papers, please refer to the original.

For technical reasons we could not place the page where the original quote is coming from.

Abstract-Summary "Rebamipide has low oral bioavailability (10%) due to its low solubility and permeability."

"Lipid nanoemulsions (LNEs) were prepared in order to improve its oral bioavailability."

"Rebamipide-loaded lipid nanoemulsions were formulated by hot homogenization and ultrasonication method."

"The optimized LNE showed 4.32-fold improvement in the oral bioavailability in comparison to a marketed tablet suspension."

"Anti ulcer activity of rebamipide LNE was studied by testing the prophylactic effect in preventing the mucosal damage in stomach region."

"Maximum prophylactic antiulcer activity was observed by per oral delivery of rebamipide as LNE."

"Our results indicated that LNEs were a promising approach for the oral delivery of rebamipide for systemic effects along with local effects in protecting gastric region, which gets damaged during peptic ulcers."

Introduction

"Rebamipide is a Biopharmaceutics Classification System (BCS) class IV drug and has low systemic availability (10%), which is due to its low solubility and low permeability [18]."

"Further, lipid nanoemulsions (LNEs) were not reported for oral delivery of rebamipide."

"LNEs are promising delivery systems composed of oil as internal phase and egg lecithin as emulsifier."

"The bioavailabilities of many drugs, such as danazol, cefuroxime axetil, and baicalin were enhanced by formulating them into LNEs [19–21]."

"BCS class II drugs are commonly prepared as LNEs, but class IV drugs are rarely reported through the LNEs."

"The enhancement in bioavailability of rebamipide is only possible when both solubility and permeability of the drug are increased."

"An effort was made to prepare LNE formulation to enhance the oral performance of rebamipide."

"The prepared LNEs were characterized and in vivo studies were conducted, i.e., anti-ulcer activity and pharmacokinetic effects."

Methodology

"About 1 ml of LNE formulation was taken in Eppendorf tubes and centrifuged in a centrifuge (Heraeus Biofuge, Germany) at a speed of 13,000 rpm for a time period of 10 min and percentage creaming volume for each formulation was determined."

"The oral bioavailabilities of the optimized LNE formulation (F4) and rebamipide tablet suspension were estimated by conducting pharmacokinetic studies in male Wistar rats with an single oral dose (10 mg/kg body weight)."

"The animals were differentiated into five groups each containing six rats, which were administered with optimized LNE formulation (F4), rebamipide tablet suspension, blank LNE (10 mg/kg body weight)."

"Group 3 Pretreatment with rebamipide tablet suspension (10 mg/kg) and then inducing ulcers with ethanol (80%) treatment."

Results and Discussion

"The cumulative percent release of drug from the formulations F1–F4 were 57.0 ± 1.72, 60.24 ± 0.89, 66.83 ± 1.59, and 71.89 ± 1.80, respectively in 24 h. In general, the LNEs released slowly in 0.1 N HCl when compared to that in pH 6.8 phosphate buffer, which could be due to the solubility difference of drug in these media."

"The PK parameters of rebamipide in individual rats for optimized LNE and marketed tablet suspension were calculated by using Kinetica software."

"The PK parameters C_{max}, $t_{max,}$ AUC_{0-24}, MRT, and $t_{1/2}$ values were calculated for optimized LNE formulation and compared with that of marketed product as a control."

"As in the case of test formulation (LNE), the drug would be slowly released from the oil globules into the aqueous environment and then get absorbed."

Conclusion

"Further, prophylactic effect of the rebamipide LNE was also studied, in comparison with tablet suspension and blank emulsion."

"The rank order of protective effect against the damage caused by ethanol (80%) was; F4 (LNE) > tablet suspension > blank LNE (placebo) > control (ethanol treated)."

"The developed rebamipide lipid nanoemulsion (F4) containing phospholipids, showed superior performance in terms of pharmacokinetic and pharmacodynamic effects in rats over the tablet suspension."

Novel Drug Delivery Approach via Self-Microemulsifying Drug Delivery System for Enhancing Oral Bioavailability of Asenapine Maleate: Optimization, Characterization, Cell Uptake, and In Vivo Pharmacokinetic Studies [22] This is a machine-generated summary of:

Patel, Mitali H.; Mundada, Veenu P.; Sawant, Krutika K.: Novel Drug Delivery Approach via Self-Microemulsifying Drug Delivery System for Enhancing Oral Bioavailability of Asenapine Maleate: Optimization, Characterization, Cell Uptake, and In Vivo Pharmacokinetic Studies [22]

Published in: AAPS PharmSciTech (2019)

Link to original: https://doi.org/10.1208/s12249-018-1212-z

If you want to cite the papers, please refer to the original.

For technical reasons we could not place the page where the original quote is coming from.

Abstract-Summary "Asenapine maleate (AM)-loaded self-microemulsifying drug delivery system (AM-SMEDDS) was prepared to increase its oral bioavailability."

"The AM-SMEDDS showed globule size and zeta potential of 21.1 ± 1.2 nm and -19.3 ± 1.8 mV, respectively."

"In vitro drug release study showed $99.2 \pm 3.3\%$ of drug release at the end of 8 h in phosphate buffer pH 6.8."

"Confocal and flow cytometry study showed that cellular uptake of coumarin-6 loaded SMEDDS was significantly enhanced by Caco-2 cells as that of coumarin-6 solution."

"Intestinal lymphatic transport study using Cycloheximide (CHX) showed that the AUC_{total} of AM-SMEDDS reduced about 35.67% compared with that without the treatment of CHX indicating involvement of lymphatic system in intestinal absorption of AM-loaded SMEDDS."

"These findings demonstrated the potential of SMEDDS for oral bioavailability improvement of AM via lymphatic uptake."

INTRODUCTION

"Lipid-based formulations are promising tools to enhance absorption of drugs as some lipid-based excipients are known absorption enhancers [23]."

"Self-micro emulsifying drug delivery system (SMEDDS) has achieved considerable attention to conquer low solubility and oral absorption of poorly water-soluble drugs."

"SMEDDS is isotropic mixture of oil, surfactant, cosurfactant, and drug which emulsify in the presence of water to form nano-sized globules in gastrointestinal tract (GIT) [24]."

"Negi and others [25] formulated SMEDDS using P-gp modulator excipients to evaluate systemic availability of irinotecan."

"Till date, authors have developed in situ nasal gel and transfersomes for transdermal delivery of AM [26, 27] and there is no marketed oral SMEDDS formulation."

"The present work was aimed to develop a SMEDDS for AM to augment oral bioavailability by decreasing its first pass metabolism."

MATERIALS AND METHODS

"Capryol 90 and Transcutol HP were received from Gattefosse India Private Limited (Mumbai, India) as gift samples."

"For surfactant screening, oil was mixed in a ratio of 1:1 with surfactant."

"For cosurfactant screening, surfactant was mixed in the ratio of 2:1 with cosurfactant."

"This mixture was mixed in the ratio of 1:1 with oil, diluted 50 times with water, and percentage transmittance was recorded [28, 29]."

"Cremophor EL was mixed with Transcutol HP in different weight ratios (1:0, 1:1, 2:1, 3:1, 1:2, and 1:3 w/w)."

"Capryol 90 was mixed with Smix and the ratio was changed from 1:9 to 9:1 (w/w)."

"Cremophor EL was mixed with Transcutol HP in the ratio of 3:1."

CHARACTERIZATION

"AM-SMEDDS and AM suspension (equivalent to 10 mg of AM) was filled in dialysis bag and placed in 250 mL of drug release media at 100 rpm at 37°C."

"Aliquots were withdrawn from release medium at specified time intervals, filtered, and analyzed by UV spectroscopy at 270 nm [30]."

"A volume equivalent to 10 mg AM of SMEDDS and suspension was added in lumen of stomach and tied at each end with a thread."

"It was placed in an organ tube containing 30 ml of phosphate buffer (pH 7.4) with continuous aeration at 37°C."

"At predetermined time intervals (0.5, 1, and 2 h), samples were withdrawn and replenished by the same volume of fresh buffer solution."

"At predetermined intervals of time (4, 6, 8, and 10 h), aliquots were withdrawn, filtered, and concentrations of AM in aliquots were analyzed by a UV spectrophotometer at 269 nm."

CELL LINE STUDIES

"Caco-2 cells were grown in 75-cm^2 tissue culture flasks containing growth medium (20% MEM, 20% FBS) and maintained in an incubator (Thermo-Fischer, Waltham, USA)."

"Caco-2 cells were seeded at a concentration of 1×10^4 in 96-well plates and placed in a CO_2 incubator for 24 h [31, 32]."

"After incubation, 100 μl of sample (AM suspension, AM-SMEDDS, and blank SMEDDS) was added to each well in the range of 10–250 μg/ml."

"This study was carried out by seeding 1×10^6 cells/well on coverslips in six-well plates for 24 h. Cells were incubated with coumarin-6 solution and coumarin-6-loaded SMEDDS for 4 h. After incubation, the cells were fixed by 70% ethanol followed by rinsing with PBS."

"Caco-2 cells were seeded at a concentration of 1×10^5 cells to each well in a six-well plate for 24 h. Coumarin-6 solution and coumarin-6-loaded SMEDDS was added to each well for 4 h. Then, the cells were washed with PBS and trypsinized."

RESULTS AND DISCUSSION

"The results showed that polyoxy 35 castor oil (Cremophor EL) had highest ability to emulsify oil (Capryol 90) as compared to other surfactants."

"The longer the length of the hydrophobic alkyl chain, the higher the molecular volume of the oil phase affecting the emulsification ability of surfactant mixtures."

"The addition of suitable cosurfactant lowers the interfacial tension, fluidizes the hydrocarbon region of the interfacial film, and decreases the bending stress of the interface which results in improvement in spontaneity of emulsification and reduction in emulsion droplet size and polydispersity [33]."

"Smaller microemulsion region was obtained at 1:0 ratio of S/Cos, as surfactant alone was insufficient to minimize the o/w interfacial tension."

"Further increase in a surfactant from 1:1 to 1:3 showed a decrease in the formation of microemulsion region indicating that optimum emulsification was achieved at 1:1 ratio of S/Cos."

CHARACTERIZATION

"These results indicate that a significant amount of drug will be carried to intestine inside the nano-sized globules when formulated as SMEDDS [34]."

"This enhanced uptake of SMEDDS can be owing to the presence of potential absorption enhancers (Cremophor EL and Transcutol HP) which may alter the epithelial barrier property and the small globule size."

"There might be various reasons in improvement of oral absorption and bioavailability by formulating AM-SMEDDS which are as follows: (i) intestinal lymphatic transport of AM entrapped in the nano-sized globules of SMEDDS, (ii) improvement in solubility of AM by SMEDDS which kept the drug in the solubilized form, and (iii) absorption-enhancing effect of oil and surfactant."

CONCLUSIONS

"Lipid-based formulations of AM, i.e., SMEDDS were successfully developed with small globule size which aided in the absorption of drug and enhanced its bioavailability."

"The cell uptake study showed that transport of AM-SMEDDS across Caco-2 cell monolayer was effectively enhanced as compared to AM suspension."

"The relative bioavailability of AM-SMEDDS was remarkably improved in rats."

"Cremophor EL showed significant CYP3A inhibitory effect which contributed in the bioavailability improvement."

Lipophilic Conjugates of Drugs: A Tool to Improve Drug Pharmacokinetic and Therapeutic Profiles [35] This is a machine-generated summary of:

Han, Sifei; Mei, Lianghe; Quach, Tim; Porter, Chris; Trevaskis, Natalie: Lipophilic Conjugates of Drugs: A Tool to Improve Drug Pharmacokinetic and Therapeutic Profiles [35]

Published in: Pharmaceutical Research (2021)

Link to original: https://doi.org/10.1007/s11095-021-03093-x

If you want to cite the papers, please refer to the original.

For technical reasons we could not place the page where the original quote is coming from.

Abstract-Summary "Lipophilic conjugates (LCs) of small molecule drugs have been used widely in clinical and pre-clinical studies to achieve a number of pharmacokinetic and therapeutic benefits."

"Such conjugation strategies have been employed to promote drug association with endogenous macromolecular carriers (e.g. albumin and lipoproteins), and this in turn results in altered drug distribution and pharmacokinetic profiles, where the changes can be 'general' (e.g. prolonged plasma half-life) or 'specific' (e.g. enhanced delivery to specific tissues in parallel with the macromolecular carriers)."

"Another utility of LCs is to enhance the encapsulation of drugs within engineered nanoscale drug delivery systems, in order to best take advantage of the targeting and pharmacokinetic benefits of nanomedicines."

"The current review provides a summary of the mechanisms by which lipophilic conjugates, including in combination with delivery vehicles, can be used to control drug delivery, distribution and therapeutic profiles."

Introduction

"LCs may be employed to achieve a range of drug delivery and pharmacokinetic benefits via different mechanisms including: (1) Long-acting injections or implants where the prodrug is dissolved in a lipid-based matrix, and maintains very slow drug release for weeks; (2) Modulation of lipophilicity via prodrug formation offers a means to enhance passive membrane permeability and in doing so to optimise the passage of drugs across biological barriers including the gastrointestinal (GI) tract for oral absorption, skin for topical absorption, and blood-brain barrier (BBB) for brain penetration; (3) Alteration of drug lipophilicity or conjugation of the drug to a lipophilic moiety at chemical sites that are metabolically labile may change drug-transporter or drug-metabolic enzyme interactions dramatically, thus changing transporter-mediated membrane permeability (and influencing absorption and/or tissue distribution) or modulating metabolic profiles; (4) Lipophilic conjugation may enhance drug binding to endogenous lipid carriers such as albumin, and this may result in altered drug disposition, metabolism and excretion; (5) A number of lipid-derivative structures can integrate into endogenous lipid metabolic/transport pathways, and therefore access targets such as the lymphatic system; (6) Alteration of a drug's physicochemical properties such as increasing lipophilicity and amphiphilicity may also improve drug encapsulation within nanoscale delivery systems, which enables the application of nanomedicine approaches to a wider range of drug 'payloads'."

To achieve very slow release or delayed absorption following injection (i.e. long-acting injections)

"To overcome this issue, modification of drug structures to prolong the half-life and oral or parenteral extend release formulations may be employed such as polymer based sustained release systems [36]."

"One such option is long acting injections (LAIs) that slowly release drug over time following a single injection given every few weeks to months."

"The aqueous solubility of many drugs (even when categorised as poorly water soluble by 'common' standards) often makes it a challenge to prevent drug partitioning out of the sustained release system, causing rapid drug leakage and absorption from the injected site."

"The LC may thus become a good candidate for sustained release from a drug-aggregate or microparticle-based LAI."

"LAI formulations of these drugs allow markedly prolonged duration of action following intramuscular administration (up to 8 weeks for aripiprazole and up to 14 weeks for paliperidone), which provides great clinical benefits for the long-term treatment required for schizophrenia."

To improve membrane permeability and thus oral absorption, topical absorption or blood brain barrier penetration

"In order to achieve sufficient oral bioavailability, drug molecules are required to have both adequate aqueous solubility (to allow dissolution in GI fluid) and cell membrane permeability (to facilitate passage across the gut epithelium)."

"Permeability may be improved by increasing the lipophilicity of the drug but sometimes this is not possible as it will alter other drug properties such as activity at target and off-target sites, metabolism etc To address this issue, LC approaches can be used to enhance permeability and this has resulted in successful launch of a number of prodrugs including tenofovir alafenamide and tenofovir disoproxil fumarate [37], sofosbuvir (where the phosphate group in the nucleotide derivative is double-capped with l-alanine-isopropyl ester and phenol moieties) [38], as well as other products and investigational candidates summarised elsewhere [39, 40]."

"Many drug delivery strategies have been explored to enhance BBB penetration by overcoming specific barriers or utilising active transportation mechanisms [41]."

"There have been several proof-of-concept preclinical studies where LC were explored to enhance BBB penetration by increasing logP, thereby increasing permeability across the lipid-bilayer of cell membrane."

To modulate influx/efflux transport or drug metabolism (for half-life management, permeability enhancement, etc.)

"In an in vitro hASBT transfected Madin-Darby Canine Kidney (MDCK) cell model (a common mammalian cell line used for drug transport studies), the prodrug showed high affinity to hASBT and resulted in 1-2 orders of magnitude higher transcellular permeability than the parent drug gabapentin [42]."

"To the examples above where the prodrugs were designed to enhance transporter-mediated influx, LC can also be employed to redirect drug permeation away from dependence on a transporter-mediated pathway to a passive diffusion dominant mechanism by enhancing lipophilicity."

"Elacytarabine improves efficacy in the treatment of cytarabine resistant tumor cells and one of the mechanisms is thought to be due to increased drug uptake by avoiding the dependency on the human equilibrative nucleoside transporter (hENT) for influx [43, 44]."

"Transporter mediated efflux is a common mechanism underlying drug resistance, poor oral bioavailability, and poor BBB penetration."

To modulate drug disposition by promoting association with endogenous macromolecular carriers (e.g. albumin)

"Drug binding to endogenous circulating macromolecular structures (such as albumin and lipoproteins) can influence drug distribution and tissue distribution via a number of mechanisms."

"This section focuses on modification of small molecule drugs to enhance binding to endogenous plasma proteins after administration, rather than administration of the drugs in association with specific nanocarriers."

"One of the functions of albumin (66.5 kDa) is to serve as a transport protein for endogenous molecules such as fatty acids, metal ions and bilirubin and there are a number of binding sites on albumin that exogenous drug molecules can also interact with."

"The protein binding behaviour of the prodrug is not influenced by Cremophor EL (a surfactant required in the formulations of both the prodrug and parent drug)."

"The binding profile of the parent drug paclitaxel has been reported to be significantly changed in the presence of Cremophor EL."

To integrate into lipid biotransformation and transport pathways (association with lipoproteins)

"Incorporation of lipophilic prodrugs into intestinal lipoproteins following oral administration provides a unique means to promote drug delivery into lymph, which may bring about a number of advantages: (a) avoidance of hepatic first-pass metabolism leading to enhanced oral bioavailability, as transport via the lymph leads to direct access into the systemic circulation without passage through the portal vein and liver (which is the main route for absorption of most drugs); (b) enhanced drug delivery to targets in the mesenteric lymphatics, including the lymphatic vessels, lymph nodes and their resident immune cells, thereby improving the efficacy of drugs such as immunotherapies."

"In this application, co-administration of lipid-based formulations or postprandial drug administration is often required to promote lymphatic uptake of the prodrugs, as lipid intake increases the production of intestinal chylomicrons and thus the association of lipophilic compounds with chylomicron transport pathways into lymph."

To facilitate encapsulation in nanoscale drug delivery systems (NanoDDS)

"The prodrugs discussed in previous sections were typically administered as the conjugate alone or after dissolving in a lipid-based formulations (e.g. oil solutions, micelles, self-emulsifying drug delivery systems (SEDDS) etc)."

"NanoDDS include systems such as liposomes, reconstituted or synthetic lipoproteins (including lipoprotein-mimicking nanoparticles), polymeric and lipid nanoparticles, self-assembled supramolecules and polymer systems (e.g. dendrimers), etc Taking advantage of nanoscale sizes (typically 10-1000 nm in diameter), nanomedicines can enable specific pharmacokinetic or targeted delivery benefits, such as promoting drug solubility or absorption, prolonging circulation half-life, achieving altered disposition and facilitating targeted drug delivery to specific tissues."

"Lipid nanoparticles (spherical nanoscale particles 'filled' with solid or liquid lipids, in contrast to the multilayered vesicles of liposomes) have been extensively studied to enhance drug delivery to tumours."

"LC have been be used to improve encapsulation within these lipid-based nanoscale drug delivery carriers, including, solid lipid nanoparticles for paclitaxel [45] and synthetic lipoprotein-mimicking particles for porphyrin [46]."

Preparation and Pharmacokinetics Evaluation of Solid Self-Microemulsifying Drug Delivery System (S-SMEDDS) of Osthole [47] This is a machine-generated summary of:

Sun, Chaojie; Gui, Yun; Hu, Rongfeng; Chen, Jiayi; Wang, Bin; Guo, Yuxing; Lu, Wenjie; Nie, Xiangjiang; Shen, Qiang; Gao, Song; Fang, Wenyou: Preparation and Pharmacokinetics Evaluation of Solid Self-Microemulsifying Drug Delivery System (S-SMEDDS) of Osthole [47]

Published in: AAPS PharmSciTech (2018)

Link to original: https://doi.org/10.1208/s12249-018-1067-3

If you want to cite the papers, please refer to the original.

For technical reasons we could not place the page where the original quote is coming from.

Abstract-Summary "The study was performed aiming to enhance the solubility and oral bioavailability of poorly water-soluble drug osthole by formulating solid self-microemulsifying drug delivery system (S-SMEDDS) via spherical crystallization technique."

"The liquid self-microemulsifying drug delivery system (L-SMEDDS) of osthole was formulated with castor oil, Cremophor RH40, and 1,2-propylene glycol after screening various lipids and emulsifiers."

"The type and amount of polymeric materials, good solvents, bridging agents, and poor solvents in S-SMEDDS formulations were further determined by single-factor study."

"In vitro release study demonstrated a sustained release of the drug from osthole S-SMEDDS."

"Comparing with osthole aqueous suspension and L-SMEDDS, osthole S-SMEDDS increased bioavailability by 205 and 152%, respectively."

"The results suggested that S-SMEDDS was an effective oral solid dosage form, which can improve the solubility and oral bioavailability of poorly water-soluble drug osthole."

INTRODUCTION

"A number of pharmaceutical methods [48–57], which are especially liquid self-microemulsifying drug delivery system (L-SMEDDS), have been developed as viable drug delivery approaches to address the abovementioned problems."

"Drug delivered by L-SMEDDS can spontaneously form microemulsion with droplet size of dozens of nanometers in the gastrointestinal tract after oral administration [58–61] and hence improve the absorption and bioavailability of poorly water-soluble drugs [62]."

"As a part of our long-term interest in solving the oral bioavailability of poorly soluble drug [62–64], we solidified the L-SMEDDS into solid self-microemulsifying drug delivery system (S-SMEDDS) by applying spherical crystallization technique, which was a one-step solidification method that was executed in the liquid phase."

"We developed the osthole S-SMEDDS from osthole L-SMEDDS by applying spherical crystallization technique."

"The pharmacokinetics parameters including oral bioavailability of osthole S-SMEDDS were determined in comparison with the drug suspension and the L-SMEDDS."

MATERIALS AND METHODS

"The prepared osthole L-SMEDDS was diluted with pH 6.8 PBS to form microemulsion."

"One milliliter of osthole L-SMEDDS was diluted to 50 mL by double-distilled water, 0.9% sodium chloride solution, hydrochloric acid (0.1 M), and pH 6.8 phosphate-buffered saline (PBS), respectively."

"One milliliter of osthole L-SMEDDS was diluted with pH 6.8 PBS to 10, 25, 50, 100, and 150 times, respectively."

"One milliliter of osthole L-SMEDDS was added into 50 mL of pH 6.8 PBS."

"The selected osthole L-SMEDDS and the polymer solution were then quickly added into the poor solvent, respectively."

"The osthole S-SMEDDS prepared via the spherical crystallization technique was evaluated by optical observation in an N-108M optical microscope (Ningbo, China) to determine whether microspheres formed after drying."

"The L-SMEDDS and S-SMEDDS of osthole were diluted 50-fold with pH 6.8 PBS to form the microemulsion before the TEM study."

RESULTS AND DISCUSSIONS

"When the dilution time was overly large, the zeta potential of the microemulsion by the osthole L-SMEDDS would go infinitely close to zero."

"The osthole S-SMEDDS prepared according to the optimized formulation and preparation parameters had a yield of 83.91 ± 3.31% and encapsulation efficiency of 78.39 ± 2.25% with good reproducibility."

"There was 95% drug release from osthole L-SMEDDS within 2 h while only about 15% release from the S-SMEDDS."

"Osthole S-SMEDDS was firstly dissolved into osthole L-SMEDDS, and then the existence of the enteric polymer also prevented the release of osthole in simulated gastric juice [65]."

"When stored at low temperature (4°C) for 10 days, the osthole S-SMEDDS showed no change in size or encapsulation efficiency."

CONCLUSIONS

"The optimal formulation consisted of EC and Eudragit S100 in ratio of 1:2 as polymers."

"Besides, the optimal volume of 0.08% SDS aqueous solution used as poor solvent was found to be 120 mL. The formulation was stirred at a screw speed of 400 rpm for 35 min under 25°C."

"Prepared osthole S-SMEDDS presented high yield and encapsulation efficiency."

"In vivo pharmacokinetics study showed that the osthole S-SMEDDS significantly increased the oral bioavailability compared with drug suspension or L-SMEDDS in rabbits."

"Based on these facts, osthole S-SMEDDS successfully addressed the low solubility and oral bioavailability of the drug, which could potentially expand the drug applications in clinical settings."

Nanoemulsion-based dissolving microneedle arrays for enhanced intradermal and transdermal delivery [66] This is a machine-generated summary of:

Nasiri, Muhammad Iqbal; Vora, Lalitkumar K.; Ershaid, Juhaina Abu; Peng, Ke; Tekko, Ismaiel A.; Donnelly, Ryan F.: Nanoemulsion-based dissolving microneedle arrays for enhanced intradermal and transdermal delivery [66]

Published in: Drug Delivery and Translational Research (2021)

Link to original: https://doi.org/10.1007/s13346-021-01107-0

Copyright of the summarized publication:

The Author(s) 2021

If you want to cite the papers, please refer to the original.

For technical reasons we could not place the page where the original quote is coming from.

Abstract-Summary "The development of dissolving microneedles (DMN) is one of the advanced technologies in transdermal drug delivery systems, which precisely deliver the drugs through a rapid dissolution of polymers after insertion into the skin."

"We fabricated nanoemulsion-loaded dissolving microneedle (DMN) arrays for intradermal and transdermal drug delivery."

"Task, model drug (amphotericin B, AmB)-loaded nanoemulsion (NE) were prepared by the probe-sonication method."

"Transdermal porcine skin permeation studies showed significantly higher permeability of AmB (29.60 ± 8.23 µg/patch) from AmB-NE-DMN compared to MN-free AmB-NE patches (5.0 ± 6.15 µg/patch) over 24 h. Antifungal studies of optimized AmB-NE-DMN, AmB-loaded discs and drug-free DMN against Candida albicans, confirmed the synergistic activity of Campul-MCM C-8, used in the nanoemulsion formulation."

"This study establishes that nanoemulsion based dissolving microneedle may serve as an efficient system for intradermal as well as transdermal drug delivery."

Introduction

"NE has many advantages, like easy fabrication, high stability, increased drug solubility, and enhanced bioavailability, particularly for hydrophobic drugs [67, 68]."

"Formerly, various studies have been reported related to the fabrication of AmB-loaded nanoemulsion to treat topical fungal infections [69–72]."

"We present a simple and innovative in-situ AmB NE generation in DMN polymeric hydrogel to prepare the AmB NE-loaded DMN arrays that could synergetically improve the intradermal delivery of AmB. Initially, AmB based NE was prepared by the probe-sonication method in polymeric hydrogel and then characterized for droplet size, PDI, and zeta potential and, subsequently, cast into DMN by single-step centrifugation method."

"This delivery system was designed, specifically focusing on determining the possibility of using a novel DMN system to facilitate intradermally and transdermal delivery of AmB loaded NE."

"The developed AmB-NE-DMN system was then evaluated for ex vivo intradermal neonatal porcine skin permeation and drug deposition studies."

Materials and methods

"AmB-NE-DMN were fabricated by pouring AmB-NE onto the top surface of the MN moulds, and the moulds were centrifuged at 3500 rpm for 20 min and then allowed to dry for 24 h at room temperature and then kept in an oven at 37 ± 2 °C for further drying of 24 h. MN arrays were then removed from the moulds and evaluated for needle formation and mechanical strength [73, 74]."

"After insertion of MN patch, a cylindrical 12.0 g stainless steel weight was placed onto the top of the MN arrays patch to prevent MN expulsion and placed inside the oven at 37 °C ± 2 °C for 24 h. The tissue sample, taken from that portion of the skin where the AmB-NE-MN had been inserted, were obtained using a scalpel."

"Permeation from control AmB-NE was performed in the same manner, except instead of inserting a DMN array, a needle-free patch of the same dimensions and formulation was placed on top of the skin, followed by the stainless-steel weight."

Results and discussion

"Several formulations (AmB-NE-F1 to AmB-NE-F5) were prepared to optimize the concentration of PVA and PVP in order to achieve a good penetration of DMN into parafilm M® as well as excised porcine skin."

"It was indicated that AmB-NE-DMN arrays displayed higher ex vivo skin permeation compared with MN-free AmB-NE patches over 24 h. The DMN containing nanosized droplet probably increases the delivery of the drug by permeating the stratum corneum barrier and through MN induced micro-conduits into the skin, thereby allowing systemic drug absorption."

"The AmB NE loaded DMN arrays were manufactured with drugs in the needle tips (to deliver intradermally) and in the baseplate (to deliver transdermally)."

"To test antifungal activities of these developed NE-DMN formulations with AmB, the study was divided into seven groups such as A, B, C, D, E, F, and G (A = tips of AmB-NE-MN-F5 arrays (approx."

Conclusion

"The NE of the highly hydrophobic drug was successfully optimized and incorporated into DMN arrays to penetrate the skin and dissolve rapidly in the skin to achieve adequate drug permeation and deposition."

"NE loaded DMN arrays were formulated with the model drug, AmB. The stability of the drug-loaded NE was confirmed by using a particle size analyzer."

"As per obtained results, AmB NE-loaded DMN could provide the synergistic antifungal effect to ensure its efficacy against Candida albicans."

"This proof-of-concept work, therefore, represents meaningful advancement in the usage of DMN technologies in combination with nanoemulsion for delivery of lipophilic drugs into the viable skin (epidermis and dermis) layers for maximum therapeutic achievement and better patient compliance."

Pharmacokinetic and Anti-inflammatory Effects of Sanguinarine Solid Lipid Nanoparticles [75] This is a machine-generated summary of:

Li, Weifeng; Li, Huani; Yao, Huan; Mu, Qingli; Zhao, Guilan; Li, Yongmei; Hu, Hua; Niu, Xiaofeng: Pharmacokinetic and Anti-inflammatory Effects of Sanguinarine Solid Lipid Nanoparticles [75]

Published in: Inflammation (2013)
Link to original: https://doi.org/10.1007/s10753-013-9779-8

If you want to cite the papers, please refer to the original.

For technical reasons we could not place the page where the original quote is coming from.

Abstract-Summary "The sanguinarine (SG) was studied for its pharmacokinetic and anti-inflammatory activities with prepared solid lipid nanoparticles (SLNs)."

"The drug release profile of SG was examined in pH 7.4 PBS and 85 % of the SG loaded in SLNs was gradually released during 24 h. We used mice endotoxin shock model which was induced by lipopolysaccharide (1 mg/kg) to examine the anti-inflammatory function of SG-SLNs."

"Healthy Kunming mice were administered orally with saline, SG (10 mg/kg), and SG-SLNs (10 mg/kg), respectively, at 12 and 1 h before lipopolysaccharide (LPS) injection."

"SG-SLNs revealed significant anti-inflammatory effects through inhibition of LPS-induced tumor necrosis factor-alpha level, interleukin 6 level, and nitric oxide production in serum."

INTRODUCTION

"Solid lipid nanoparticles (SLNs) are colloidal particles, consisting of a matrix composed of lipids being solid at both room and body temperatures, dispersed in an aqueous surfactant solution [76]."

"Methods usually used in preparing SLNs include high-pressure homogenization, high-shear homogenization and ultrasound, emulsion solvent/evaporation, solvent injection, and microemulsion [77]."

"SLNs can be prepared for multiple routes of administration including oral administration, and the technology could increase the solubility and dissolution rate of the drugs [78]."

"Studies have indicated that poorly hydrophilic drugs that incorporated into SLNs can enhance the oral bioavailability."

"We successfully prepared SG-SLNs by film-ultrasonic dispersion method."

"The aim of this article is to study the pharmacokinetic and anti-inflammatory activities of SG after single dose intragastric administration of SG-SLNs in comparison with SG solution."

MATERIALS AND METHODS

"The total drug content of SG-SLNs was determined by acetonitrile emulsion breaking, and the amount of drug that was free in the SLNs was measured by the low temperature speed centrifugation (15,000 rpm, 4 °C, 1 h)."

"Quality control samples were prepared in similar way at three concentration levels of 0.5, 5, and 10 μg/ml for SG."

"Male Kunming mice were randomly divided into four groups (n = 24 per group): the SG treatment group and the SG-SLNs treatment group, which were administered orally with SG (10 mg/kg) and SG-SLNs (10 mg/kg), respectively, at 12 and 1 h-post injection of LPS; the LPS group, which received an intraperitoneal injection of 1 mg/kg LPS; and the control group, which receive the equal volume of saline."

"Blood samples from four groups of mice (saline, LPS, LPS + SG, LPS + SG-SLNs) were extracted at 1 and 6 h after injection of LPS."

RESULT AND DISCUSSION

"SG-SLNs were prepared by film-ultrasonic dispersion method as it results in small particles of 117 nm and a negative zeta potential of −31.3 mV. It was reported that a zeta potential value of nanoparticles of more than −30 mV is enough for good stability [79]."

"A key issue analyzed in this study was the feasibility of using solid lipid nanoparticles to deliver SG and the ability of nanoparticles to deliver SG was measured by determining the drug release."

"We measured TNF-α level at 1 h after LPS injection and NO and IL-6 levels at 6 h after LPS injection to test whether SG-SLNs could be implicated in the inhibition of inflammatory cytokines."

"SG-SLNs suppressed LPS-induced TNF-α, NO, and IL-6 release."

CONCLUSION

"The present results demonstrate that SG-SLNs have a better solubility as nanodispersion and SLNs delivery system can enhance the anti-inflammation and pharmacokinetic activity of SG."

"SG-SLNs system is a promising delivery system for enhancing the bioavailability."

PEGylated Lipid Nanocontainers Tailored with Sunseed-Oil-Based Solidified Reverse Micellar Solution for Enhanced Pharmacodynamics and Pharmacokinetics of Metformin [75] This is a machine-generated summary of:

Kenechukwu, Franklin Chimaobi; Nnamani, Daniel Okwudili; Nmesirionye, Bright Ugochukwu; Isaac, God'spower Tochukwu; Momoh, Mumuni Audu; Attama, Anthony Amaechi: PEGylated Lipid Nanocontainers Tailored with Sunseed-Oil-Based Solidified Reverse Micellar Solution for Enhanced Pharmacodynamics and Pharmacokinetics of Metformin [75]

Published in: Journal of Pharmaceutical Innovation (2022)

Link to original: https://doi.org/10.1007/s12247-022-09654-w

Copyright of the summarized publication:

If you want to cite the papers, please refer to the original.

For technical reasons we could not place the page where the original quote is coming from.

Abstract-Summary "The aim of this study was to formulate and evaluate sunseed-oil-based PEGylated nanostructured lipid carriers (PEG-NLC) for enhanced delivery and prolonged antidiabetic activity of metformin."

"The PEG-NLC and non-PEGylated NLC were formulated by high shear homogenization and thereafter characterized by scanning electron microscopy, mean particle size determination, photon correlation spectroscopy, differential scanning calorimetry (DSC), and Fourier transform infrared (FT-IR) spectroscopy."

"DSC results showed reduced crystallinity and hence greater possibility of enhanced drug solubility and entrapment, while FTIR results showed drug-excipient compatibility."

"The PEG-NLCs showed enhanced drug release in simulated biorelevant media and prolonged antidiabetic activity compared with both non-PEGylated NLC and controls."

"Batch D_{40} containing the highest amount of PEG-4000 (optimized formulation) gave sixfold increase in pharmacokinetics properties than marketed sample (Glucophage®)."

"Sunseed-oil-based PEGylated NLC has proven to be a stable and safe carrier system for enhanced delivery and prolonged antidiabetic activity of metformin."

Introduction

"The novelty embodied in this study is the use of sunseed-oil-based PEGylated nanoengineered lipid carrier to enhance the antidiabetic activity of metformin via improved half-life, reduced dosage and dosing frequency as well as through enhanced permeability across the biological membrane."

"Although we had earlier utilized sunseed oil in the development of PEGylated formulations for enhanced delivery of antifungal agents[80], to the best of our knowledge, there is currently a dearth of information in the literature on the use of sunseed-oil-based PEGylated nanoengineered lipid carrier for oral delivery of metformin for enhanced treatment of diabetes mellitus."

"The investigation of sunseed-oil-based PEGylated nanolipid carrier for enhanced metformin delivery to optimize the efficacy, intestinal absorption, bioavailability, biological half-life, and reduce drug dose, dosing frequency and drug adverse effects, informs the aim of this study."

"The objective of this study was to formulate, characterize and evaluate in vitroand in vivo pharmacodynamic properties of metformin-loaded sunseed-oil-based PEGylated nanostructured lipid carrier (PEG-NLC) for improved delivery of metformin hydrochloride."

Materials and Methods

"Metformin hydrochloride pure sample was obtained as a gift from May and Baker PLC (Ikeja, Lagos State, Nigeria)."

"Phospholipon® 90H (P90H) (Phospholipid GmbH, Köln, Germany), sorbitol (Caesar & Loretz, Hilden, Germany), polyethylene glycol 4000 (PEG 4000) (Ph."

"Carl Roth GmbH + Co. KG Karlsruhe, Germany), beeswax (Carl Roth, Karlsruhe, Germany), Polysorbate 80 (Tween® 80) (Acros Organics, Geel, Belgium), sunseed oil (double refined), vitamin A fortified (Kewalram Chanrai Group, Lagos, Nigeria), Alloxan (Merck KGaA, Darmstadt, Germany), distilled water (Lion water, University of Nigeria, Nsukka, Nigeria), and other solvents and reagents were used as procured from their manufacturers without further purification."

"Adult albino Wistar rats of both sexes were procured from the Faculty of Veterinary Medicine, University of Nigeria, Nsukka."

Preparation of PEGylated Lipid Matrices

"The homogenous mixture of the lipid matrix (LM_1) was stirred further at room temperature and then allowed to cool and solidify."

"After 24 h, this lipid matrix was melted in the thermoregulated bath at a temperature of 80 °C, followed by the addition of 9.0 g (9.017 ml) of super-refined sunseed oil."

"The homogenous mixture of the lipid matrix (LM_2) was stirred at room temperature until solidification."

"More so, after 24 h, various quantities (90, 80, and 60%w/w) of the prepared lipid matrix (LM_2) were melted together with corresponding amounts of polyethylene glycol (PEG 4000) (10, 20, and 40%w/w) incorporated at 80 °C over the oil bath to give PEGylated lipid matrices containing 1:9, 2:8, 4:6 ratios of PEG: lipid matrix, respectively, which were appropriately stirred and allowed to solidify."

Preparation of Drug-Loaded PEGylated Lipid Matrices

"Representative drug-loaded PEGylated lipid matrices were prepared by fusion using the PEGylated lipid matrices and metformin."

"With target PEGylated lipid concentration of 5.0%w/w and target drug concentrations of 1.0%w/w of metformin in the PEGylated nanostructured lipid carriers to be developed, 2.5 g of each of the lipid matrices was melted in the thermoregulated oil bath at a temperature of 80 °C followed by the addition of 0.5 g of metformin."

"Each mixture was stirred continuously until a homogenous, transparent white melt was obtained."

"The drug-loaded lipid matrices were allowed to cool and solidify at room temperature."

Differential Scanning Calorimetry (DSC) Analysis of Plain and Drug-Loaded PEGylated Lipid Matrices

"About 5 mg of each sample was weighed into an aluminum pan and hermetically sealed, and the thermal behavior was determined in the range of 20–350 °C at a heating rate of 5 °C/min."

"The temperature was held at 80 °C for 10 min and, after that, cooled at the rate of 5 to 10 °C/min."

"Baselines were determined using an empty pan, and all the thermograms were baseline-corrected."

Fourier Transform Infrared (FT-IR) Spectroscopic Analysis of Drug-Loaded PEGylated Lipid Matrices

"The spectrum was recorded in the wavelength region of 4000 to 400 cm^{-1} with a threshold of 1.303, sensitivity of 50, and resolution of 2 cm-1."

"A smart attenuated total reflection (SATR) accessory was used for data collection."

"The pellet was placed in the light path, and the spectrum was obtained."

"Spectra were collected in 60 s using Gram A1 spectroscopy software, and the chemometrics was performed using TQ Analyzer1."

Preparation of PEGylated Nanostructured Lipid Carriers

"PEGylated nanostructured lipid carriers encapsulating metformin (D_0, D_{10}, D_{20}, D_{40}) were prepared using the drug, PEGylated lipid matrices, Polysorbate® 80 (Tween® 80) (mobile surfactant), sorbitol (cryoprotectant), and distilled water (vehicle) by the high shear hot homogenization method [81, 82]."

"The PEGylated lipid matrix was melted at 80 °C in the thermoregulated heater (IKA instrument), and metformin was introduced into the melted lipid and stirred thoroughly."

"India), and the mixture was further dispersed using an Ultra-Turrax T25 (IKA-Werke, Staufen, Germany) homogenizer at 1000 rpm for 5 min."

"The obtained pre-emulsion was homogenized at 15,000 rpm for 30 min and allowed to cool and re-crystallize at room temperature."

Characterization of the Non-PEGylated and PEGylated NLCs

"Mean diameter, Z. Ave (nm), and polydispersity indices (PDI) of the formulations were measured using a zeta sizer nano-ZS (Malvern Instrument, Worcestershire, UK) equipped with a 10-mw He–NE laser employing the wavelength of 633 nm and a backscattering angle of 173° at 25 °C."

"The stability of the formulations was ascertained using zeta potential measurement."

"Each sample was diluted with deionized water to avoid multiple scattering and to maintain the number of counts per second in the region of 600, and measured at angle of 90° and temperature of 25 °C."

Compatibility Study by Fourier Transform Infrared (FTIR) Spectroscopy

"Fourier transform infrared (FT-IR) spectroscopic analysis was conducted using a Shimadzu FT-IR 8300 Spectrophotometer (Shimadzu, Tokyo, Japan), and the spectrum was recorded in the wavelength region of 4000 to 400 cm^{-1} with a threshold of 1.303, the sensitivity of 50, and resolution of 2 cm^{-1} range."

"A smart attenuated total reflection (SATR) accessory was used for data collection."

"The pellet was placed in the light path, and the spectrum was obtained."

"Spectra were collected in 60 s using Gram A1 spectroscopy software, and the chemometrics was performed using TQ Analyzer1."

Determination of Encapsulation Efficiency

"The encapsulation efficiency of each formulation was determined."

"A 5 ml volume of each formulation was placed in a centrifuge tube and centrifuged for 30 min at an optimized speed of 4000 rpm to obtain two phases (the aqueous and lipid phases)."

"A 1 ml volume of the aqueous phase was measured with the aid of a syringe and then diluted 10, 000-fold using distilled water."

"The absorbance of the dilutions was taken using a UV spectrophotometer (Jenway 6405, UK) at a wavelength of 231.5 nm, and the EE % was calculated using the formula below:"

Loading Capacity

"Loading capacity of the drug in lipid carriers depends on the type of lipid matrix, solution of drug in melted lipid, miscibility of drug melt and lipid melt, chemical and physical structure of solid lipid matrix, and the polymorphic state of the lipid material [83]."

"Loading capacity (LC) is expressed as the ratio between the entrapped drug by the lipid and the total quantity of the lipids used in the formulation."

In Vitro Drug Release Studies

"The dissolution medium consisted of 500 mL of freshly prepared phosphate buffer maintained at 37 ± 1 °C utilizing a thermoregulated water bath."

"The polycarbonate dialysis membrane used as a release barrier was pre-treated by soaking it in the dissolution medium for 24 h before the commencement of each release experiment."

"2 ml of the drug-loaded PEGylated NLC was placed in the dialysis membrane, securely tied with a thermoresistant thread, and then immersed in the dissolution medium under agitation provided by the paddle at 200 rpm."

In Vivo Antidiabetic Studies

"Wistar strain albino rats of both sexes weighing between 150 and 200 g were bred in our institution."

"The animals were housed in standard environmental conditions, kept at a body temperature of 37 °C using warming lamps, and left for one week to acclimatize with the new laboratory environment while being fed a standard laboratory low chow diet."

"All the animals were fasted for 12 h but were allowed free access to water before the commencement of the experiments."

Induction of Diabetes

"The rats were divided into groups of six animals, and each group of animals was housed in a separate cage."

"Diabetes was induced by a single intraperitoneal injection of a freshly prepared solution of alloxan (150 mg/ kg) in normal saline for all the groups."

"The diabetic rats (glucose level above 200 mg/dl) were separated into ten groups of six animals (n = 6)."

Evaluation of Antidiabetic Activity

"The different formulations of the metformin-loaded PEGylated NLC were administered orally to the animals according to their weight as follows: Group A:

received the test formulation (batch D_0) equivalent to 100 mg/kg of metformin hydrochloride."

"Group B: received the test formulation (batch D_{10}) equivalent to 100 mg/kg of metformin hydrochloride."

"Group C: received the test formulation (batch D_{20}) equivalent to 100 mg/kg dose of metformin hydrochloride."

"Group D: received the test formulation (batch D_{40}) equivalent to 100 mg/kg dose of metformin hydrochloride."

"Group F: received pure metformin equivalent to 100 mg/kg dose of metformin hydrochloride."

In Vivo Bioavailability Study

"The formulations were administered orally, and thereafter blood samples were collected from the tail vein of the rats at time intervals of 0, 1, 3, 6, 12, and 24 h using heparinized hematocrit tubes, centrifuged at 5,000 rpm for 5 min to separate the plasma and then stored at -4 °C till analyzed."

"Approximately 0.2 ml of the plasma sample was diluted with equal volume of acetonitrile and centrifuged at 2,000 rpm for 5 min after which 0.1 ml of the supernatant was diluted in distilled water and assayed spectrophotometrically (Unico 2102 PC UV/Vis Spectrophotometer, New York, USA) for drug content."

"The plasma concentration–time curve was then plotted and further evaluated to obtain the pharmacokinetic parameters such as the maximum plasma concentration (C_{max}) and the corresponding time (T_{max}) using Phoenix® WinNonlin (version 6.3; Pharsight, St Louis, MO, USA), based on the average blood drug concentration."

Safety Studies

"Three groups of animals were selected from the antidiabetic studies in the previous section, i.e., one animal from each group (optimized formulation group (G_{40}), diabetic but untreated rat group and non-diabetic/normal rabbit group) was utilized for the purpose of histopathological studies in line with earlier reports [84, 85]."

"Small pieces of liver tissues in each group were collected in 10% neutral buffered formalin for proper fixation for 24 h. These tissues were processed and embedded in paraffin wax."

"Sections of 5–6 um in thickness were cut and stained with hematoxylin and eosin (H & E)."

"These sections were examined photomicroscopically for necrosis, steatosis and fatty changes of hepatic cell [84, 85]."

Statistical Analysis

"All experiments were performed in replicate for the validity of the statistical analysis."

"Analysis of variance (ANOVA) using LSD post hoc multiple comparison test was performed on the data sets generated using SPSS 16.0."

"Differences were considered significant for p-values < 0.05."

Results and Discussion

"The DSC thermograms of beeswax, PEG 4000, metformin, Phospholipon® 90H, and lipid matrix (LM_1) showed single sharp endothermic melting peaks at 73,

74.5, 265.5, and 122.1 °C with enthalpies of -55.3, -3.8, -28.4, and -38.9 mW/mg, respectively, indicating the highly crystalline nature of these materials."

"Lipid matrix (LM_1) and PEG lipid matrix (LM_2) showed endothermic melting peaks at 88.6 and 93.1 °C with enthalpies of -32.5 and -3.9 mW/mg, respectively, and the reduced enthalpy indicates that the lipid matrices are less crystalline than the individual components used in the formulation."

"The metformin-loaded PEGylated lipid matrix (drug-loaded PEG-LM_2) also showed two endothermic peaks at 89.0 and 258.0 °C with enthalpy of -0.7 and -0.4 mW/mg, indicating molecular dispersion of metformin in the lipid matrix."

Fourier Transform Infrared Spectroscopy of Drug and Drug-Loaded Lipid Matrices

"The spectrum of the metformin-loaded LM_2 shows principal absorption bands at 3819.18 cm^{-1} (O–H stretching), 3471.98 cm^{-1} (N–H stretching), 2916.47 cm^{-1} ($(CH_3)_2$-N absorption), 2854.74 cm^{-1} ($(CH_3)_2$-N absorption), 2314.66 cm^{-1} (-C = C- stretching), 1735.99 cm^{-1} (-C = O vibration), 1643.41 cm^{-1} (N–H deformation), 1458.23 cm^{-1} (symmetric N–H deformation), 1396.51 cm^{-1} (N–H deformation), 1165.04 cm^{-1} (C-N stretching), 1095.60 cm^{-1} (C-N stretching), 1049.31 cm^{-1} (C-N stretching), 717.54 cm^{-1} (N–H wagging), 601.81 cm^{-1} (C-H out-of-plane bending) and 547.80 cm^{-1}(C-N–C deformation)."

"The spectrum of the PEGylated metformin-loaded lipid matrix (drug-loaded PEG-LM_2) shows principal absorption bands at 3433.41 cm^{-1} (N–H stretching), 2916.47 cm^{-1} ($(CH_3)_2$-N absorption), 2854.74 cm^{-1} ($(CH_3)_2$-N absorption), 2314.66 cm^{-1} (-C = C- stretching), 1975.17 cm^{-1} (-C = O vibration), 1735.99 cm^{-1} (conjugated C = C bond vibration), 1651.12 cm^{-1} (N–H deformation), 1458.23 cm^{-1} (symmetric N–H deformation), 1226.77 cm^{-1} (C-O vibration), 1157.33 cm^{-1} (C-N stretching), 964.44 cm^{-1} (N–H out-of-plane bending), 840.99 cm^{-1} (NH_2 rocking), 717.54 cm^{-1} (C-H out-of-plane bending), and 586.38 cm^{-1} (C-N–C deformation)."

Mean Particle Size, Polydispersity Indices, and Surface Charges of the NLC Formulations

"The formulation with the most miniature particle size was D_{20} (290.6 nm), and formulation D_{10} showed maximum particle size (880.60 nm)."

"The PDI value of batch D_0 and D_{10} was 0.477 and 0.625, showing the minimum and maximum PDI in the formulation, respectively."

"These results showed a unimodal size distribution of the different particle sizes, indicating stable formulations [86]."

Fourier Transform Infrared Spectroscopy of Drug-Loaded NLC Formulations

"The FTIR spectrum of D_0 shows principal absorption peaks at 3363.97 cm^{-1} (N–H stretching), 2090.91 cm^{-1} (carboxylic acid C = O vibration), 1643.41 cm^{-1} (N–H deformation), 1103.32 cm^{-1} (C-N stretching), and 578.66 cm^{-1} (C-N–C deformation)."

"The FTIR spectrum of D_{20} shows principal absorption peaks at 3363.97 cm^{-1} ($(CH_3)_2$-N absorption), 2924.18 cm^{-1}, 2854.74 cm^{-1} (N–H stretching), 2121.77 cm^{-1} (carboxylic acid C = O vibration), 1643.41 cm^{-1} (N–H deformation), 1512.24 cm^{-1}

(asymmetric N–H deformation), 1087.89 cm^{-1} (C-N stretching), and 532.37 cm^{-1} (C-N–C deformation)."

"The FTIR spectrum of D_{20} shows principal absorption peaks at 3873.19 cm^{-1} (N–H stretching), 3302.24 cm^{-1} (asymmetric N–H stretching), 2924.18 cm^{-1}, 2507.54 cm^{-1} ($(CH_3)_2$-N absorption), 2360.95 cm^{-1} (-C = C- stretching), 2083.19 cm^{-1} (carboxylic acid C = O vibration), 1620.26 cm^{-1} (N–H deformation), 1535.39 cm^{-1} (asymmetric N–H deformation), 1411.94 cm^{-1} (N–H deformation), 1033.88 cm^{-1} (C-N stretching), 794.7 cm^{-1} (N–H wagging) and 493.79 cm^{-1} (C-N–C deformation)."

Encapsulation Efficiency and Loading Capacity of PEGylated and Non-PEGylated NLCs

"The drug loading efficiency could be affected by the drug's molecular weight, the volumetric size of the carrier, chemical interactions between the drug and the carrier, solubility of the drug in the carrier, miscibility in the lipid matrix, and the lipid phase polymorphic state [87–89]."

"Formulations with smaller particle sizes had higher encapsulation efficiency and drug loading capacity, and this is consistent with previous reports."

"The encapsulation efficiencies were in the range of 98.78–99.83% for the NLCs, with the formulation D_0 showing the highest EE%, while D_{20} had the least."

"The loading capacity (LC) of the NLC batches was in the range of 16.50–16.64 g of metformin per 100 g of lipid."

In Vitro Drug Release From the Metformin-Loaded PEGylated and Non-PEGylated NLCs

"The in vitro release kinetics of nanoparticles provides vital information regarding the ability to modify the drug release; when performed correctly, they can be correlated to in vivo behaviors through predictive mathematical models [90, 91]."

"Many mathematical models have been proposed to predict release-kinetic models, such as the zero-order, first-order, Higuchi, Korsmeyer–Peppas, Weibull, Gompertz, Peppas–Sahlin, Bakers–Lonsdale, Hopfenberg, Hixson–Crowell [92]."

"Applying the kinetic model is critical in clarifying the release mechanism, which helps design the drug release control [93]."

"Five different mathematical equations were used to predict the release kinetics: zero-order equation, first-order equation, Hixson–Crowell, Higuchi, and Korsmeyer–Peppas models to describe the kinetics of metformin release from the PEG-NLC formulations."

"The formulations predominantly followed the diffusion mechanism of drug release."

"The release rate of the drug was higher in the formulations than the release in Glucophage."

"PEGylation enhanced the dissolution and release rate of the drug from the formulations."

In Vivo Antidiabetic Activity

"The effect of the PEGylated nanostructured lipid carrier (PEG-NLC) on the blood glucose level of non-diabetic rats was assessed and compared to both the pure drug sample and commercially available product (glucophage®)."

"In the animal groups treated with the batches D_0, D_{10}, $D_{20,}$ and D_{40}, the blood glucose reduction effectively commenced within an hour of oral administration with maximum blood glucose reduction (T_{max}) achieved in 24 h. The blood glucose level produced by D_0, $D_{10,}$ and D_{40} formulations reduced from 100% to 44.36%, 43.51%, and 44.41%, respectively, indicating that batch D_{40} was the best formulation, which was optimized for further in vivo bioavailability study."

"With the pure drug and commercial sample, the formulations generally maintained the blood glucose level of the rats within the normoglycemic level for 12–24 h. Thus, it depicts that metformin could be effectively and efficiently delivered as sunseed-oil-based PEGylated NLCs."

In Vivo Oral Bioavailability

"The mean AUC_{-24} values for optimized metformin-loaded non-PEGylated and PEGylated NLC indicate approximately threefold and sixfold increase in systemic bioavailability of metformin from metformin-loaded non-PEGylated and PEGylated NLC, respectively."

"Metformin-loaded non-PEGylated and PEGylated NLC formulations maintained a steady slow decrease or gradual clearance of drug throughout the study compared with the rapid exponential decrease in the marketed sample (Glucophage®)."

"The decrease in metformin concentration in the blood by the formulation could be attributed to an increase in the circulating half-life of metformin when administered as metformin-loaded PEGylated NLC."

"By implication, there was enhancement in the circulation longevity of metformin in the developed NLC formulations, but the effect was highest with metformin-loaded PEGylated NLC."

Safety of the Metformin-Loaded PEGylated NLC Formulations

"The observation from the histological studies conducted on the liver of the rabbits from various groups showed significant hepatitis (arrow) on the liver of the rats in diabetic but untreated group (b)."

"Photomicrograph of liver section of rats from experimental diabetic group that received metformin-loaded PEGylated NLC (optimized batch D_{40}) (a) showed reduced periportal mononuclear infiltration of cells-periportal hepatitis (arrow), while the non-diabetic and untreated group (c) showed normal portal area and hepatocytes (H and E, mag."

"The results of histological study showed that short-term administration of developed metformin-loaded PEGylated NLC formulation did not cause any significant changes in the body regarding the relative liver morphology of the treated rats."

"By implication, short-term oral administration of sunseed-oil-based metformin-loaded PEGylated NLC will have no negative effects on somatic growth."

Conclusions

"This study evaluates the potential of metformin-loaded beeswax-based PEGylated nanostructured lipid carrier (PEG-NLC) for enhanced metformin delivery to treat diabetes mellitus."

"Metformin-loaded beeswax-based PEGylated nanostructured lipid carriers (PEG-NLCs) were successfully prepared by high shear hot homogenization method using structured lipids (beeswax and Phospholipon 90H), liquid lipids (sunseed oil), and PEG-4000 and evaluated for improved treatment of diabetes mellitus."

"Using sunseed-oil-based PEGylated NLC as a delivery system for metformin achieves a steady-state drug concentration in the blood that is therapeutically effective, safe, and remains for an extended period."

"Sunseed-oil-based PEG-NLCs present a promising approach for improving metformin delivery and oral bioavailability."

4.3 Pharmacokinetics and Pharmacodynamics of Vesicular Nanocarriers

Introduction by the Editor

Vesicular nanocarriers, including liposomes, niosomes, and ethosomes, represent a remarkable frontier in drug delivery and encapsulation. These lipid-based nanoscale structures consist of bilayer membranes that encapsulate therapeutic agents, offering a biocompatible and versatile platform for controlled drug release and targeted delivery. Vesicular nanoparticles involve various lipid classes such as phospholipids (e.g., phosphatidylcholine), glycerides (e.g., triglycerides), and surfactants (e.g., Tween, Span). Phospholipids contribute to structural stability and biocompatibility, forming the basis for liposomes and lipid bilayers in lipid nanoparticles. Glycerides provide a hydrophobic core for drug encapsulation and controlled release, while surfactants aid in particle stabilisation and dispersion. Liposomes are composed of phospholipid bilayers, mimic biological membranes, and can encapsulate both hydrophobic and hydrophilic compounds. Niosomes, composed of non-ionic surfactants, exhibit enhanced stability and adaptability for encapsulating various drug types. Ethosomes, enriched with ethanol, facilitate deeper penetration through the skin, making them valuable for transdermal drug delivery. These vesicular nanoparticles hold tremendous potential for modifying drug release kinetics, enhancing solubility, and optimising the delivery of drugs to specific tissues.

Machine generated summaries

Disclaimer: The summaries in this chapter were generated from Springer Nature publications using extractive AI auto-summarization: An extraction-based summarizer aims to identify the most important sentences of a text using an algorithm and uses those original sentences to create the auto-summary (unlike generative AI). As the constituted sentences are machine selected, they may not fully reflect the body of the work, so we strongly advise that the original content is read and cited. The auto generated summaries were curated by the editor to meet Springer Nature publication standards.

To cite this content, please refer to the original papers.

Machine generated keywords: dispersion, liposome, breast, cancer, optimize, lipid, aqueous, gel, bioavailability, permeability, transdermal, breast cancer, death, anticancer, ocular

Response surface optimization of biocompatible elastic nanovesicles loaded with rosuvastatin calcium: enhanced bioavailability and anticancer efficacy [94] This is a machine-generated summary of:

Elsayed, Ibrahim; El-Dahmy, Rania Moataz; El-Emam, Soad Z.; Elshafeey, Ahmed Hassen; El Gawad, Nabaweya Abdelaziz Abd; El-Gazayerly, Omaima Naim: Response surface optimization of biocompatible elastic nanovesicles loaded with rosuvastatin calcium: enhanced bioavailability and anticancer efficacy [94]

Published in: Drug Delivery and Translational Research (2020)
Link to original: https://doi.org/10.1007/s13346-020-00761-0

If you want to cite the papers, please refer to the original.

For technical reasons we could not place the page where the original quote is coming from.

Abstract-Summary "The goal of this study was to statistically optimize novel elastic nanovesicles containing rosuvastatin calcium to improve its transdermal permeability, bioavailability, and anticancer effect."

"The optimized elastic nanovesicular dispersion is composed of 20% cetyl alcohol, 53.47% Tween 80, and 26.53% clove oil."

"Carboxy methylcellulose was utilized to convert the optimized elastic nanovesicular dispersion into elastic nanovesicular gels."

"The cell viability assay of the optimized gel on MCF-7 and Hela cell lines showed significant antiproliferative and potent cytotoxic effects when compared to the drug gel."

"The optimized gel accomplished a significant increase in rosuvastatin bioavailability upon comparison with the drug gel."

"The optimized gel could be considered as a promising nanocarrier for statins transdermal delivery to increase their systemic bioavailability and anticancer effect."

Introduction

"Transdermal delivery of ROS can be a reasonable alternative to achieve sustained and controlled effects, reduce the incidence of adverse effects, and escape from the first-pass metabolism [95, 96]."

"ROS was previously prepared in the form of solid lipid nanoparticles (RC-SLNs) which achieved 4.6-fold bioavailability enhancement when compared to the drug suspension after oral administration by Wistar rats [97]."

"Permeation enhancers such as essential oils are frequently utilized to improve the transdermal drug delivery [98]."

"Among five essential oils of clove, chuanxiong, angelica, cinnamon, and Cyprus oils, only clove oil was found to be the optimum permeation enhancer for the transdermal delivery of ibuprofen where the drug bioavailability was increased by 2.4-fold upon comparison with the control [99, 100]."

"This would enhance ROS bioavailability, minimize its hepatic uptake, and consequently, increase its therapeutic concentration available at the site of action and potentiate its anticancer effect."

Materials and methods

"Following drying at 37 °C, the stained drop was examined under TEM at a voltage of 100 kV. The ROS permeation through newly born rat skin was investigated from both the OEND and the OENG formulae and compared to its permeation from the equivalent drug suspension and gel."

"The transdermal permeation enhancement ratio (ER) was estimated utilizing the following equation [101]: Samples from both the OEND and OENG formulae were applied onto newly born rat skin membranes for 1 h under the same conditions applied in both the in vitro release and the ex vivo permeation studies."

"After the application of both OENG and the drug gel, blood samples (3 mL) were taken from the rabbits' ear vein into pre-heparinized glass tubes at different time intervals of 0.5, 1, 2, 3, 4, 6, 8, and 24 h. The gathered samples underwent centrifugation with a speed of 4000 rpm at 4 °C for 15 min."

Results and discussion

"The enormous surface area of the formed nanovesicles and the incorporation of T80 enhanced ROS diffusion from the prepared OEND to the medium leading to a significant increase in ROS release rate and extent when compared to the drug suspension with f_2 value of 31 [102, 103]."

"The EN gel containing 2% CMC was chosen as the OENG due to having the closest PS, PDI, ZP, and EE values to those of the OEND formula and due to having the highest release $T_{50\%}$ and viscosity when compared to the other EN gels."

"The high zeta potential value of the OENG formula (− 42.51 mV) enhanced its penetration through the cancer cells as it was previously stated that ZP values above + 10 mV or − 30 mV could boost the drug accumulation and prevent its diffusion out of tumor cells [104, 105]."

Conclusion

"It was converted to gel (OENG), utilizing 2% CMC, to increase its physical stability and facility its application on the skin."

"The optimized EN gel (OENG) was found to be biocompatible and non-irritant upon application onto newborn rat skin."

"ROS transdermal permeation from the OENG was significantly higher than the drug gel."

"ROS bioavailability in the case of the OENG was significantly higher than the drug gel."

A review on multivesicular liposomes for pharmaceutical applications: preparation, characterization, and translational challenges [106] This is a machine-generated summary of:

Chaurasiya, Akash; Gorajiya, Amruta; Panchal, Kanan; Katke, Sumeet; Singh, Ajeet Kumar: A review on multivesicular liposomes for pharmaceutical applications: preparation, characterization, and translational challenges [106]

Published in: Drug Delivery and Translational Research (2021)
Link to original: https://doi.org/10.1007/s13346-021-01060-y

If you want to cite the papers, please refer to the original.

For technical reasons we could not place the page where the original quote is coming from.

Abstract-Summary "The usage of MVL for sustained drug delivery has seen progression over the last decade due to successful clinical and commercial applications."

"Drug molecules are encapsulated in internal aqueous compartments of MVL, separated by lipid bilayer septa to form polyhedral structures."

"Despite the frequent use of unilamellar liposomes, characterization of MVLs is critical due to different puzzling problems, such as real-time size evaluation, initial burst, and in vivo performance."

"Available regulatory guidelines on liposomal drug product development are insufficient to assure ample in vitro-in vivo behavior of MVL."

"This review hereby highlights the innovations pertaining to development and manufacturing procedures, drug release mechanisms, and characterization techniques."

"Despite the intricacies involved in the development of MVL, establishing steadfast characterization techniques and regulatory paths could pave the way to its extensive clinical use."

Introduction

"This necessity led to the development and clinical application of multivesicular liposomes (MVLs)."

"The structure and conformation of MVLs are the underlying reason for its higher stability and extended drug release [107]."

"Drug release from MVLs combines the diffusion of Active Pharmaceutical Ingredient (API) through the MVL membrane, the gradual erosion of the same, and rearrangement of "polyhedral" vesicles [108, 109]."

"Besides using the amphipathic phospholipid (AP), it is mandatory to use a neutral lipid such as triglycerides in the production of MVLs; else, this would lead to the production of unilamellar or multilamellar liposomes."

"Despite existing techniques used to characterize conventional liposomes, the study of MVLs becomes difficult due to their size, complex structure, and unconventional release pattern demonstrating complex kinetics [108, 110]."

"Considering the importance of MVL in sustained release application and delivery of therapeutic agents, we have reviewed various opportunities and challenges associated with MVL-based product development."

Mechanism of MVL formation and drug release

"The selection of these excipients and its optimization can help design a stable MVL system with desired drug release characteristics."

“Proper optimization of AP is essential to develop MVL with maximum drug entrapment, desired drug release, and required stability.”

“Various factors facilitating the functioning of CP during and after MVL preparation, like buffer type, ionic strength, and CP concentration, needed to be optimized to develop a stable drug product [111].”

“Different process conditions like time, shear, temperature, and solvent evaporation can impact the preparation of stable particles with desired properties.”

“Even MVL in finished products is so delicate and fragile that particles readily undergo physicochemical degradation in unwanted conditions of high temperature, humidity, etc During MVL preparation, various intermediates are formed (like primary emulsion, secondary emulsion, spherules), which are stable for a limited time.”

Characterization techniques

“Various quality attributes like particle size, surface charge, membrane rigidity, entrapment efficiency, and drug release characteristics can impact the fate of drug products during storage and usage [112].”

“These techniques not only help to understand the structure of MVLs, but also help to speculate the mechanism and release rate of drug molecules from the carrier as they undergo agglomeration, collapse, and rearrangement of honeycomb structures in the release media.”

“Drug encapsulation in MVL can be measured by separating and processing free drug from entrapped drug and analyzing drug content using techniques like spectroscopy and chromatography [113, 114].”

“The measurement of pH in the external and internal phases will help understand the stability and drug release characteristics of the developed MVL product.”

“Ionic concentration across the lipid membrane highly impacts the drug entrapment and stability of MVL particles.”

Commercial and clinical status

“Exparel® is a multivesicular liposomal injectable formulation of bupivacaine, a non-opioid analgesic based on DepoFoam® technology, developed and marketed by Pacira Pharmaceuticals Inc. Exparel® is used as a local anesthetic for the treatment of infiltration/field block and nerve block.”

“For sustained delivery of TXA during intended usage, it is formulated into an MVL using DepoFoam® technology as DepoTXA by Pacira Pharmaceuticals.”

“A formulation was developed to release the drug in a sustained manner over a more extended time by encapsulating leridistim in multivesicular liposomes (DepoLeridistim).”

“Sustained pharmacodynamic effects by elevated neutrophil counts were observed with this novel formulation resulting in reduced dosing frequency in contrast to daily injections with unencapsulated leridistim [115].”

“Similar to small molecules, MVLs also demonstrated high drug loading and sustained release of therapeutic proteins and peptides.”

“Insulin-like growth factor-I (IGF-I) is rapidly cleared from circulation after subcutaneous or intravenous injections; therefore, a lipid based sustained release system using DepoFoam® technology (DepoIGF-I) was developed.”

Applications of MVL

"To overcome this issue, cytarabine was formulated into MVL for sustained drug delivery to the eye."

"DDC (2′, 3′-dideoxycytidine) was encapsulated into the MVL; resulting in a 20-fold increase in the half-life of the drug in-vivo and thus offering the possibility of intrathecal drug delivery for molecules that are unable to cross the blood–brain barrier [116]."

"The MVL formulation of ropivacaine released the drug over a prolonged period and showed improved pharmacokinetic behavior, significantly reducing its toxicity on subcutaneous injection to rats [117]."

"Similar to this study, bleomycin-loaded MVL product was developed to maintain drug levels for a prolonged period and reduce the toxicity related to frequent dosing or continuous administration."

"The results from the in-vivo study demonstrated reduced peptide degradation and exhibited its sustained release when formulated into MVL [118]."

Translational challenges

"The development and manufacturing of complex drug products like MVL are complicated and require a thorough understanding of science involved and expertise over the parameters that affect the critical quality attributes (CQAs) of the finished product."

"We will discuss various challenges and roadblocks associated with MVL-based product development, large-scale manufacturing, intellectual property (IP), and regulatory approval."

"Proper design of experiments is needed to optimize formulation and process variables, which plays an important role in obtaining stable MVL products with desirable characteristics like particle size, encapsulation efficiency, and drug release."

"Scarcity of the subject matter experts possessing sufficient knowledge and experience in MVL-based product development/manufacturing have led to very few successful formulations entering the commercial market."

"Due to the lack of general guidelines on the development of MVL based drug products, the industries face challenges in understanding regulatory bodies' expectations for the design of such products."

Liposomes as a Novel Ocular Delivery System for Brinzolamide: In Vitro and In Vivo Studies [119] This is a machine-generated summary of:

Li, Huili; Liu, Yongmei; Zhang, Ying; Fang, Dailong; Xu, Bei; Zhang, Lijing; Chen, Tong; Ren, Ke; Nie, Yu; Yao, Shaohua; Song, Xiangrong: Liposomes as a Novel Ocular Delivery System for Brinzolamide: In Vitro and In Vivo Studies [119]

Published in: AAPS PharmSciTech (2015)

Link to original: https://doi.org/10.1208/s12249-015-0382-1

If you want to cite the papers, please refer to the original.
For technical reasons we could not place the page where the original quote is coming from.

Abstract-Summary "The liposomes of Brz (Brz-LPs) were produced by the thin-film dispersion method with a particle size of 84.33 ± 2.02 nm and an entrapment efficiency of 98.32 ± 1.61%."

"The corneal permeability was measured using modified Franz-type diffusion cells, and Brz-LPs showed 6.2-fold increase in the apparent permeability coefficient when compared with the commercial available formulation (B rz-Sus)."

"Brz-LPs (1 mg/mL Brz) showed a more sustained and effective intraocular pressure reduction (5–10 mmHg) than Brz-Sus (10 mg/mL Brz) in white New Zealand rabbits."

"Brz-LPs were a hopeful formulation of Brz for glaucoma treatment and worthy of further investigation."

INTRODUCTION

"Brinzolamide (Brz), one of the carbonic anhydrase inhibitors, is effective in reducing IOP for glaucoma therapy mainly through decreasing the production of the aqueous humor [120]."

"Brz-Sus, with obvious granular sensation, usually has poor patient compliance and low bioavailability caused by the uncomfortable feeling and subsequently undesirable tears wash to wipe the drug off rapidly after instillation."

"Our group has successfully developed a solution formulation of Brz using the inclusion complex of hydroxypropyl-β-cyclodextrin, in which just half dosage of Brz achieved an equivalent IOP-lowing efficiency of Brz-Sus in vivo [121]."

"The aim of this work was to develop liposomes as a novel ocular delivery system for Brz (Brz-LPs) to improve its retention time and IOP-lowering efficacy."

"The in vitro corneal permeability and in vivo IOP-lowering efficiency of Brz-LPs were evaluated by comparison with Brz-Sus."

MATERIALS AND METHODS

"The unentrapped Brz was firstly removed from the liposomes by centrifuging at 50,000 rpm for 30 min at a temperature of 4°C followed by removal of the supernatant."

"Around 5~6 mg samples (b-LPs, physical mixture of Brz and b-LPs (PM), pure Brz and Brz-LPs) were sealed in standard aluminum pans, respectively, and heated at a constant rate of 10°C/min in a nitrogen atmosphere with a temperature range of 50~350°C."

"The powder samples (pure Brz, PM, b-LPs, and Brz-LPs) were examined with an X-ray power diffractometer (XRD; X'Pert Pro Philips, Netherlands) operated with Cu Kα X radiation, a voltage of 40 kV, and a current of 40 mA. The scans were conducted at a scanning rate of 10°/min in the 2θ range from 50° to 300°."

RESULTS AND DISCUSSION

"Brz-LPs were successfully prepared [122], and the process parameters were systemically optimized according to our preliminary screening study, including the lipid-to-cholesterol (L/C) molar ratio, lipid-to-drug (L/D) molar ratio, and ultrasonic power."

"Further increase in the cholesterol content (when the L/C molar ratio was set at 7:7) did not correspondingly enhance the Brz entrapment into liposomes."

"The optimal Brz-LPs achieved an EE of 98.32 ± 1.61% (n = 3)."

"Three batches of Brz-LPs had no significant difference in EE, size, and zeta potential, which demonstrated that the preparation process was reproducible."

"After 12 h, there almost 100% Brz was released from Brz-LPs."

"Brz-LPs showed a 6.2-fold increase in P_{app} when compared with Brz-Sus."

"The higher ocular hypotensive effect produced by Brz-LPs attributed to the fact that liposomes had a higher binding affinity to the corneal surface than Brz-Sus [123], which caused better absorption of Brz."

CONCLUSIONS

"Brz-LPs were successfully prepared by a modified thin-film dispersion method."

"TEM, DSC, XRD, and FTIR results revealed the successful formation of Brz-LPs."

"Ex vivo corneal transport experiment indicated that Brz-LPs had better trans-membrane permeation ability than Brz-Sus and could improve the transport of Brz across the cornea by sixfold."

QbD-based optimization of raloxifene-loaded cubosomal formulation for transdemal delivery: ex vivo permeability and in vivo pharmacokinetic studies [124] This is a machine-generated summary of:

Gupta, Tanushree; Kenjale, Prathmesh; Pokharkar, Varsha: QbD-based optimization of raloxifene-loaded cubosomal formulation for transdemal delivery: ex vivo permeability and in vivo pharmacokinetic studies [124]

Published in: Drug Delivery and Translational Research (2022)

Link to original: https://doi.org/10.1007/s13346-022-01162-1

If you want to cite the papers, please refer to the original.

For technical reasons we could not place the page where the original quote is coming from.

Abstract-Summary "The objective of the present study was to develop an alternative transdermal delivery of RLX to improve its absorption, bypass first pass metabolism, and subsequently improve bioavailability."

"RLX-loaded cubosomes were prepared using the ethanol injection method followed by microfluidization technique and optimized using the QbD-based 2^3 factorial design."

"In vitro dissolution study indicated that the RLX-loaded cubosomes released 98.26% of the drug compared to pure RLX dispersion (58.6%)."

"Accelerated stability study as per ICH guidelines displayed no significant change in the formulation characteristics and drug-related performance of the developed formulation."

"Ex vivo permeability studies demonstrated a prolonged release from cubosomal formulation."

"Pharmacokinetic studies revealed that the relative bioavailability of the optimized transdermal RLX-loaded cubosomes increased by 2.33-fold and 1.22-fold when compared with the oral RLX dispersion and transdermal RLX hydro-ethanolic solution respectively."

"The developed RLX-loaded cubosomes may have potential to overcome the problems associated with the existing marketed oral dosage forms of RLX."

Introduction

"The self-assembly of amphiphilic lipids as a result of the hydrophobic effect could potentially lead to some well-defined, thermodynamically stable structures such as lamellar (La), hexagonal (HII), and bicontinuous (QII) cubic phases, all of which have a sufficient average degree of molecular orientation and structural symmetry, collectively known as lyotropic liquid crystal (LLC) systems [125]."

"Liposomes are the most common nanostructures created by dispersion of these poorly water-soluble structures in aqueous media, as a result of the dispersion of a large lamellar phase; cubosomes, formed by reversed bicontinuous cubic phase; and hexosomes, formed by reversed bicontinuous cubic phase."

"Cubosomes are thus bicontinuous cubic liquid phases that include two distinct water regions separated by surfactant-controlled bilayers."

"The present strategy was to explore cubosomes to improve drug solubility, drug loading capacity, and effective delivery for RLX via transdermal route."

Materials and method

"One milliliter of freshly manufactured RLX-loaded cubosomal dispersions was diluted in purified water to make 10 mL, and 3 mL of the diluted samples was deposited in an Eppendorf tube and centrifuged at 4000 rpm for 15 min."

"Samples were mixed with KBr and FTIR of pure RLX, RLX + GMO physical mixture, RLX + F127 physical mixture, and physical mixture, and optimized formulation of RLX-loaded cubosome (F8) was scanned in transmission mode over a wavelength range of 3600–600 cm^{-1}."

"Developed cubosomal formulation of RLX was applied for 2 h and 24 h respectively to observe the histopathological differences on excised female Wistar rat skin."

"Whereas, the second and third group of rats got a single transdermal dosage of 2 mg/kg of RLX hydro-ethanolic solution and 2 mg/kg of optimized RLX-loaded cubosomes respectively applied to the dorsal skin above the spine."

Results and discussion

"The examination of the three-dimensional response surface diagrams indicated that the formulation variables amount of RLX and amount of Pluronic F127 have a strong influence on %EE, particle size distribution, and zeta potential."

"The relative bioavailability of the developed RLX-loaded cubosomal formulation was obtained by comparing the bioavailability with the pure RLX dispersion given orally and RLX hydro-ethanolic solution applied transdermally to the male Wistar rats in the in vivo pharmacokinetic studies."

"The pharmacokinetic parameters of optimized RLX-loaded cubosome formulation were enhanced with a C_{max} value of 39.97 ± 0.901 μg/mL, $t_{½}$ of 17.89 ± 1.1 h, and $AUC_{0\text{-}72}$ of 1104.76 ± 1.72 h μg/mL. Thus, the administration of RLX-loaded cubosomes transdermally showed improved bioavailability when compared with pure RLX given orally and transdermally due to its nanoparticle size, unique structure of cubosomes, and its structural similarity to the skin which provides high flexibility in transdermal drug delivery for the lipophilic drugs [126]."

Conclusion

"Transdermal administration of RLX-loaded cubosomes with improved pharmacokinetics was examined for the first time in this study."

"The QbD approach was successfully used to optimize the composition of the RLX-loaded cubosomal formulation."

"Pharmacokinetic study clearly established the potential of the cubosomal formulation of RLX in enhancing its bioavailability."

"In vivo pharmacodynamic studies are in progress to establish the potential of RLX-loaded cubosomes in the treatment of breast cancer."

4.4 Pharmacokinetics and Pharmacodynamics of Metallic Nanocarriers

Introduction by the Editor

Metallic nanoparticles have emerged as versatile carriers for drug delivery and theranostic applications, seamlessly merging therapeutic and diagnostic functions. By precisely engineering their size, shape, and surface properties, these nanoparticles offer targeted drug delivery, optimizing treatment efficacy and minimizing side effects. Their ability to encapsulate drugs, peptides, nucleic acids and provide controlled drug release aligns with personalized medicine goals. Moreover, the inherent physicochemical properties of metallic nanoparticles, such as gold and silver, facilitate multimodal imaging, enabling real-time visualization of disease sites and therapeutic responses. Engineered at the nanoscale, these particles exhibit unique physicochemical properties owing to their small size and high surface area-to-volume ratio. Gold nanoparticles, for instance, are known for their tunable optical properties and potential in diagnostics, imaging, and targeted drug delivery. Silver nanoparticles possess antimicrobial capabilities with applications in wound healing and infections. Iron oxide nanoparticles exhibit superparamagnetic behaviour, rendering them valuable for magnetic resonance imaging and targeted drug delivery.

Machine Generated Summaries

Disclaimer: The summaries in this chapter were generated from Springer Nature publications using extractive AI auto-summarization: An extraction-based summarizer aims to identify the most important sentences of a text using an algorithm and uses those original sentences to create the auto-summary (unlike generative AI). As the constituted sentences are machine selected, they may not fully reflect the body of the work, so we strongly advise that the original content is read and cited. The auto generated summaries were curated by the editor to meet Springer Nature publication standards.

To cite this content, please refer to the original papers.

Machine generated keywords: biomedical, nanoparticle, nanostructure, gold, diameter, toxicity, mgkg, charge, polyethylene, average, biomolecule, systemic toxicity, core, narrow, depend

Pharmacokinetics, clearance, and biosafety of polyethylene glycol-coated hollow gold nanospheres [127] This is a machine-generated summary of:

You, Jian; Zhou, Jialin; Zhou, Min; Liu, Yang; Robertson, J David; Liang, Dong; Van Pelt, Carolyn; Li, Chun: Pharmacokinetics, clearance, and biosafety of polyethylene glycol-coated hollow gold nanospheres [127]

Published in: Particle and Fibre Toxicology (2014)

Link to original: https://doi.org/10.1186/1743-8977-11-26

If you want to cite the papers, please refer to the original.

For technical reasons we could not place the page where the original quote is coming from.

Abstract-Summary "Hollow gold nanospheres (HAuNS) is a class of photothermal conducting agent that have shown promises in photoacoustic imaging, photothermal ablation therapy, and drug delivery."

"We investigated the cytotoxicity, complement activation, and platelet aggregation of polyethylene glycol (PEG)-coated HAuNS (PEG-HAuNS, average diameter of 63 nm) in vitro and their pharmacokinetics, biodistribution, organ elimination, hematology, clinical chemistry, acute toxicity, and chronic toxicity in mice."

"PEG-HAuNS did not induce detectable activation of the complement system and did not induce detectable platelet aggregation."

"The single effective dose of PEG-HAuNS in photothermal ablation therapy was determined to be 12.5 mg/kg."

"Quantitative analysis of the muscle, liver, spleen, and kidney revealed that the levels of Au decreased 45.2%, 28.6%, 41.7%, and 40.8%, respectively, from day 14 to day 90 after the first intravenous injection, indicating that PEG-HAuNS was slowly cleared from these organs in mice."

"Our data support the use of PEG-HAuNS as a promising photothermal conducting agent."

Background

"Gold nanoparticles (AuNPs) show several features that make them well suited for biomedical applications, including straightforward synthesis, stability, and the ability to selectively incorporate recognition molecules such as peptides or proteins [128]."

"Hollow gold nanospheres (HAuNS) are a novel class of AuNPs composed of a thin Au shell with a hollow interior."

"Unlike solid AuNPs, HAuNS have plasmon absorption in the near-infrared (NIR) region and display strong photothermal conducting properties suitable for photothermal ablation therapy."

"HAuNS' unique combination of small size (30–50 nm in diameter) and a strong, tunable absorption band (520–950 nm) suggests that HAuNS are a promising mediator for a variety of biomedical applications, including imaging and cancer therapy [129, 130]."

"There has been no detailed study on the toxicity of HAuNS."

Results and discussion

"The major microscopic observation in the PEG-HAuNS-treated groups was deposition of a brown-black pigment in the liver, spleen, lungs, heart, adrenal cortex, and injection site."

"An increase in follicular lymphoid hyperplasia and extramedullary hematopoiesis was observed in PEG-HAuNS-treated animals compared to the respective controls, but the incidence and severity was minimal, considered a normal biologic response to injection of foreign material, and not adverse."

"The adrenal gland in only the female mice had an increase in vacuolation of the cortex and pigment deposition in the cortex at the 14-day time point in the treated animals but not in their respective controls."

"Deposition of a dark brown pigment similar to that observed in other tissues and identified as HAuNS occurred in the perivascular area and the wall of vessels at the injection site at the tail vein in 3/24 treated animals — 2 in the 14-day group and 1 at 90 days."

Conclusion

"Under the conditions of this study and on the basis of clinical signs, survivability, clinical pathology, and total body and relative organ weights and pathology, the administration of PEG-HAuNS was not considered adverse in female or male mice at an accumulated injection dose of 125 mg/kg, which was 10-fold higher than the effective therapeutic dose."

"Pathologic and quantitative analysis for Au in formalin-fixed tissues indicated that the levels of deposited HAuNS were decreased at 90 days, but HAuNS was not completely eliminated."

"The primary test-substance-related microscopic observation in this study was the deposition of pigment, primarily in the macrophages of the liver and spleen that was identified as HAuNS based on TEM and Au quantification."

"Quantitative analysis for Au in the liver, spleen, and kidney revealed that the levels of deposited PEG-HAuNS gradually decreased from day 14 to day 90 after the first injection, indicating that PEG-HAuNS was gradually cleared from the body, albeit slowly."

Methods

"To study the nanoparticles' effects on platelet aggregation, whole human blood was centrifuged 8 min at 200 g to obtain platelet-rich plasma, which was then treated with PEG-HAuNS (0.008, 0.04, 0.2, and 1.0 mg/mL) or collagen (positive control; Helena Laboratories, Beaumont, TX) for 15 min at 37°C."

"For the pharmacokinetic study, 8 healthy female Swiss mice (22–25 g; Charles River Laboratories, Wilmington, MA) were each injected intravenously with 0.125 mL of ^{64}Cu-labeled PEG-HAuNS (activity: 20 μCi; 6.25 mg/kg of 50 OD PEG-HAuNS)."

"Uptake of nanoparticles in various tissues was calculated as %ID/g. The antitumor activity of PEG-HAuNS at different doses was investigated in female nude mice (Charles River Laboratories) bearing human ovarian tumors."

"Mice in groups 1 to 3 were injected intravenously with a single dose of PEG-HAuNS (5 mL/kg of 25, 50, or 100 OD nanoparticles, corresponding to 3.13 mg/kg, 6.25 mg/kg, or 13.5 mg/kg, respectively)."

Pharmacokinetics of magnetic iron oxide nanoparticles for medical applications [131] This is a machine-generated summary of:

Nowak-Jary, Julia; Machnicka, Beata: Pharmacokinetics of magnetic iron oxide nanoparticles for medical applications [131]

Published in: Journal of Nanobiotechnology (2022)

Link to original: https://doi.org/10.1186/s12951-022-01510-w

If you want to cite the papers, please refer to the original.

For technical reasons we could not place the page where the original quote is coming from.

Abstract-Summary "Magnetic iron oxide nanoparticles (MNPs) have been under intense investigation for at least the last five decades as they show enormous potential for many biomedical applications, such as biomolecule separation, MRI imaging and hyperthermia."

"The uniqueness of MNPs is due to their nanometric size and unique magnetic properties."

"After digesting MNPs in lysosomes, iron ions are incorporated into the natural circulation of this element in the body, which reduces the risk of excessive storage of nanoparticles."

"One of the key issues for the therapeutic applications of magnetic nanoparticles is their pharmacokinetics which is reflected in the circulation time of MNPs in the bloodstream."

"These characteristics depend on many factors, such as the size and charge of MNPs, the nature of the polymers and any molecules attached to their surface, and other."

"Since the pharmacokinetics depends on the resultant of the physicochemical properties of nanoparticles, research should be carried out individually for all the nanostructures designed."

"Almost every year there are new reports on the results of studies on the pharmacokinetics of specific magnetic nanoparticles, thus it is very important to follow the achievements on this matter."

Introduction

"Magnetic iron oxide nanoparticles (MNPs) have been intensively developed and widely adopted for a range of biomedical applications such as tumors imaging (MRI) [132, 133], hyperthermia [134, 135], drug delivery [136, 137], gene therapy [138] and magnetic separation of cells or biomolecules [139, 140]."

"Irrespective of the specific applications, all magnetic nanostructures following in vivo administration are recognized by the host immunological mechanisms and eliminated from the body [141, 142]."

"Iron ions contained in magnetic nanoparticles are trace elements in the body."

"The progress in the research on the use of magnetic nanoparticles for biomedical applications has shown that their pharmacokinetics and biodistribution are influenced by the size, shape, charge and, above all, surface chemistry of the nanostructures [143]."

"It is suggested that all these parameters must be considered in order to develop magnetic nanostructures particularly useful in biomedical applications."

Intravenously injected MNPs

"Developing the method of atherosclerotic lesions imaging using magnetic iron oxide nanoparticles coated with PEGylated lipids, they demonstrated that MNPs with the iron core size of 10 nm exhibited longer half-life in the mouse bloodstream ($t_{1/2} = 1.41$ h) compared to analogous nanoparticles, but with a larger core diameter ($d = 35$ nm, $t_{1/2} = 1.01$ h)."

"Since the charge of the nanoparticles' surface has an impact on the degree of the proteins' absorption and circulation time in the bloodstream, the types of functional groups displayed on the MNPs surface play an important role in determining the half-life of the nanostructures [144]."

"The results showed that SPIO-alginate was eliminated at a high rate from the serum (half-life of 0.27 h) at a dose of 6.12 mg Fe/kg and accumulated mainly in the liver and the spleen after injection, whereas the $t_{1/2}$ of MNP-alginate at a high dose was 0.59 h. Prospero and others [145] investigated the circulation time of citrate coated magnetic nanoparticles in rats depending on their mode of administration."

Pharmacokinetics and clearance of MNPs in other administration methods

"As is turned out, almost 90% of the naked MNPs dosage (with the size of 20–30 nm in diameter) were still present in the lung even two weeks after the administration [146]."

"Other studies have shown that after 28 days of silica coated MNPs (d_H=50 nm) inhalation, a significant amount of them was accumulated in the liver, kidneys and testes, whereas the percentage of the nanoparticles remaining in the lungs was similar to other tissues (for example the heart or the brain) [147, 148]."

"When it comes to the oral administration of the MNPs, there are several biological barriers that must be overcome in order to successfully deliver therapeutic agents immobilized on the nanoparticles."

"The liver is the major clearance organ in the oral administration of nanoparticles, unless special MNPs surface modifications make them resistant to these macrophages."

The methods of determining the pharmacokinetics and biodistribution of MNPs

"To the ex vivo imaging techniques depicted above, the Magnetic Resonance Imaging (MRI) method can be applied in order to image MNPs distribution in vivo."

"Depending on the magnetic properties of the studied particles, it considers both relaxation pathways: the first structures are those that reduce the longitudinal (T_1) relaxation time and cause positive contrast enhancement (Gd^{3+} complexes), and the other group includes particles called negative contrast agents which are based on magnetic iron oxide nanoparticles resulting in a darker condition in the T_2-weighted image [149, 150]."

"Magnetic Particle Imaging (MPI) is an in vivo imaging method which makes use of the nonlinear magnetic response of magnetic iron oxide nanoparticles [151, 152]."

"Magnetic Susceptibility Measurement (MSM) [153] and the technique with the use of Superconducting Quantum Interference Device (SQUID) [154] are magnetometry techniques exploiting the magnetic properties of iron oxide nanoparticles for detection."

"There is a broad range of methods for determining the half-life and biodistribution of magnetic iron oxide nanoparticles, but despite this, research on broadening the spectrum of techniques used is still ongoing."

Nonstandard methods of extending the circulation time of nanoparticles in the blood

"The results of the study consisting in T_1 NMR measurements on blood samples showed that R_1 relaxation rate of circulating blood of mice treated with ferucarbotran-loaded RBCs was approximately 1.5-fold higher than blood R_1 of mice treated with free ferucarbotran, whereby the half-life of bulk ferucarbotran nanoparticles was less than 1 h, whereas blood half-life of ferucarbotran loaded RBCs was 48 h. In recent years cell membrane-based nanosystems have engaged attention due to their superior biocompatibility and functionality [155]."

"The new studies indicate that MNPs coating with erythrocyte membrane can significantly extend their blood circulation time."

"The above results indicate that coating MNPs with erythrocyte membrane can significantly extend their blood circulation time, even more effectively than the traditional PEGylation method."

Conclusion and perspectives

"MNPs are often used as contrast agents in MRI imaging, hyperthermia, the diagnosis and treatment of cancerous tumors, but also countless studies focus on designing magnetic nanoparticles as drug carriers in targeted therapy as well as nanoparticles for the separation of malignant cells."

"The success of properly designing nanoparticles in the laboratory is determined by their pharmacokinetics, especially in vivo, which, in turn, is determined by many parameters characterizing nanoparticles such as charge and size."

"The surface charge of MNPs play an important role in the physical stability and influence their interaction with the biological system."

"The ligands and functional covering molecules often significantly increase the hydrodynamic the MNPs which leads to the macrophage and systemic clearance of designed nanoparticles."

"Despite some general rules, such as the influence of nanoparticle size on the way of their clearance, studies on pharmacokinetics should be performed individually for each designed nanostructure."

4.5 Pharmacokinetics and Pharmacodynamics of Implants

Introduction by the Editor

Drug-loaded implants, a pioneering avenue in controlled drug delivery, leverage biocompatible polymers to sustainably release therapeutic agents. Polylactic-co-glycolic acid (PLGA) is a widely employed polymer due to its biodegradability and tuneable degradation rate. Implants, often fashioned as rods or discs, encapsulate drugs within the polymer matrix, enabling controlled release over extended periods. Additionally, polyethylene glycol (PEG) coatings enhance biocompatibility and minimise host responses. This innovative approach offers advantages in maintaining steady drug levels, reducing dosing frequency, and improving patient compliance. It holds immense potential for diverse medical applications, from contraception to orthopaedic therapies. In-situ gelling implants are drug delivery systems where liquid formulations transform into gels upon administration, providing sustained drug release at the implantation site. These implants are typically composed of biocompatible polymers like thermosensitive Pluronic F127 or chitosan derivatives, which undergo sol-to-gel transitions due to temperature or pH changes. Upon injection, the liquid implants rapidly form gels, adhering to tissue surfaces and allowing for localised and prolonged drug release. This innovative approach bypasses the need for surgical implantation, offers minimally invasive administration, and enhances patient comfort. In-situ gelling implants hold promise in diverse applications, including post-operative pain management, ocular treatments, and wound healing, revolutionising the landscape of controlled drug delivery.

Machine Generated Summaries

Disclaimer: The summaries in this chapter were generated from Springer Nature publications using extractive AI auto-summarization: An extraction-based summarizer aims to identify the most important sentences of a text using an algorithm and uses those original sentences to create the auto-summary (unlike generative AI). As the constituted sentences are machine selected, they may not fully reflect the body of the work, so we strongly advise that the original content is read and cited. The auto generated summaries were curated by the editor to meet Springer Nature publication standards.

To cite this content, please refer to the original papers.

Machine generated keywords: implant, inflammation, eye, rabbit, surgery, degradation, polymer, matrix, ocular, degrade, soluble, topical, compliance, period, molecular weight

Pharmacokinetics of the Carmustine Implant [156] This is a machine-generated summary of:

Fleming, Alison B.; Saltzman, W. Mark: Pharmacokinetics of the Carmustine Implant [156]

Published in: Clinical Pharmacokinetics (2002)

Link to original: https://doi.org/10.2165/00003088-200241060-00002

If you want to cite the papers, please refer to the original.

For technical reasons we could not place the page where the original quote is coming from.

Abstract-Summary "Given the localised nature of the drug in the brain tissue, no direct pharmacokinetic measurements have been made in humans after implantation of a carmustine wafer."

"Drug distribution and clearance have been extensively studied in both rodent and non-human primate brains at various times after implantation."

"Studies to characterise the degradation of the polymer matrix, the release kinetics of carmustine and the metabolic fate of the drug and polymer degradation products have been conducted both in vitro and in vivo."

"GLIADEL®[1] wafers have been shown to release carmustine in vivo over a period of approximately 5 days; when in continuous contact with interstitial fluid, wafers should degrade completely over a period of 6 to 8 weeks."

"Pharmacokinetic studies in animals and associated modelling have demonstrated the capability of this modality to produce high dose-delivery (millimolar concentrations) within millimetres of the polymer implant, with a limited penetration distance of carmustine from the site of delivery."

"In non-human primates, the presence of significant doses in more distant regions of the brain (centimetres away from the implant) has been shown to persist over the course of a week."

"The drug in this region was presumed to be transported from the implant site by either cerebral blood flow or cerebrospinal fluid flow, suggesting that although drug is able to penetrate the blood-brain barrier at the site of delivery, it may re-enter within the confines of the brain tissue."

1. Chemistry, Degradation, Release Kinetics and Metabolism of the Carmustine Implant

"The hydrophobic nature of p(CPP:SA) would also provide the encapsulated drug with some protection from hydrolytic degradation before release, an important feature in the case of carmustine which has a half-life of only 52 minutes in phosphate buffer, pH 7.4, at 37°C.[157] The degradation of p(CPP:SA) 20: 80 in aqueous environments has been characterised both qualitatively, by SEM,[158] and quantitatively, by measuring the decrease in polymer molecular weight, the wafer mass loss, the release of soluble low molecular weight oligomers, and the crystallinity change of the polymer.[159, 160] Quantitative experiments demonstrated that the degradation of p(CPP:SA) occurs in two phases.[159] During the first phase, known as the induction period, water penetrates into the matrix and hydrolyses the bonds of the copolymer."

"Morphological studies using SEM demonstrated that the wafers degrade in a layerwise fashion, whereby a porous erosion zone starts on the outside of the wafer and gradually extends to the inner core.[158] The release of CPP and SA from wafers does not occur at the same rate, presumably due to the differences in stability of anhydride linkages as well as the solubility of the monomers."

2. Distribution of Carmustine and Pharmacokinetic Model

"The concentration of bound drug, B, is defined analogously to the concentration of unbound drug (equation 5): For the case of carmustine, the following assumptions can be made:[161] (a) carmustine is neither eliminated nor bound in the membrane phase; (b) the concentration of bound carmustine is directly proportional to the concentration of free carmustine in both ECS and ICS; (c) carmustine is eliminated by first-order processes in the ECS and ICS; and (d) local equilibrium is achieved between the ECS, ICS and CM."

"Considering diffusion as the sole mechanism of transport (i.e. neglecting convection), equation (12a) becomes the diffusion/elimination equation (equation 12b): Analytical solutions to equations (12a) and (12b) have been compared to the concentration profile of carmustine in the brains of animals following implantation of a drug-loaded polymer matrix."

"The observation of relatively long penetration distances of carmustine and measurements of high concentrations of drug in the CSF at early times for both rats and non-human primates suggests that this may be an important transport mechanism immediately following the implantation of the carmustine wafer."

3. Computer Simulations of Carmustine Distribution

"To model drug delivery to the rat brain, a finite mesh was applied to a normal coronal slice in two dimensions, accounting for the presence of three features within the mesh domain: white matter, ventricles and the polymer implant.[162] The white matter was delineated from the brain parenchyma because of its effect on transport properties; the regular arrangement of nerve fibres within it provides a low resistance pathway for flow relative to the ECS of the grey matter.[163] The ventricles, which are essentially an internal boundary, provide a route out of the parenchyma into the CSF for drug and interstitial fluid."

"In another set of simulations, MRI images of a PNET were used to construct a finite element mesh that captured the geometric features of a human tumour.[164] Unlike simulations conducted using the anatomy of the normal rat brain, these tumour simulations did not differentiate white matter or ventricular space in the mesh geometry; the anatomical structure of the tumour itself was assumed to be more important because of the size of a human neoplasm (approximately 4cm in diameter) relative to the anticipated drug penetration."

4. Summary and Clinical Implications

"Drugs must be transported far enough from their source to be effective in eliminating cells at the growing tumour margin; on the other hand, high concentrations of drug in normal brain tissue can have adverse affects."

"Pharmacokinetic studies in animals and associated modelling have demonstrated the ability of carmustine wafers to achieve delivery of high doses (millimolar concentrations) within millimetres of the implant, while exposing more distant locations in the brain to only low concentrations (micromolar concentrations or less)."

"In the clinical case, where oedema is typical for postoperative craniotomy patients,[165–167] the resulting convective flow probably plays a role in distributing drugs into the tissue surrounding the wafer-lined tumour bed."

"The penetration of drug into the region surrounding the implant is limited by the fact that carmustine is highly lipophilic, such that elimination through capillary walls or ependymal barriers is rapid."

[Section 5]

"Of these devices, controlled release polymers offer the theoretical advantage of continuous delivery that is not subject to clogging by tissue debris and protection of labile drugs prior to their release.[168] In 1996, the carmustine implant (GLIADEL®), a biodegradable, chemotherapy agent-loaded polymer matrix, was approved by the US Food and Drug Administration after it had been demonstrated to be a well tolerated and effective treatment for recurrent malignant glioma.[166] This was the first new treatment to be approved for malignant gliomas in 20 years, since the introduction of systemic chemotherapy."

"GLIADEL®is comprised of the anticancer drug carmustine (BCNU) homogeneously distributed in a solid copolymer disk, and is currently approved for the treatment of recurrent glioblastoma multiforme."

"Pharmacokinetic measurements were not obtained in patients during clinical trials due to the localisation of drug in the brain tissue."

"During the development of the carmustine implant, studies were conducted both in vitro and in animal models in order to determine the release kinetics of carmustine, the degradation properties of the copolymer disc, the metabolic fate of the drug and copolymer degradation products, and the interstitial distribution of carmustine."

Pharmacokinetics and Efficacy of Bioerodible Dexamethasone Implant in Concanavalin A-induced Uveitic Cataract Rabbit Model [169] This is a machine-generated summary of:

Chennamaneni, Srinivas Rao; Bohner, Austin; Bernhisel, Ashlie; Ambati, Balamurali K.: Pharmacokinetics and Efficacy of Bioerodible Dexamethasone Implant in Concanavalin A-induced Uveitic Cataract Rabbit Model [169]

Published in: Pharmaceutical Research (2014)

Link to original: https://doi.org/10.1007/s11095-014-1410-7

If you want to cite the papers, please refer to the original.

For technical reasons we could not place the page where the original quote is coming from.

Abstract-Summary "To advance therapy for the treatment of concurrent uveitis and post-cataract surgical inflammation; we evaluated pharmacokinetics and pharmacodynamics of Bioerodible Dexamethasone Implant (BDI) containing 0.3 mg of dexamethasone (DXM) in Concanavalin A (Con A) induced uveitis followed by phacoemulsification in New Zealand White (NZW) rabbits."

"In Con A-induced inflammatory uveitic cataract model the BDI controlled anterior and posterior segment inflammation as well as retinal thickening more effectively than topical drops."

"The exposure (AUC_{0-t}) of DXM with BDI is superior in all ocular tissues, while topical drops did not achieve therapeutic posterior segment levels and did not control inflammation nor prevent retinal edema and architectural disruption."

"Our results demonstrate the superiority of the BDI in suppressing Con A-induced inflammation and retinal edema in NZW rabbits and highlight the need for sustained bidirectional delivery of potent anti-inflammatory agents for 5 to 6 weeks to optimize clinical outcomes."

Introduction

"Cataract surgery is often performed in older patients who have comorbidities, e.g., diabetes or uveitis [170], which increase the risk of CME or inflammation."

"Ocular inflammation after cataract surgery can prolong recovery time and increase the likelihood of CME, synechiae formation, and posterior capsule opacification (PCO) [171, 172]."

"To prevent short and long-term postoperative complications, corticosteroids and non-steroidal anti-inflammatory drugs (NSAID's) are administered topically after cataract surgery."

"The lack of compliance, along with complexity of topical eye drop regimens, can delay recovery from surgery to several weeks or months if CME or uveitis occurs."

"In order to mitigate poor compliance and to enhance therapeutic effectiveness, sustained release of steroidal and antiviral drugs loaded in bioerodible or non-bioerodible implants have been developed for the implantation in the vitreous cavity for the treatment of various ocular diseases in the posterior segment."

Materials and Methods

"Approximately 2.0 mg of microspheres were dispersed in 5.0 ml of 0.2% (w/v) PVA solution, diluted 5 times with deionized water and used for particle size analysis."

"Approximately 1.0 ml of whole blood was collected (before injecting euthanasia solution), placed into labeled micro-centrifuge tubes, and allowed to clot and centrifuged at 6,000 rpm for 6.0 min; serum was collected and stored in a freezer until bio-analysis."

"Twenty seven female NZW rabbits (Western Oregon Rabbit Co, USA) weighing 2.5 to 3 kg were anesthetized with a 0.8–1.2 ml intramuscular injection of a 4:1 mixture of 100 mg/ml xylazine and 50 mg/ml ketamine."

"To each tissue sample 100 μl of BSS was added and the sample homogenized in an ice bath with sonic dismembrator (Fisher Scientific, USA) at low speed for 1.0 min followed by liquid-liquid extraction using tert-Butyl methyl ether."

Results

"After 15 days of treatment PCO was minimal in the BDI group; in contrast, there was 1.5+ PCO in eyes which received topical DXM drops."

"The loss in body weight in the topical drops group may be due to systemic exposure of DXM (100.0 vs. <5.0 ng/ml, $P < 0.001$) [173, 174]."

"With topical drops tissue DXM concentrations declined after week 4 and were below quantifiable limits (BQL) in iris/ciliary body and retina/choroid on day 42."

"Time to reach highest observed concentration (T_{max}) in the iris/ciliary body and retina/choroid after BDI implantation was faster (7–14 days) compared to topical drops (14–23 days)."

"In the topical drop group, retinal thickness increased significantly ($P < 0.05$) by week 1 and persisted up to week 6 in comparison to both normal control and BDI groups."

Discussion

"Unlike topical drops, BDI was effective in limiting inflammation in both the anterior and posterior segments of the Con A-injected eyes following uveitic cataract surgery."

"Since uncontrolled inflammation in the posterior segment heralds CME, which can lead to permanent vision loss, we believe that BDI-based drug delivery could be effective in therapeutic management of severe posterior segment disorders."

"We observed that DXM pharmacokinetics in healthy (phacoemulsification only) [175] and inflamed eyes (Con A intravitreal injection followed by phacoemulsification, current study) are significantly different, suggesting that the underlying inflammatory status of the eye influences implant degradation and pharmacokinetics."

"The concentration-time profiles for DXM were different between healthy rabbit eyes and inflamed eyes for the aqueous humor, vitreous humor, iris/ciliary body and retina/choroid."

"In the topical drops group, vitreous humor concentrations were negligible, and the levels of DXM decreased tremendously in inflamed eyes in iris/ciliary body (>18 times) and retina/choroid (~40 times)."

4.6 Pharmacokinetics and Pharmacodynamics of Topically Administered Novel Drug Delivery Systems

Introduction by the Editor

Topically administered novel drug delivery systems and nanoparticles have revolutionised the field of dermatology and beyond. These advanced systems are designed to efficiently deliver therapeutic agents to the skin, offering targeted treatment while minimising systemic side effects. By encapsulating drugs within nanocarriers, such as lipid-based nanoparticles, polymeric micelles, and dendrimers, these delivery systems enhance drug solubility, stability, and penetration into the skin layers. Whether treating skin disorders, wounds, or cosmetic concerns, these technologies ensure controlled and sustained drug release, optimising therapeutic outcomes and patient compliance. This paradigm shift in drug delivery opens doors to precision medicine in dermatology, wound care, and cosmetic applications, promising safer, more efficient, and personalised solutions for a diverse range of conditions.

Machine Generated Summaries

Disclaimer: The summaries in this chapter were generated from Springer Nature publications using extractive AI auto-summarization: An extraction-based summarizer aims to identify the most important sentences of a text using an algorithm and uses those original sentences to create the auto-summary (unlike generative AI). As the constituted sentences are machine selected, they may not fully reflect the body of the work, so we strongly advise that the original content is read and cited. The auto generated summaries were curated by the editor to meet Springer Nature publication standards.

To cite this content, please refer to the original papers.

Machine generated keywords: skin, transdermal, microneedle, iontophoresis, mns, corneum, transdermal delivery, stratum, stratum corneum, permeation, charge, delivery, field, injection, technology

Microneedle-based insulin transdermal delivery system: current status and translation challenges [176] This is a machine-generated summary of:

Zhao, Jing; Xu, Genying; Yao, Xin; Zhou, Huirui; Lyu, Boyang; Pei, Shuangshuang; Wen, Ping: Microneedle-based insulin transdermal delivery system: current status and translation challenges [176]

Published in: Drug Delivery and Translational Research (2021)

Link to original: https://doi.org/10.1007/s13346-021-01077-3

Copyright of the summarized publication:

Controlled Release Society 2021

If you want to cite the papers, please refer to the original.

For technical reasons we could not place the page where the original quote is coming from.

Abstract-Summary "For patients with type 1 and advanced type 2 diabetes mellitus, insulin therapy is essential."

"Non-invasive insulin delivery technologies are pursued because of their benefits of decreasing patients' pain, anxiety, and stress."

"Microneedle (MN) technology is one of the most promising tactics, which can effectively deliver insulin through skin stratum corneum in a minimally invasive and painless way."

"This article will review the research progress of MNs in insulin transdermal delivery, including hollow MNs, dissolving MNs, hydrogel MNs, and glucose-responsive MN patches, in which insulin dosage can be strictly controlled."

"The clinical studies about insulin delivery with MN devices have also been summarized and grouped based on the study phase."

"There are still several challenges to achieve successful translation of MNs-based insulin therapy."

Introduction

"MNs can painlessly penetrate SC and access skin epidermis and dermis to release drug."

"The micro-channels caused by MNs are temporarily presented for drug delivery and can be recovered in a short time after MNs removal, which will prevent long-term skin damage [177]."

"Solid MNs are drug-free that are designed only to pierce the skin and generate micro-channels."

"After removing solid MNs, the drug formulation is administrated on the pre-treated skin and the drug substance will enter into the skin through the micro-channels [178]."

"The resealing time of the micro-channels after solid MNs removed varies with the skin hydration state and skin thickness, which will affect the absorption time of drugs [179]."

"Insulin administration requires a precise dose to improve glycemic control while reducing hypoglycemic events; thus, these two types of MNs may not be suitable for insulin delivery."

MNs used for insulin transdermal delivery

"Hollow MNs are indirect auxiliary administration, in which hydrogel and dissolving MN can be directly loaded with insulin for administration, and both can significantly increase the amount of insulin transdermal delivery."

"To avoid the high temperature for polymer dissolution in micro-molding method, researchers have extensively explored the water-soluble polymers in the fabrication of heat-sensitive drugs loaded dissolving MNs, including HA [180], γ-PGA, mixture of starch and gelatin [181], and mixture of fish gelatin and sucrose [182]."

"Compared with dissolving MNs, hydrogel MNs can achieve a more consistent and controlled insulin release profile without polymer deposition in the skin."

"The closed-loop glucose-responsive MN patch which can detect blood glucose level and adjust the release of insulin in time has become a hotspot in MNs delivery of insulin [183]."

"A porous polymer layer, which encapsulated glucose-responsive formulation (composed of insulin, GOx, and sodium bicarbonate ($NaHCO_3$)) in the pores, was coated out of the MNs needles by dip-coating method [184]."

Clinical progress of MNs-based insulin delivery

"In 2016, Kochba and others [185] reported an improved insulin pharmacokinetics (PK) profile of patients with type 2 diabetes using Micronjet®, a hollow MN device containing four needles, by comparing with conventional SC injection administration (ClinicalTrials.gov Identifier: NCT01684956)."

"The patients were given lispro (IL) or regular human insulin (RHI) by MN-based intradermal (ID) injections and SC injections after standard liquid meals (ClinicalTrials.gov Identifier: NCT00553488) [186]."

"Another result of the same research team showed that compared with SC injection, the secondary insulin PK endpoint showed faster ID availability in different doses and diets (ΔT_{max} −16 min, ΔT50 rising −7 min, ΔT50falling −30 min, all $p < 0.05$) [187] (ClinicalTrials.gov Identifier: NCT01120444)."

"These findings indicate that, compared with traditional SC injection, MN-based ID insulin administration has a beneficial effect on insulin PK and PPG for both IL and RHI versus SC."

Challenges

"This type of MNs is more suitable for drug delivery that does not require frequent administration, such as vaccines, but it is undesirable for insulin long-term use."

"Of all, drug loading is the most important factor affecting the effectiveness of insulin MNs."

"For dissolving MNs to be completely inserted into the skin, insulin is only loaded on the needle tip, which causes another problem, that is, low load capacity."

"The successful insertion of MNs into the skin without breakage or bending is fundamental for effective insulin delivery, which requires sufficient strength and stiffness of MNs."

"An optimal MN should be designed with low insertion force, the force required to insert MNs into living the skin and high breaking force, the force needles can withstand before fracturing."

"While when the effective dose in animals was converted to humans, the amount of insulin required greatly exceeded the drug-loaded amount of the MNs."

Conclusion and perspective

"MN technology shows great advantages in pain relief from frequent injections and administration convenience, which can ease diabetic patients from needle phobia under traditional SC insulin injection and therefore improve their life quality."

"Many efforts have been made to design and fabricate MN-based insulin delivery systems and numerous great research works have been reported."

"The production of insulin MNs involves multiple steps, continuous and automatic online control under GMP scale will help to ensure the products consistency, reproducibility, and quality."

"Numerous researchers are devoted to developing glucose-responsive MNs for insulin delivery."

"With the continuous development of MNs fabricating technology, the ingenious combination of MNs with other medical devices or intelligent drug delivery systems, and the continuous improvement of the MNs evaluation and regulatory system, MNs will be widely used for the transdermal delivery of insulin and other active molecules."

"Glucose-responsive MNs have become a research trend that combines novel materials for biosensors or intelligent drug delivery systems."

Microneedle Mediated Transdermal Delivery of Protein, Peptide and Antibody Based Therapeutics: Current Status and Future Considerations [188] This is a machine-generated summary of:

Kirkby, Melissa; Hutton, Aaron R.J.; Donnelly, Ryan F.: Microneedle Mediated Transdermal Delivery of Protein, Peptide and Antibody Based Therapeutics: Current Status and Future Considerations [188]

Published in: Pharmaceutical Research (2020)

Link to original: https://doi.org/10.1007/s11095-020-02844-6

If you want to cite the papers, please refer to the original.

For technical reasons we could not place the page where the original quote is coming from.

Abstract-Summary "The intrinsic properties of biopharmaceuticals has restricted the routes available for successful drug delivery."

"One such strategy involves the use of microneedles (MNs), which are able to painlessly penetrate the stratum corneum barrier to dramatically increase transdermal drug delivery of numerous drugs."

"This review reports the wealth of studies that aim to enhance transdermal delivery of biopharmaceutics using MNs."

"The true potential of MNs as a drug delivery device for biopharmaceuticals will not only rely on acceptance from prescribers, patients and the regulatory authorities, but the ability to upscale MN manufacture in a cost-effective manner and the long term safety of MN application."

Introduction

"Dissolving MNs use biocompatible polymers mixed with the drug to form the needle tips."

"The main limitation associated with dissolving MNs is the deposition of polymer, alongside the drug, into the skin."

"The primary disadvantage associated with hollow MNs is the potential for drug flow resistance to occur – either by clogging of needle openings with skin tissue during insertion [189], or by compression of the MNs by dense dermal tissue [190]."

"This removes the advantage associated with other MN types, for example, hydrogel-forming MNs, whereby the drug may sit in a compressed tabled or lyophilised wafer above the array until the array is applied."

"Hydrogel-forming MNs have been made from light responsive polymer materials to control drug release [191]."

"As briefly discussed above, the type of material used to make MNs in dissolving and hydrogel-forming systems will influence the drugs ability to diffuse into the skin."

MN Mediated Transdermal Delivery of Protein, Peptide and Antibody Based Therapeutics

"In vitro permeation of FITC-BSA loaded MNs (1000 μm, 64 needles/cm^2) across full thickness porcine skin following 5 min MN insertion was found to be 45% at 24 h. FITC-BSA appeared to be concentrated in the epidermis, upper layers of the dermis, and around sites of microneedle penetration, with little fluorescent signal observed in the lower dermis."

"Polymyxin B loaded MNs were applied to porcine ear skin for 30 s. The rate of drug delivery was found to be greater than the control (drug loaded disc without MNs) for the first 4 h post MN application, after which rate of permeation was equal to the control, but the percentage of drug delivered transdermally was significantly greater."

"After assuring stability of FITC-BSA remained intact in the formulation, in vitro studies were used to assess transdermal delivery of FITC-BSA from the novel dissolving MNs (600 μm needle length)."

Safety and Clinical Translation of MN Based Products for Protein, Peptide and Antibody Based Therapeutics

"Delivery of high molecular weight, high dose and low potency protein and peptide based therapies has become more commonplace, allowing MNs to be considered for delivery of drugs which was previously thought unlikely or even impossible."

"Dissolving MNs may be better placed for vaccine delivery, as their infrequent use reduces the issue of repeated polymer deposition within the skin."

"Studies revealed that repeat application of both dissolving (once daily for 5 weeks) and hydrogel-forming (twice daily for 3 weeks) MNs did not alter skin appearance or barrier function and caused no measurable disturbance of serum biomarkers of infection, inflammation or immunity [192]."

"Whilst it seems unwanted immunogenic effects from protein and peptide drugs delivered via MNs are unlikely, long-term studies exploring the immune response following repeated application of MNs containing biological drugs are required before clinical acceptance can be assured, and will likely have to be shown on a drug-by-drug basis."

Conclusion

"Numerous studies have illustrated the ability to deliver therapeutic doses of protein, peptide and antibody based therapies using MNs as an alternative drug delivery device."

"It is vitally important to change this mind set by proving that MNs offer far more than acting as simple vaccine delivery devices."

"This field has huge potential, with the possibility of MN sampling devices being delivered to patients' homes and returned to the laboratory for analysis, without the patient having to enter a clinical setting."

"As this MN design could be CE marked as a medical device rather than a drug product, reduced regulatory requirements could mean that pharmaceutical companies may be willing to first invest in a device such as this."

"As the first drug containing MN product to potentially achieve commercialisation, it appears many pharmaceutical companies are willing to delay MN development until its success becomes apparent."

Non-dermal applications of microneedle drug delivery systems [193] This is a machine-generated summary of:

Panda, Apoorva; Matadh, V. Anusha; Suresh, Sarasija; Shivakumar, H. N.; Murthy, S. Narasimha: Non-dermal applications of microneedle drug delivery systems [193]

Published in: Drug Delivery and Translational Research (2021)

Link to original: https://doi.org/10.1007/s13346-021-00922-9

If you want to cite the papers, please refer to the original.

For technical reasons we could not place the page where the original quote is coming from.

Abstract-Summary "Microneedles (MNs) are micron-scaled needles measuring 100 to 1000 μm that were initially explored for delivery of therapeutic agents across the skin."

"Considering the success in transcutaneous drug delivery, the application of microneedles has been extended to different tissues and organs."

"The review captures the application of microneedles to the oral mucosa, the eye, vagina, gastric mucosa, nail, scalp, and vascular tissues for delivery of vaccines, biologics, drugs, and diagnostic agents."

"The technology has created easy access to the poorly accessible segments of eye to facilitate delivery of monoclonal antibodies and therapeutic agents in management of neovascular disease."

"The technology has been successful to overcome the delivery hurdles and enable direct delivery of drug to the target sites, thus maximizing the efficacy thereby reducing the required dose."

"This review is an attempt to capture the non-dermatological applications of microneedles being explored and provides an insight on the future trends in the field of microneedle technology."

Introduction

"It was observed that microneedle pretreatment followed by application of ultrasound on topically applied nanoparticle formulation demonstrated a multi-modal delivery of antibody conjugated PEGylated gold nanoparticles that enhanced the contrast of OCT images of the oral cancer."

"In the same study, it was noted that pilocarpine-coated microneedles were able to increase the drug concentration by nearly 45-folds when compared with topical application."

"The studies indicated that hollow microneedles have been useful in the intrascleral delivery of drugs."

"Application of microneedles for vaginal drug delivery is still being explored at this stage with insights still needed on mucosal irritation and toxicity along with user feasibility studies that would enhance this microneedle application furthermore."

"The study clearly demonstrated the ability of microneedles to improve the bioavailability of orally administered drug."

Conclusion

"Although the extensive studies are carried out on the skin application of microneedles, investigations are still in progress to deploy the technology for delivering drug to other tissues and organs."

"With the success of technology in drug delivery to the skin, applications of microneedles have been explored for oral mucosa, ocular, muscular, cardiac, nail, scalp, and vascular delivery."

"It is necessary to have a thorough understanding of the barrier disruption and its subsequent restoration when microneedles are directly applied to other tissues and organs."

"Usage of microneedles from a drug delivery perspective into other tissues and organs is still at very early stages."

"In case the safety concerns of the microneedle technology are duly addressed, drug delivery to the remote inaccessible affected organs and tissues may become a true reality in future."

Iontophoresis application for drug delivery in high resistivity membranes: nails and teeth [194] This is a machine-generated summary of:

Martins Andrade, Jayanaraian F.; da Cunha Miranda, Thamires; Cunha-Filho, Marcílio; Taveira, Stephânia Fleury; Gelfuso, Guilherme M.; Gratieri, Taís: Iontophoresis application for drug delivery in high resistivity membranes: nails and teeth [194]

Published in: Drug Delivery and Translational Research (2022)

Link to original: https://doi.org/10.1007/s13346-022-01244-0

If you want to cite the papers, please refer to the original.

For technical reasons we could not place the page where the original quote is coming from.

Abstract-Summary "Iontophoresis has been vastly explored to improve drug permeation, mainly for transdermal delivery."

"Nails and teeth are accessible structures for target drug delivery but possess low water content compared to the skin and impose significant barriers to drug permeation."

"Iontophoresis application in nail and teeth structures may be a safe and effective way to improve drug transport across the nail and drug distribution through dental structures, making treatments more effective and comfortable for patients."

"Iontophoresis presents a promising option to enhance drug permeation through the nail and dental tissues, and further developments in these areas could lead to widespread clinical use."

Introduction

"In simple terms, iontophoresis is the application of a mild electric current to a pharmaceutical formulation with the objective of enhancing drug penetration towards a biological tissue."

"For long, the skin has been the most studied biological target of iontophoresis, which can be attributed not only to the demand for targeted treatments with lower side effects and easier patient compliance compared to oral administration, but also the wide range of drug candidates with undesirable physicochemical properties for optimal gastrointestinal absorption, particularly regarding an ionized state at physiological pH [195]."

"Even though trustable in vitro models for ocular and mucosal drug delivery research could also be obtained from porcine [196, 197], their processing is much more laborious, as these tissues must not be frozen before the experiment, a facility granted in skin research."

"We provide an overview of the iontophoretic application to these "hard structures," summarize recent studies, and discuss the current limitations of drugs delivered to these highly resistant membranes."

Nail

"The high degree of cross-linking in the intermediate layer due to cystine amino acid disulfide bonds makes the nail plate a significant barrier to drug permeation [198]."

"The main clinical presentation of onychomycosis is distal subungual onychomycosis which infects the nail bed, the thin layer of epidermal tissue right above the vascularized tissue."

"Onychomycosis is the leading cause of nail diseases."

"The topical approach also has limitations, i.e., antifungals do not present excellent permeability properties due its physicochemical characteristics, such as high lipophilicity, which are associated with a very keratinized nail plate that limits drug permeation, altogether it lowers cure rate chance [199–201]."

"There is no standardized protocol to treat nail psoriasis, and the available treatment consists mainly of painful intralesional or topical steroids with low permeation properties and systemic immunosuppressants, which can cause various side effects."

Teeth

"Non-mineralized tissue includes pulp, composed of nerves and blood vessels and mineralized tissue includes, from the outermost to the innermost, the enamel, the dentine, and the cementum [202]."

"Dentine is less mineralized than enamel, contains more fluids, and is more permeable than enamel due to dentinal tubules on the dentine's surface [203]."

"The cementum is a thin layer of mineralized tissue covering the teeth roots, with a composition very similar to the bones [203]."

"To the nails, hard dental tissues contains similar water content, i.e., enamel has 3–4% of aqueous content [204] and dentine has an average of 10% [205]."

Basic iontophoresis principles

"The relative contributions of EO and EM depend on the physicochemical properties of the molecule, the formulation characteristics, the iontophoretic conditions, and, naturally, the tissue structure to which it is applied."

"The nail has an isoelectric point (pI) of around 5, thus, at a formulation pH higher than 5, solvent flow can be expected from the anode-to-cathode direction [206]."

"An absent or even lower EO flow in the anode-to-cathode direction would be expected when the drug solutions were applied."

"At pH 3, the nail plate is positively charged as well, and even though EO flows from cathode to anode, EM is stronger and able to increase drug delivery of these drugs to deep-seated tissues."

"In the same sense, at pHs higher than the nail's isoelectric point, i.e., 5, cathodal iontophoresis should only benefit transungual transportation of drugs bearing negative charges, being, in this case, EM the sole iontophoretic component."

Nail iontophoresis

"When the current is applied directly to the nail, the maximum tolerable current is usually lower due to the higher current density caused by the reduction in the applied area."

"The lack of appendages in the nail plate and proximal nail fold makes the paracellular and transcellular pathways the only option for electrical current, in contrast with the skin, where the appendageal pathway is the lower resistance path not only to current passage but to molecules and colloidal drug delivery systems [207]."

"Nail hydration is also associated with lower pain sensation during iontophoresis application, mainly at the beginning, when electrical resistance is higher, as shown in a study with six patients that went under nail hydration for 3 min before current application of 0.2 mA at 50–60 V to reduce pain and tickling sensation [208]."

Dental iontophoresis

"As previously described, the enamel is an avascular and acellular tissue; hence, it cannot regenerate itself, therefore, the first iontophoresis applications involved ion transportation [209, 210] in an attempt to remineralize the enamel layer to avoid further damage to this structure, which is indeed a promising approach due to the high electrical mobility of such ions."

"A previous study also demonstrates an enhanced delivery of F^-, Ca^{2+}, K^+, and Na^+ ions into whole enamel depth by electrokinetic flows generated by a direct current application to avoid enamel deterioration and assist remineralization in early carious lesions."

"One of the first things to be considered is that enamel is the densest and electrically isolating part of the teeth, thus applying iontophoresis through this tissue to deliver larger molecules may require higher electric currents, which should be increased slowly to avoid discomfort and pain to the patients or should be done with lower current, but longer duration [211]."

Concluding remarks and perspective

"Other than dentine sclerosis, which is responsible for this reduced permeability in carious teeth, it is also possible that in caries-affected dentin, the tissue impairment increases the involvement of more mobile ionic species, which compete for carrying the electrical current and, therefore, decrease the drug delivery efficiency."

"The tissue is impaired, and the resistivity is compromised, possibly higher current densities can be tolerated, which could compensate for the lower delivery efficiency."

"Unless the tissue is mechanically impaired, neither the nail nor the dentine structure presents any less resistive gateways for the transient channels to be created, as is the case of the follicular pathway regarding the cutaneous route."

"A noticeable advantage of ungueal iontophoresis observed in practically all studies is the maintenance of tissue physiological characteristics after iontophoresis completion compared to other more aggressive physical enhancement methods for drug permeation or even nail removal."

Transdermal Delivery of Salmon Calcitonin Using a Dissolving Microneedle Array: Characterization, Stability, and In vivo Pharmacodynamics [212] This is a machine-generated summary of:

Zhang, Lu; Li, Yingying; Wei, Fang; Liu, Hang; Wang, Yushuai; Zhao, Weiman; Dong, Zhiyong; Ma, Tao; Wang, Qingqing: Transdermal Delivery of Salmon Calcitonin Using a Dissolving Microneedle Array: Characterization, Stability, and In vivo Pharmacodynamics [212]

Published in: AAPS PharmSciTech (2020)

Link to original: https://doi.org/10.1208/s12249-020-01865-z

If you want to cite the papers, please refer to the original.

For technical reasons we could not place the page where the original quote is coming from.

Abstract-Summary "To overcome the limitations of the conventional routes, two new dissolving microneedle arrays (DMNAs) based on transdermal sCT delivery systems were developed, namely sCT-DMNA-1 (sCT/Dex/K90E) and sCT-DMNA-2 (sCT/Dex-Tre/K90E) with the same dimension, meeting the requirements of suitable mechanical properties."

"The stability study exhibited that the addition of trehalose could improve the stability of sCT in DMNA under high temperature and humidity."

"Further, in vivo pharmacodynamic study revealed that DMNA patch could significantly enhanced relative bioavailability to approximately 70%, and the addition of trehalose was found to be beneficial for sCT transdermal delivery."

"SCT-DMNA is expected to replace traditional dosage form, providing a secure, efficient, and low-pain therapeutic strategy for bone disorders."

INTRODUCTION

"Although the nasal delivery avoids the impact of the adverse environment of the gastrointestinal tract, the poor bioavailability of sCT could limit its curative effects."

"Tas and others fabricated microneedles (MNs) with only five-needle arrays from stainless steel sheets, where each MN was 730 μm long, 180 μm wide at the base, 50 μm in thickness, and coated the surface of MNs with sCT to design the delivery system sCT-MNs [213]."

"The authors demonstrated that the AUC (area under the curve) of sCT-MNs was not statistically different from intravenous injection, suggesting that the use of microneedle patch for sCT delivery could be a viable alternative to injection of sCT and other peptides or protein therapeutics."

"Based on the previous research [214] of optimizing the fabrication process of DMNA with improved needle drug proportion, we have prepared two formulations of sCT-dissolving microneedle array (sCT-DMNA-1 and sCT-DMNA-2) and thoroughly investigated their stability as well as in vivo bioavailability."

MATERIALS AND METHODS

"While sCT-DMNA-1 solution was prepared by mixing 2.5 mg sCT and 25 mg Dextran in 50 μL DI, sCT-DMNA-2 solution was prepared by dissolving 12.5 mg trehalose into the needle formulation of sCT-DMNA-1."

"An appropriate amount of sCT-DMNA-1/sCT-DMNA-2 solution was added onto the surface of PDMS female mold and centrifuged at 3000×g force for 10 min to make the solution enter into the micropores of the mold."

"The needle and base part of sCT-DMNAs were separated with a surgical blade and dissolved in 1 mL PBS solution, which was prepared by mixing 0.87 g dipotassium hydrogen phosphate and 8.34 g potassium dihydrogen phosphate in 1000 mL ultrapure water."

RESULTS

"Contrastingly, at the storage conditions of room temperature with high humidity and high temperature with low humidity, the drug loadings of sCT-DMNA-1 were drastically reduced (even lower than 60% of the initial drug loading), with statistically significant difference ($p < 0.05$)."

"At room temperature with high humidity and high temperature with low humidity, the drug loadings of sCT-DMNA-2 were only slightly decreased, in comparison with those of sCT-DMNA-1 which were drastically reduced, the differences were statistically significant ($p < 0.05$), indicating that the trehalose could enhance the chemical stability of sCT encapsulated in DMNA."

"The hypocalcemic effect (D%) of the gel group was significantly lower than that of the positive control groups and sCT-DMNAs groups ($p < 0.01$), with relative bioavailability (PA%) only 11%, which was approximately one-sixth times less than that of sCT-DMNA-2."

DISCUSSION

"During the preparation and solidification process of sCT-DMNAs, the drug, encapsulated in the needle portion diffusing [215, 216] to the base, directly leads to the reduction of NDP."

"DI water was used as the needle solvent and highly volatile ethanol as the base solvent for two-step injection in mold, followed by rapid drying process, leading to significantly increase in the NDP of sCT-DMNAs."

"Trehalose is found in natural organisms, where it can not only replace water molecules by forming hydrogen bonds with polar residues of lipid and/or protein molecules (water substitution hypothesis) [217, 218] but also act as a protector against various stress conditions including freeze, osmotic stress, oxidation, and changes in temperature [219]."

"The reason may be that the existence of trehalose accelerates the needle dissolution speed of sCT-DMNA-2 when the microneedles penetrated the epidermis, thus improving the drug release amount, due to the high hydrophilicity [220] of trehalose (solubility in water: trehalose ≈ 689 mg/mL, dextran ≈ 50 mg/mL, 20°C)."

CONCLUSION

"To investigate the feasibility and in vivo pharmacological activity of DMNA patches for non-injection administration of biomacromolecule drugs, two formulations of sCT-DMNA patches (sCT-DMNA-1 and sCT-DMNA-2) were prepared."

"The results demonstrated that the mechanical properties and the drug distributions of the DMNA patches met all the requirements for transdermal administration."

"Comparing the results of the two sCT-DMNA formulations, it was observed that sCT-DMNA-2 containing trehalose was more conducive to the stability and transdermal delivery of sCT."

Enhancement of the transdermal delivery of zidovudine by pretreating the skin with two physical enhancers: microneedles and sonophoresis [221] This is a machine-generated summary of:

Martínez-Segoviano, Irene de Jesús; Ganem-Rondero, Adriana: Enhancement of the transdermal delivery of zidovudine by pretreating the skin with two physical enhancers: microneedles and sonophoresis [221]

Published in: DARU Journal of Pharmaceutical Sciences (2021)

Link to original: https://doi.org/10.1007/s40199-021-00402-y

If you want to cite the papers, please refer to the original.

For technical reasons we could not place the page where the original quote is coming from.

Abstract-Summary "The use of permeation enhancers is necessary to favor the passage of this drug through the skin, due to its physicochemical properties and to the natural permeation barrier imposed by the skin."

"To evaluate the effect of two permeation enhancers, sonophoresis and microneedles, on the permeability of AZT through the skin."

"Sonophoresis was applied under different conditions (i.e., amplitude, duty cycle and application time), selected according to an experimental design, where the response variables were the increase in temperature of the skin surface and the increase in transepidermal water loss."

"Pretreatment of the skin with sonophoresis increased AZT transport significantly, reducing the lag time."

"The maximum flux (27.52 $\mu gcm^{-2} h^{-1}$) and the highest total amount permeated (about 624 $\mu g/cm^2$) were obtained when applying sonophoresis in continuous mode, with an amplitude of 20%, and an application time of 2 min."

"The use of microneedles further increased the flux (30.41 $\mu gcm^{-2} h^{-1}$) and the total amount permeated (about 916 $\mu g/cm^2$), relative to sonophoresis."

"The results are encouraging in terms of promoting AZT transport through the skin using sonophoresis or microneedles as permeation enhancers."

Introduction

"It was the first agent approved for clinical use to treat AIDS, and for years it has shown to be effective, becoming the most widely used drug for antiretroviral therapy."

"In the case of AZT, the main reported adverse effects include hematological complications, headache, nausea, vomiting, myalgia, nail pigmentation, and insomnia."

"Although AZT is administered orally, several drawbacks are associated with this route of administration, including its short biological half-life (approximately 1 h) and the fact that it suffers from significant first-pass metabolism [222]."

"Most of these strategies are intended to favor the passage of AZT through the skin, as an alternative route of administration."

"The main objective of this work was to study the capability of sonophoresis and microneedles (two minimally invasive penetration promoters) to enhance the permeability of AZT into and through the skin, comparing the results with previous reports that make use of permeation enhancers."

Materials and methods

"Permeation experiments were conducted with porcine ear skin using Franz-type cells (diffusion area 0.78 cm^2) for a period of 24 h. The pigskin was removed from the cartilage, was cut with the aid of a dermatome (Zimmer 901, Ohio, USA) to a thickness of approximately 750 μm, and was stored at -20 °C until its use."

"Once the permeation study was completed, the drug solution was removed from the donor, rinsing the skin profusely with PBS pH 7.4."

"Once the skin was mounted in Franz's cells, 1 ml of an aqueous solution of SLS 1% (w/v), used as a coupling medium, was added to the donor compartment, placing the sonotrode at a distance of 1 cm above the skin surface, and operating the ultrasound equipment under the conditions previously determined, according to a 2^3 factorial design."

"The skin samples were mounted in Franz diffusion cells, filling the receptor compartment with 2 mL of PBS pH 7.4."

Results and discussion

"Kumar Narishetty & Panchagnula [223] and Thomas & Panchagnula [224], reported fluxes of 11.7 $\mu g cm^{-2} h^{-1}$ and 23.4 $\mu g cm^{-2} h^{-1}$, as well as lag times of 14.2 h and 46.9 h, respectively, for AZT aqueous solutions of 50 mg/mL. In our case, it was not possible to obtain a flux value, since as already mentioned, AZT was only quantified in the receptor medium at 24 h. Low-frequency ultrasound or sonophoresis has shown to be effective in promoting drug transport through the skin, reducing the lag time."

"Although the flux obtained in this work by pretreating the skin with microneedles or with sonophoresis, is far from the values reported by these authors, it should be expected to further increase the flux by: (i) Using a higher concentration of the drug; (ii) modifying the application scheme of sonophoresis; (iii) combining sonophoresis with some formulation containing chemical enhancers to achieve a synergistic effect; (iv) formulating AZT in a transdermal delivery system with microneedles, in order to promote the continuous passage of the drug through the skin."

Conclusions

"AZT hardly permeates through intact skin, as a result of its physicochemical properties."

"Pretreatment of the skin with sonophoresis increased AZT transport, reducing significantly the lag time, founding that an amplitude of 20%, and an application time of 2 min, in continuous mode, presented the highest flux for the first 7 h, and also the highest total amount permeated."

"Although the results of this work are encouraging in terms of promoting AZT transport through the skin using sonophoresis or microneedles as permeation enhancers, further studies must be carried out, considering the factors that affect their performance, in order to achieve flux values that guarantee effective plasma concentrations."

Lidocaine Transdermal Patch: Pharmacokinetic Modeling and In Vitro–In Vivo Correlation (IVIVC) [225] This is a machine-generated summary of:

Kondamudi, Phani Krishna; Tirumalasetty, Phani Prasanth; Malayandi, Rajkumar; Mutalik, Srinivas; Pillai, Raviraj: Lidocaine Transdermal Patch: Pharmacokinetic Modeling and In Vitro–In Vivo Correlation (IVIVC) [225]

Published in: AAPS PharmSciTech (2015)

Link to original: https://doi.org/10.1208/s12249-015-0390-1

If you want to cite the papers, please refer to the original.

For technical reasons we could not place the page where the original quote is coming from.

Abstract-Summary "The present study aims to develop the correlation between in vitro and in vivo skin permeation of lidocaine in its transdermal patch."

"Pharmacokinetic (PK) parameters such as AUC_{last} and C_{max} were predicted with the established in vitro–in vivo correlation (IVIVC) models."

"The minimum prediction errors in NDC method for C_{max} were found to be −30.9 and −25.4% for studies I (in vivo study in human volunteers with one batch of Lidoderm patch; internal validation) and II (in vivo study in human volunteers with another batch of Lidoderm patch; external validation), respectively, whereas minimum prediction errors in NCA method were relatively low (3.9 and 0.03% for studies I and II, respectively) compared to those in NDC method."

"The established method in this study could be a potential approach for predicting the bioavailability and/or bioequivalence for transdermal drug delivery systems."

INTRODUCTION

"Most of the transdermal patches intended for local therapy exhibit plasma drug concentrations below the limit of quantification and hence pose low risk for systemic toxicities."

"There is limited role of plasma drug concentration with respect to localized analgesic effect of LTP."

"Even though evaluation of plasma drug concentration is not an accurate measure for drug availability at the site of action for topical products, in the absence of well-defined clinical methodology, bioequivalence (BE) with pharmacokinetic (PK) end point serves as a surrogate for establishing the equivalence between generic and branded LTP [226]."

"Establishing bioequivalence between RLD and test products is cumbersome due to low systemic bioavailability of LTP, when compared with other dosage forms."

"Like other dosage forms, in vitro drug release testing (IVRT) methodologies are traditionally used for the optimization of formula and process for transdermal patches."

MATERIALS AND METHODS

"Franz diffusion cells were used to investigate the ex vivo skin permeation of lidocaine from transdermal patches."

"Preparation of skin for ex vivo permeation study has been performed by thawing the skin in 0.9% NaCl for not more than 1 h at room temperature and cutting to appropriate Franz cell size ($\approx$5 cm^2)."

"Blood samples of 7 mL were collected periodically at −1, 0, 1.5, 3, 4, 6, 8, 9, 10, 11, 12, 13, 14, 15, 16, 20, 24, 30, and 36 h of post-patch-application, and plasma drug concentrations were determined by a validated liquid chromatography–tandem mass spectrometry (LC-MS/MS) method."

"PK analysis for both studies was done by using Phoenix software version 6.3 (Pharsight™, Certara L.P.)."

"PK modeling of lidocaine was performed by using plasma drug concentration–time profile of intravenous bolus (50 mg dose) data obtained from the literature [227]."

RESULTS AND DISCUSSION

"Two-compartmental and non-compartmental PK parameters were used for establishing IVIVC by NDC and NCA methods, respectively."

"Level A correlation between normalized percent cumulative in vitro permeation and in vivo permeation was established using non-compartmental analysis (NCA) and numerical deconvolution (NDC) method."

"The advantages of this IVIVC method over conventional methods are (a) minimal compartmental modeling error, (b) easy to calculate in vivo permeation, and (c) simple mathematical expressions."

"The percent prediction error of C_{max} in study I, using internal validation method, was found to be −30.9 and 3.9% for NDC and NCA methods, respectively."

"We attempted to establish a preliminary IVIVC method by NCA method to demonstrate the correlation between in vitro and in vivo skin permeation of lidocaine."

"This further confirms the superiority of NCA method in establishing IVIVC over NDC method in this study."

CONCLUSION

"The present study successfully develops the correlation between in vitro and in vivo skin permeation of lidocaine in its transdermal patch."

"Of the present study, the NCA method is a superior approach over NDC method in predicting the C_{max} and AUC_{last} after administration of LTP."

"The study successfully shows that NCA method may be used as a biorelevant quality control (BRQC) in vitro tool for quality assurance of transdermal products."

Enhanced Transdermal Drug Delivery by Sonophoresis and Simultaneous Application of Sonophoresis and Iontophoresis [228] This is a machine-generated summary of:

Park, Juhyun; Lee, Hyowon; Lim, Guei-Sam; Kim, Nayoung; Kim, Dongwon; Kim, Yeu-Chun: Enhanced Transdermal Drug Delivery by Sonophoresis and Simultaneous Application of Sonophoresis and Iontophoresis [228]

Published in: AAPS PharmSciTech (2019)

Link to original: https://doi.org/10.1208/s12249-019-1309-z

If you want to cite the papers, please refer to the original.

For technical reasons we could not place the page where the original quote is coming from.

Abstract-Summary "We devised an improved device that can perform not only the single application of sonophoresis or iontophoresis but also the simultaneous application."

"The enhancement effect of sonophoresis was evaluated for various cosmeceutical drugs using a Franz diffusion cell."

"The relationship was found between the enhancement effect of sonophoresis and the physicochemical properties of drugs."

"The simultaneous treatment of sonophoresis and iontophoresis enhanced skin penetration of glutamic acid to 240% using the fabricated device."

"The simultaneous application showed significantly higher enhancement ratio than application of sonophoresis or iontophoresis alone."

"The improved device achieved skin penetration enhancement of various cosmeceutical drugs with lower intensity and a short application time."

"The miniaturized device with sonophoresis and iontophoresis is a promising approach due to enhanced transdermal drug delivery and feasibility of self-administration in cosmetic and therapeutic fields."

INTRODUCTION

"There are two ways to enhance skin permeability: chemical enhancement methods including azone derivatives, fatty acids, alcohols, esters, sulfoxides, pyrrolidones, glycols, surfactants, and terpenes and physical enhancement methods including electroporation, ultrasound, iontophoresis, and microneedles [229–231]."

"Sonophoresis increases the permeability of the skin and transfer drugs effectively."

"Ahmet and others have reported that skin permeability of drug linearly depends on the ultrasound intensity and exposure time, and the enhancement is directly proportional to the total ultrasound energy density [232]."

"The repulsive forces between the same charge push the molecule into the skin; thus, iontophoresis promotes drug delivery with a low current density [233]."

"Shirouzu and others [234] reported that the combination of sonophoresis and iontophoresis synergistically enhances the skin penetration of vitamin B_{12}."

"One of the proposed mechanisms is that enhancement effect is mainly caused by the stratum corneum diffusivity of molecules increased by sonophoresis and the electro-osmotic water flow by iontophoresis application [235]."

MATERIALS AND METHODS

"The device was applied to the porcine dorsal skin with sonophoresis and iontophoresis."

"The dorsal skin of mini-pig was supplied by Cronex (South Korea) without drying and stored at – 80°C and thawed at room temperature prior to the experiment."

"The porcine skin was mounted on the Franz diffusion cell and then 0.3 ml of 1 mM fluorescein (Sigma-Aldrich) solution (PBS buffer based) was applied into donor chamber."

"The mini-pig is promising model animal for percutaneous drug absorption because mini-pig and human skin are similar in consideration of cellular composition, morphology, and epidermal thickness [236, 237]."

"From previous research, there is not a quantitative, but a qualitative similarity between percutaneous drug absorption across porcine and human skin."

"Various cosmetic ingredients were prepared for the skin penetration experiments."

"All hydrophilic drugs were dissolved in PBS for the skin penetration experiment."

"The treated skin was mounted on the Franz diffusion cell."

RESULTS AND DISCUSSION

"The aim of skin penetration experiment was to evaluate the enhancement effect of sonophoresis for various drugs."

"Niacinamide with the lowest molecular weight showed the most enhanced skin permeability than the other hydrophilic drugs."

"In all group, it was also observed that the enhancement ratio from 0 to 5 h was significantly higher than that from 5 h to 24 h. This result indicated that sonophoresis could induced the fast transport of hydrophobic drugs through an intracellular pathway."

"The skin penetration of drug after 5 h was also significantly enhanced compared to the control group."

"The experiment was carried out at different frequencies (280 and 350 kHz) and intensities (levels 1, 2, and 3) with the simultaneous application as well as single application of ultrasound and iontophoresis to evaluate the skin permeation enhancement effect of sonophoresis and iontophoresis."

CONCLUSION

"We have designed the miniaturized device for this study, which can operate not only the single application but also the simultaneous application of sonophoresis and iontophoresis."

"In spite of a short application time and lower intensity level than previous studies, the improved device achieved significant skin penetration enhancement of various hydrophilic and hydrophobic drugs."

"The simultaneous application is a promising approach because this strategy achieves transdermal enhancement at the low-energy density."

"Considering safety, this simultaneous application has advantage of reduced the skin irritation because this strategy enhances transdermal drug delivery at the low-energy density."

"This simultaneous application of sonophoresis and iontophoresis can be extended for therapeutic fields as well as cosmetic fields."

Iontophoresis to Overcome the Challenge of Nail Permeation: Considerations and Optimizations for Successful Ungual Drug Delivery [238] This is a machine-generated summary of:

Chen, Kevin; Puri, Vinam; Michniak-Kohn, Bozena: Iontophoresis to Overcome the Challenge of Nail Permeation: Considerations and Optimizations for Successful Ungual Drug Delivery [238]

Published in: The AAPS Journal (2021)

Link to original: https://doi.org/10.1208/s12248-020-00552-y

If you want to cite the papers, please refer to the original.

For technical reasons we could not place the page where the original quote is coming from.

Abstract-Summary "Iontophoresis is a widely used drug delivery technique that has been used clinically to improve permeation through the skin for drugs and other actives in topical formulations."

"It is however not commonly used for the treatment of nail diseases despite its potential to improve transungual nail delivery."

"It also includes relevant details about the nail structure, the mechanisms of iontophoresis, and the associated in vitro and in vivo studies which have been used to investigate the optimal characteristics for a transungual iontophoretic drug delivery system."

"Iontophoresis is undoubtedly a promising option to treat nail diseases, and the use of this technique for clinical use will likely improve patient outcomes."

INTRODUCTION

"In the case of onychomycosis, there are both systemic and topical treatments."

"Available topical lacquers such as Loceryl® and Penlac® have less side effects and better patient compliance but suffer from low cure rates due to poor penetration of drug through the nail plate [239, 240]."

"Nail psoriasis is also treated using systemic or topical treatments, both of which suffer problems similar to the treatment methods for onychomycosis."

"Immunosuppressant drugs (such as corticosteroids, methotrexate, infliximab) are used to treat nail psoriasis, which have a wide variety of side effects depending on which drug is used."

"Topical antipsoriatic drugs suffer similar problems to that of anti-onychomycotic ones, featuring poor penetration of drug through the nail [241]."

"This work aims to elucidate areas in need of more research in this field so that iontophoresis may be utilized clinically to enhance topical treatments of nail disease."

THE NAIL AND NAIL MODELS IN TRANSUNGUAL IONTOPHORESIS

"Many nail models have been used in order to study transungual drug delivery and emulate in vivo conditions as closely as possible, each featuring their own pros and cons."

"Human nail clippings are the most frequently used in vitro models for studying transungual iontophoresis because they are relatively easy to obtain over full nail plates and are more representative of the in vivo nail than animal hooves."

"There was a high correlation ($r^2 = 0.93$) between the drug load into the hoof membrane and nail plate, but there was a much lower correlation in the permeation of the two models ($r^2 = 0.56$) [242]."

"A major obstacle in studying transungual drug delivery is inter-nail variability, presenting as statistically significant differences between nails despite no other differences being present [201]."

"A transungual drug delivery system should be able to address the differing nail barrier functionalities present in an in vivo setting."

MECHANISMS OF IONTOPHORESIS

"Iontophoresis enhances transungual drug delivery primarily by two mechanisms: electromigration and electroosmosis."

"The solvent's flux is also proportional to the applied voltage and therefore current density [243], making a higher current density favorable for electroosmosis, similar to electromigration."

"When cathodal iontophoresis was used instead, the transport of neither ion was significantly different from the passive transport, as the electroosmotic solvent flow was opposing permeation through the nail."

"Using a low pH means that the nail will become anion permselective, and electroosmosis will flow from cathode to anode, which is the opposite direction of electromigration."

"The only situation in which electromigration and electroosmosis will flow in the same direction is if a drug is positively charged at a high pH or negatively charged at a low pH. As aforementioned, the contributions of electromigration are much more significant than that of electroosmosis in the nails [244–246]."

PRACTICALITY OF TRANSUNGUAL IONTOPHORESIS

"All transungual iontophoresis studies have reported a significant increase of both permeation and drug loading into the active site of diffusion, but therapy cannot be successful unless the drug is delivered to the entire nail unit that is affected by disease."

"Like the increase in delivery in the active diffusion area, an increase in current density results in a proportional increase in drug loading into the peripheral area, which was seen when a twofold increase in current density resulted in a twofold increase in peripheral nail drug loading (0.25 mA/cm^2, 0.020 ± 0.014 μg/mg; 0.5 mA/cm^2, 0.039 ± 0.015 μg/mg) [247]."

"These release profiles are likely related to the fact that most of the drug in passive delivery are confined in the dorsal layer of the nail, while iontophoresis effectively delivers drug into all layers of the nail."

FORMULATION DEVELOPMENT AND DESIGN FOR A TRANSUNGUAL IONTOPHORETIC DRUG DELIVERY SYSTEM

"While the drug experiences considerable difficulty in permeating the nail barrier, its mycological effectiveness makes it a promising choice for topical delivery, and its ability to assume a positive charge at reasonable pHs makes it suitable for iontophoresis."

"The latter is advantageous for transungual iontophoresis because it carries a charge at reasonable pH values (itraconazole's pK_a is 3.7, requiring extremely low pHs to achieve protonation), has better transungual permeation, and has improved aqueous and organic solubility, while not having significantly different antifungal activity."

"The main effect of drug concentration in transungual iontophoresis is on electromigration, as the transport number is in part determined by the proportion of the charge carried by the drug."

"At this pH, the transport of all drugs was very low relative to normal pHs, most likely due to the excessive competition for electromigration from hydronium ions [248]."

NAIL PRETREATMENTS

"The most important effect of hydration is the increase of pore size in the nails, allowing for relatively large drug molecules to better permeate the nail [245]."

"Using a hydrating solution with modified pH can also change the permselectivity of the nail, which affects the direction of electroosmosis, one of the mechanisms of iontophoresis, and the ability for either cations or anions to permeate through the membrane more easily."

"While urea is keratolytic and causes hydration and swelling, the data suggested that the pores created by urea were too small to enhance permeation of molecules larger than water, and glycolic acid primarily weakens lipid binding, which would not help much in permeation of the nail plate."

CLINICAL USE OF TRANSUNGUAL IONTOPHORESIS

"Four controlled clinical studies regarding the efficacy of iontophoresis to treat nail diseases have been reported: two for onychomycosis and two for nail psoriasis."

"While the results of the nail growth study are not fully validated [249], there was a significantly lower amount of fungal elements in the nail in group A by the end of the 4 weeks (study group, 74% of patients negative for fungal elements; control group, 50% of patients negative for fungal elements)."

"Another clinical study on forty patients used onychomycosis severity index (a scale from 0 to 35 that represents how much of the nail is affected by the disease), a visual analog scale (a scale from 0 to 10 based on patient-reported pain while walking with shoes on), and potassium hydroxide microscopy (to evaluate presence of fungal elements) to evaluate the efficacy of Lamisil® with iontophoresis."

CONCLUSIONS

"Despite the apparent effectiveness of iontophoresis in improving drug delivery through the nails, the technique is still not widely used in the clinic, and many patients with nail diseases are not treated due to lack of safety and efficacy of existing treatment methods."

"The technique has been tested in many different nail models and has been shown to significantly improve drug permeation in every model it has been tested in, including in vivo application."

"Transungual iontophoresis is a promising approach to address the issue of poor treatment options for nail diseases."

"Some areas in the field that could benefit from further studies are ionto-keratolysis, the application of a greater variety of permeation enhancers, the use of disease model keratin films to both emulate disease states and improve on inter-nail variability, and the use of a nail gel formulation for the treatment of nail psoriasis (the way in which transungual treatment of onychomycosis is typically treated)."

Influencing factors and drug application of iontophoresis in transdermal drug delivery: an overview of recent progress [250] This is a machine-generated summary of:

Wang, Yu; Zeng, Lijuan; Song, Wenting; Liu, Jianping: Influencing factors and drug application of iontophoresis in transdermal drug delivery: an overview of recent progress [250]

Published in: Drug Delivery and Translational Research (2021)

Link to original: https://doi.org/10.1007/s13346-021-00898-6

If you want to cite the papers, please refer to the original.

For technical reasons we could not place the page where the original quote is coming from.

Abstract-Summary "Transdermal drug delivery is limited by the stratum corneum of skin, which blocks most molecules, and thus, only few molecules with specific physicochemical properties (molecular weight < 500 Da, adequate lipophilicity, and low melting point) are able to penetrate the skin."

"Iontophoresis technology, which uses a small current to improve drug permeation through skin, is one of the effective ways to circumvent the stratum corneum."

"The focus will be on the latest advancements in iontophoretic transdermal drug delivery and application of iontophoresis with other enhancing technologies."

"The challenges of this technology for drug administration have also been highlighted, and some iontophoretic systems approved for clinical use are described."

Introduction

"Since skin serves as a natural barrier that protects the body from xenobiotics and hazardous substances, drug transdermal permeation is restrained [251]."

"SC, the outermost layer of skin, is the main barrier for drug permeation."

"Transdermal delivery of macromolecular drugs is a big challenge."

"Several technologies have been developed to overcome the barrier of SC and promote skin permeability of drugs, including physical technologies such as sonophoresis, microneedles, and iontophoresis, as well as chemical technologies like penetration enhancers [252, 253]."

"Of all the technologies adopted to enhance drug transdermal permeation, iontophoresis not only provides ease application but also ensures the delivery of drugs in programmed and controlled manner [254]."

"Iontophoresis refers to the use of low-density electric current for transdermal drug delivery enhancement [255]."

Transport mechanisms

"Iontophoresis is a method of introducing ionized and neutral drugs into the skin by using electric current [256]."

"Charged drugs are forced across the skin by electronic repulsion of similar charges."

"Cationic drugs can permeate through skin by using a positively charged working electrode."

"Anionic drugs can cross skin by negatively charged electrodes [254]."

Factors influencing iontophoretic process

"According to the Faraday's law, the iontophoretic flux of a drug molecule (J_{DRUG}) is the sum of the fluxes due to electromigration (J_{EM}), electroosmosis (J_{EO}), and passive delivery (J_p) [195, 257]."

"Equation, it would appear that iontophoretic flux should be mainly affected by applied current and drug concentration."

"They analyzed the pharmacokinetic parameters, which suggested the concentration-dependent iontophoretic tacrine delivery at lower concentrations and the plateau in tacrine permeation flux at higher concentration."

"The effect of current density on iontophoretic flux is similar to drug concentration."

"The pH of the drug-containing solution is another important factor which can impact on iontophoretic transport."

"The iontophoretic flux of midazolam was 18 ± 3.2 nmol·h^{-1} at the pH 4.5 donor solution, which was significantly higher than that with the pH 3 donor (6.0 ± 3.0 nmol·h^{-1}) [258]."

Iontophoretic device

"An iontophoretic device consists of four parts: power source, control circuit, electrodes, and reservoirs."

"An electrotransport drug delivery system typically consists of a power supply connected to a pair of electrodes in contact with ionically conductive reservoirs that, in turn, are in contact with the skin."

"Phoresor iontophoretic drug delivery system from Iomed is used to deliver iontocaine (lidocaine and epinephrine combination) for local dermal anesthesia [254, 255]."

"Wearable electronic disposable drug delivery (WEDD) developed by BirchPoint Medical Inc. is a portable, disposable patch having a thin, flexible battery having capability to supply variable voltages for versatility in drug delivery and expands the range of drugs which can be delivered by iontophoresis [259]."

"Another electrode is placed in the receptor compartment filling with buffer, and a small current is applied to delivering the drug through the skin."

Application for drug delivery

"They found that the application of iontophoresis dramatically increased the flux of both drugs compared with passive delivery."

"More significantly, the results indicated that iontophoresis technology could considerably reduce the interspecies differences in transdermal drug delivery."

"Talbi and others designed a controllable transdermal patch for lidocaine delivery and investigated lidocaine permeation amounts at different current density through in vitro experiments by iontophoresis [260]."

"To transdermal drug delivery system, iontophoresis is also applied to cosmetics in order to promote the entry of cosmetic ingredients into skin."

"KPV permeation increased to 35.2 μg/cm^2/h by iontophoresis treatment and the skin retention level of KPV by iontophoresis increased tenfold as compared with passive delivery."

"The electrorepulsion in cathodal iontophoresis might be the only driving force for ODN delivery, since the skin was negatively charged and cation selective at physiological pH. Considering the limited transfer of macromolecule and negatively charged molecules across intact skin, combination with microneedles or other approaches might be necessary to achieve transdermal delivery of ODNs [261, 262]."

Combination application with other enhancing technologies

"Tokumoto and others used iontophoresis combined with low frequency ultrasound to enhance transdermal delivery of mannitol across mouse skin [263]."

"Higher transdermal drug permeation can be achieved by the combination of chemical penetration enhancer and iontophoresis."

"Studies indicated that a therapeutic amount of diclofenac potassium could be transdermally delivered by the combination of iontophoresis with enhancers like geraniol or l-menthol."

"The skin was pretreated with the microneedle device for 1 min, and then a constant current of 0.5 mA/cm^2 was applied for 4 h. The results demonstrated that the combination of microneedle and iontophoresis showed synergetic effects, and the enhancement effect was much greater than the method used alone."

"Considering that combination strategy could obtained a high permeation-enhancing effect and overcame the limitations of iontophoresis that could not disrupt the skin barrier, the combination of iontophoresis and microneedles was a potential strategy for transdermal delivery of macromolecules."

Clinical use

"While a number of researches have been conducted and significant progresses have been made in clinical trials of iontophoresis, only three products (Ionsys®, LidoSite®, and Zecuity®) have been approved by FDA [264]."

"LidoSite®, an iontophoretic drug device, was used to rapidly deliver lidocaine for local anesthesia."

"Reverse iontophoresis, which is used for the extraction of a molecule from the body rather than its delivery into the body, has been widely applied in devices for diagnostic application and has shown tremendous potential in glucose monitoring."

"Glucowatch® is a device that provides a needle-free means of monitoring blood glucose levels in diabetic patients [259]."

Challenges

"One of the major challenges regarding transdermal iontophoresis is developing low cost and stable devices that can provide efficient transdermal drug delivery."

"In iontophoretic delivery, the rate of drug delivery scales with the electric current [265]."

"Under low current density, iontophoresis does not disrupt the skin structure that may cause skin damage [266], but the maximum delivery rate is limited [4]."

"Only molecules approximately 13 kDa molecular weight can permeate through intact skin by iontophoresis [267]."

"The contradiction between drug delivery rate and safety needs to be balanced."

"The transdermal delivery of negative charged drugs such as insulin and hyaluronic acid [268] is limited under physiological conditions."

Conclusion

"Although significant progress in iontophoresis has been recorded and important breakthroughs have been made in the clinic field, there are still some challenges, such as the transdermal delivery of low potent efficiency drugs and biological drugs."

"Combination of iontophoresis with sonophoresis, chemical enhancers, and microneedles has been utilized by scientists to solve the limitation."

"With more thorough understanding and in-depth research of iontophoresis, more transdermal iontophoretic products will be developed for the benefit of patients."

A comparison of dissolving microneedles and transdermal film with solid microneedles for iloperidone in vivo: a proof of concept [269] This is a machine-generated summary of:

Bhadale, Rupali S.; Londhe, Vaishali Y.: A comparison of dissolving microneedles and transdermal film with solid microneedles for iloperidone in vivo: a proof of concept [269]

Published in: Naunyn-Schmiedeberg's Archives of Pharmacology (2022)

Link to original: https://doi.org/10.1007/s00210-022-02309-0

Copyright of the summarized publication:

The Author(s), under exclusive licence to Springer-Verlag GmbH Germany, part of Springer Nature 2022

If you want to cite the papers, please refer to the original.

For technical reasons we could not place the page where the original quote is coming from.

Abstract-Summary "The current research aimed to compare the antipsychotic activity and pharmacokinetics of ILO-loaded dissolving microneedles (DMNs) and transdermal film with a solid microneedle (STF)."

"Studies were compared with transdermal film (TF) on untreated skin as a passive control."

"STF and DMNs had considerably greater AUC and C_{max} ($p \leq 0.001$) than transdermal film."

"In pharmacodynamic tests, STF and DMNs demonstrated significant ($p \leq 0.001$) forelimb retraction time (FRT) and hindlimb retraction time (HRT) delay responses as compared to control and TF."

"In the skin irritation test, no adverse effects such as erythema or edema were observed at the end of the 48 h. Thus, antipsychotic activity (paw test) and pharmacokinetics studies revealed sustained action of DMN and STF."

"This research revealed that improved efficacy of DMN and STF for antipsychotic drug delivery may be an alternative to the existing dosage form."

Introduction

"Long-term use of antipsychotic drugs has been related to extrapyramidal symptoms, despite the fact that these treatments have been shown to be effective at reducing symptoms (EPS)."

"The US Food and Drug Administration approved the atypical antipsychotic drug iloperidone (ILO) in May 2009 for the treatment of schizophrenia in adults."

"This confirms the theory that, despite its antipsychotic properties, long-term oral use of ILO causes oxidative stress and, as a result, EPS (Shirzadi & Ghaemi [270]; Pillai and others [271])."

"A transdermal patch for the treatment of schizophrenia was recently licensed by the US Food and Drug Administration (FDA) as SECUADO ® (asenapine) transdermal system in 2019 ("Dosage & Application | SECUADO® (asenapine) Patch", [272])."

"Microneedles (MNs), a type of physical enhancer, are an emerging technology that is being used in a rising prevalence of transdermal drug delivery."

"By 2025, the worldwide transdermal drug delivery market is expected to reach $95.57 billion (Schizophrenia Facts and Statistics, [273])."

Material and method

"The inclusion complex of ILO was used further while preparing DMNs and TF."

"The solution was placed in a vacuum desiccator for 2 h to eliminate air bubbles, then cast onto a glass slide and placed in a 60 °C preheated oven for 4–5 h. Finally, the film was removed and kept in a desiccator at room temperature until further use."

"The TF (passive control) and DMNs (1 mg/kg) were placed uniformly at the back of the neck of rats in groups 1 and 2 respectively."

"ILO and carbamazepine standard solutions were prepared by dissolving 5 mg each in 5 mL methanol separately (1000 ppm)."

"Blank rat plasma was spiked with ILO and carbamazepine to provide plasma calibration standards."

"The blank brain tissue was homogenized with ILO and carbamazepine to construct brain calibration standards."

Result and discussion

"Following TF (passive control) of ILO, STF, and DMNs, the average peak rat plasma concentrations (C_{max}) were 180.66 ± 8.47 ng/ml, 954.66 ± 2.81 ng/ml, and 927.17 ± 4.57 ng/ml respectively."

"The development of micropores in the top layer of the skin significantly improves systemic absorption (10 times) of ILO through DMNs and STF as compared to the TF (passive control, $p \leq 0.001$)."

"The C_{max} values of TF (passive control), DMNs, and STF were compared in brain tissue."

"In the paw test, the STF (group III) and DMNs (group IV) exhibited a significant ($p \leq 0.001$) delay after 12 h to 48 h in HRT and FRT when compared to the control group and TF (passive control, group II)."

"The DMNs and STF have enhanced the release of ILO, whereas TF (passive control) had a longer release time (C_{max} of 48 h)."

Conclusion

"This study investigated the use of microneedles for antipsychotic drug delivery (ILO) for the first time."

"Biopolymers such as HPβCD, PVA, and PVPK-30 have been used to develop the sustained-release formulation of ILO, which includes dissolving MNs and a transdermal film."

"The findings revealed that dissolving MNs and a combination of solid MNs with TF resulted in effective plasma and brain delivery as demonstrated in the pharmacokinetic and pharmacodynamic study."

Transdermal Delivery of Baclofen Using Iontophoresis and Microneedles [274] This is a machine-generated summary of:

Junaid, Mohammad Shajid Ashraf; Banga, Ajay K.: Transdermal Delivery of Baclofen Using Iontophoresis and Microneedles [274]

Published in: AAPS PharmSciTech (2022)

Link to original: https://doi.org/10.1208/s12249-022-02232-w

Copyright of the summarized publication:

If you want to cite the papers, please refer to the original.

For technical reasons we could not place the page where the original quote is coming from.

Abstract-Summary "A transdermal baclofen delivery system might be the solution to this problem."

"This research focuses on evaluating microneedles, iontophoresis, and a combination of microneedles-iontophoresis as transdermal delivery enhancement strategies for baclofen."

"Anodal iontophoresis was applied at a current density of 0.5 mA/cm^2, and transdermal delivery was assessed from pH 4.5 (45.51±0.76 μg/cm^2) and pH 7.4 (68.84±10.13 μg/cm^2) baclofen solutions."

"Iontophoresis enhanced baclofen delivery but failed to reach target delivery."

"Maltose microneedles were used to create hydrophilic microchannels on the skin, and this technique enhanced baclofen delivery by 89-fold."

"Both microneedles (447.88±68.06 μg/cm^2) and combination of microneedles – iontophoresis (428.56±84.33 μg/cm^2) reached the target delivery range (222–1184 μg/cm^2) for baclofen."

"The findings of this research suggest that skin could be a viable route for delivery of baclofen."

INTRODUCTION

"The transdermal route offers painless drug delivery, greater convenience, and patient compliance by increasing dose intervals, ease of termination in case of toxicity, bypassing first-pass metabolism, and avoidance of the peak and valley effect (typically caused by conventional oral drug delivery)."

"Since baclofen is hydrophilic (log P −0.78) in nature, therapeutically relevant drug levels might not be achieved through passive transdermal delivery [275]."

"For enhancing the transdermal delivery of non-ideal drug molecules like baclofen, employing physical enhancement techniques such as microneedles and iontophoresis could be a viable strategy."

"This research focuses on exploring physical enhancement techniques for transdermal delivery of baclofen at therapeutically relevant levels."

"Iontophoresis has been explored as a potential physical enhancement strategy for the delivery of baclofen, and the effect of donor solution pH on anodal iontophoretic baclofen delivery has been investigated."

"Combination of iontophoresis and microneedles has been applied and assessed as a delivery strategy for baclofen."

MATERIALS AND METHOD

"Baclofen was obtained from T.C.I. America (Portland, OR, USA)."

"Acetonitrile was procured from Pharmco-aaper (Brookfield, CT, USA), and sodium phosphate dibasic was obtained from Sigma Aldrich, USA."

"Silver wire was obtained from Fisher Scientific (Fair Lawn, NJ, USA)."

"Silver/Silver chloride electrodes were procured from A-M systems (Sequim, WA, USA)."

METHOD

"Skin pieces with an electrical resistance higher than 10 kΩ were selected for the in vitro permeation studies."

"Dermatomed porcine ear skin was placed on the Franz diffusion cell exposing 0.64 cm^2 of the skin to the donor and receptor chambers."

"Anodal iontophoretic delivery of baclofen through dermatomed porcine ear skin was evaluated by conducting in vitro permeation studies on vertical Franz diffusion cells as described in "Design of In Vitro Permeation Study.""

"After skin was microporated by the application of microneedles, in vitro permeation study was conducted as described in "Design of In Vitro Permeation Study.""

"To evaluate the combined effect of microneedles and iontophoresis on transdermal delivery of baclofen, skin was first microporated using microneedles as described in the "Microneedle-aided Delivery of Baclofen" section under "Design of In Vitro Permeation Study.""

RESULTS

"Solubility of baclofen was evaluated in PBS, CBS, propylene glycol, and methanol."

"The PBS and CBS solutions containing baclofen delivered 4.20±0.55 μg/cm^2 and 3.64±1.24 μg/cm^2 into the receptor, respectively."

"There was no significant difference ($p > 0.05$) between the amounts of drug delivered into the receptor between the PBS and CBS solutions."

DISCUSSION

"This can be explained by the fact that skin becomes net positively charged when pH of the applied solution (on skin) drops below pH 4.8 (isoelectric point of skin) compared to net negative charge at physiological pH. This phenomenon has a negative impact on the contribution of electroosmosis in drug delivery enhancement of anodal iontophoresis [276, 277]."

"With microneedles, by creating hydrophilic microchannels through epidermis, this lipophilic barrier is bypassed, resulting in higher systemic delivery of baclofen."

"The enhancement caused by iontophoresis (16-fold enhancement over passive delivery) is considerably lower when compared to that caused by microneedles (89-fold enhancement over passive delivery and 5.5-fold enhancement over iontophoretic delivery), resulting in an insignificant difference in transdermal drug delivery between microneedles only and combination group."

"These results are in line with findings of another study conducted by Garland and others, where it is reported that microneedles significantly enhanced transdermal delivery of small hydrophilic compounds, but a combination of microneedles and iontophoresis did not further enhance delivery [278]."

CONCLUSION

"The goal of this study was to investigate a more convenient and non-invasive route of baclofen administration as an alternative to the currently available routes."

"Microneedles effectively enhanced transdermal delivery of baclofen and delivered drug within the target delivery range for the treatment of multiple sclerosis."

"This research successfully demonstrates the potential for delivering therapeutic doses of baclofen in multiple sclerosis treatment through the transdermal route using physical enhancement techniques."

The Relationship between the Drug Delivery Properties of a Formulation of Teriparatide Microneedles and the Pharmacokinetic Evaluation of Teriparatide Administration in Rats [279] This is a machine-generated summary of:

Oh, Yu-Jeong; Kang, Nae-Won; Jeong, Hye-Rin; Sohn, Seo-Yeon; Jeon, Yae-Eun; Yu, Na-Young; Hwang, Yura; Kim, Sunkyung; Kim, Dae-Duk; Park, Jung-Hwan: The Relationship between the Drug Delivery Properties of a Formulation of Teriparatide Microneedles and the Pharmacokinetic Evaluation of Teriparatide Administration in Rats [279]

Published in: Pharmaceutical Research (2022)

Link to original: https://doi.org/10.1007/s11095-022-03254-6

If you want to cite the papers, please refer to the original.

For technical reasons we could not place the page where the original quote is coming from.

Abstract-Summary "This study examines the relationship between the drug delivery properties of the solid formulation with teriparatide and the pharmacokinetic properties of teriparatide in vivo."

"Teriparatide microneedles with different dissolution rates were prepared using sucrose and carboxymethylcellulose (CMC)."

"There were three aspects of this study: (1) The dissolution rate of teriparatide from both formulations (sucrose and CMC) was measured in vitro. (2) After administration into porcine skin ex vivo, the diffusion rate of FITC-dextran was observed using a confocal microscope. (3) Pharmacokinetic studies were performed in rats and pharmacokinetic data compared with the release rate and the diffusion pattern."

"In the in vitro dissolution experiment, 80% of teriparatide was released within 30 min from the CMC MNs, whereas 80% of teriparatide was released within 10 min from the sucrose MNs."

"When teriparatide is delivered into the skin using microneedles, the release rate from the solid formulation determines the drug's pharmacokinetic properties."

Introduction

"Human recombinant parathyroid hormone (1–34) (PTH) (CosMED Pharmaceutical Co, Ltd Kyoto, Japan) was delivered using dissolving MNs, and the pharmacokinetic profile of MN administration was found to be compatible with that of the subcutaneous administration of teriparatide [280]."

"Daddona and others conducted a clinical study using teriparatide-coated MNs and found that pharmacokinetic data from MNs were comparable to those from SC administration [281]."

"The pharmacokinetic properties of teriparatide MNs have been reported in previous preclinical and clinical model studies."

"Two types of coated MNs with rapid or slow release rates of teriparatide were prepared to obtain the correlation between the dissolution rate of teriparatide in vitro and the diffusion rate into the skin ex vivo and the pharmacokinetic results in vivo."

Materials and Methods

"Te-Su and Fl-Su MNs were prepared by dissolving sucrose in distilled water (DW) at 60°C overnight, followed by the addition of teriparatide or FITC-dextran."

"To observe the shape of the coating layer on the microneedles using an optical microscope, a coating layer was prepared by adding FITC-dextran (Sigma Aldrich, St. Louis, MO, USA) which has the same molecular weight (3–4 kDa) as teriparatide."

"Trypan blue–coated microneedles were placed on porcine skin (CRONEX, Seoul, Korea) and pressed with a force of 5 kg for 10 s and then attached to the skin using adhesive tape for 1 min."

"FITC-dextran MNs were administered to ex vivo porcine skin, and fluorescence images of FITC-dextran were observed with z-stacks at 5, 15, 30, 50, 80, 120, 180, and 240 min using a confocal microscope (ECLIPSE TE2000-E, Nikon, Osaka, Japan)."

Results and Discussion

"The release rate of teriparatide from the coated layer was much faster with sucrose MNs than with CMC MNs."

"The sucrose MN group exhibited a much higher C_{max} value (6809.0 ± 1717.3 pg/mL) than the CMC MN group (961.5 ± 194.0 pg/mL) despite similar drug content, which might be explained by the faster dissolution of the sucrose-coated layer and its quicker diffusion rate into the skin."

"The fast dissolution and diffusion of the sucrose MNs in the skin allowed the teriparatide to be absorbed rapidly into the bloodstream, thereby showing a lower T_{max} value in the sucrose MN group (0.4 ± 0.1 h) than in the CMC MN group (1.0 ± 1.3 h)."

"When the drug was delivered into the skin using microneedles, the in vitro dissolution rate of the coating layer was a critical variable in determining its pharmacokinetic properties, as with other drug administration methods."

Conclusion

"The pharmacokinetic properties of microneedles have not been clearly evaluated in previous studies of this method of teriparatide delivery."

"Previous studies of teriparatide-coated microneedles aimed at intradermal delivery did not consider the drug formulation."

"A quickly dissolving sucrose layer and a slowly dissolving CMC coating layer were designed, and their pharmacokinetic properties and correlations were observed."

"This study found that sucrose MNs dissolved quickly, and the model drug also diffuses rapidly in the skin."

"The difference in drug delivery properties of the two MNs also made a difference in C_{max}, T_{max}, and AUC_{last} values."

"An understanding of the drug delivery properties of the MN coating layer is necessary to obtain the desired pharmacokinetic properties of coated MNs and thus improve this method of drug delivery."

Bibliography

1. Abdifetah O, Na-Bangchang K (2019) Pharmacokinetic studies of nanoparticles as a delivery system for conventional drugs and herb-derived compounds for cancer therapy: a systematic review. Int J Nanomed:5659–5677
2. Chidambaram M, Manavalan R, Kathiresan K (2011) Nanotherapeutics to overcome conventional cancer chemotherapy limitations. J Pharm Pharm Sci 14(1):67–77
3. Alai M, Lin WJ (2014) Novel lansoprazole-loaded nanoparticles for the treatment of gastric acid secretion-related ulcers: in vitro and in vivo pharmacokinetic pharmacodynamic evaluation. AAPS J 16:361–372. https://doi.org/10.1208/s12248-014-9564-0
4. Prausnitz MR, Langer R (2008) Transdermal drug delivery. Nat Biotechnol 26(11):1261–1268
5. Shaker R, Castell DO, Schoenfeld PS, Spechler SJ (2003) Nighttime heartburn is an underappreciated clinical problem that impacts sleep and daytime function: the results of a Gallup survey conducted on behalf of the American Gastroenterological Association. Am J Gastroenterol 98:1487–1493
6. Katz PO, Koch FK, Ballard ED, Bagin RG, Gautille TC, Checani GC et al (2007) Comparison of the effects of immediate-release omeprazole oral suspension, delayed-release lansoprazole capsules and delayed-release esomeprazole capsules on nocturnal gastric acidity after bedtime dosing in patients with night-time GERD symptoms. Aliment Pharmacol Ther 25:197–205
7. Wang H, Jia Y, Hu W, Jiang H, Zhang J, Zhang L (2013) Effect of preparation conditions on the size and encapsulation properties of mPEG-PLGA nanoparticles simultaneously loaded with vincristine sulfate and curcumin. Pharm Dev Technol 18:694–700
8. Cetin M, Atila A, Sahin S, Vural I (2013) Preparation and characterization of metformin hydrochloride loaded-Eudragit®RSPO and Eudragit®RSPO/PLGA nanoparticles. Pharm Dev Technol 18:570–576
9. Wu ZM, Zhou LY, Guo XD, Jiang W, Ling L, Qian Y et al (2012) HP55-coated capsule containing PLGA/RS nanoparticles for oral delivery of insulin. Int J Pharm 425:1–8
10. Cartiera MS, Johnson KM, Rajendran V, Caplan MJ, Saltzman WM (2009) The uptake and intracellular fate of PLGA nanoparticles in epithelial cells. Biomaterials 30:2790–2798
11. Song XR, Zhao Y, Wu WB, Bi YQ, Cai Z, Chen QH et al (2008) PLGA nanoparticles simultaneously loaded with vincristine sulfate and verapamil hydrochloride: systematic study of particle size and drug entrapment efficiency. Int J Pharm 350:320–329
12. He W, Yang M, Fan JH, Feng CX, Zhang SJ, Wang JX et al (2010) Influences of sodium carbonate on physicochemical properties of lansoprazole in designed multiple coating pellets. AAPS PharmSciTech 11:1287–1293
13. Kaur A, Nigam K, Tyagi A, Dang S (2022) A preliminary pharmacodynamic study for the management of Alzheimer's disease using memantine-loaded PLGA nanoparticles. AAPS PharmSciTech 23(8):298. https://doi.org/10.1208/s12249-022-02449-9
14. Abd Razak M, Julianto T, Majeed ABA (2019) Enhancement of memantine uptake in the brain by incorporation with nanoparticles and given intranasally. IBRO Reports 6:S169
15. Kaur A, Nigam K, Bhatnagar I, Sukhpal H, Awasthy S, Shankar S, Tyagi A, Dang S (2020) Treatment of Alzheimer's diseases using donepezil nanoemulsion: an intranasal approach. Drug Deliv Transl Res 10(6):1862–1875

16. Bali ZK, Bruszt N, Tadepalli SA, Csurgyók R, Nagy LV, Tompa M, Hernádi I (2019) Cognitive enhancer effects of low memantine doses are facilitated by an alpha7 nicotinic acetylcholine receptor agonist in scopolamine-induced amnesia in rats. Front Pharmacol 10:73
17. Narala A, Guda S, Veerabrahma K (2019) Lipid nanoemulsions of rebamipide: formulation, characterization, and in vivo evaluation of pharmacokinetic and pharmacodynamic effects. AAPS PharmSciTech 20:1–9. https://doi.org/10.1208/s12249-018-1225-7
18. Kim KT, Lee J-Y, Park J-H, Cho H-J, Yoon I-S, Kim D-D (2017) Capmul MCM/Solutol HS15-based microemulsion for enhanced oral bioavailability of rebamipide. J Nanosci Nanotechnol 17:2340–2344
19. Devalapally H, Silchenko S, Zhou F, McDade J, Goloverda G, Owen A et al (2013) Evaluation of a nanoemulsion formulation strategy for oral bioavailability enhancement of danazol in rats and dogs. J Pharm Sci 102(10):3808–3815
20. Patel Y, Poddar A, Sawant K (2012) Formulation and characterization of cefuroxime Axetil nanoemulsion for improved bioavailability. J Pharm Bioallied Sci 4(Suppl 1):S4–S5
21. Zhao L, Wei Y, Huang Y, He B, Zhou Y, Fu J (2013) Nanoemulsion improves the oral bioavailability of baicalin in rats: in vitro and in vivo evaluation. Int J Nanomed 8:3769–3779
22. Patel MH, Mundada VP, Sawant KK (2019) Novel drug delivery approach via self-microemulsifying drug delivery system for enhancing oral bioavailability of asenapine maleate: optimization, characterization, cell uptake, and in vivo pharmacokinetic studies. AAPS PharmSciTech 20:1–8. https://doi.org/10.1208/s12249-018-1212-z
23. Bajaj H, Bisht S, Yadav M, Singh V (2011) Bioavailability enhancement: a review. Int J Pharm Bio Sci 2(2):202–216
24. Kohli K, Chopra S, Dhar D, Arora S, Khar RK (2010) Self-emulsifying drug delivery system: an approach to enhance oral bioavailability. Drug Discov Today 15(21/22):958–965
25. Negi LM, Tariq M, Talegaonkar S (2013) Nano scale self-emulsifying oil based carrier system for improved oral bioavailability of camptothecin derivative by P-glycoprotein modulation. Colloids Surf B 111:346–353
26. Kulkarni JA, Avachat AA (2017) Pharmacodynamic and pharmacokinetic investigation of cyclodextrin mediated asenapine maleate in situ nasal gel for improved bioavailability. Drug Dev Ind Pharm 43(2):234–245
27. Shreya AB, Managuli RS, Menon J, Kondapalli L, Hegde AR, Avadhani K et al (2016) Nano-transfersomal formulations for transdermal delivery of asenapine maleate: in vitro and in vivo performance evaluations. J Liposome Res 26(3):221–232
28. Prajapati ST, Joshi H, Patel CN (2013) Preparation and characterization of self microemulsifying drug delivery system of olmesartan medoxomil for bioavailability improvement. Aust J Pharm 2013:1–9
29. Singh S, Pathak K, Bali V (2012) Product development studies on surface-adsorbed nanoemulsion of olmesartan medoxomil as a capsular dosage form. AAPS PharmSciTech 13(4):1212–1221
30. Balakumar K, Raghavan CV, Selvana NT, Hari Prasad R, Abdu S (2013) Self nanoemulsifying drug delivery system (SNEDDS) of rosuvastatin calcium: design, formulation, bioavailability and pharmacokinetic evaluation. Colloids Surf B 112:337–343
31. Bandyopadhyay S, Katare OP, Singh B (2012) Optimized self nano-emulsifying systems of ezetimibe with enhanced bioavailability potential using long chain and medium chain triglycerides. Colloids Surf B 100:50–61
32. Calcagno AM, Ludwig JA, Fostel JM, Gottesman MM, Ambudkar SV (2006) Comparison of drug transporter levels in normal colon, colon cancer, and Caco-2 cells: impact on drug disposition and discovery. Mol Pharmacol 3(1):87–93
33. Nepal PR, Han HK, Choi HK (2010) Preparation and in vitro–in vivo evaluation of Witepsol® H35 based self-nanoemulsifying drug delivery systems (SNEDDS) of coenzyme Q10. Eur J Pharm Sci 39:224–232
34. Zhang P, Liu Y, Feng N, Xu J (2008) Preparation and evaluation of self microemulsifying drug delivery system of oridonin. Int J Pharm 355:269–276

35. Han S, Mei L, Quach T, Porter C, Trevaskis N (2021) Lipophilic conjugates of drugs: a tool to improve drug pharmacokinetic and therapeutic profiles. Pharm Res 38(9):1497–1518. https://doi.org/10.1007/s11095-021-03093-x
36. Siegel RA, Rathbone MJ (2012) Overview of controlled release mechanisms. In: Siepmann J, Siegel RA, Rathbone MJ (eds) Fundamentals and applications of controlled release drug delivery. Springer, pp 19–43
37. Ray AS, Fordyce MW, Hitchcock MJM (2016) Tenofovir alafenamide: a novel prodrug of tenofovir for the treatment of Human Immunodeficiency Virus. Antivir Res 125:63–70
38. Sofia MJ, Bao D, Chang W, Du J, Nagarathnam D, Rachakonda S, Reddy PG, Ross BS, Wang P, Zhang H-R, Bansal S, Espiritu C, Keilman M, Lam AM, Steuer HMM, Niu C, Otto MJ, Furman PA (2010) Discovery of a β-d-2′-Deoxy-2′-α-fluoro-2′-β-C-methyluridine nucleotide prodrug (PSI-7977) for the treatment of Hepatitis C virus. J Med Chem 53(19):7202–7218
39. Beaumont K, Webster R, Gardner I, Dack K (2003) Design of ester prodrugs to enhance oral absorption of poorly permeable compounds: challenges to the discovery scientist. Curr Drug Metab 4(6):461–485
40. Wiemer AJ, Wiemer DF (2015) Prodrugs of phosphonates and phosphates: crossing the membrane barrier. Top Curr Chem 360:115–160
41. Pandit R, Chen LY, Gotz J (2020) The blood-brain barrier: physiology and strategies for drug delivery. Adv Drug Deliv Rev 165-166:1–14
42. Rais R, Fletcher S, Polli JE (2011) Synthesis and in vitro evaluation of gabapentin prodrugs that target the human apical sodium-dependent bile acid transporter (hASBT). J Pharm Sci 100(3):1184–1195
43. Burke AC, Giles FJ (2011) Elacytarabine - lipid vector technology overcoming drug resistance in acute myeloid leukemia. Expert Opin Investig Drugs 20(12):1707–1715
44. Bergman AM, Kuiper CM, Voorn DA, Comijn EM, Myhren F, Sandvold ML, Hendriks HR, Peters GJ (2004) Antiproliferative activity and mechanism of action of fatty acid derivatives of arabinofuranosylcytosine in leukemia and solid tumor cell lines. Biochem Pharmacol 67(3):503–511
45. Duivenvoorden R, Tang J, Cormode DP, Mieszawska AJ, Izquierdo-Garcia D, Ozcan C, Otten MJ, Zaidi N, Lobatto ME, van Rijs SM, Priem B, Kuan EL, Martel C, Hewing B, Sager H, Nahrendorf M, Randolph GJ, Stroes ESG, Fuster V et al (2014) A statin-loaded reconstituted high-density lipoprotein nanoparticle inhibits atherosclerotic plaque inflammation. Nat Commun 5:3065
46. Rajora MA, Ding L, Valic M, Jiang W, Overchuk M, Chen J, Zheng G (2017) Tailored theranostic apolipoprotein E3 porphyrin-lipid nanoparticles target glioblastoma. Chem Sci 8(8):5371–5384
47. Sun C, Gui Y, Hu R, Chen J, Wang B, Guo Y, Lu W, Nie X, Shen Q, Gao S, Fang W (2018) Preparation and pharmacokinetics evaluation of solid self-microemulsifying drug delivery system (S-SMEDDS) of osthole. AAPS PharmSciTech 19:2301–2310. https://doi.org/10.1208/s12249-018-1067-3
48. Zhang Y, He L, Yue S, Huang Q, Zhang Y, Yang J (2017) Characterization and evaluation of a self-microemulsifying drug delivery system containing tectorigenin, an isoflavone with low aqueous solubility and poor permeability. Drug Deliv 24(1):632–640
49. Wu L, Qiao Y, Wang L, Guo J, Wang G, He W et al (2015) A self-microemulsifying drug delivery system (SMEDDS) for a novel medicative compound against depression: a preparation and bioavailability study in rats. AAPS PharmSciTech 16(5):1051–1058
50. Zhang JB, Lv Y, Zhao S, Wang B, Tan MQ, Xie HG et al (2014) Effect of lipolysis on drug release from self-microemulsifying drug delivery systems (SMEDDS) with different core/shell drug location. AAPS PharmSciTech 15(3):731–740
51. Bachhav YG, Patravale VB (2009) SMEDDS of glyburide: formulation, in vitro evaluation, and stability studies. AAPS PharmSciTech 10(2):482–487
52. Pawar SK, Vavia PR (2012) Rice germ oil as multifunctional excipient in preparation of self-microemulsifying drug delivery system (SMEDDS) of tacrolimus. AAPS PharmSciTech 13(1):254–261

53. Singh AK, Chaurasiya A, Awasthi A, Mishra G, Asati D, Khar RK et al (2009) Oral bioavailability enhancement of exemestane from self-microemulsifying drug delivery system (SMEDDS). AAPS PharmSciTech 10(3):906–916
54. Xing Q, Song J, You XH, Xu DL, Wang KX, Song JQ et al (2016) Microemulsions containing long-chain oil ethyl oleate improve the oral bioavailability of piroxicam by increasing drug solubility and lymphatic transportation simultaneously. Int J Pharm 511(2):709–718
55. Lu J, Obara S, Liu F, Fu W, Zhang W, Kikuchi S (2017) Melt extrusion for a high melting point compound with improved solubility and sustained release. AAPS PharmSciTech 19:1–13
56. Ochi M, Kimura K, Kanda A, Kawachi T, Matsuda A, Yuminoki K et al (2016) Physicochemical and pharmacokinetic characterization of amorphous solid dispersion of meloxicam with enhanced dissolution property and storage stability. AAPS PharmSciTech 17(4):932–939
57. Wang TR, Hu QB, Zhou MY, Xue JY, Luo YC (2016) Preparation of ultra-fine powders from polysaccharide-coated solid lipid nanoparticles and nanostructured lipid carriers by innovative nano spray drying technology. Int J Pharm 511(1):219–222
58. Ding W, Hou X, Cong S, Zhang Y, Chen M, Lei J et al (2016) Co-delivery of honokiol, a constituent of Magnolia species, in a self-microemulsifying drug delivery system for improved oral transport of lipophilic sirolimus. Drug Deliv 23(7):2513–2523
59. Benival DM, Devarajan PV (2015) In situ lipidization as a new approach for the design of a self microemulsifying drug delivery system (SMEDDS) of doxorubicin hydrochloride for oral administration. J Biomed Nanotechnol 11(5):913–922
60. Sangsen Y, Wiwattanawongsa K, Likhitwitayawuid K, Sritularak B, Graidist P, Wiwattanapatapee R (2016) Influence of surfactants in self-microemulsifying formulations on enhancing oral bioavailability of oxyresveratrol: studies in Caco-2 cells and in vivo. Int J Pharm 498(1–2):294–303
61. Baek MK, Lee JH, Cho YH, Kim HH, Lee GW (2013) Self-microemulsifying drug-delivery system for improved oral bioavailability of pranlukast hemihydrate: preparation and evaluation. Int J Nanomed 8:167–176
62. Li F, Hu RF, Wang B, Gui Y, Cheng G, Gao S et al (2017) Self-microemulsifying drug delivery system for improving the bioavailability of huperzine A by lymphatic uptake. Acta Pharm Sin B 7(3):353–360
63. Cheng G, Hu RF, Ye L, Wang B, Gui Y, Gao S et al (2016) Preparation and in vitro/in vivo evaluation of puerarin solid self-microemulsifying drug delivery system by spherical crystallization technique. AAPS PharmSciTech 17(6):1336–1346
64. Hu R, Zhu J, Ma F, Xu X, Sun Y, Mei K et al (2006) Preparation of sustained-release silyb in microspheres by spherical crystallization technique. J Chin Pharm Sci 15(2):83–91
65. Maghsoodi M, Sadeghpoor F (2010) Preparation and evaluation of solid dispersions of piroxicam and Eudragit S100 by spherical crystallization technique. Drug Dev Ind Pharm 36(8):917–925
66. Nasiri MI, Vora LK, Ershaid JA, Peng K, Tekko IA, Donnelly RF (2022) Nanoemulsion-based dissolving microneedle arrays for enhanced intradermal and transdermal delivery. Drug Deliv Transl Res 12(4):881–896. https://doi.org/10.1007/s13346-021-01107-0
67. Singh Y, Meher JG, Raval K, Khan FA, Chaurasia M, Jain NK et al (2017) Nanoemulsion: concepts, development and applications in drug delivery. J Control Release 252:28–49
68. Chavda VP, Shah D (2016) A review on novel emulsification technique: a nanoemulsion. Trends Drug Deliv 3:25–34
69. Sosa L, Clares B, Alvarado HL, Bozal N, Domenech O, Calpena AC (2017) Amphotericin B releasing topical nanoemulsion for the treatment of candidiasis and aspergillosis. Nanomedicine 13:2303–2312. https://doi.org/10.1016/j.nano.2017.06.021
70. Hussain A, Singh S, Webster TJ, Ahmad FJ (2017) New perspectives in the topical delivery of optimized amphotericin B loaded nanoemulsions using excipients with innate anti-fungal activities: a mechanistic and histopathological investigation. Nanomedicine 13:1117–1126. https://doi.org/10.1016/j.nano.2016.12.002

71. Hussain A, Singh VK, Singh OP, Shafaat K, Kumar S, Ahmad FJ (2016) Formulation and optimization of nanoemulsion using antifungal lipid and surfactant for accentuated topical delivery of amphotericin B. Drug Deliv 23:3101–3110. https://doi.org/10.3109/10717544.2016.1153747
72. Caldeira LR, Fernandes FR, Costa DF, Frézard F, Afonso LCC, Ferreira LAM (2015) Nanoemulsions loaded with amphotericin B: a new approach for the treatment of leishmaniasis. Eur J Pharm Sci Off J Eur Fed Pharm Sci 70:125–131. https://doi.org/10.1016/j.ejps.2015.01.015
73. Ershaid JA, Vora L, Donnelly RF (2021) Novel fluphenazine decanoate nanoemulsion loaded dissolving microneedles for transdermal delivery. In: Proc Control Release Soc Virtual Annu Meet
74. Li M, Vora LK, Peng K, Donnelly RF (2021) Trilayer microneedle array assisted transdermal and intradermal delivery of dexamethasone. Int J Pharm 612:121295. https://doi.org/10.1016/j.ijpharm.2021.121295
75. Li W, Li H, Yao H, Mu Q, Zhao G, Li Y, Hu H, Niu X (2014) Pharmacokinetic and anti-inflammatory effects of sanguinarine solid lipid nanoparticles. Inflammation 37:632–638. https://doi.org/10.1007/s10753-013-9779-8
76. Joshi MD, Müller RH (2009) Lipid nanoparticles for parenteral delivery of actives. Eur J Pharm Biopharm 71:161–172
77. Mehnert W, Mäder K (2001) Solid lipid nanoparticles: production, characterization and applications. Adv Drug Deliv Rev 47:165–196
78. Song X, Zhao Y, Hou S, Xu F, Zhao R, He J, Chen Q (2008) Dual agents loaded PLGA nanoparticles: systematic study of particle size and drug entrapment efficiency. Eur J Pharm Biopharm 69:445–453
79. Riddick T (1968) Control of colloid stability through zeta potential: with a closing chapter on its relationship to cardiovascular disease. Zeta-Meter Inc, Wynnewood
80. Kenechukwu FC, Attama AA, Ibezim EC, Nnamani PO, Umeyor CE, Uronnachi EM et al (2018) Novel intravaginal drug delivery system based on molecularly PEGylated lipid matrices for improved antifungal activity of miconazole nitrate. BioMed Res Int 20:1–18
81. Pingale A, Gondkar S, Saudagar R (2018) Nanostructured lipid carrier (NLC): a modern approach for intranasal drug delivery. World J Pharm Res 7(9):1574–1588
82. Uner M (2005) Preparation, characterization and physico-chemical properties of solid lipid nanoparticles (SLN) and nanostructured lipid carriers (NLC): their benefits as colloidal drug carrier systems. Pharmazie 61:375–386
83. Barenholz Y, Anselem S (1993) In: Gregoriades G (ed) Quality control assays in the development and clinical use of liposome-based formulation. CRC Press, Boca Raton, pp 527–616
84. Momoh M, Adikwu M, Ibezim E, Attama A (2011) Effect of metformin and Vernonia amygdalina leaf extract loaded PEGylated-mucin formulation on haematological, kidney and liver indices of healthy and diabetes rats. J Pharm Res 4(10):3455–3459
85. Momoh M, Adedokun M, Adikwu M, Kenechukwu F, Ibezim E, Ugwoke E (2013) Design, characterization and evaluation of PEGylated-mucin for oral delivery of metformin hydrochloride. Afr J Pharm Pharmacol 7(7):347–355
86. Momoh M, Kenechukwu F, Attama A (2013) Formulation and evaluation of novel solid lipid microparticles as a sustained release system for the delivery of metformin hydrochloride. Drug Deliv 20(3–4):102–111
87. Rostamkalaei SS, Akbari J, Saeedi M, Morteza-Semnani K, Nokhodchi A (2019) Topical gel of metformin solid lipid nanoparticles: a hopeful promise as a dermal delivery system. Colloids Surf B 175:150–157
88. Attama A, Momoh M, Builders P (2012) Lipid nanoparticulate drug delivery systems: a revolution in dosage form design and development. In: Sezer AD (ed) Recent advances in novel drug carrier systems. InTech, Rijeka, pp 107–140. Available from: http://www.intechopen.com/books/recent-advances-in-novel-drug-carrier-systems/lipid-nanoparticulate-

drug-delivery-systems-a-revolution-in-dosage-form-design-and-development. Accessed 9 Jun 2021.
89. Sharma G, Parchur AK, Jagtap JM, Hansen CP, Joshi A (2019) Hybrid nanostructures in targeted drug delivery. In: Hybrid nanostructures for cancer theranostics. Elsevier, pp 139–158. Available from: https://linkinghub.elsevier.com/retrieve/pii/B9780128139066000081. Accessed 26 Feb 2021
90. Cardot J, Beyssac E, Alric M (2007) In vitro–in vivo correlation: importance of dissolution in IVIVC. Dissolution Technol 14:15
91. Sjögren E, Abrahamsson B, Augustijns P, Becker D, Bolger M, Brewster M et al (2014) In vivo methods for drug absorption - Comparative physiologies, model selection, correlations with in vitro methods (IVIVC), and applications for formulation/API/excipient characterization including food effects. Eur J Pharm Sci 57:23–31
92. Bruschi ML (2015) Mathematical models of drug release. In: Bruschi ML (ed) Strategies to modify the drug release from pharmaceutical systems. Woodhead Publishing, pp 63–86. Available from: https://www.sciencedirect.com/science/article/pii/B9780081000922000059. Accessed 12 Aug 2019
93. Suvi K, Belma M, Pouya S, Paraskevi S (eds) (2019) IDF Diabetes Atlas, 9th edn. International Diabestes Federation. Available from: http://www.diabetesatlas.org. Accessed 14 Jun 2020
94. Elsayed I, El-Dahmy RM, El-Emam SZ, Elshafeey AH, El Gawad NA, El-Gazayerly ON (2020) Response surface optimization of biocompatible elastic nanovesicles loaded with rosuvastatin calcium: enhanced bioavailability and anticancer efficacy. Drug Deliv Transl Res 10:1459–1475. https://doi.org/10.1007/s13346-020-00761-0
95. Chein YW (1987) Transdermal controlled system medication, vol 9. Marcel Dekkar, New York, pp 697–703
96. Shamma RN, Elsayed I (2013) Transfersomal lyophilized gel of buspirone HCl: formulation, evaluation and statistical optimization. J Liposome Res 23:244–254
97. Dudhipala N, Veerabrahma K (2017) Improved anti-hyperlipidemic activity of rosuvastatin calcium via lipid nanoparticles: pharmacokinetic and pharmacodynamic evaluation. Eur J Pharm Biopharm 110:47–57
98. Herman A, Herman AP (2014) Essential oils and their constituents as skin penetration enhancer for transdermal drug delivery: a review. J Pharm Pharmacol 67:473–485
99. Shen Q, Li W, Li W (2007) The effect of clove oil on the transdermal delivery of ibuprofen in the rabbit by in vitro and in vivo methods. Drug Dev Ind Pharm 33:1369–1374
100. Chen J, Jiang Q, Wu Y, Liu P, Yao J, Lu Q et al (2015) Potential of essential oils as penetration enhancers for transdermal administration of ibuprofen to treat dysmenorrhoea. Molecules 20:18219–18236
101. Al-Mahallawi AM, Khowessah OMSR (2014) Nano-transfersomal ciprofloxacin loaded vesicles for non-invasive trans-tympanic ototopical delivery: in-vitro optimization, ex-vivo permeation studies, and in-vivo assessment. Int J Pharm 472:304–314
102. Lawrence MJ (1994) Surfactant systems: their use in drug delivery. Chem Soc Rev 23:417–424
103. Polli JE, Rekhi GS, Augsburger LLSV (1997) Methods to compare dissolution profiles and a rationale for wide dissolution specifications for metoprolol tartrate tablets. J Pharm Sci 86:690–700
104. Lieleg O, Baumgärtel R, Bausch A (2009) Selective filtering of particles by theextracellular matrix: an electrostatic bandpass. Biophys J 97:1569–1577
105. Nomura T, Koreeda N, Yamashita F, Takakura Y, Hashida M (1998) Effect of particle sizeand charge on the disposition of lipid carriers after intratumoral injection into tissue-isolated tumors. Pharm Res 15:128–132
106. Chaurasiya A, Gorajiya A, Panchal K, Katke S, Singh AK (2022) A review on multivesicular liposomes for pharmaceutical applications: preparation, characterization, and translational challenges. Drug Deliv Transl Res:1–9. https://doi.org/10.1007/s13346-021-01060-y
107. Philippot JR, Schuber F (1994) Liposomes as tools in basic research and industry. CRC Press

108. Sun L, Wang T, Gao L et al (2013) Multivesicular liposomes for sustained release of naltrexone hydrochloride: design, characterization and in vitro/in vivo evaluation. Pharm Dev Technol 18:828–833. https://doi.org/10.3109/10837450.2012.700934
109. Jain SK, Gupta Y, Jain A et al (2007) Multivesicular liposomes bearing celecoxib-beta-cyclodextrin complex for transdermal delivery. Drug Deliv 14:327–335. https://doi.org/10.1080/10717540601098740
110. Katre NV, Asherman J, Schaefer H et al (1998) Multivesicular liposome (DepoFoam) technology for the sustained delivery of insulin-like growth factor-I (IGF-I). J Pharm Sci 87:1341–1346. https://doi.org/10.1021/js980080t
111. Schutt EG, McGuire RW, Walters PA et al (2019) Method for formulating large diameter synthetic membrane vesicles (U.S. Patent No. US20110250264A1). U.S. Patent and Trademark Office. https://patft.uspto.gov/netacgi/nph-Parser?Sect1=PTO2&Sect2=HITOFF&p=1&u=%2Fnetahtml%2FPTO%2Fsearch-bool.html&r=15&f=G&l=50&co1=AND&d=PTXT&s1=Pacira.ASNM.&OS=AN/Pacira&RS=AN/Pacira. Accessed 24 August 2021
112. Lu B, Ma Q, Zhang J et al (2021) Preparation and characterization of bupivacaine multivesicular liposome: a QbD study about the effects of formulation and process on critical quality attributes. Int J Pharm 598:120335. https://doi.org/10.1016/j.ijpharm.2021.120335
113. Vafaei SY, Dinarvand R, Esmaeili M et al (2015) Controlled-release drug delivery system based on fluocinolone acetonide-cyclodextrin inclusion complex incorporated in multivesicular liposomes. Pharm Dev Technol 20:775–781. https://doi.org/10.3109/10837450.2014.920358
114. Zhao Y, Liu J, Sun X et al (2010) Sustained release of hydroxycamptothecin after subcutaneous administration using a novel phospholipid complex—DepoFoam™ technology Drug. Dev Ind Pharm 36:823–831. https://doi.org/10.3109/03639040903520975
115. Langston MV, Ramprasad MP, Kararli TT et al (2003) Modulation of the sustained delivery of myelopoietin (Leridistim) encapsulated in multivesicular liposomes (DepoFoam). J Control Release 89:87–99. https://doi.org/10.1016/s0168-3659(03)00073-7
116. Kim S, Scheerer S, Geyer MA et al (1990) Direct cerebrospinal fluid delivery of an antiretroviral agent using multivesicular liposomes. J Infect Dis 162:750–752. https://doi.org/10.1093/infdis/162.3.750
117. Shen Y, Ji Y, Xu S et al (2011) Multivesicular liposome formulations for the sustained delivery of ropivacaine hydrochloride: preparation, characterization, and pharmacokinetics. Drug Deliv 18:361–366. https://doi.org/10.3109/10717544.2011.557788
118. Li N, Shi A, Wang Q et al (2019) Multivesicular liposomes for the sustained release of angiotensin I-converting enzyme (ACE) inhibitory peptides from peanuts: design, characterization, and in vitro evaluation. Molecules 24:1746. https://doi.org/10.3390/molecules24091746
119. Li H, Liu Y, Zhang Y, Fang D, Xu B, Zhang L, Chen T, Ren K, Nie Y, Yao S, Song X (2016) Liposomes as a novel ocular delivery system for brinzolamide: in vitro and in vivo studies. AAPS PharmSciTech 17:710–717. https://doi.org/10.1208/s12249-015-0382-1
120. Siesky B, Harris A, Cantor LB, Kagemann L, Weitzman Y, McCranor L et al (2008) A comparative study of the effects of brinzolamide and dorzolamide on retinal oxygen saturation and ocular microcirculation in patients with primary open-angle glaucoma. Brit J Ophthalmol 92(4):500–504. https://doi.org/10.1136/bjo.2007.125187
121. Zhang Y, Ren K, He Z, Li H, Chen T, Lei Y et al (2013) Development of inclusion complex of brinzolamide with hydroxypropyl-beta-cyclodextrin. Carbohydr Polym 98(1):638–643. https://doi.org/10.1016/j.carbpol.2013.06.052
122. Hironaka K, Inokuchi Y, Tozuka Y, Shimazawa M, Hara H, Takeuchi H (2009) Design and evaluation of a liposomal delivery system targeting the posterior segment of the eye. J Control Release 136(3):247–253. https://doi.org/10.1016/j.jconrel.2009.02.020
123. Natarajan JV, Ang M, Darwitan A, Chattopadhyay S, Wong TT, Venkatraman SS (2012) Nanomedicine for glaucoma: liposomes provide sustained release of latanoprost in the eye. Int J Nanomed 7:123–131. https://doi.org/10.2147/IJN.S25468

124. Gupta T, Kenjale P, Pokharkar V (2022) QbD-based optimization of raloxifene-loaded cubosomal formulation for transdemal delivery: ex vivo permeability and in vivo pharmacokinetic studies. Drug Deliv Transl Res 12(12):2979–2992. https://doi.org/10.1007/s13346-022-01162-1
125. Yang D, O'Brien DF, Marder SR (2002) Polymerized bicontinuous cubic nanoparticles (cubosomes) from a reactive monoacylglycerol. J Am Chem Soc 124:13388–13389
126. Noor AH, Ghareeb MM (2020) Formulation and evaluation of ondansetron HCl nanoparticles for transdermal delivery. Iraqi J Pharm Sci Drug Deliv Transl Res 29:70–79
127. You J, Zhou J, Zhou M, Liu Y, Robertson JD, Liang D, Van Pelt C, Li C (2014) Pharmacokinetics, clearance, and biosafety of polyethylene glycol-coated hollow gold nanospheres. Part Fibre Toxicol 11:1–4. https://doi.org/10.1186/1743-8977-11-26
128. Pissuwan D, Valenzuela SM, Cortie MB (2006) Therapeutic possibilities of plasmonically heated gold nanoparticles. Trends Biotechnol 24:62–67
129. Lee S, Chon H, Lee M, Choo J, Shin SY, Lee YH, Rhyu IJ, Son SW, Oh CH (2009) Surface-enhanced Raman scattering imaging of HER2 cancer markers overexpressed in single MCF7 cells using antibody conjugated hollow gold nanospheres. Biosens Bioelectron 24:2260–2263
130. Melancon MP, Zhou M, Li C (2011) Cancer theranostics with near-infrared light-activatable multimodal nanoparticles. Acc Chem Res 44:947–956
131. Nowak-Jary J, Machnicka B (2022) Pharmacokinetics of magnetic iron oxide nanoparticles for medical applications. J Nanobiotechnol 20(1):305. https://doi.org/10.1186/s12951-022-01510-w
132. Lu AH, Zhang XQ, Sun Q, Zhang Y, Song Q, Schüth F et al (2016) Precise synthesis of discrete and dispersible carbon-protected magnetic nanoparticles for efficient magnetic resonance imaging and photothermal therapy. Nano Res 9(5):1460–1469
133. Zhou Q, Wei Y (2017) For better or worse, iron overload by superparamagnetic iron oxide nanoparticles as a mri contrast agent for chronic liver diseases. Chem Res Toxicol 30(1):73–80
134. Yin PT, Shah BP, Lee KB (2014) Combined Magnetic nanoparticle-based MicroRNA and hyperthermia therapy to enhance apoptosis in brain cancer cells. Small 10(20):4106–4112
135. Obaidat IM, Narayanaswamy V, Alaabed S, Sambasivam S, Muralee Gopi CVV (2019) Principles of magnetic hyperthermia: a focus on using multifunctional hybrid magnetic nanoparticles. Magnetochemistry 5(4):67
136. Arruebo M, Fernández-Pacheco R, Ibarra MR, Santamaría J (2007) Magnetic nanoparticles for drug delivery. Nano Today 2(3):22–32
137. Wong J, Prout J, Seifalian A (2017) Magnetic nanoparticles: new perspectives in drug delivery. Curr Pharm Des 23(20):2908–2917
138. Mu X, Li J, Yan S, Zhang H, Zhang W, Zhang F et al (2018) siRNA delivery with stem cell membrane-coated magnetic nanoparticles for imaging-guided photothermal therapy and gene therapy. ACS Biomater Sci Eng 4(11):3895–3905
139. Luo B, Zhou X, Jiang P, Yi Q, Lan F, Wu Y (2018) PAMA–Arg brush-functionalized magnetic composite nanospheres for highly effective enrichment of phosphorylated biomolecules. J Mater Chem B 6(23):3969–3978
140. Uskoković V, Tang S, Wu VM (2019) Targeted magnetic separation of biomolecules and cells using earthicle-based ferrofluids. Nanoscale 11(23):11236–11253
141. Boraschi D, Italiani P, Palomba R, Decuzzi P, Duschl A, Fadeel B et al (2017) Nanoparticles and innate immunity: new perspectives on host defence. Semin Immunol 34:33–51
142. Fadeel B (2019) Hide and seek: nanomaterial interactions with the immune system. Front Immunol 10:133
143. Stepien G, Moros M, Pérez-Hernández M, Monge M, Gutiérrez L, Fratila RM et al (2018) Effect of surface chemistry and associated protein corona on the long-term biodegradation of iron oxide nanoparticles in vivo. ACS Appl Mater Interf 10(5):4548–4560
144. Mahmoudi M, Sheibani S, Milani AS, Rezaee F, Gauberti M, Dinarvand R et al (2015) Crucial role of the protein corona for the specific targeting of nanoparticles. Nanomedicine 10(2):215–226

145. Prospero AG, Fidelis-de-Oliveira P, Soares GA, Miranda MF, Pinto LA, dos Santos DC et al (2020) AC biosusceptometry and magnetic nanoparticles to assess doxorubicin-induced kidney injury in rats. Nanomedicine 15(5):511–525
146. Al Faraj A, Lacroix G, Alsaid H, Elgrabi D, Stupar V, Robidel F et al (2008) Longitudinal3He and proton imaging of magnetite biodistribution in a rat model of instilled nanoparticles. Magn Reson Med 59(6):1298–1303
147. Kwon J, Hwang S, Jin H, Kim D, Minai-Tehrani A, Yoon H et al (2008) Body distribution of inhaled fluorescent magnetic nanoparticles in the mice. J Occup Health 50(1):1–6
148. Kwon J, Kim D, Minai-Tehrani A, Hwang S, Chang S, Lee E et al (2009) Inhaled fluorescent magnetic nanoparticles induced extramedullary hematopoiesis in the spleen of mice. J Occup Health. 51(5):423–431
149. Khalkhali M, Rostamizadeh K, Sadighian S, Khoeini F, Naghibi M, Hamidi M (2015) The impact of polymer coatings on magnetite nanoparticles performance as MRI contrast agents: a comparative study. DARU J Pharm Sci 23(1):45
150. Corot C, Robert P, Idee J, Port M (2006) Recent advances in iron oxide nanocrystal technology for medical imaging☆. Adv Drug Deliv Rev 58(14):1471–1504
151. Pablico-Lansigan MH, Situ SF, Samia ACS (2013) Magnetic particle imaging: advancements and perspectives for real-time in vivo monitoring and image-guided therapy. Nanoscale 5(10):4040
152. Weizenecker J, Gleich B, Rahmer J, Dahnke H, Borgert J (2009) Three-dimensional real-time in vivo magnetic particle imaging. Phys Med Biol 54(5):L1–L10
153. Levy M, Luciani N, Alloyeau D, Elgrabli D, Deveaux V, Pechoux C et al (2011) Long term in vivo biotransformation of iron oxide nanoparticles. Biomaterials 32(16):3988–3999
154. Škrátek M, Dvurečenskij A, Kluknavský M, Barta A, Bališ P, Mičurová A et al (2020) Sensitive SQUID bio-magnetometry for determination and differentiation of biogenic iron and iron oxide nanoparticles in the biological samples. Nanomaterials 10(10):1993
155. Fang RH, Kroll AV, Gao W, Zhang L (2018) Cell membrane coating nanotechnology. Adv Mater 30(23):1706759
156. Fleming AB, Saltzman WM (2002) Pharmacokinetics of the carmustine implant. Clin Pharmacokinet 41:403–419. https://doi.org/10.2165/00003088-200241060-00002
157. Loo TL, Dion RT, Dixon L et al (1966) The antitumor agent, 1,3-Bis(2-chloroethyl)-1-nitrosourea. J Pharm Sci 55(5):492–497
158. Dang WB, Daviau T, Brem H (1996) Morphological characterization of polyanhydride biodegradable implant GLIADEL® during in vitro and in vivo erosion using scanning electron microscopy. Pharm Res 13(5):683–691
159. Dang WB, Daviau T, Ying P et al (1996) Effects of GLIADEL® wafer initial molecular weight on the erosion of wafer and release of BCNU. J Control Release 42(1):83–92
160. Wu MP, Tamada JA, Brem H et al (1994) In-vivo versus in-vitro degradation of controlled-release polymers for intracranial surgical therapy. J Biomed Mater Res 28(3):387–395
161. Fung LK, Shin M, Tyler B et al (1996) Chemotherapeutic drugs released from polymers: distribution of 1,3-bis(2-chloroethyl)-1-nitrosourea in the rat brain. Pharm Res 13(5):671–682
162. Kalyanasundaram S, Leong KW (1997) Intracranial drug delivery systems. STP Pharma Sci 7(1):62–70
163. Reulen HJ, Graham R, Spatz M et al (1977) Role of pressure gradients and bulk flow in dynamics of vasogenic brain edema. J Neurosurg 46(1):24–35
164. Wang CH, Li J, Teo CS et al (1999) The delivery of BCNU to brain tumors. J Control Release 61(1–2):21–41
165. Englehard HH (2000) The role of interstitial BCNU chemotherapy in the treatment of malignant glioma. Surg Neurol 53(5):458–464
166. Brem H, Piantadosi S, Burger PC et al (1995) Placebo-controlled trial of safety and efficacy of intraoperative controlled delivery by biodegradable polymers of chemotherapy for recurrent Gliomas. Lancet 345(8956):1008–1012

167. Castillo M, Ewend MG, Cush S et al (1998) Magnetic resonance imaging appearance of carmustine-impregnated implantable wafers. Int J Neurol 4(5):380–384
168. Walter KA, Tamargo RJ, Olivi A et al (1995) Intratumoral chemotherapy. Neurosurgery 37(6):1129–1145
169. Chennamaneni SR, Bohner A, Bernhisel A, Ambati BK (2014) Pharmacokinetics and efficacy of Bioerodible Dexamethasone implant in Concanavalin A-induced uveitic cataract rabbit model. Pharm Res 31:3179–3190. https://doi.org/10.1007/s11095-014-1410-7
170. Adamis AP, Shima DT (2005) The role of vascular endothelial growth factor in ocular health and disease. Retina 25(2):111–118
171. Tejwani S, Murthy S, Sangwan VS (2006) Cataract extraction outcomes in patients with Fuchs' heterochromic cyclitis. J Cataract Refract Surg 32(10):1678–1682
172. Roesel M, Tappeiner C, Heinz C, Koch JM, Heiligenhaus A (2009) Comparison between intravitreal and orbital floor triamcinolone acetonide after phacoemulsification in patients with endogenous uveitis. Am J Ophthalmol 147(3):406–412
173. Kyuki K, Shibuya T, Tsurumi K, Fujimura H (1981) Topical anti-inflammatory activity of dexamethasone 17-valerate (author's transl). Nihon Yakurigaku Zasshi 77(1):73–85
174. Michel C, Cabanac M (1999) Effects of dexamethasone on the body weight set point of rats. Physiol Behav 68(1–2):145–150
175. Chennamaneni SR, Mamalis C, Archer B, Oakey Z, Ambati BK (2013) Development of a novel bioerodible dexamethasone implant for uveitis and postoperative cataract inflammation. J Control Release 167(1):53–59
176. Zhao J, Xu G, Yao X, Zhou H, Lyu B, Pei S, Wen P (2022) Microneedle-based insulin transdermal delivery system: current status and translation challenges. Drug Deliv Transl Res 1:1–25. https://doi.org/10.1007/s13346-021-01077-3
177. Prausnitz MR (2017) Engineering microneedle patches for vaccination and drug delivery to skin. Annu Rev Chem Biomol Eng 8:177–200. https://doi.org/10.1146/annurev-chembioeng-060816-101514
178. Pradeep Narayanan S, Raghavan S (2017) Solid silicon microneedles for drug delivery applications. Int J Adv Manuf Technol 93(1):407–422. https://doi.org/10.1007/s00170-016-9698-6
179. Gupta J, Gill HS, Andrews SN, Prausnitz MR (2011) Kinetics of skin resealing after insertion of microneedles in human subjects. J Control Release 154(2):148–155. https://doi.org/10.1016/j.jconrel.2011.05.021
180. Liu S, Jin MN, Quan YS, Kamiyama F, Katsumi H, Sakane T et al (2012) The development and characteristics of novel microneedle arrays fabricated from hyaluronic acid, and their application in the transdermal delivery of insulin. J Control Release 161(3):933–941. https://doi.org/10.1016/j.jconrel.2012.05.030
181. Zhang Y, Wu M, Tan D, Liu Q, Xia R, Chen M et al (2021) A dissolving and glucose-responsive insulin-releasing microneedle patch for type 1 diabetes therapy. J Mater Chem B 9(3):648–657. https://doi.org/10.1039/d0tb02133d
182. Vassilieva EV, Kalluri H, McAllister D, Taherbhai MT, Esser ES, Pewin WP et al (2015) Improved immunogenicity of individual influenza vaccine components delivered with a novel dissolving microneedle patch stable at room temperature. Drug Deliv Transl Res 5(4):360–371. https://doi.org/10.1007/s13346-015-0228-0
183. Chen G, Yu J, Gu Z (2019) Glucose-responsive microneedle patches for diabetes treatment. J Diabetes Sci Technol 13(1):41–48. https://doi.org/10.1177/1932296818778607
184. Ullah A, Choi HJ, Jang M, An S, Kim GM (2020) Smart microneedles with porous polymer layer for glucose-responsive insulin delivery. Pharmaceutics 12(7):606. https://doi.org/10.3390/pharmaceutics12070606
185. Kochba E, Levin Y, Raz I, Cahn A (2016) Improved insulin pharmacokinetics using a novel microneedle device for intradermal delivery in patients with type 2 diabetes. Diabetes Technol Ther 18(9):525–531. https://doi.org/10.1089/dia.2016.0156
186. Pettis RJ, Hirsch L, Kapitza C, Nosek L, Hövelmann U, Kurth HJ et al (2011) Microneedle-based intradermal versus subcutaneous administration of regular human insulin or insulin lis-

pro: pharmacokinetics and postprandial glycemic excursions in patients with type 1 diabetes. Diabetes Technol Ther 13(4):443–450. https://doi.org/10.1089/dia.2010.0183
187. McVey E, Hirsch L, Sutter DE, Kapitza C, Dellweg S, Clair J et al (2012) Pharmacokinetics and postprandial glycemic excursions following insulin lispro delivered by intradermal microneedle or subcutaneous infusion. J Diabetes Sci Technol 6(4):743–754. https://doi.org/10.1177/193229681200600403
188. Kirkby M, Hutton AR, Donnelly RF (2020) Microneedle mediated transdermal delivery of protein, peptide and antibody based therapeutics: current status and future considerations. Pharm Res 37(6):117. https://doi.org/10.1007/s11095-020-02844-6
189. Gardeniers HJGE, Luttge R, Berenschot EJW, De Boer MJ, Yeshurun SY, Hefetz M et al (2003) Silicon micromachined hollow microneedles for transdermal liquid transport. J Microelectromech Syst 12:855–862
190. Martanto W, Moore JS, Couse T, Prausnitz MR (2006) Mechanism of fluid infusion during microneedle insertion and retraction. J Control Release 112:357–361
191. Hardy JG, Larrañeta E, Donnelly RF, McGoldrick N, Migalska K, McCrudden MTC, Irwin NJ, Donnelly L, McCoy CP (2016) Hydrogel-forming microneedle arrays made from light-responsive materials for on-demand transdermal drug delivery. Mol Pharm 13:907–914
192. Vicente-Perez EM, Larrañeta E, McCrudden MTC, Kissenpfennig A, Hegarty S, McCarthy HO et al (2017) Repeat application of microneedles does not alter skin appearance or barrier function and causes no measurable disturbance of serum biomarkers of infection, inflammation or immunity in mice in vivo. Eur J Pharm Biopharm 117:400–407
193. Panda A, Matadh VA, Suresh S, Shivakumar HN, Murthy SN (2022) Non-dermal applications of microneedle drug delivery systems. Drug Deliv Transl Res 1:1–2. https://doi.org/10.1007/s13346-021-00922-9
194. Martins Andrade JF, da Cunha MT, Cunha-Filho M, Taveira SF, Gelfuso GM, Gratieri T (2023) Iontophoresis application for drug delivery in high resistivity membranes: nails and teeth. Drug Deliv Transl Res 13(5):1272–1287. https://doi.org/10.1007/s13346-022-01244-0
195. Kalaria DR, Singhal M, Patravale V, Merino V, Kalia YN (2018) Simultaneous controlled iontophoretic delivery of pramipexole and rasagiline in vitro and in vivo: transdermal polypharmacy to treat Parkinson's disease. Eur J Pharm Biopharm 127:204–212
196. Gratieri T, Gelfuso GM, Rocha EM, Sarmento VH, de Freitas O, Lopez RFV (2010) A poloxamer/chitosan in situ forming gel with prolonged retention time for ocular delivery. Eur J Pharm Biopharm 75:186–193
197. Pereira MN, Reis TA, Matos BN, Cunha-Filho M, Gratieri T, Gelfuso GM (2017) Novel ex vivo protocol using porcine vagina to assess drug permeation from mucoadhesive and colloidal pharmaceutical systems. Colloids Surf B Biointerfaces 158:222–228
198. Baswan SM, Li SK, Kasting GB (2016) Diffusion of uncharged solutes through human nail plate. Pharm Dev Technol 21:255–260
199. Monti D, Egiziano E, Burgalassi S, Tampucci S, Terreni E, Tivegna S et al (2018) Influence of a combination of chemical enhancers and iontophoresis on in vitro transungual permeation of nystatin. AAPS PharmSciTech 19:1574–1581
200. Manda P, Sammeta SM, Repka MA, Murthy SN (2012) Iontophoresis across the proximal nail fold to target drugs to the nail matrix. J Pharm Sci 101:2392–2397
201. Benzeval I, Bowen CR, Guy RH, Delgado-Charro MB (2013) Effects of iontophoresis, hydration, and permeation enhancers on human nail plate: infrared and impedance spectroscopy assessment. Pharm Res 30:1652–1662
202. Wanasathop A, Li S (2018) Iontophoretic drug delivery in the oral cavity. Pharmaceutics 10:121
203. Shahmoradi M, Bertassoni LE, Elfallah HM, Swain M (2014) Fundamental structure and properties of enamel, dentin and cementum. In: Ben-Nissan B (ed) Advances in calcium phosphate biomaterials. Berlin, Heidelberg, Springer, pp 511–547
204. Zheng J, Weng LQ, Shi MY, Zhou J, Hua LC, Qian LM et al (2013) Effect of water content on the nanomechanical properties and microtribological behaviour of human tooth enamel. Wear 301:316–323

205. Ito S, Saito T, Tay FR, Carvalho RM, Yoshiyama M, Pashley DH (2005) Water content and apparent stiffness of non-caries versus caries-affected human dentin. J Biomed Mater Res 72B:109–116
206. Chen K, Puri V, Michniak-Kohn B (2021) Iontophoresis to overcome the challenge of nail permeation: considerations and optimizations for successful ungual drug delivery. AAPS J 23:25
207. Santos GA, Angelo T, Andrade LM, Silva SMM, Magalhães PO, Cunha-Filho M et al (2018) The role of formulation and follicular pathway in voriconazole cutaneous delivery from liposomes and nanostructured lipid carriers. Colloids Surf B Biointerfaces 170:341–346
208. Dutet J, Delgado-Charro MB (2009) In vivo transungual iontophoresis: effect of DC current application on ionic transport and on transonychial water loss. J Control Release 140:117–125
209. Abla N, Naik A, Guy RH, Kalia YN (2005) Effect of charge and molecular weight on transdermal peptide delivery by iontophoresis. Pharm Res 22:2069–2078
210. Abla N, Naik A, Guy RH, Kalia YN (2006) Topical iontophoresis of valaciclovir hydrochloride improves cutaneous aciclovir delivery. Pharm Res 23:1842–1849
211. Pita M, Halámek J, Chinnapareddy S, White DJ, Gartstein V, Katz E (2012) Permeability of human tooth surfaces studied in vitro by electrochemical impedance spectroscopy. Electroanalysis 24:1033–1038
212. Zhang L, Li Y, Wei F, Liu H, Wang Y, Zhao W, Dong Z, Ma T, Wang Q (2021) Transdermal delivery of salmon calcitonin using a dissolving microneedle array: characterization, stability, and in vivo pharmacodynamics. AAPS PharmSciTech 22:1–9. https://doi.org/10.1208/s12249-020-01865-z
213. Tas C, Mansoor S, Kalluri H, Zarnitsyn VG, Choi SO, Banga AK et al (2012) Delivery of salmon calcitonin using a microneedle patch. Int J Pharm 423(2):257–263. https://doi.org/10.1016/j.ijpharm.2011.11.046
214. Wang Q, Yao G, Dong P, Gong Z, Li G, Zhang K et al (2015) Investigation on fabrication process of dissolving microneedle arrays to improve effective needle drug distribution. Eur J Pharm Sci 66:148–156. https://doi.org/10.1016/j.ejps.2014.09.011
215. Lock JY, Carlson TL, Carrier RL (2018) Mucus models to evaluate the diffusion of drugs and particles. Adv Drug Deliv Rev 124(15):34–49. https://doi.org/10.1016/j.addr.2017.11.001
216. Okamura E (2019) Solution NMR to quantify mobility in membranes: diffusion, protrusion, and drug transport processes. Chem Pharm Bull 67(4):308–315. https://doi.org/10.1248/cpb.c18-00946
217. Crowe LM (2002) Lessons from nature: the role of sugars in anhydrobiosis. Comp Biochem Physiol 131A:505–513. https://doi.org/10.1016/S1095-6433(01)00503-7
218. Crowe JH, Hoekstra FA, Crowe L (1992) Anhydrobiosis. Annu Rev Physiol 54:579–599. https://doi.org/10.1146/annurev.ph.54.030192.003051
219. Elbein AD, Pan YT, Pastuszak I, Carroll D (2003) New insights on trehalose: a multi functional molecule. Glycobiology 13:17–27. https://doi.org/10.1093/glycob/cwg047
220. Laskowska E, Kuczyńska-Wiśnik D (2020) New insight into the mechanisms protecting bacteria during desiccation. Curr Genet 66(2):313–318. https://doi.org/10.1007/s00294-019-01036-z
221. Martínez-Segoviano ID, Ganem-Rondero A (2021) Enhancement of the transdermal delivery of zidovudine by pretreating the skin with two physical enhancers: microneedles and sonophoresis. DARU J Pharm Sci 29(2):279–290. https://doi.org/10.1007/s40199-021-00402-y
222. Sosnik A, Chiappetta DA, Carcaboso AM (2009) Drug delivery systems in HIV pharmacotherapy: what has been done and the challenges standing ahead. J Controlled Release 138:2–15. https://doi.org/10.1016/j.jconrel.2009.05.007
223. Kumar Narishetty ST, Panchagnula R (2005) Effect of L-menthol and 1,8-cineole on phase behavior and molecular organization of SC lipids and skin permeation of zidovudine. J Controlled Release 102:59–70. https://doi.org/10.1016/j.jconrel.2004.09.016
224. Thomas NS, Panchagnula R (2003) Transdermal delivery of zidovudine: effect of vehicles on permeation across rat skin and their mechanism of action. Eur J Pharm Sci 18:71–79

225. Kondamudi PK, Tirumalasetty PP, Malayandi R, Mutalik S, Pillai R (2016) Lidocaine transdermal patch: pharmacokinetic modeling and in vitro–in vivo correlation (IVIVC). AAPS PharmSciTech 17:588–596. https://doi.org/10.1208/s12249-015-0390-1
226. Response to Citizen Petition 2006. http://www.fda.gov/ohrms/dockets/dockets/06p0522/06p-0522-cp00001-01-vol1.pdf. Accessed 17 July 2015
227. Collinsworth KA, Kalman SM, Harrison DC (1974) The clinical pharmacology of lidocaine as an antiarrhythymic drug. Circulation 50(6):1217–1230. https://doi.org/10.1161/01.cir.50.6.1217
228. Park J, Lee H, Lim GS, Kim N, Kim D, Kim YC (2019) Enhanced transdermal drug delivery by sonophoresis and simultaneous application of sonophoresis and iontophoresis. AAPS PharmSciTech 20:1–7. https://doi.org/10.1208/s12249-019-1309-z
229. Prausnitz MR, Mitragotri S, Langer R (2004) Current status and future potential of transdermal drug delivery. Nat Rev Drug Discov 3:115–124
230. Kim Y-C, Late S, Banga AK, Ludovice PJ, Prausnitz MR (2008) Biochemical enhancement of transdermal delivery with magainin peptide: modification of electrostatic interactions by changing pH. Int J Pharm 362:20–28
231. Kang S-M, Song J-M, Kim Y-C (2012) Microneedle and mucosal delivery of influenza vaccines. Expert Rev Vaccines 11:547–560
232. Tezel A, Sens A, Tuchscherer J, Mitragotri S (2001) Frequency dependence of sonophoresis. Pharm Res-Dordr 18:1694–1700
233. Anderson CR, Morris RL, Boeh SD, Panus PC, Sembrowich WL (2003) Effects of iontophoresis current magnitude and duration on dexamethasone deposition and localized drug retention. Phys Ther 83:161–170
234. Shirouzu K, Nishiyama T, Hikima T, Tojo K (2008) Synergistic effect of sonophoresis and iontophoresis in transdermal drug delivery. JCEJ 41:300–305
235. Hikima T, Ohsumi S, Shirouzu K, Tojo K (2009) Mechanisms of synergistic skin penetration by sonophoresis and iontophoresis. Biol Pharm Bull 32:905–909
236. Mahl JA, Vogel BE, Court M, Kolopp M, Roman D, Nogués V (2006) The minipig in dermatotoxicology: methods and challenges. Exp Toxicol Pathol 57:341–345
237. Yoshimatsu H, Ishii K, Konno Y, Satsukawa M, SJIjop Y. (2017) Prediction of human percutaneous absorption from in vitro and in vivo animal experiments. Int J Pharm 534:348–355
238. Chen K, Puri V, Michniak-Kohn B (2021) Iontophoresis to overcome the challenge of nail permeation: considerations and optimizations for successful ungual drug delivery. AAPS J 23:1–5. https://doi.org/10.1208/s12248-020-00552-y
239. Del Rosso JQ (2014) The role of topical antifungal therapy for onychomycosis and the emergence of newer agents. J Clin Aesthet Dermatol 7(7):10–18
240. Shahi S, Deshpande S (2017) Iontophoresis: an approach to drug delivery enhancement. Int J Pharm Sci Res 8(10):4056
241. Thatai P, Khan AB (2020) Management of nail psoriasis by topical drug delivery: a pharmaceutical perspective. Int J Dermatol 59(8):915–925
242. Kushwaha A, Jacob M, Shiva Kumar HN, Hiremath S, Aradhya S, Repka MA et al (2015) Trans-ungual delivery of itraconazole hydrochloride by iontophoresis. Drug Dev Ind Pharm 41(7):1089–1094
243. Dragicevic N, Maibach HI (2017) Percutaneous penetration enhancers physical methods in penetration enhancement. Springer, Berlin
244. Delgado-Charro MB (2012) Iontophoretic drug delivery across the nail. Expert Opin Drug Deliv 9(1):91–103
245. Hao J, Li SK (2008) Transungual iontophoretic transport of polar neutral and positively charged model permeants: effects of electrophoresis and electroosmosis. J Pharm Sci 97(2):893–905
246. Hao J, Smith KA, Li SK (2008) Chemical method to enhance transungual transport and iontophoresis efficiency. Int J Pharm 357(1–2):61–69

247. Nair AB, Kim HD, Chakraborty B, Singh J, Zaman M, Gupta A et al (2009) Ungual and trans-ungual iontophoretic delivery of terbinafine for the treatment of onychomycosis. J Pharm Sci 98(11):4130–4140
248. Smith KA, Hao J, Li SK (2010) Influence of pH on transungual passive and iontophoretic transport. J Pharm Sci 99(4):1955–1967
249. Oon HH, Tan HH (2010) Iontophoretic terbinafine delivery in onychomycosis: questionable nail growth. Br J Dermatol 162(3):699–700
250. Wang Y, Zeng L, Song W, Liu J (2021) Influencing factors and drug application of iontophoresis in transdermal drug delivery: an overview of recent progress. Drug Deliv Transl Res 23:1–2. https://doi.org/10.1007/s13346-021-00898-6
251. Lane ME (2013) Skin penetration enhancers. Int J Pharm 447(1–2):12–21
252. Pastore MN, Kalia YN, Horstmann M, Roberts MS (2015) Transdermal patches: history, development and pharmacology. Br J Pharmacol 172(9):2179–2209
253. Thotakura N, Kaushik L, Kumar V, Preet S, Babu PV (2018) Advanced approaches of bioactive peptide molecules and protein drug delivery systems. Curr Pharm Des 24(43):5147–5163
254. Dixit N, Bali V, Baboota S, Ahuja A, Ali J (2007) Iontophoresis - an approach for controlled drug delivery: a review. Curr Drug Deliv 4(1):1–10
255. Ajay KB, Peter CP (2017) lontophoretic devices: clinical applications and rehabilitation medicine. Crit Rev Phys Rehabil Med 29(1–4):247–279
256. Ita K (2017b) Percutaneous transport of psychotropic agents. J Drug Deliv Sci Technol 39:247–259
257. Gratieri T, Kalia YN (2013) Mathematical models to describe iontophoretic transport in vitro and in vivo and the effect of current application on the skin barrier. Adv Drug Deliv Rev 65:315–329
258. Djabri A, Guy RH, Delgado-Charro MB (2019) Potential of iontophoresis as a drug delivery method for midazolam in pediatrics. Eur J Pharm Sci 128:137–143
259. Subramony JA, Sharma A, Phipps JB (2006) Microprocessor controlled transdermal drug delivery. Int J Pharm 317:1–6
260. Talbi Y, Campo E, Brulin D, Fourniols JY (2018) Controllable and re-usable patch for transdermal iontophoresis drug delivery. Electron Lett 54(12):739–740
261. Ita K (2017d) Dermal/transdermal delivery of small interfering RNA and antisense oligonucleotides- advances and hurdles. Biomed Pharmacother 87:311–320
262. Tezel A, Dokka S, Kelly S, Hardee GE, Mitragotri S (2004) Topical delivery of anti-sense oligonucleotides using low-frequency sonophoresis. Pharm Res 21:2219–2225
263. Tokumoto S, Higo N, Todo H, Sugibayashi K (2016) Effect of combination of low-frequency sonophoresis or electroporation with iontophoresis on the mannitol flux or electroosmosis through excised skin. Biol Pharm Bull 39(7):1206–1210
264. Ita K (2016) Transdermal iontophoretic drug delivery: advances and challenges. J Drug Target 24(5):386–391
265. Jijie R, Barras A, Boukherroub R, Szunerits S (2017) Nanomaterials for transdermal drug delivery: beyond the state of the art of liposomal structures. J Mater Chem B 5(44):8653–8675
266. Zhang Y, Yu J, Kahkoska AR, Wang J, Buse JB, Gu Z (2019) Advances in transdermal insulin delivery. Adv Drug Deliv Rev 139:51–70
267. Noh G, Keum T, Seo JE, Bashyal S, Eum NS, Kweon MJ, Lee S, Sohn DH, Lee S (2018) Iontophoretic transdermal delivery of human growth hormone (hGH) and the combination effect of a new type microneedle, Tappy Tok Tok®. Pharmaceutics 10(3):153
268. Kim KT, Lee J, Kim MH, Park JH, Lee JY, Song JH, Jung M, Jang MH, Cho HJ, Yoon IS, Kim DD (2017) Novel reverse electrodialysis-driven iontophoretic system for topical and transdermal delivery of poorly permeable therapeutic agents. Drug Deliv 24(1):1204–1215
269. Bhadale RS, Londhe VY (2023) A comparison of dissolving microneedles and transdermal film with solid microneedles for iloperidone in vivo: a proof of concept. Naunyn-Schmiedeberg's Arch Pharmacol 396(2):239–246. https://doi.org/10.1007/s00210-022-02309-0

270. Shirzadi A, Ghaemi N (2006) Side effects of atypical antipsychotics: extrapyramidal symptoms and the metabolic syndrome. Harv Rev Psychiatry 14(3):152–164. https://doi.org/10.1080/10673220600748486
271. Pillai A, Parikh V, Terry A, Mahadik S (2007) Long-term antipsychotic treatments and crossover studies in rats: differential effects of typical and atypical agents on the expression of antioxidant enzymes and membrane lipid peroxidation in rat brain. J Psychiatr Res 41(5):372–386. https://doi.org/10.1016/j.jpsychires.2006.01.011
272. Dosage & application I SECUADO® (asenapine) Patch. Secuado (2022) https://www.secuado.com/hcp/prescribing-secuado/. Accessed 23 Apr 2022
273. Schizophrenia facts and statistics. Schizophrenia.com (2022) http://schizophrenia.com/szfacts.htm. Accessed 23 Jan 2022
274. Junaid MS, Banga AK (2022) Transdermal delivery of baclofen using iontophoresis and microneedles. AAPS PharmSciTech 23(3):84. https://doi.org/10.1208/s12249-022-02232-w
275. Sznitowska M, Janicki S, Gos T (1996) The effect of sorption promoters on percutaneous permeation of a model zwitterion baclofen. Int J Pharm 137:125–132
276. Volpato NM, Santi P, Colombo P (1995) Iontophoresis enhances the transport of acyclovir through nude mouse skin by electrorepulsion and electroosmosis. Pharm Res 12:1623–1627
277. Saepang K, Li SK, Chantasart D (2021) Effect of pH on iontophoretic transport of pramipexole dihydrochloride across human epidermal membrane. Pharm Res 38:657–668
278. Garland MJ, Caffarel-Salvador E, Migalska K, Woolfson AD, Donnelly RF (2012) Dissolving polymeric microneedle arrays for electrically assisted transdermal drug delivery. J Controlled Release 159:52–59
279. Oh YJ, Kang NW, Jeong HR, Sohn SY, Jeon YE, Yu NY, Hwang Y, Kim S, Kim DD, Park JH (2022) The relationship between the drug delivery properties of a formulation of teriparatide microneedles and the pharmacokinetic evaluation of teriparatide administration in rats. Pharm Res 39(5):989–999. https://doi.org/10.1007/s11095-022-03254-6
280. Naito C, Katsumi H, Suzuki T, Quan Y-S, Kamiyama F, Sakane T, Yamamoto A (2018) Self-dissolving microneedle arrays for transdermal absorption enhancement of human parathyroid hormone (1–34). Pharmaceutics 10(4):215
281. Daddona PE, Matriano JA, Mandema J, Maa Y-F (2011) Parathyroid hormone (1–34)-coated microneedle patch system: clinical pharmacokinetics and pharmacodynamics for treatment of osteoporosis. Pharm Res 28(1):159–165

Chapter 5
Clinical Applications of Pharmacokinetic and Pharmacodynamic Studies of Targeted Novel Drug Delivery Systems

Sankalp A. Gharat, Munira M. Momin, and Tabassum Khan

Introduction by the Editor

The clinical applications of pharmacokinetic (PK) and pharmacodynamic (PD) studies of nanoparticles are diverse and offer significant advancement in drug delivery and personalized medicine. Researchers can design treatment regimens by utilizing the knowledge gained from PK and PD studies to enhance therapeutic outcomes, and minimize adverse effects, ultimately advancing patient care and overall well-being. Based on preclinical and clinical data, PK-PD modeling can potentially direct formulation design and dosage regimen selection in the development of novel drug delivery systems. This method relates the properties of the physiological system and drug delivery system to the dose of the drug and the physiological response. A schematic representation of PK-PD modeling in the development of nanoparticle-based drug delivery system is depicted in Fig. 5.1.

PK-PD modeling is applied to various aspects of drug delivery, including brain targeting, cancer targeting, pulmonary targeting, parenteral targeting, delivery of proteins and peptides, and gene delivery.

1. Brain Targeting [1–3]:

    Blood-Brain Barrier (BBB) Penetration: PK-PD modeling predicts the ability of nanoparticles and drugs to cross the BBB based on their physicochemical

S. A. Gharat · M. M. Momin (✉)
Department of Pharmaceutics, SVKM's Dr. Bhanuben Nanavati College of Pharmacy, Mumbai, Maharashtra, India
e-mail: sankalp.gharat@bncp.ac.in; munira.momin@bncp.ac.in

T. Khan
Department of Pharmaceutical Chemistry and Quality Assurance, SVKM's Dr. Bhanuben Nanavati College of Pharmacy, Mumbai, Maharashtra, India
e-mail: tabassum.khan@bncp.ac.in

S. A. Gharat et al. (eds.), *Pharmacokinetics and Pharmacodynamics of Novel Drug Delivery Systems: From Basic Concepts to Applications*,
https://doi.org/10.1007/978-981-99-7858-8_5

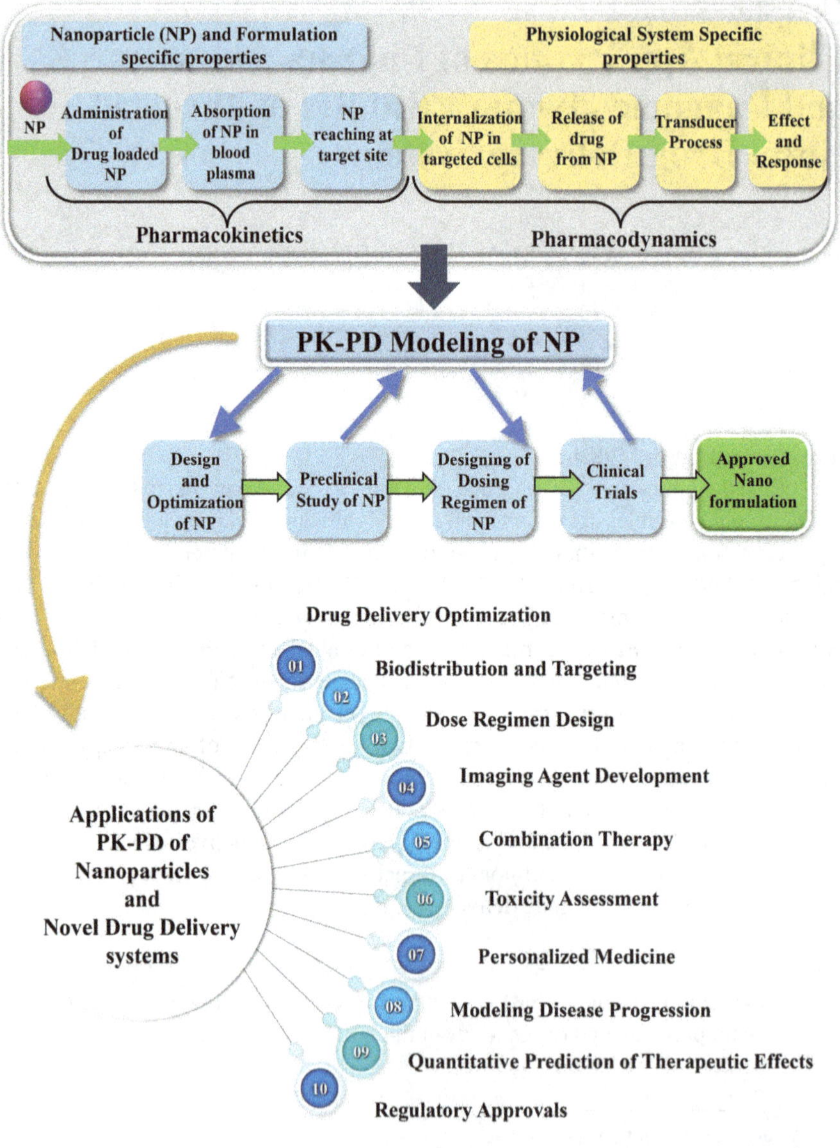

Fig. 5.1 Schematic representation of PK-PD modeling in the nanoparticle-based drug delivery system development

properties. This helps in selecting and designing drug delivery systems that can effectively reach the brain.

Optimal Dosage: PK-PD modeling helps to determine the optimal dose required to achieve effective therapeutic concentration in the brain, accounting for factors like drug-receptor interactions and elimination kinetics.

Time Course of Action: Modeling the time course of drug concentration in the brain assists in scheduling dose to achieve sustained therapeutic effects.

2. Cancer Targeting [4–6]:

    Tumor Accumulation: PK-PD modeling helps estimate the accumulation of therapeutic agents in tumor tissues through the EPR effect. This information guides the design of nanoparticles with appropriate characteristics for enhanced tumor targeting.
    Therapeutic Efficacy: PK-PD modeling assists in predicting the therapeutic efficacy by modeling drug-receptor interactions and tumor growth kinetics of targeted therapies and optimizing dosing regimens.
    Combination Therapies: PK-PD modeling can evaluate the synergistic effects of combining different drugs delivered using nanoparticles, aiding in the development of effective combination therapies.

3. Pulmonary Targeting [7–9]:

    Deposition and Distribution: PK-PD modeling predicts the deposition pattern of inhaled nanoparticles in the respiratory tract. This information helps in the design of inhalation therapies for targeted lung delivery.
    Local Effects: Modeling drug distribution within pulmonary tissues helps understand how the drug interacts with lung cells, leading to optimized treatment strategies for lung diseases.

4. Ocular Drug Delivery [10–13]:

    Formulation Optimization: PK-PD modeling guides the optimization of nanoparticle formulations for ocular administration by predicting their distribution, clearance, and therapeutic efficacy.
    Safety Assessment: Modeling systemic exposure and potential off-target effects helps in evaluating the safety of ocular drug delivery systems.

In these applications, PK-PD modeling provides insight into the pharmacokinetics, biodistribution, target engagement, and therapeutic effects of drug delivery systems. This enables researchers and clinicians to make informed decisions about formulation design, dosing regimens, and treatment strategies, ultimately leading to more effective and targeted therapies with reduced side effects. This chapter focuses on biodistribution of nanoparticles in case of cancer, brain drug delivery, pulmonary drug delivery, parenteral drug delivery and gene delivery.

5.1 Targeted Delivery to Central Nervous System and Brain

Introduction by the Editor

The brain presents a complex array of molecular targets crucial for various physiological and pathological processes. Neurotransmitter receptors, such as glutamate, dopamine, and serotonin receptors, play pivotal roles in neuronal communication and mood regulation. Enzymes like acetylcholinesterase are targets for cognitive disorders like Alzheimer's disease. Inflammatory markers, including cytokines and chemokines, are implicated in neuroinflammatory conditions. For brain tumors, growth factors like epidermal growth factor receptor (EGFR) and vascular endothelial growth factor (VEGF) are significant. Moreover, specific receptors like transferrin receptors and low-density lipoprotein receptors are exploited for drug delivery across the blood-brain barrier. These molecular targets offer opportunities for both diagnostic and therapeutic interventions in neurology and neuroscience. Nanoparticles hold immense promise in targeting specific molecular markers within the brain, offering a novel approach to addressing neurological disorders. Surface modification with ligands enables nanoparticles to recognise and bind to receptors like transferrin receptors or integrins, facilitating their transport across the blood-brain barrier. For instance, nanoparticles are designed to deliver therapeutic agents that target amyloid plaques in Alzheimer's disease, or they can carry anticancer drugs to brain tumor sites by exploiting overexpressed receptors. By precisely modifying the properties of nanoparticles, such as size, surface charge, and ligand specificity, researchers can enhance their affinity for specific molecular targets, unlocking new possibilities for precision medicine in the realm of neurological conditions.

Machine generated summaries

Disclaimer: The summaries in this chapter were generated from Springer Nature publications using extractive AI auto-summarization: An extraction-based summarizer aims to identify the most important sentences of a text using an algorithm and uses those original sentences to create the auto-summary (unlike generative AI). As the constituted sentences are machine selected, they may not fully reflect the body of the work, so we strongly advise that the original content is read and cited. The auto generated summaries were curated by the editor to meet Springer Nature publication standards.

To cite this content, please refer to the original papers.

Machine generated keywords: brain, cns, nanoparticle, bbb, alzheimer, alzheimer disease, brain target, cancer, pulmonary, lung, antimicrobial, bloodbrain, blood-brain barrier, biomolecule, pkpd, brain, cns, bbb, brain target, nanoparticle, blood-brain, bloodbrain barrier, intranasal, alzheimer, alzheimer disease, central nervous, polymer, nasal, tumor, parkinson

In Vitro to In Vivo Extrapolation Linked to Physiologically Based Pharmacokinetic Models for Assessing the Brain Drug Disposition [14] This is a machine-generated summary of:

Murata, Yukiko; Neuhoff, Sibylle; Rostami-Hodjegan, Amin; Takita, Hiroyuki; Al-Majdoub, Zubida M.; Ogungbenro, Kayode: In Vitro to In Vivo Extrapolation Linked to Physiologically Based Pharmacokinetic Models for Assessing the Brain Drug Disposition [14]

Published in: The AAPS Journal (2022)

Link to original: https://doi.org/10.1208/s12248-021-00675-w

If you want to cite the papers, please refer to the original.

For technical reasons we could not place the page where the original quote is coming from.

Abstract-Summary "Several in vitro brain systems have been evaluated, but the ultimate use of these data in terms of translation to human brain concentration profiles remains to be fully developed."

"Linking up in vitro-to-in vivo extrapolation (IVIVE) strategies to physiologically based pharmacokinetic (PBPK) models of brain is a useful effort that allows better prediction of drug concentrations in CNS components."

"Such models may overcome some known aspects of inter-species differences in CNS drug disposition."

"Due to the inability to directly measure brain concentrations in humans, compound-specific (drug) parameters are often obtained from in silico or in vitro studies."

"This report summarizes the state of IVIVE-PBPK-linked models and discusses shortcomings and areas of further research for better prediction of CNS drug disposition."

INTRODUCTION

"Whilst a large part of this relates to lack of good experimental models mimicking relevant mechanisms of the disease, the difficulties associated with the location of the drug effect, namely central nervous system (CNS), cannot be dismissed."

"BBB has many features that makes establishing relationship between the drug concentrations in systemic circulation and in CNS more challenging."

"PBPK models based on human physiology allow prediction of drug concentrations in target tissues, which has been well documented and have become a critical tool in nonclinical and clinical study design and regulatory review."

"CNS PBPK models have been reported based on various types of model structures and parameter acquisition methods."

"We describe the structure of the CNS, factors that determine the central distribution of drugs, and methods for experimental evaluation."

STRUCTURE OF THE CNS

"The spaces between the arachnoid membrane and spinal cord or brain, including the ventricles, are filled with medium called spinal or cranial cerebrospinal fluid (CSF), respectively."

"Interstitial fluid (ISF) occupies intercellular space of the brain (20% of the total brain volume of around 1250 mL in humans [15, 16]), mediating the exchange of drugs between brain cells and CSF."

"Since predominantly protein-unbound unionized drugs penetrate these barriers, the distribution of drugs in the brain is determined by factors like the pH (pH; 7.3 [17]) or protein content ($\approx$ 0.2 g/L [16, 18]) of CSF/ISF and plasma."

"After reaching the brain, drugs distribute throughout the CNS by the flow of CSF (0.2–0.4 mL/min in human [16, 19])."

"For drugs targeting CNS diseases, it is also important to consider the influence of the diseases on the physiology of the brain; CNS diseases such as stroke, brain tumor, and meningitis, as well as aging, may change barrier function (BBB, BCSFB) or composition and/or flow rate of the CSF and/or ISF."

TECHNIQUES TO ESTIMATE BRAIN DISTRIBUTION OF CNS AND TRANSPORT ACROSS THE BBB

"The brain slice method was originally developed by Kakee and others [20], and further refined by Friden [21]."

"The high-throughput brain slice method is a precise and robust technique for estimating the overall uptake of drugs into brain tissue through determination of the unbound volume of distribution in the brain ($V_{u,brain}$; ml·g brain^{-1}) [22]."

"The brain slice method is more physiologically based than the brain homogenate method with respect to the assessment of drug distribution in the brain since active transport systems, pH gradients, and cell-cell interactions are conserved."

"The brain slice method is suitable for estimation of target-site PK in the early drug discovery process and fundamental pharmacological studies."

"The brain homogenate binding method measures the intracellular binding by equilibrium dialysis of diluted brain homogenates and allows estimation of $V_{u,brain}$."

"A widely used method to measure the permeability of the BBB in vivo is the in situ brain perfusion technique."

PBPK MODELS FOR BRAIN

"The model development was supported by physiological brain parameters obtained from rats (volumes), and in vitro binding (plasma and brain) parameters, while other brain drug disposition parameters were estimated."

"Westerhout and others [23], developed a multi-compartment model (including five brain and CSF) for paracetamol in rats using data obtained by serial sampling of blood and microdialysis probes at different regions of CSF and brain."

"Yamamoto and others [24, 25], in the latest update of their PBPK model for brain disposition, also proposed a workflow for the use of in silico and in vitro data to inform active transport parameters across the BBB and BCSFB in the context of their model."

"Ball and others [26, 27], described PBPK models for CSF and brain disposition of drugs using plasma, CSF and brain ECF concentration data from in situ or microdialysis in rats."

TRANSPORTERS IN BLOOD-BRAIN BARRIER

"Expression of Mdr1a or MDR1 was the highest in rats (mean, 18.4 pmol/mg of total protein), while bcrp or BCRP was the highest in human (mean, 4.26 pmol/mg of total protein), and in both species SLC family transporter (Glut1/GLUT1) were the most abundant (mean in rat, 77.3 pmol/mg of total protein; mean in human 188 pmol/mg of total protein, respectively)."

"Among the most highly expressed transporters, more monocarboxylate transporter 1 (Mct1/MCT1) was present in the rats compared to humans (mean, 9.4 versus 2.7 pmol/mg total protein, respectively)."

"More amino acid transporter (Lat1/LAT1) was expressed in the rats (mean, 2.6 pmol/mg protein) compared to the humans (mean, 0.63 pmol/mg total protein)."

"Protein expression levels of MDR1 in the human brains were reported not to be statistically different between all studies (<2.5-fold difference)."

"The reports on most transporters' expression are comparable between studies; except for expression of 4f2hc, Fatp1, ABC2, CTL2, RFC ranging from 3- and 35.7-fold difference."

DISCUSSION AND CONCLUSIONS

"The IVIVE of permeability by scaling with absolute transporter abundance to the PBPK model is a fundamental aspect in translational abilities to humans whether from in vitro experiments or from non-clinical data."

"With the increasing number of reports on absolute transporter abundance in recent years, this leaves an opportunity to refine these IVIVE-PBPK models so they can be the next generation of tools for a more successful CNS drug development."

"Model-informed drug development (MIDD) is no longer just an aspirational idea."

"As MIDD enters mainstream use during drug development by many pharmaceutical companies, community assessment of various models applied to a certain problem and settling on some selected models that can be used repeatedly by a mass of users with assurance on reproducibly of results become inevitable."

Nanosized Transferosome-Based Intranasal In Situ Gel for Brain Targeting of Resveratrol: Formulation, Optimization, In Vitro Evaluation, and In Vivo Pharmacokinetic Study [28] This is a machine-generated summary of:

Salem, Heba F.; Kharshoum, Rasha M.; Abou-Taleb, Heba A.; Naguib, Demiana M.: Nanosized Transferosome-Based Intranasal In Situ Gel for Brain Targeting of Resveratrol: Formulation, Optimization, In Vitro Evaluation, and In Vivo Pharmacokinetic Study [28]

Published in: AAPS PharmSciTech (2019)

Link to original: https://doi.org/10.1208/s12249-019-1353-8

If you want to cite the papers, please refer to the original.
For technical reasons we could not place the page where the original quote is coming from.

Abstract-Summary "The aim of the study was to enhance RES bioavailability through developing intranasal transferosomal mucoadhesive gel."

"Reverse evaporation–vortexing sonication method was employed to prepare RES-loaded transferosomes."

"Mucoadhesive gels were prepared and evaluated, then optimized RES transferosomes were incorporated into the selected gel and characterized using FTIR spectroscopy, in vitro release, and ex vivo permeation study."

"In vitro drug release from transferosomal gel was 65.87 ± 2.12% and ex vivo permeation was 75.95 ± 3.19%."

"Histopathological study confirmed the safety of the optimized formula."

"The C_{max} of RES in the optimized RES trans-gel was 2.15 times higher than the oral RES suspension and $AUC_{(0-\infty)}$ increased by 22.5 times."

"The optimized RES trans-gel developed intranasal safety and bioavailability enhancement through passing hepatic and intestinal metabolism."

INTRODUCTION

"Nasal drug delivery has showed great interest recently as a promising route since it provides numerous advantages over oral or parenteral administration [29, 30]."

"Studies have been recently reported that intranasal drug delivery can be effective for the management of several central nervous system diseases such as Alzheimer's diseases [31], Parkinson's disease [32], brain tumors, sleep disorders [33], and schizophrenia [34]."

"The majority of recent studies oriented toward the development of mucoadhesive nasal formulations also the use permeation enhancers to enhance the nasal absorption and increase the nasal residence time [35]."

"The aim of the present study is to enhance RES bioavailability and to accomplish direct nose to brain targeting through an optimum transferosomal formulation."

"Mucoadhesive gels were prepared and evaluated, then the optimum transferosomes were incorporated into the selected gel and subjected to in vitro release, ex vivo permeation, histopathological examination for nasal mucosa tolerability, and finally were applied to in vivo pharmacokinetics study."

MATERIALS AND METHODS

"The obtained film was hydrated with 10 ml of simulated nasal fluid (SNF, pH 5.5) [36] containing permeation enhancer (PE) except for oleic acid, which was added with soya lecithin; the obtained vesicles were allowed to swell for 2 h at room temperature then vortexed for 20 min."

"The optimized loaded RES transferosomes were prepared and analyzed for vesicle size, PDI, EE%, and % of RES released as prescribed previously."

"Stability testing of the optimized loaded RES transferosomes was determined through storing the optimized formula (5 mg RES/5 ml) in tightly closed amber glass vials in the refrigerator at 4°C and at room temperature of 25°C up to 3 months."

"The optimized RES-TRS gel was tested for permeation behavior against RES suspension gel through sheep nasal mucosa instead of the dialysis cellophane membrane, used formerly throughout the dialysis method utilized during in vitro release studies with the same test conditions."

RESULTS AND DISCUSSION

"As to optimize the RES-loaded transferosome, relationships among the studied independent variables were A (ratio of PC to PE), B (ratio of PC and PE to surfactants), C (type of surfactant), and D (type of PE), at three levels (− 1, 0, + 1)."

"The higher in vitro release percentage is attributed to the dual effect of PE and surfactant, as they increase the partitioning of RES from transferosomes through improving vesicular bilayer fluidity and enhancing RES solubility [37]."

"The results of the pharmacokinetic parameters were highly significant at P < 0.05) for the optimized RES-trans gel compared with oral RES suspension, indicating that the absorption of RES was significantly increased by intranasal transferosome administration."

"The sustained plasma concentration of RES was obtained up to 24 h after intranasal administration of the optimized RES-trans gel, and the results go with the data obtained for in vitro release and ex vivo permeation study."

CONCLUSION

"The optimized transferosomes containing Cremophor RH 40, ethanol, and soya lecithin can be well thought out as an efficient nanocarrier for bioavailability enhancement of RES through the nasal route."

"The optimized transferosomes were found to be safe and tolerable to the sheep nasal mucosa; they displayed vesicle size of 83.79 ± 2.54 nm, entrapment efficiency (EE%) up to 72.58 ± 4.51%, and ex vivo permeation up to 75.95 ± 3.19%."

"The intranasal RES-transferosomal mucoadhesive gel showed gorgeous elevation in $AUC_{(0-\infty)}$ and C_{max} by 22.5 and 2.15 times, respectively, which is counted as a valuable enhancement in the bioavailability of RES compared with oral RES suspension."

"The developed transferosomes could be counted as an elegant nanocarrier for nasal delivery of drugs that have a low oral bioavailability."

Chitosan Engineered PAMAM Dendrimers as Nanoconstructs for the Enhanced Anti-Cancer Potential and Improved In vivo Brain Pharmacokinetics of Temozolomide [38] This is a machine-generated summary of:

Sharma, Ashok Kumar; Gupta, Lokesh; Sahu, Hitesh; Qayum, Arem; Singh, Shashank K.; Nakhate, Kartik T.; Ajazuddin, None; Gupta, Umesh: Chitosan Engineered PAMAM Dendrimers as Nanoconstructs for the Enhanced Anti-Cancer Potential and Improved In vivo Brain Pharmacokinetics of Temozolomide [38]

Published in: Pharmaceutical Research (2018)

Link to original: https://doi.org/10.1007/s11095-017-2324-y

Copyright of the summarized publication:

Springer Science+Business Media, LLC, part of Springer Nature 2017

If you want to cite the papers, please refer to the original.

For technical reasons we could not place the page where the original quote is coming from.

Abstract-Summary "To establish a platform for the possibility of effective and safe delivery of Temozolomide (TMZ) to brain via surface engineered (polyamido-amine) PAMAM dendrimer for the treatment of glioblastoma."

"The present study aims to investigate the efficacy of PAMAM-chitosan conjugate based TMZ nanoformulation (PCT) against gliomas in vitro as well as in vivo."

"PCT was more efficacious in terms of IC_{50} values compared to pure TMZ against U-251 and T-98G glioma cell lines."

"This study exhibits the potential applicability of dendrimer and CS in improving the anticancer activity and delivery of TMZ to brain."

"The attractive ex vivo cytotoxicity against two glioma cell lines; U-251 and T-98G and phase solubility studies of TMZ revealed remarkable results."

"Studies of prepared nanoformulation were significant and promising that explored the double concentration of TMZ in brain due to surface functionality of dendrimer."

"The reported work is novel and non- obvious as none of such approaches using chitosan anchored dendrimer for TMZ delivery has been reported earlier."

Introduction

"It comprises of the high-grade tumors such as malignant gliomas [39] and the glioblastoma [40].The current challenges in brain therapeutics are associated with the drug delivery through blood brain barrier (BBB) in a safe and effective manner."

"Use of nano-vectors and their recognition by tumor is the foremost approach for targeting brain tumor."

"Chitosan is a natural origin polymer obtained from chitin."

"Many drugs such as temozolomide, teniposide, cisplatin, docetaxel, carmustine etc have been approved by FDA for the treatment of brain tumor."

"For TMZ it is prerequisite that it must be administered in higher systemic doses to achieve therapeutic levels in the brain due to its short half-life of about 1.8 h (in plasma)."

"We attempted to combine the nanotechnological polymer therapeutics approach using natural origin polymer such as chitosan to establish an improved delivery of TMZ to brain."

Materials and Methods

"After 24 h, the solution was twice dialyzed for 10 min against sodium acetate buffer (pH 5.0) maintaining perfect sink conditions so as to remove free drug from the formulation (TMZ loaded PCS i.e. Temozolomide loaded PAMAM-chitosan conjugate or PCT)."

"The MTT assay was carried out using 96 well plates (Tarsons, India) and formulations (and other test samples) were seeded in concentration ranges equivalent to 10–100 μg/mL equivalents to TMZ."

"Pure drug solution (TMZ) and conjugated formulations PCT (600 µg equivalent TMZ) were administered intravenously through tail vein to 3rd, 4th and 5th groups of animals, respectively each day for a week."

"At predetermined time intervals the blood sample (0.2 mL) were collected from the animals through Retro Orbital Plexus (ROP) using micro blood collecting tubes and diluted in appropriate solvent then centrifuged (REMI C-24BL, India) at 10000 rpm at 4°C."

Results and Discussion

"The low pdi values of blank formulation (PCS) compared to drug loaded one (PCT), provides information of monodispersity means mono modal particle size distributions with uniform particle size."

"The higher average size of the PCT can be inferred to conjugation between PAMAM and CS as well as due to drug loading, which is due increased chain of polymer length and accordingly the size of formulation."

"The drug loaded PCT showed more hemolysis compared to others due to cumulative effect of TMZ, PAMAM and CS."

"The conjugated formulation, PCT, was found significantly safer ($p < 0.05$) and biocompatible with slightest effect on hematological parameters compared to drug."

"Drug distribution observed for the formulation (i.e. PCT) was always higher compared to pure TMZ."

Conclusion

"TMZ loaded PCS conjugated formulation (PCT) was developed and characterized that displayed sustained release behavior of the encapsulated drug."

"Studies performed were significant and promising in terms of biocompatibility as well safety profile of TMZ in conjugated formulation of PAMAM dendrimers."

"These PAMAM dendrimer based novel conjugated formulation provides a remarkable delivery of TMZ for targeting the brain."

"The study is one of its own kinds as it has explored the possibility of dendrimers as nanocarriers for the delivery of drugs to brain."

"It is novel in the sense that conjugated approach for the delivery of drug using dendrimers was followed so as to make the surface of dendrimers more compatible in the biological milieu."

"The present work will be continuously explored up to other possible in vivo studies and molecular level to make an ideal drug delivery approach for brain tumor."

Emerging Insights for Translational Pharmacokinetic and Pharmacokinetic-Pharmacodynamic Studies: Towards Prediction of Nose-to-Brain Transport in Humans [41] This is a machine-generated summary of:

Ruigrok, Mitchel J. R.; de Lange, Elizabeth C. M.: Emerging Insights for Translational Pharmacokinetic and Pharmacokinetic-Pharmacodynamic Studies: Towards Prediction of Nose-to-Brain Transport in Humans [41]

Published in: The AAPS Journal (2015)

Link to original: https://doi.org/10.1208/s12248-015-9724-x

Copyright of the summarized publication:

If you want to cite the papers, please refer to the original.

For technical reasons we could not place the page where the original quote is coming from.

Abstract-Summary "To investigate the potential added value of intranasal drug administration, preclinical studies to date have typically used the area under the curve (AUC) in brain tissue or cerebrospinal fluid (CSF) compared to plasma following intranasal and intravenous administration to calculate measures of extent like drug targeting efficiencies (%DTE) and nose-to-brain transport percentages (%DTP)."

"CSF does not necessarily provide direct information on the target site concentrations, while total brain concentrations are not specific to that end either as non-specific binding is not explicitly considered."

"To predict nose-to-brain transport in humans, the use of descriptive analysis of preclinical data does not suffice."

"Nose-to-brain research should be performed translationally and focus on preclinical studies to obtain specific information on absorption from the nose, and distinguish between the different transport routes to the brain (absorption directly from the nose to the brain, absorption from the nose into the systemic circulation, and distribution between the systemic circulation and the brain), in terms of extent as well as rate."

INTRODUCTION

"Numerous candidate drugs for CNS diseases were efficacious during in vitro and preclinical in vivo studies."

"This review aims to provide insight in advanced experimental and mathematical modeling approaches using preclinical data, and proposed steps to be taken for translation between conditions and ultimately to species translatability for nose-to-brain transport in humans."

"To that end, the impact of the blood–brain barriers on drug distribution into the CNS is shortly discussed, followed by a summary on the knowledge of the nasal anatomy, histology, and physiology and their species differences; direct nose-to-brain drug transport mechanisms; evidence for direct nose-to-brain drug and drug delivery systems transport in animals; and evidence for direct nose-to-brain drug transport in humans."

"This information finally feeds into considerations and suggestions for future studies on translation of preclinical nose-to-brain PK and PK-PD data to the human situation."

INTRANASAL ADMINISTRATION TO CIRCUMVENT THE IMPACT OF THE BLOOD–BRAIN BARRIERS ON DRUG DISTRIBUTION INTO THE CNS

"Most CNS-active drugs tend to enter the brain mainly by passing through the BBB."

"For the more lipophilic drugs that can pass cell membranes readily, transcellular passage of the BBB may be counteracted by the action of efflux transporter proteins, such as P-glycoprotein (Pgp) and multidrug resistance-related proteins (MRPs) that are present on the cell membranes of the brain capillary endothelial cells."

"There is a need for safer, easier, and less invasive brain drug delivery techniques which bypass the BBB."

"Researchers answered to this need by exploring IN drug administration as a method to enhance the delivery of drugs into the brain while bypassing the BBB [42]."

"There is a potential for direct nose-to-brain delivery as drugs could bypass the BBB."

"Direct nose-to-brain delivery could be a promising drug administration technique for patients who suffer from CNS diseases."

NASAL ANATOMY, HISTOLOGY, AND PHYSIOLOGY AND SPECIES DIFFERENCES

"The nasal cavity contains two functional regions which are concerned with (a) the conditioning and filtration of inhaled air before it enters the lungs (respiratory region) and (b) the sense of smell (olfactory region)."

"Four types of epithelium can be found in the nasal cavity: squamous, transitional, respiratory, and olfactory [43]."

"These types of epithelium line the surface of distinct regions in the nasal cavity."

"Respiratory epithelium lines the main chamber of the nasal cavity and the nasopharynx."

"Olfactory epithelium is present on the cranial side of the nasal cavity."

"Olfactory epithelium is the most relevant for direct nose-to-brain delivery as it provides a direct link between the nasal cavity and the CNS which bypasses the BBB [44]."

"Filtration of air occurs as harmful agents are deposited into mucus which is present on the surface epithelium of the nasal cavity."

NOSE-TO-BRAIN DRUG TRANSPORT MECHANISMS

"Drugs can enter the CNS via the olfactory bulb by transport along the olfactory nerve."

"These two drug transport mechanisms circumvent the BBB, resulting in direct nose-to-brain transport."

"Direct nose-to-brain drug delivery can also occur to a lesser extent via transport along the trigeminal nerve."

"After passing through the respiratory and olfactory epithelium, drugs can move along the trigeminal nerve via intracellular or extracellular transport mechanisms where they can enter the brain through either the cribriform plate or the pons."

"Drugs can also reach the CNS via initial absorption into the systemic circulation, followed by blood-brain transport."

"It is important to realize that in direct nose-to-brain transport, drug transporters and metabolizing capacity of nasal mucosa should be considered as well."

"Wong and Zuo (2010) highlighted the importance and implications of how nasal metabolism might influence the transport of drugs via direct nose-to-brain transport."

EVIDENCE FOR NOSE-TO-BRAIN DRUG AND DRUG DELIVERY SYSTEM TRANSPORT IN ANIMALS

"The extent was based on AUC values in brain and plasma following IV and IN drug administration, and expressed as drug targeting efficiency percentage (%DTE) and the nose-to-brain direct transport percentage (%DTP)."

"Several IN administered small molecule drugs have been shown to be delivered into the CNS via direct nose-to-brain transport."

"This indicates that other processes seem to be involved that favor direct nose-to-brain transport, especially as the IN administration of the moderately lipophilic drug acetaminophen did not lead to brain distribution enhancement [45]."

"These results show several CNS-active small molecule drugs are transported into the CNS of rats via direct nose-to-brain transport."

"Aside from small molecule drugs, peptides have also been shown to enter the CNS via direct nose-to-brain drug transport."

"Proteins have also been shown to enter the CNS after IN drug administration in rats via direct nose-to-brain transport."

"Direct nose-to-brain transport of venlafaxine loaded alginate nanoparticles has been shown in rats [46]."

EVIDENCE FOR NOSE-TO-BRAIN DRUG TRANSPORT IN HUMANS

"Although direct nose-to-brain drug transport has obtained increased attention, only one study so far has collected quantitative PK data which confirmed this type of transport in humans [47]."

"Born and others (2002) obtained PK evidence of direct nose-to-brain transport in humans after IN administration of the peptides melanocortin [48, 49], vasopressin, and insulin [47]."

"Merkus and others (2003) was unable to confirm direct nose-to-brain transport of melatonin and hydroxocobalamin, on the basis of plasma and CSF concentrations obtained in humans, when comparing the extent of drug transport enhancement after IN and IV administration [50]."

"Direct nose-to-brain drug transport in humans remains an unsolved issue."

"Direct nose-to-brain drug transport in humans can also be assessed indirectly, via a non-quantitative approach by measuring drug-specific PD."

"Collecting quantitative PK evidence for direct nose-to-brain transport in humans is difficult for several reasons."

TRANSLATIONAL PK-PD OF NOSE-TO-BRAIN REMOXIPRIDE TRANSPORT IN RATS

"The PK model developed by Stevens and others (2011) shows the added value of separation and quantitation of systemic and direct nose-to-brain transport after IN administered remoxipride in freely moving rats, in terms of extent and rates [51]."

"Two absorption compartments were identified in this model to describe (1) the absorption rate constant of remoxipride from the nasal cavity into the central compartment (ka_{13}) and (2) the absorption rate constant via direct nose-to-brain transport (ka_{24})."

"This study demonstrated successful separation and quantitation of systemic and direct nose-to-brain transport in rats after IN administration of remoxipride, in terms of extent as well as rate."

"The PK-PD relationship of remoxipride following IN administration could be adequately predicted by simulations of the PK-PD model, demonstrating successful translation of remoxipride PK in rats between the two distinct routes of drug administration."

TRANSLATIONAL PK AND PK-PD OF IV REMOXIPRIDE FROM RATS TO HUMANS

"Successful mathematical modeling of unbound remoxipride PK in plasma and brain ECF in rats after IV and IN drug administration, as shown in the previous study example [51], provided the basis for the "humanized" PK and finally the PK-PD model [52]."

"For translation of the preclinical PK model to humans, allometric scaling was applied and the "humanized" PK model successfully predicted existing plasma PK of remoxipride as measured in humans [53]."

"This PD information could be used to see if the PD obtained in humans would be adequately predicted by the "humanized" PK-PD model for which drug-specific and biological-system-specific parameters were obtained from the literature."

"It was demonstrated that the humanized PK-PD model could satisfactorily predict PK-PD relationship of remoxipride in humans."

TOWARDS PREDICTION OF NOSE-TO-BRAIN TRANSPORT IN HUMANS

"Translational PK and PK-PD models developed on the basis of preclinical data on CNS active drugs is a promising approach to improve prediction of CNS target site concentrations in human and associated effects."

"For the translation of animal to human PK and PK-PD following IV administered drugs on the basis of preclinical data, preclinical experiments will substantially improve if we include: unbound drug concentrations, as it is the unbound drug concentration that drives transport processes (BBB transport, intra-brain distribution, unbound brain concentrations) and target interactions that lead to drug effects [54, 55]."

"To further work on the prediction of human PK and PK-PD following IN administration, in addition to the aforementioned points, we need to consider the: explicit distinction between the different absorption/transport routes (i.e., absorption directly from nose to the brain, absorption from the nose into the systemic circulation, and distribution between the systemic circulation and the brain), as the rate and extent of each absorption/transport pathway may independently differ between rats and humans."

CONCLUSIONS

"Numerous IN administered substances, such as small molecules, biologics, and specialized drug delivery systems, have been shown to enter the CNS of animals via direct nose-to-brain transport while bypassing the BBB."

"Circumvention of the BBB is facilitated by extracellular and intracellular transport processes of drugs along the olfactory nerve and the trigeminal nerve which provide direct entry points to the CNS."

"Preclinical animal studies show encouraging results, confirming direct nose-to-brain transport."

"It is of great importance to investigate the predictive value of preclinical animal models within the context of direct nose-to-brain transport."

Pharmacokinetic and pharmacodynamic evaluation of nasal liposome and nanoparticle based rivastigmine formulations in acute and chronic models of Alzheimer's disease [56] This is a machine-generated summary of:

Rompicherla, Sampath Kumar L.; Arumugam, Karthik; Bojja, Sree Lalitha; Kumar, Nitesh; Rao, C. Mallikarjuna: Pharmacokinetic and pharmacodynamic evaluation of nasal liposome and nanoparticle based rivastigmine formulations in acute and chronic models of Alzheimer's disease [56]

Published in: Naunyn-Schmiedeberg's Archives of Pharmacology (2021)

Link to original: https://doi.org/10.1007/s00210-021-02096-0

If you want to cite the papers, please refer to the original.

For technical reasons we could not place the page where the original quote is coming from.

Abstract-Summary "Rivastigmine, a reversible dual cholinesterase inhibitor, is a more tolerable and widely used choice of drug for AD."

“Nanoformulations including liposomes and PLGA nanoparticles can encapsulate hydrophilic drugs and deliver them efficiently to the brain.”

“The present study attempts to evaluate the pharmacokinetic and pharmacodynamic properties of nasal liposomal and PLGA nanoparticle formulations of rivastigmine in acute scopolamine-induced amnesia and chronic colchicine induced cognitive dysfunction animal models, and validate the best formulation by employing pharmacokinetic and pharmacodynamic (PK-PD) modeling.”

“Nasal liposomal rivastigmine formulation showed the best pharmacokinetic features with rapid onset of action (Tmax = 5 min), higher Cmax (1489.5 ± 620.71), enhanced systemic bioavailability (F = 118.65 ± 23.54; AUC = 35,921.75 ± 9559.46), increased half-life (30.92 ± 8.38 min), and reduced clearance rate (Kel (1/min) = 0.0224 ± 0.006) compared to oral rivastigmine (Tmax = 15 min; Cmax = 56.29 ± 27.05; F = 4.39 ± 1.82; AUC = 1663.79 ± 813.54; t1/2 = 13.48 ± 5.79; Kel (1/min) = 0.0514 ± 0.023).”

“Further, the liposomal formulation significantly rescued the memory deficit induced by scopolamine as well as colchicine superior to other formulations as assessed in Morris water maze and passive avoidance tasks.”

“PK-PD modeling demonstrated a strong correlation between the pharmacokinetic parameters and acetylcholinesterase inhibition of liposomal formulation.”

Introduction

“Cholinesterase inhibitors represent the significant fraction of currently available drugs as they primarily enhance the cholinergic neurotransmission in the brain by delaying the degradation of acetylcholine available in the synaptic clefts (Sharma [57]).”

“Donepezil, galantamine, and rivastigmine are the FDA-approved cholinesterase inhibitors recommended for treatment of AD.”

“Rivastigmine is a second-generation carbamate derivative and reversible, noncompetitive cholinesterase inhibitor widely used in mild to moderate AD cases, and studies propose maximal therapeutic benefits with early and continuous treatment.”

“The nasal route can be exploited in the present study to deliver rivastigmine efficiently to the brain.”

“Despite the direct access to the brain, limited drug absorption and nasal permeability remain a challenge with nasal delivery of hydrophilic drugs.”

“To overcome this, nanocarriers such as PLGA nanoparticles and liposomes represent efficient vehicle systems to deliver the hydrophilic cargo (rivastigmine) to the brain (Vieira and Gamarra [58]).”

Materials and methods

“The drug EE and loading efficiency (LE) was calculated by the following equations: (Li [59]) Lyophilized rivastigmine-loaded nanoparticles were suspended in water/saline (0.1% w/v), and solution was vortexed for 5 min.”

“The overnight fasted rats were administered with different rivastigmine formulations 30 min following scopolamine treatment.”

“The whole brain was rapidly weighed and homogenized in PBS (10%w/v), and 500 μL of this whole homogenate was taken in a 1.5-mL centrifuge tube, centrifuged at 10,000 rpm for 5 min at − 10 °C, and clear supernatant 100 μL was taken for the AChE activity."

"Rats treated with colchicine ICV were treated with different rivastigmine formulations and sampled at different time intervals until 2 h on the 12th day post-colchicine administration."

"Following rivastigmine administration, 15-min lag time for assessment was chosen based on pilot PK and PD studies where Cmax was observed."

Results

"PLGA nanoparticles showed significantly lower systemic bioavailability and larger volume of distribution when compared to liposomes as well as nasal pure drug."

"Rivastigmine-loaded liposomes had a lower clearance rate compared to nanoparticles as well as nasal pure drug."

"The absolute bioavailability of rivastigmine was significantly higher for liposomes compared to pure drug and nanoparticles."

"On day 14 (1st retention test) and day 21 (2nd retention test), the COL group displayed significantly increased escape latency and reduced time spent in the target quadrant over the SC and ACSF groups indicating the memory deficit induced in the colchicine-alone-treated rats."

"In the 2nd and 3rd retention test assessed on the 17th and 23rd-day post-colchicine administration, COL showed a significantly higher number of crossings and increase in time spent in dark chamber compared to sham and ACSF indicating the memory deficit induced by colchicine, as it failed to remember the electric shock."

Discussion

"Intranasally administered rivastigmine liposomes had significantly improved pharmacokinetic parameters compared to the nasal pure rivastigmine, nasal rivastigmine nanoparticles, and oral rivastigmine."

"The present study compared the effects of a cholinesterase inhibitor, rivastigmine between pure form and the novel formulations administered via the intranasal route in the scopolamine-induced memory-impaired rats in the MWM and PAT tasks."

"Rivastigmine at a dose of 2.5 mg/kg via the oral route inhibited the cholinesterase enzyme in the cortex and hippocampus by 20–30% and significantly reduced the effects of scopolamine on reference and working memory in MWM and PAT (Emerich and Walsh [60])."

"INRL treatment significantly reversed the memory deficit induced by scopolamine superior to other formulations as assessed in MWM and PAT tasks."

"The intranasal liposomal rivastigmine (INRL) treatment has a beneficial effect over the amnesia induced by scopolamine superior to the conventional oral treatment and also intranasal pure rivastigmine administration."

Conclusion

"Nasal delivery of rivastigmine liposomes showed ideal pharmacokinetic characteristics with rapid absorption, enhanced systemic bioavailability, half-life, and mean residence time superior to other formulations."

"Besides, it reversed the memory deficit induced by acute (scopolamine) as well as chronic (colchicine) models and exhibited a good agreement between pharmacokinetic and pharmacodynamic activities."

"The nasal route of rivastigmine liposomal formulation can be an ideal choice for treating dementia related to Alzheimer's disease."

Brain Targeted Curcumin Loaded Turmeric Oil Microemulsion Protects Against Trimethyltin Induced Neurodegeneration in Adult Zebrafish: A Pharmacokinetic and Pharmacodynamic Insight [61] This is a machine-generated summary of:

More, Suraj; Pawar, Atmaram: Brain Targeted Curcumin Loaded Turmeric Oil Microemulsion Protects Against Trimethyltin Induced Neurodegeneration in Adult Zebrafish: A Pharmacokinetic and Pharmacodynamic Insight [61]

Published in: Pharmaceutical Research (2023)

Link to original: https://doi.org/10.1007/s11095-022-03467-9

If you want to cite the papers, please refer to the original.

For technical reasons we could not place the page where the original quote is coming from.

Abstract-Summary "Its effect as oral brain targeted formulation for neuroprotection has not yet reported."

"The objective of the study was to investigate the pharmacokinetic of curcumin loaded turmeric oil microemulsion for brain targeting and probing the protective effect against trimethyltin induced neurodegeneration in adult zebrafish."

"In vivo plasma and brain pharmacokinetics was performed to determine improvement in relative bioavailability in rats followed by biodistribution and histopathological evaluation."

"The neuroprotective effect of the formulation was assessed in trimethyltin induced neurodegeneration model using adult zebrafish by behavioral analysis and biochemical analysis."

"Histopathological evaluation confirmed neuroprotective effect on zebrafish brains."

"Results showed a great potential of curcumin microemulsion for brain targeting in the effective treatment of neurological ailments."

Introduction

"Curcumin (CUR) and curcuminoids, polyphenolic compounds of turmeric, exhibited many biological activities and found effective on wide range of diseases [62–66]."

"The results of previously reported in vitro and in vivo preclinical studies showed the neuroprotective effect of CUR on AD [67]."

"In line with these reports, our group has previously disclosed the first time formulation development of turmeric oil for brain targeting where improved preliminary brain pharmacokinetics in adult zebrafish was reported [68]."

"The aim of present investigation was probing the pharmacokinetics and brain targeting efficiency of curcumin loaded turmeric oil microemulsion (CUR ME) and exploring the neuroprotective activity of the same in experimental model of neurodegeneration."

"It is worth mentioning that the present study explored the neuroprotective effect of CUR ME for the first time by using in novel TMT induced neurodegeneration in adult zebrafish."

Materials and Methods

"Oil was mixed with the S:CoS (2:1 w/w Kolliphor RH40:ethanol) mixture using vortex mixer followed by addition of CUR at a dose of 25 mg/mL. The CUR was dissolved by vortex mixing and sonication for 15 min."

"Upon solubilization of the CUR, water was added in remaining amount followed by vortex mixing for 15 min."

"Brain homogenates were stored at − 70°C prior to RP-HPLC analysis."

"Zebrafish were anaesthetized using cold anaesthesia and CUR ME and CUR solution were administered using micropipette as per our previously reported method [68]."

"The Pearless C18 (5 μm, 250 × 4.6 mm, Chromatopak, India) column was used for analysis with mobile phase comprising of 2% acetic acid (pH 3): acetonitrile in a ratio of 40:60 v/v. Flow rate was kept 1 mL per min with detection wavelength of 428 nm and retention time about 9 min."

Results

"The complete elimination of the drug from both the formulations was observed at the end of 3 h. The plasma concentration observed with CUR ME was significantly higher than CUR solution ($p < 0.001$) with relative bioavailability of 200%."

"CUR ME treated group revealed statistically nonsignificant difference with control group (ns, $p > 0.05$) but statistically highly significant difference with disease control group (###, $p < 0.001$) showing the similar performance as that of control group which indicates the recollection of spatial memory and learning behavior."

"CUR solution treated group revealed statistically highly significant difference with control group (***, $p < 0.001$) and statistically significant difference with disease control group (##, $p < 0.01$)."

"CUR Solution treated group was found to be improving GSH levels greater than control group ($p < 0.001$) and other two treatment groups ($p < 0.01$)."

Discussion

"We presented a simple microemulsion formulation administered by oral route for enhancing the drug delivery to the brain."

"In previous study we have reported formulation development and preliminary in vivo evaluation in adult zebrafish whereas this investigation consisted of in vitro drug release, in vivo pharmacokinetics in wistar rats and pharmacodynamics in experimental model of neurodegeneration using adult zebrafish."

"Spatial memory assessment of CUR in TMT induced AD in adult zebrafish was tried for the first time."

"Results showed deprived levels of MDA and improved levels of GSH and SOD which indicated reduction in oxidative stress in the zebrafish brain in disease control group when treated with CUR ME."

"Results showed promising use of CUR ME for passive drug delivery to the brain with neuroprotective effects desired for effective treatment of neurodegenerative disorders."

Conclusion

"The current investigation comprised of in vivo pharmacokinetic and pharmacodynamic assessment of CUR ME which was previously developed and reported by our group."

"Pharmacodynamics assessment in adult zebrafish revealed improved spatial memory when treated with developed formulation than compared with disease control and other treatment groups."

"These encouraging results highlighted the promising use of CUR ME for the treatment of neurodegenerative disorders."

"Use of CUR ME can be considered as promising approach for treatment of neurodegenerative disorders after clinical evaluation for translation to humans."

Chondroitin sulfate conjugation facilitates tumor cell internalization of albumin nanoparticles for brain-targeted delivery of temozolomide via CD44 receptor-mediated targeting [69] This is a machine-generated summary of:

Kudarha, Ritu R.; Sawant, Krutika K.: Chondroitin sulfate conjugation facilitates tumor cell internalization of albumin nanoparticles for brain-targeted delivery of temozolomide via CD44 receptor-mediated targeting [69]

Published in: Drug Delivery and Translational Research (2020)

Link to original: https://doi.org/10.1007/s13346-020-00861-x

If you want to cite the papers, please refer to the original.

For technical reasons we could not place the page where the original quote is coming from.

Abstract-Summary "Temozolomide (TMZ) loaded chondroitin sulfate conjugated albumin nanoparticles (CS-TNPs) were fabricated by desolvation method were chondroitin sulfate (CS) was used as the surface exposed ligand to achieve CD44 receptor mediated targeting of brain tumor."

"Cell viability assay data demonstrated higher cytotoxicity of CS-TNPs as compared with pure TMZ."

"The CD44 receptor blocking assay and receptor poisoning assay in U87 MG cells confirmed the CD44 receptor and endocytosis-mediated (caveolae pathway) uptake of CS-TNPs."

"Results revealed significant enhancement in pharmacokinetic profile of CS-TNPs as compared with TMZ alone."

"Biodistribution results demonstrated higher accumulation of TMZ in the brain by CS-TNPs as compared with the pure drug that confirmed the brain targeting ability of nanoparticles."

"From all obtained results, it may be concluded that CS-TNPs are promising carrier to deliver TMZ to the brain for targeted therapy of brain tumor."

Introduction

"In surface-modified albumin nanocarriers, various ligands have been used for either altering various properties of therapeutic moiety (like pharmacokinetic properties, stability, circulation half-life, and release behavior) or as a targeting moiety [70, 71]."

"CD44 receptors are highly overexpressed in various cancers like breast cancer, brain tumor, hepatic and cervical cancers, etc which can be targeted using various targeting moieties [72–75]."

"The present research work focused on fabrication of temozolomide-loaded CS-conjugated bovine serum albumin (BSA) nanoparticles (CS-TNPs) and its investigation in brain targeting for enhancing the therapeutic concentration of TMZ in the brain."

"The cytotoxic potential of TMZ after encapsulation in CS-TNPs was assessed by cell viability assay and BBB passage was estimated by in vitro monolayer and co-culture model."

"Pharmacokinetic and biodistribution studies were performed to confirm brain targeting and biochemical parameters were estimated to assess the toxicity potential of prepared CS-TNPs in the rat model."

Experimental

"U87 MG cells at a density of 1×10^5 cells/well were seeded on plates and incubated at 37 °C for 24 h. After 24 h, the old medium was discarded, replaced with a fresh medium containing TMZ, TNPs, and CS-TNPs at a concentration equivalent to 100 µg/ml of pure drug and incubated for another 24 h. Then cells were detached using trypsin-EDTA, washed with PBS, and fixed using 70% ethanol."

"For that, the cells were pretreated with 10 mg/ml free HA polymer (175–350 kDa and hydrated overnight in serum- and antibiotic-free medium) for 1 h before addition of CS-TNPs for an additional 12 h and 24 h. After incubation, the cells were lysed using 1% Triton X-100 and the amount of TMZ inside the cells was analyzed by HPLC [76]."

"The cells were seeded in a 96-well plate and incubated for 24 h. After 24 h, the medium was replaced with a fresh medium containing TMZ, TNPs, and CS-TNPs (5-100 μg/ml)."

Results and discussion

"As compared with TMZ, the suppression in the cell viability was 1.27-fold and 1.68-fold higher for TNPs and CS-TNPs which indicated higher cytotoxic potential of prepared nanoparticles as compared with pure drug."

"CS-TNPs showed enhanced cell cytotoxicity as compared with TNPs which may be correlated with its higher cellular uptake and CD44 receptor targeting ability that led to higher suppression of cell growth."

"This higher uptake of CS-TNPs led to increased concentration of loaded TMZ inside the cells which caused enhanced cell cycle arrest in the G2/M phase and indicated higher suppression of cell growth."

"This may be due to higher uptake of CS-TNPs via the CS-mediated CD44 receptor which led to increased TMZ concentration in the cells that ultimately caused increased ROS generation that led to oxidative stress to the cells, inhibited the cell proliferation, and caused cell death."

Conclusions

"Temozolomide-loaded chondroitin sulfate-modified albumin nanoparticles (CS-TNPs) were fabricated for efficient tumor internalization to target brain tumor."

"The results demonstrated that TNPs and CS-TNPs both have the ability to improve the therapeutic efficacy of TMZ after encapsulation in nanoparticles."

"As compared with TNPs, CS-TNPs showed better tumor-targeting efficiency in cell line studies."

"It may be concluded that CS-TNPs are promising carriers to deliver TMZ to the brain for targeted therapy of brain tumor."

Nanoparticles for drug delivery in Parkinson's disease [77] This is a machine-generated summary of:

Baskin, Jonathan; Jeon, June Evelyn; Lewis, Simon J. G.: Nanoparticles for drug delivery in Parkinson's disease [77]

Published in: Journal of Neurology (2020)

Link to original: https://doi.org/10.1007/s00415-020-10291-x

If you want to cite the papers, please refer to the original.

For technical reasons we could not place the page where the original quote is coming from.

Abstract-Summary "Although effective symptomatic treatments for Parkinson's disease (PD) have been available for some time, efficient and well-controlled drug delivery to the brain has proven to be challenging."

"Several exciting strategies including drug carrier nanoparticles targeting specific intracellular pathways and structural reconformation of tangled proteins as well as introducing reprogramming genes have already shown promise and are likely to deliver more tailored approaches to the treatment of PD in the future."

Introduction

"Current treatment strategies for Parkinson's disease (PD) largely focus on relieving motor symptoms by increasing dopamine levels within the central nervous system (CNS) or by stimulating dopamine receptors."

"We will focus on nanoparticles that may impact on future PD therapies."

"We will first describe the composition and structure of the various subtypes of nanoparticles that have been developed for PD, highlighting polymer-based, lipid-based and inorganic agents."

"We will then outline how nanoparticles may be applied in the future treatment of PD from a functional perspective."

Nanoparticle composition

"Polymeric nanoparticles can be composed of synthetic or naturally occurring polymers and are organised into a variety of structures, which have been shown to be biocompatible, biodegradable, and non-toxic [78]."

"Of the clinically approved nanoparticles, PEG is the most widely used polymer as it is used to coat nanoparticles as well as to stabilise proteins in more conventional therapies [79]."

"There are few other examples of polymeric nanoparticles in clinical use with the majority of current trials being conducted in cancer therapies [80]."

"Despite the proof of concept in these clinical trials, polymeric nanoparticles have not progressed beyond preclinical studies in Parkinson's disease."

"Newer liposomal drugs are available, but there is limited clinical availability of the other types of lipid nanoparticles."

"Despite the emerging clinical use of exosomes and the wide clinical usage of liposomes, research into lipid-based nanoparticles for PD is currently limited to preclinical studies."

Application to Parkinson's Disease

"Nanoparticles have been trialled around a number of different rationales including providing the sustained release of conventional PD therapies, evading the immune system and facilitating entry into the CNS [81]."

"Potential benefits of nanotherapies extend beyond simply improving the pharmacokinetic properties of conventional treatments to the delivery of novel therapies that could target particular intracellular pathways (e.g. oxidative stress, inflammation) or even specific genes [82]."

Sustained release

"The earliest research into nanoparticles for PD involved improving the release profile of dopamine and then other dopaminergic medications, such as levodopa, dopamine agonists and monoamine oxidase B inhibitors."

"These liposomes released dopamine for 40 days with levels that remained elevated for 25 days following implantation."

"A progression of studies resulted in a PLGA-PLA nanosphere being developed in 2012 that allowed for a once weekly administration of subcutaneous levodopa, resulting in elevated plasma dopamine levels that were sustained for 20 days in 6-OHDA lesioned rats."

"In an effort to avoid subcutaneous administration, more recent research has focused on intranasal formulations of levodopa-loaded PLGA nanoparticles, which appear to offer motor benefits that can persist for at least a week following administration [83]."

"Most of the recent research into extending the release of conventional PD therapies has been with polymeric nanoparticles rather than liposomes, which make up the majority of clinically approved nanoparticles in other diseases."

Immunoevasion

"Coating nanoparticles with polyethylene glycol (PEG) can bestow a 'stealth property', which prolongs peripheral circulation [80]."

"PEG is used as an adjunct with other nanoparticles such as liposomes or as components of polymeric micelles rather than directly conjugated to PD drugs."

"Pegylated formulations of conventional PD therapies might be destined to follow the same fate as pegylated insulin, whose development was ceased despite completion of Phase III studies."

"This does not mean that pegylation is not a viable strategy, as evidenced by the numerous pegylated therapies available in other conditions, but financial incentives might favour research into disease modifying therapies in PD rather than adding to the numerous symptomatic treatments that are currently available."

Mucosal entry

"Nanoparticles can optimise intranasal delivery via the use of mucoadhesive polymers, such as chitosan or by conjugating nanoparticles with ligands capable of binding to the nasoepithelial surface, such as lectins."

"Chitosan nanoparticles have been used for intranasal formulations of bromocriptine [84], levodopa [85], pramipexole [86], selegeline [86], rotigotine [87] and ropinirole [88]."

"Conjugating nanoparticles with ligands capable of binding to the nasoepithelium can also enhance nasal absorption and have been used to deliver anti-PD drugs intranasally."

"In one study, fluorescence labelling was able to show that odorranalectin conjugation increased the uptake of urocortin-containing nanoparticles in the brain (sustained and continued to increase over 8 h) compared to those without odorranalectin [89]."

"Another method of improving intranasal drug absorption has been the incorporation of nanoparticles into thermosensitive gels, which improve the intranasal residence time of the drug due to increased viscosity and hence absorption."

Blood–brain barrier entry

"Enhanced entry is usually achieved by functionalising the surface of nanoparticles with ligands known to target receptors on the BBB."

"Intranasally administered rotigotine-loaded nanoparticles conjugated with lactoferrin have shown increased concentration in the striatum over other areas in the brain such as the hippocampus [90, 91]."

"One study has demonstrated that urocortin-loaded Lf-conjugated PEG-PLGA nanoparticles were capable of increasing brain uptake by threefold, when compared to unconjugated nanoparticles [92]."

"These lipoproteins, such as apolipoprotein B and/or E, then facilitate entry into the brain by receptor-mediated endocytosis."

"PS80-coated chitosan nanoparticles have been used to enhance the entry of ropinirole, which typically crosses the BBB poorly due to its hydrophilicity [93]."

"This suggests that PS80-coated nanoparticles not only improved CNS entry across the BBB but also reduced opsonisation by the mononuclear phagocytic system, allowing more effective drug delivery."

Cell-targeting

"The conjugation of specific ligands on nanoparticles can be used to promote selective uptake into dopaminergic neurons."

"The Rabies Virus Glycoprotein (RVG29) is thought to facilitate internalisation into neurons by specifically binding to the nicotinic acetylcholine receptor (nAchR), which is found on dopaminergic neurons, as well as the extracellular surface of the brain's microvascular endothelial cells."

"Studies have demonstrated that RVG29 can be conjugated with liposomes to deliver a dopamine derivative to the substantia nigra and striatum in PD rodents with a higher specificity than other neurons and in greater concentrations than was achieved with non-conjugated liposomes [94]."

"This technique of conjugating nanoparticles with ligands capable of being internalised within specific neurons offers advantages when looking to target pathogenic intracellular processes (see below)."

Triggered release

"To combat off-target toxicity and adverse effects, nanoparticles can be modified to release drug in selective environments, such as inside intracellular organelles."

"The boronate ester undergoes dynamic changes in acidic environments, so that the drug is protected from release in the circulation but is rapidly released when taken into the acidic environment of the endolysosomes within neurons [95]."

"A triggered release nanoparticle ThermoDox®, which is a thermosensitive liposome that releases doxorubicin in response to temperatures at 40 °C, is currently in a phase III study for the treatment of hepatocellular carcinoma in combination with radiofrequency ablation, which provides the external heat required for activation [96, 97]."

Pathological targets

"Other examples of anti-α-synuclein aggregation are reasonably diverse but aggregates of several to hundreds of gold atoms called 'gold nanoclusters' have been reported to inhibit α-synuclein fibrillation, reverse dopaminergic neuronal loss and improve motor behaviour in MPTP-induced mice [98], whilst melatonin-loaded polymeric nanoparticles made from dopamine polymers (polydopamine) have demonstrated a cytoprotective effect and suppress α-synuclein phosphorylation in vitro [99]."

"Early results in MPTP-induced mice have demonstrated a reduction in striatal iron, a reduction in α-synuclein levels, a reversal of dopaminergic neuronal loss and an improvement in motor deficits [100]."

"The same group utilised a synthetic peptide called angiopep conjugated to a poly-L-lysine dendrimer to successfully deliver GDNF intravenously to the CNS of PD-induced rats resulting in improved locomotor activity and reduced dopaminergic neuronal loss [101]."

"The combination of FUS with microbubbles complexed to liposomes carrying GDNF significantly improved behaviour of MPTP-induced mice and reduced dopaminergic neuronal loss when compared with liposomal delivery alone."

"This resulted in reduced α-synuclein expression, dopaminergic neuronal loss and improvements in behaviour in MPTP-induced mice."

The Future

"As our understanding of the pathogenic mechanisms of PD improves, nanoparticles could be designed to replace or repair specific pathological processes."

"Various mechanisms of sensing the environment and triggered release are already being studied including using focused ultrasound, external heat or local pH. It is foreseeable that techniques to respond to local dopamine levels or deep brain stimulation cues are possible."

"Despite the enormous potential demonstrated in preclinical studies, there is still a paucity of clinical translation for even the most basic applications of nanoparticles to PD therapy [80]."

"Addressing these challenges is essential before the potential benefit of nanoparticles in the treatment of PD can be realised."

Thin film hydration versus modified spraying technique to fabricate intranasal spanlastic nanovesicles for rasagiline mesylate brain delivery: Characterization, statistical optimization, and in vivo pharmacokinetic evaluation [102] This is a machine-generated summary of:

Ali, Mohamed Mahmoud; Shoukri, Raguia Aly; Yousry, Carol: Thin film hydration versus modified spraying technique to fabricate intranasal spanlastic nanovesicles for rasagiline mesylate brain delivery: Characterization, statistical optimization, and in vivo pharmacokinetic evaluation [102]

Published in: Drug Delivery and Translational Research (2022)

Link to original: https://doi.org/10.1007/s13346-022-01285-5

If you want to cite the papers, please refer to the original.

For technical reasons we could not place the page where the original quote is coming from.

Abstract-Summary "This study aims to form RM-loaded spanlastic vesicles for intranasal (IN) administration to overcome its hepatic metabolism and permit its direct delivery to the brain."

"RM-loaded spanlastics were prepared using thin film hydration (TFH) and modified spraying technique (MST)."

"The optimized system prepared using MST (MST 2) has shown higher desirability factor with smaller PS and higher EE%; thus, it was selected for further in vivo evaluation where it revealed that the extent of RM distribution from the intranasally administered spanlastics to the brain was comparable to that of the IV drug solution with significantly high brain-targeting efficiency (458.47%)."

"These results suggest that the IN administration of the optimized RM-loaded spanlastics could be a promising, non-invasive alternative for the efficient delivery of RM to brain tissues to exert its pharmacological activities without being dissipated to other body organs which subsequently may result in higher pharmacological efficiency and better safety profile."

Introduction

"The intranasal (IN) route for drug administration has become far superior to other conventional routes in improving the delivery of drugs into the central nervous system (CNS) through BBB [103]."

"The intranasally administered drug may reach the brain directly via the olfactory pathway or after absorption into the lymphatic system and then to the cerebrovasculature to reach the brain [104]."

"Among their numerous advantages, their ability to pass through BBB after intranasal administration is considered one of the gold standard applications of nanoparticles development to study the pathogenesis and alleviate the symptoms of many CNS disorders."

"Nanoparticles are also able to overcome the rapid intranasal mucocilliary clearance due to their rapid absorption aided by their small particle size and elasticity that permit the direct passage of the intranasally administered nanoparticles through the olfactory pathway to brain [105]."

Methods

"The obtained spanlastic vesicular systems were subjected to four consecutive freeze–thaw cycles at –8 °C for 8 h and 25 °C for 1 h aiming to enhance the entrapment of RM inside the nanosystem [106]."

"RM levels in plasma samples and brain homogenates were measured by a validated liquid chromatography-mass spectrometry (LC–MS/MS) method."

"Specific volumes of RM stock solution were mixed with one hundred microliter of clonazepam stock solution (100 ng/mL) and then spiked into 0.5 mL plasma or brain homogenate to obtain the following concentrations: 0.1, 1, 10, 20, 100, and 1000 ng/mL. For RM extraction from the samples, 0.5 mL of plasma samples or brain homogenates was mixed with 100 μL of clonazepam solution as IS and 4 mL of ethyl acetate to precipitate the proteins followed by vortexing (Stuart SA8, BiBBY Sterlin Ltd., UK) for 5 min and centrifugation for 10 min at 3000 rpm."

Results and discussion

"We managed to prepare RM-loaded spanlastics using 2 different techniques, TFH and MST, aiming to produce nanovesicles with small PS and high EE%."

"In our preliminary trials, TFH was successfully adopted to formulate RM-loaded spanlastics with high EE%; however, it was incapable of producing vesicles with PS less than 100 nm; thus, MST was applied to produce vesicles with smaller PS."

"RM-loaded spanlastic nanovesicles were prepared by TFH and MST in which 2^3 factorial design was constructed to study the effects of the different formulation variables on the characteristics of the formulated RM-spanlastic nanovesicles using both techniques."

"ANOVA results have shown that the PS of the formulated spanlastics was significantly affected by the type of Span used with p values of 0.0041 and 0.0017 for TFH and MST, respectively."

Conclusion

"Two different preparation techniques, namely, MST and TFH, were utilized to formulate RM-loaded spanlastic vesicles."

"The investigated formulation variables affecting each technique were statistically characterized, compared, and optimized to prepare physically stable spanlastic vesicular system with small PS (< 100 nm) and high EE%."

"The in vivo pharmacokinetic results revealed that the extent of RM distribution to the brain was comparable for the intranasally administered MST 2 and the IV drug solution; however, the optimized spanlastic vesicular system has shown a significantly high BTE% which indicates the higher proportion of drug reaching the brain relative to the plasma after IN administration compared to the IV route."

Lipid Nanoparticles Improve the Uptake of α-Asarone Into the Brain Parenchyma: Formulation, Characterization, In Vivo Pharmacokinetics, and Brain Delivery [107] This is a machine-generated summary of:

Ramalingam, Prakash; Ganesan, Palanivel; Prabakaran, D. S.; Gupta, Pardeep K.; Jonnalagadda, Sriramakamal; Govindarajan, Karthivashan; Vishnu, Revuri; Sivalingam, Kalaiselvi; Sodha, Srushti; Choi, Dong-Kug; Ko, Young Tag: Lipid Nanoparticles Improve the Uptake of α-Asarone Into the Brain Parenchyma: Formulation, Characterization, In Vivo Pharmacokinetics, and Brain Delivery [107]

Published in: AAPS PharmSciTech (2020)

Link to original: https://doi.org/10.1208/s12249-020-01832-8

If you want to cite the papers, please refer to the original.

For technical reasons we could not place the page where the original quote is coming from.

Abstract-Summary "Our research primarily focuses on the delivery of natural therapeutic compound, α-asarone, for the treatment of brain-related diseases."

"This study aims at formulating a lipid nanoparticulate system of α-asarone (A-LNPs) that could be used as a brain drug delivery system."

"After intravenous administration of A-LNPs or free α-asarone, significantly higher levels of α-asarone from the A-LNPs were detected in murine plasma and brain parenchyma fractions, confirming the ability of A-LNPs to not only maintain a therapeutic concentration of α-asarone in the plasma, but also transport α-asarone across the blood-brain barrier."

"These findings confirm that lipid nanoparticulate systems enable penetration of natural therapeutic compound α-asarone through the blood-brain barrier and may be a candidate for the treatment of brain-related diseases."

INTRODUCTION

"α-Asarone is a highly lipophilic compound that showed higher bioavailability in the brain at the dosage level of lesser than 50 mg/kg for various disease treatments, including anxiety, stress, AD, and PD [108, 109]."

"Dissolution, oral bioavailability, and brain delivery of α-asarone were improved by using various oral drug delivery vehicles like solid dispersions, solid lipid nanoparticles, and self-microemulsifying systems [110–112]."

"Nano-lipid formulations, with their ability to cross BBB, are currently the novel trend in the development of natural nanoparticle system for the treatment of diverse central nervous disorders, such as PD, AD, and other brain-related disorders [113]."

"Among diverse lipid-based nanotechnology, lipid nanoparticles (LNPs) showed better biocompatibility, solubility, and improved delivery of bioactive compounds to the brain [112]."

"Lipid nanoparticulate systems are highly compatible, more stable, and very efficient in the delivery of various lipophilic natural therapeutic compounds to various target sites within the body, including the brain [114]."

MATERIALS AND METHODS

"The A-LNPs containing mouse blood serum samples were incubated at 37°C for 8 h. At predetermined time intervals, the samples were collected and the A-LNPs were separated from mouse blood serum proteins by centrifugation at 14,000 rpm for 10 min and redispersed in phosphate buffer."

"A 100-μL aliquot of the α-asarone solution (100 μg/mL in a mixture of Cremophor EL and ethanol (1:1, v/v)) or A-LNPs (100 μg/mL in distilled water) was injected in mice via the tail vein."

"The volume of distribution (Vd) for brain parenchyma, capillary fraction, and whole brain was calculated by the ratio of the concentration of α-asarone from the brain fractions (C_{BR}) (whole brain, brain parenchyma, and capillary fraction, respectively) to plasma concentration (C_{PL}) of asarone at different time intervals, i.e., Vd (l/kg) = C_{BR} / $C_{PL.}$ Prior to these pharmacokinetic and brain distribution studies, an

LC-MS/MS bioanalytical method was developed and validated to quantify α-asarone from biological matrices [115].”

RESULTS

“The A-LNPs with LC (3.68 ± 0.08%) displayed the smaller particle size, which was suitable for brain drug delivery of α-asarone.”

“The final cumulative releases of A-LNPs and the free α-asarone solution were 50.95 ± 4.85% and 76.74 ± 3.92% after 24 h, respectively.”

“A-LNPs displayed a significantly (3.9-fold higher) increased transport across the BBB, compared to the free α-asarone solution at all time points.”

“The plasma concentration of α-asarone from the A-LNPs was higher compared to the free α-asarone solution.”

“The A-LNPs always had higher Vd of α-asarone at 5, 15, and 30 min in the brain parenchyma than the free α-asarone solution.”

“The Vd of α-asarone of A-LNPs was 4-fold higher than that in the free α-asarone solution at all time points after intravenous administration.”

DISCUSSION

“Wang and others reported that α-asarone-loaded solid lipid nanoparticles significantly improve the α-asarone pharmacokinetics, absorption, and tissue distribution (brain and lung) after oral administration [112].”

“Intranasal administration of PLA-α-asarone nanoparticles showed significantly stronger brain targeting of α-asarone than intravenous administration [116].”

“Our LNPs could provide more opportunities to increase the α-asarone circulation time in the blood compared to conventional drug formulations.”

“Lu J and others reported that the concentrations of α-asarone were not detectable in the brain, whether through oral, intravenous, or intranasal administration [117].”

“It could be seen that the Vd of α-asarone of A-LNPs was from 3- to 4-fold higher than that in the free α-asarone solution following intravenous administration at all time points.”

“LNPs enhance the brain capillary permeability leading to an augment in the parenchymal α-asarone concentration.”

“The LNPs dramatically promoted the availability and retention time of α-asarone in the brain.”

CONCLUSIONS

“Our in vitro results demonstrated increased stability at room and refrigerated conditions for 3 months in addition to the controlled release of α-asarone at physiological conditions for 24 h. Our in vivo studies further confirmed the improved bioavailability of α-asarone in the plasma.”

“We demonstrated that α-asarone readily penetrated the brain parenchyma by crossing the brain capillary endothelium, which forms the BBB.”

“These findings confirm that lipid nanoparticulate systems enable penetration of natural therapeutic compound α-asarone through the blood-brain barrier and may be a candidate for the treatment of brain-related diseases.”

5.2 Targeted Delivery to Respiratory System

Introduction by the Editor

The respiratory system presents a diverse array of molecular targets that are crucial for its functioning and play an important role in addressing various respiratory disorders. Beta-adrenergic receptors regulate bronchial smooth muscle tone and are targeted in asthma and chronic obstructive pulmonary disease (COPD) therapies. Muscarinic receptors play a role in bronchoconstriction and are targeted for delivering bronchodilator drugs. Epithelial sodium channels (ENaCs) are implicated in maintaining airway surface hydration and are potential targets for cystic fibrosis treatment. Cytokines like interleukins and tumor necrosis factor-alpha (TNF-alpha) are involved in airway inflammation and are targeted in asthma. In infections, respiratory viruses often target angiotensin-converting enzyme 2 (ACE2) receptors in the airways. These molecular targets offer avenues for therapeutic intervention and diagnostic approaches in respiratory disorders. In the context of nanoparticles targeting molecular targets in the respiratory system, specific polymers and ligands can play crucial roles in achieving effective drug delivery. Polymers like poly(lactic-co-glycolic acid) (PLGA), chitosan, and polyethylene glycol (PEG) can be utilized to formulate nanoparticles that enhance drug stability, control release, and improve biocompatibility. Ligands such as glycyrrhetinic acid or folic acid can be conjugated to nanoparticles, allowing them to selectively bind to overexpressed receptors on lung cells. Targeting β2-adrenergic receptors using ligands enhances bronchodilator delivery to treat respiratory diseases like asthma. Similarly, surface modification with hyaluronic acid facilitates the interaction of nanoparticles with CD44 receptors in inflamed airways. By combining suitable polymers and ligands, nanoparticles can be engineered to home in on specific molecular targets in the respiratory system, enhancing drug delivery precision and therapeutic outcomes.

Machine generated summaries

Disclaimer: The summaries in this chapter were generated from Springer Nature publications using extractive AI auto-summarization: An extraction-based summarizer aims to identify the most important sentences of a text using an algorithm and uses those original sentences to create the auto-summary (unlike generative AI). As the constituted sentences are machine selected, they may not fully reflect the body of the work, so we strongly advise that the original content is read and cited. The auto generated summaries were curated by the editor to meet Springer Nature publication standards.

To cite this content, please refer to the original papers.

Machine generated keywords: pulmonary, lung, antimicrobial, infection, tract, attractive, lung cancer, anticancer, plga, cancer, nanoparticle, high bioavailability, emerge, modify, intracellular

Nanoparticle-mediated pulmonary drug delivery: state of the art towards efficient treatment of recalcitrant respiratory tract bacterial infections [118] This is a machine-generated summary of:

Huang, Zheng; Kłodzińska, Sylvia Natalie; Wan, Feng; Nielsen, Hanne Mørck: Nanoparticle-mediated pulmonary drug delivery: state of the art towards efficient treatment of recalcitrant respiratory tract bacterial infections [118]

Published in: Drug Delivery and Translational Research (2021)

Link to original: https://doi.org/10.1007/s13346-021-00954-1

If you want to cite the papers, please refer to the original.

For technical reasons we could not place the page where the original quote is coming from.

Abstract-Summary "Successful aerosolized antimicrobial therapy is still challenged by the diverse biological barriers in infected lungs."

"Nanoparticle-mediated pulmonary drug delivery is gaining increasing attention as a means to overcome the biological barriers and accomplish site-specific drug delivery by controlling release of the loaded drug(s) at the target site."

"With the aim to summarize emerging efforts in combating respiratory tract infections by using nanoparticle-mediated pulmonary delivery strategies, this review provides a brief introduction to the bacterial infection-related pulmonary diseases and the biological barriers for effective treatment of recalcitrant respiratory tract infections."

"This is followed by a summary of recent advances in design of inhalable nanoparticle-based drug delivery systems that overcome the biological barriers and increase drug bioavailability."

Introduction

"Respiratory tract infections that may occur as a result of such exposure, if pathogenic microorganisms are not cleared from the lungs upon inhalation, can be categorized into upper respiratory tract infections and lower respiratory tract infections."

"This article focuses on the most recent developments in nanoparticle-mediated pulmonary drug delivery aiming at efficient treatment of respiratory tract bacterial infections."

"Encapsulating antimicrobial agents into nanoparticles intended for inhalation offers (i) protection of the antimicrobial agents from deactivation caused by the harsh local microenvironment in lungs with chronic bacterial infections (e.g., pH value, enzymes); (ii) decreased risk of adverse effects by reducing the drug exposure to the rest of body; (iii) controlled and potentially sustained drug release (i.e., prolonged residence time in lungs, which ultimately will impact patient compliance."

"This review summarizes emerging efforts towards combating bacterial infections in the respiratory tract by using nanoparticle-based pulmonary delivery strategies, mainly focusing on lipid- and polymer-based nanoparticles owing to their good biocompatibility and translational perspective."

Bacterial infection-related pulmonary diseases

"Development of effective nanoparticle-based pulmonary drug delivery strategies to combat respiratory tract infections necessitates delicate consideration of the type of pathogens, the affected area within the respiratory tract, the pathophysiological progression and local microenvironment that the disease associates with."

"Preventing or postponing chronic pulmonary colonization by P. aeruginosa is among the primary aims in early CF treatment [119]."

"Bacterial infections caused by H. influenzae, Moraxella catarrhalis, Streptococcus pneumoniae and P. aeruginosa are clearly associated with the acute exacerbations of COPD [120, 121], constituting the main cause of mortality among COPD patients [122]."

"The experience learnt from treating CF patients might also be useful for developing new approaches for the prevention and treatment of P. aeruginosa infections in COPD."

"It is noteworthy that biofilms (e.g., of P. aeruginosa) will gradually form on the inner surface of the endotracheal tube and ventilator cycling can propel the biofilms and secretions to the distal airways, leading to persistent bacterial infections [123]."

Biological barriers to effectively targeted delivery of antimicrobials

"Lung lining fluid is distributed continuously throughout the respiratory tract and is heterogeneous regarding its molecular composition and thus properties depending on whether the localization is the conducting parts (trachea, bronchi, and bronchioles) or the alveoli."

"The pH value of lung lining fluid in the proximal part (on the surface of the mucus matrix) as well as the distal part of the respiratory tract under normal conditions is close to neutral."

"The respiratory tract diseases induce overproduction and dehydration of mucus with important impact on the interaction between mucus and the drug molecule or the drug delivery system administered to the lungs [124]."

"In addition to the non-cellular barriers (e.g., respiratory tract mucus), nanoparticle delivery systems have to overcome the cellular and intracellular barriers, including host cell membrane, efflux pumps, exocytosis and endosomal degradation, to improve the penetration into and retention of antimicrobials inside host cells [125]."

Advances in design of nanoparticle delivery systems

"It is debatable if the pH-stimulated reversible PEGylation of liposomes is optimal for a combination of mucus-penetrating properties with targeted drug delivery to bacteria and infected cells."

"PH-sensitive, surface charge-switching nanoparticles made of PLGA, poly-L-histidine (PLH) and PEG have been developed and investigated to promote the targeting to bacteria through electrostatic interactions [126]."

"In spite of the great promise demonstrated in in vitro cultured bacterial biofilms, the surface charge-adaptive approach based on pH sensitive polymers may be not applicable for aerosol antimicrobial delivery owing to the fact that the pathological microenvironment in lungs of the patients with chronic respiratory tract infections, such as the acidic pH level (approximately 5.5–6.5) and the elevated salt concentration of lung lining fluid, could lead to mistargeting of the nanoparticles."

"Lipid bilayer-enveloped polymeric nanoparticles demonstrated intensive interaction with bacterial biofilm and effective mucus penetration."

General consideration on the translation from exploratory research to clinical application

"The drug loading capacity of nanoparticles and how much of the drug delivery system can be inhaled determines how high doses of the antimicrobials can be delivered to the site of action."

"For sustained-release nanoparticles, sufficient drug loading is thus a prerequisite for not only reaching but also maintaining therapeutic concentrations of antimicrobials within a longer period of time, and if successful, this may naturally reduce the dosing frequency."

"A recent work demonstrated that pulmonary delivery of fluticasone propionate formulated in mucus-penetrating nanoparticles achieved a higher local exposure in lungs of rodents compared to that achieved with both non-formulated drug and with a mucoadhesive formulation with similar particle size and in vitro drug release profile [127]."

"To effectively and rationally optimize the drug release kinetics, standardized in vitro dissolution and release testing methods for inhalable formulations are highly necessary."

Conclusions

"Nanoparticle-mediated aerosol antimicrobial therapies may pave the way for breakthroughs, yet sufficient improvements in efficacy requires their effective penetration through the mucus and localization adequately close to the bacteria, followed by release of sufficient amounts of antimicrobials to maintain a favorable pharmacokinetics/pharmacodynamics (PK/PD) profile at the site of action."

"The progress in the fields of materials science and nanotechnology has led to a variety of innovative nanoparticle-based drug delivery systems with controllable properties, which potentially allow for effectively overcoming the delivery barriers and improving the PK/PD at the site of action."

"The majority of studies are still in the early stages of the drug development process and translation to both industrial production scale and in vivo testing needs addressing."

"Overcoming the aforementioned obstacles will lead to safer and more efficient nanoparticle-mediated aerosol antimicrobial therapy entering the clinical phases of drug development."

Novel Silibinin Loaded Chitosan-Coated PLGA/PCL Nanoparticles Based Inhalation Formulations with Improved Cytotoxicity and Bioavailability for Lung Cancer [128] This is a machine-generated summary of:

Raval, Mihir; Patel, Priya; Airao, Vishal; Bhatt, Vaibhav; Sheth, Navin: Novel Silibinin Loaded Chitosan-Coated PLGA/PCL Nanoparticles Based Inhalation Formulations with Improved Cytotoxicity and Bioavailability for Lung Cancer [128]

Published in: BioNanoScience (2020)

Link to original: https://doi.org/10.1007/s12668-020-00797-z

If you want to cite the papers, please refer to the original.
For technical reasons we could not place the page where the original quote is coming from.

Abstract-Summary "The aim of the present investigation was to develop, characterize, and evaluate chitosan-coated PLGA/PCL nanoparticles containing Silibinin (SB) intended for pulmonary delivery for treating lung cancer."

"The prepared nanoparticles (NPs) were evaluated for their physicochemical characteristic along with DSC, FTIR, and SEM."

"The anticancer activity of SB-loaded NPs was assessed in the human A549 lung cancer cell line utilizing MTT assay and anti-cancer potential assessed by clonogenic assay."

"The pharmacokinetics and tissue distribution studies of SB-loaded NPs were assessed in comparison with the SB solution."

"Chitosan-coated NPs showed the sustain release effect up to 48 h with an aerodynamic particle size of 1.82 μm."

"A tremendous increase in cell inhibition was determined by chitosan-coated PLGA NPs."

"Pharmacokinetics study showed that chitosan coating onto PLGA nanoparticles promoted to release drug in the lungs and increased vivo residence time."

"These nanoparticles coated with chitosan could open a new avenue for effective treatment of lung cancer."

Introduction

"Chemotherapy as a traditional treatment for lung cancer has inconsistent results due to drug resistance and it is very important that these therapies be modified using modern methods [129]."

"An alternative method was developed to modify the approach such as pulmonary and dosage type like nanoparticles for successful lung cancer treatment."

"Jiang and others prepared PCL composed nanoparticles and surface modified with chitosan for oral administration in lung cancer."

"They investigated that due to the mucoadhesive property of chitosan improved the therapeutic effect by interacting with mucin expressed which is present in cancer cells [130]."

"An attempt was made to prepared chitosan-coated PLGA/PCL nanoparticles in inhalation powder form to improved anticancer activity as well as the maximum amount of drug targeting to the lung."

"The cytotoxicity of CS-coated SB-loaded PLGA and PCL NPs was investigated in the A549 lung cancer cell line."

Materials and Methods

"The cells were incubated with samples for 24 h. Later the formulations were replaced with MTT (5 mg/mL) and cells were incubated for an additional 4 h at 37 °C."

"2-mL blood sample was withdrawn from rats and centrifuged at 10000 rpm for 10 min, and the plasma layer was removed."

"2 µg/mL was selected as a targeted concentration for SB nanoparticles to maintain plasma concentration above the minimum inhibitory concentration of 48 h. Dose for rat was calculated from the human equivalent dose."

"Targeted concentration for SB nanoparticle formulations selected as 2 µg/mL to maintain plasma concentration above minimum inhibitory concentration (MIC) of 48 h. The estimation of the equivalent dose for rat had been performed from the human dose."

Results and Discussion

"CS is more hydrophilic than PLGA, which results in more hydration of the SB-NPs matrix and more drugs released during the same time compared to PCL."

"Both CS-SB-PLGA NPs (loaded with SB F2) and CS-SB-PCL NPs (loaded with SB F4) formulations showed significantly higher cytotoxicity in A549 cells at 48-h treatment period, as compared to the uncoated NPs and pure SB."

"The drug level was observed at different time intervals up to 48 h. Our formulation has shown that the plasma concentration of SB exceeds the minimum inhibitory concentration required to treat lung cancer effectively."

"In both plasma and lung, mean residence time (MRT) and the half-life (t1/2) of SB were found to increase for chitosan-coated nanoparticles as compared with free drug."

"This may be attributed to the slower release and comparatively longer MRT of drug from the chitosan-coated formulation in plasma and lung tissue."

Conclusion

"Optimized SB nanoparticles further modified to inhalation form with a fine particle fraction 80.2% and signified its suitability in effective delivery for pulmonary administration."

"Nanoparticles penetrated the target site in deep lung tissue and transmitted drugs for both local and systemic acts by encouraging cellular adhesion and the delivery system retention."

"The in vivo pharmacokinetic study in rats strongly indicates a marvelous improvement in both the rate and extent of SB bioavailability from chitosan-coated PLGA nanoparticles."

"Chitosan coating on PLGA nanoparticles are promising for improving bioavailability for efficient pulmonary delivery and can be useful in treating lung cancer."

Clinical Pharmacokinetics of Inhaled Antimicrobials [131] This is a machine-generated summary of:

Stockmann, Chris; Roberts, Jessica K.; Yellepeddi, Venkata K.; Sherwin, Catherine M. T.: Clinical Pharmacokinetics of Inhaled Antimicrobials [131]

Published in: Clinical Pharmacokinetics (2015)

Link to original: https://doi.org/10.1007/s40262-015-0250-x

If you want to cite the papers, please refer to the original.
For technical reasons we could not place the page where the original quote is coming from.

Abstract-Summary "These unique pharmacokinetic characteristics make inhaled antimicrobial delivery attractive for the treatment of many pulmonary diseases."

"This review examines recent pharmacokinetic trials with inhaled antibacterials, antivirals and antifungals, with an emphasis on the clinical implications of these studies."

"Many no vel inhaled antimicrobial therapies are currently under investigation that will require detailed pharmacokinetic studies, including combination inhaled antimicrobial therapies, inhaled nanoparticle formulations of several antibacterials, inhaled non-antimicrobial adjuvants, inhaled antiviral recombinant protein therapies and semi-synthetic inhaled antifungal agents."

"The development of new inhaled delivery devices, particularly for mechanically ventilated patients, will result in a pressing need for additional pharmacokinetic studies to identify optimal dosing regimens."

Introduction

"In both circumstances, the anatomy of the airway, the physical conditions that the drug will encounter in the airway (e.g. humidity), the clearance mechanisms of the lung (e.g. mucociliary clearance and alveolar macrophages) and the pathophysiological effects of acute and chronic diseases (e.g. pulmonary tissue scarring in chronic obstructive pulmonary disease [COPD]) must be considered when an inhaled medication is being developed [132]."

"Many lower respiratory tract infections feature both purulent tracheobronchitis and alveolar disease, which require deposition of the inhaled antimicrobial throughout the lungs [133]."

"Following deposition, absorption of the inhaled drug from a solution on the airway surface can be described quantitatively with the following equation: In this 'irreversible transfer relationship', it is assumed that the systemic concentration of the drug of interest is negligible [134]."

"It is now increasingly recognized that the development of a safe and effective inhaled antimicrobial requires consideration of optimization of the whole system, including the drug, the formulation and the aerosol delivery device [135, 136]."

Antibacterials

"In a phase I study of an inhaled fosfomycin and tobramycin combination, a limited noncompartmental pharmacokinetic analysis presented at the 2008 Annual North American Cystic Fibrosis Conference identified a target dose of fosfomycin 160 mg and tobramycin 40 mg in a 4 mL volume, which was advanced into phase II trials; however, more detailed results have yet to be published [137]."

"Two parallel, randomized, placebo-controlled, phase II trials were conducted to evaluate the pharmacokinetics, safety and efficacy of once daily inhaled liposomal amikacin for the treatment of P. aeruginosa infections in patients ≥6 years of age with cystic fibrosis [138]."

"Several clinical pharmacokinetic studies have been conducted with inhaled aztreonam lysine in patients with cystic fibrosis, including a phase Ib dose-escalation trial, a phase II trial and two phase III trials [139–141]."

Antivirals

"A clinical pharmacokinetic study involving healthy adult volunteers evaluated four zanamivir dosing regimens, including inhalation via a nebulizer and a DPI [142]."

"Laninamivir octanoate has been approved for the treatment and prevention of influenza infection in Japan and is currently in phase II clinical trials in the USA (ClinicalTrials.gov study identifier NCT01793883) [143, 144]."

"In phase I clinical pharmacokinetic studies, laninamivir octanoate was administered to healthy male adult volunteers in (1) a series of single ascending doses; (2) a single high dose; and (3) a series of multiple ascending doses [145]."

"The doses administered ranged from 5 to 120 mg, with a proportional increase in the half-life ranging from 6 to 81 h. In further studies, it was observed that patients with impaired renal function had increased AUCs following administration of inhaled laninamivir octanoate [146]."

Antifungals

"Amphotericin B is a first-line agent for the treatment of life-threatening fungal infections, including those caused by Candida albicans and Aspergillus fumigatus [147]."

"The pharmacokinetics of Abelcet® (a lipid-based formulation of amphotericin B) were examined in 35 lung transplant recipients [148]."

"Concentrations of amphotericin B in bronchoalveolar lavage specimens remained high enough to inhibit the growth of most Aspergillus species over a 14-day treatment period [149]."

"Similar to findings from studies involving amphotericin B deoxycholate and Abelcet®, serum concentrations of amphotericin B following aerosolized administration of AmBisome® were undetectable in all but one patient, and that patient's concentration was 0.1 μg/mL (slightly above the lower limit of quantitation) [149]."

"These results demonstrate the superior pharmacokinetic profile of aerosolized amphotericin B formulations as compared with parenterally administered amphotericin B for the treatment of pulmonary fungal infections."

Aerosolized Delivery Devices

"These nebulization methods require very little patient coordination or skill, although pulmonary deposition is limited to approximately 10 % of the total dose [150]."

"These devices have also been customized for use in mechanically ventilated patients, and, when paired with a valved spacer, have been reported to result in up to 70 % of the inhaled dose being deposited in the lungs [151]."

"Because of the high velocity and large particle size of the drug aerosol, only 10–20 % of the emitted dose is deposited in the lungs."

"It has also been clearly demonstrated that the pattern of pulmonary deposition with an MDI depends on the patient's breathing pattern and inspiratory flow rate [152]."

"This technology has been successfully used for inhaled tobramycin, in which the use of a DPI achieved more than 3-fold higher pulmonary deposition than a jet nebulizer [153]."

Conclusions and Future Directions

"The clinical pharmacokinetics of many inhaled antimicrobials have been rigorously studied using noncompartmental methods."

"Sputum pharmacokinetics are marked by high variability for many inhaled antimicrobials, which makes it challenging to interpret the results of sputum pharmacokinetic studies."

"Despite these challenges, clinical pharmacokinetic studies of inhaled antimicrobials have informed the development of safe and therapeutic dosing regimens, led to the creation of novel aerosolized formulations and aided in designing next-generation aerosolization devices."

"Over the course of the next 5–10 years, many vitally important clinical pharmacokinetic studies will be needed to evaluate novel inhaled nanoparticle formulations of several antibacterials, including amikacin, ciprofloxacin, tobramycin and vancomycin [154, 155]."

"As we look to the future, there are a number of inhaled antimicrobials currently under investigation for which clinical pharmacokinetic studies will be needed to establish evidence of safety and efficacy."

5.3 Targeted Delivery in Cancer

Introduction by the Editor

Cancer encompasses a complex landscape of molecular targets crucial for tumor development, progression, and treatment. Some notable targets include oncogenes like Human epidermal growth factor receptor 2 (HER2), Epidermal Growth Factor Receptor (EGFR), and Kirsten rat sarcoma gene (KRAS), which are often mutated or overexpressed in various cancers. Tumor suppressor genes such as tumor protein 53 (TP53) are commonly mutated, leading to unchecked cell growth. Angiogenesis-related factors like Vascular endothelial growth factor (VEGF) and its receptors are targeted to inhibit blood vessel formation in tumors. Immune checkpoint proteins like programmed cell death protein 1 (PD-1) and cytotoxic T-lymphocyte–associated antigen 4 (CTLA-4) regulate immune responses and are targeted to enhance anti-tumor immunity. Additionally, DNA repair enzymes like poly-ADP ribose polymerase (PARP) are targeted in certain cancers with defects in DNA repair pathways. These molecular targets serve as critical points for precision medicine, allowing for tailored therapies aimed at disrupting the molecular machinery of cancer and improving patient outcomes. In the realm of nanoparticles targeting molecular targets in cancer tumors, specific polymers and ligands are pivotal for achieving effective and targeted drug delivery. Ligands such as antibodies or peptides can be conjugated to nanoparticles, enabling specific recognition and binding to overexpressed receptors on cancer cells. For instance, epidermal growth factor (EGF) ligands are utilized to target EGFR-overexpressing tumors. Similarly, ligands targeting folate receptors or integrins can facilitate nanoparticles' interaction with cancer cells. The combination of these tailored polymers and ligands empowers nanoparticles to navigate the complexities of the tumor microenvironment, ensuring precise drug delivery and potentially enhancing therapeutic outcomes in cancer treatment.

Machine generated summaries

Disclaimer: The summaries in this chapter were generated from Springer Nature publications using extractive AI auto-summarization: An extraction-based summarizer aims to identify the most important sentences of a text using an algorithm and uses those original sentences to create the auto-summary (unlike generative AI). As the constituted sentences are machine selected, they may not fully reflect the body of the work, so we strongly advise that the original content is read and cited. The auto generated summaries were curated by the editor to meet Springer Nature publication standards.

To cite this content, please refer to the original papers.

Machine generated keywords: cancer, solid, patient solid, paclitaxel, nanoparticulate, albumin, effort, inhibitor, aspect, utility, serum, japan, pkpd, expand, cycle

Pharmacokinetic aspects of the clinically used proteasome inhibitor drugs and efforts toward nanoparticulate delivery systems [156] This is a machine-generated summary of:

Kwon, Seungbin; Kim, Kyung Bo; Yeo, Yoon; Lee, Wooin: Pharmacokinetic aspects of the clinically used proteasome inhibitor drugs and efforts toward nanoparticulate delivery systems [156]

Published in: Journal of Pharmaceutical Investigation (2021)

Link to original: https://doi.org/10.1007/s40005-021-00532-0

If you want to cite the papers, please refer to the original.

For technical reasons we could not place the page where the original quote is coming from.

Abstract-Summary "There are three clinically used proteasome inhibitor drugs, namely bortezomib, carfilzomib, and ixazomib."

"Efforts are ongoing to overcome the drawbacks of the existing proteasome inhibitor drugs, optimize their pharmacokinetic aspects, and expand their clinical utility beyond the current indications, in particular for solid cancer therapy."

"This review summarizes the pharmacokinetic aspects of the clinically used proteasome inhibitor drugs and the notable findings from the recent reports on the novel nanoparticulate delivery systems of bortezomib and carfilzomib."

"With the help of novel nanoparticulate delivery systems, the therapeutic utility of the proteasome inhibitor drugs is likely to expand to various types of cancer and other pathological conditions including neurodegenerative and inflammatory diseases."

Introduction

"Such efforts have brought about the successful development of proteasome inhibitor (PI) drugs, which have transformed the treatment landscape for multiple myeloma (MM) and prolonged the survival of MM patients."

"Several PI drug candidates are under preclinical and clinical development and efforts are ongoing to develop therapies targeting other UPS components such as ubiquitin ligases and deubiquitinases (Di Costanzo and others [157])."

"Encouraged by the remarkable clinical success of the PI drugs against MM, considerable effort has continued to expand the utility of PI therapies to other types of cancer, in particular solid cancers."

"The lack of therapeutic benefits of PI drugs in patients with solid cancers has been attributed in part to their poor pharmacokinetic (PK) profiles, such as short circulation time in vivo and insufficient distribution of active drugs to the proteasome target in solid cancer (Huang and others [158])."

Proteasome inhibitor drugs in clinical use

"This concern was cleared up during the development of the first-in-class PI drug BTZ [cancer cells especially derived from MM were more responsive to proteasomal inhibition than non-malignant cells, likely due to the elevated proteotoxic stress in rapidly proliferating cells (Hanahan and Weinberg [159]; Oakes [160]); detailed account of the journey to the US FDA approval of BTZ with record efficiency was previously documented (Sanchez-Serrano [161])]."

"The clinical success of BTZ reaching a blockbuster drug status has prompted the subsequent development of CFZ and IXZ."

"Such concern was however cleared up when CFZ showed minimal interactions with non-proteasomal targets and random nucleophiles, and much improved safety profiles in patients over BTZ (Arastu-Kapur and others [162])."

"As there was a clear need for an oral PI drug and a body of knowledge gained by the successful development of BTZ, IXZ with encouraging preclinical results rapidly advanced to clinical development, leading to the US FDA approval with a record efficiency (only 6 years after the initiation of the first phase I clinical trials) (Kupperman and others [163])."

Novel nanoparticulate drug delivery systems for BTZ and CFZ

"Is a summary of notable findings from the recent reports on novel nanoformulations for BTZ and CFZ. (In the case of the orally administered IXZ, there has been no report on nanoformulation development efforts.) Earlier studies investigated the feasibility of liposomes and polymeric nanoparticles for BTZ delivery (Ashley and others [164, 165]; Swami and others [166]; Shen and others [167]), but one of the major limitations was the low drug loading content, posing a potential difficulty in clinical application (i.e., administration of therapeutic doses requiring a large amount of polymeric carrier molecules)."

"Despite encouraging in vitro results, the in vivo performances of the polymeric micelle showed no notable improvements over the CD-based formulation (in vivo tumor growth suppression, plasma PK profiles, proteasome inhibition in residual tumor tissues), except for a trend of improved tolerability (Park and others [168])."

"A take-home message from this study was that the in vivo assessment of CFZ nanoparticles requires careful examination of drug biodistribution, drug release kinetics, target engagement, and anticancer efficacy."

Conclusion

"In the quest to identify the optimal nanoformulations of PI drugs, it would be important to assess the PK and biodistribution profiles of the active drug that can engage with the proteasome target."

"In the nanomedicine field, the ongoing challenge has been how to distinguish and compare the fractions of encapsulated, unencapsulated, and unbound drugs in assessing the PK and PD profiles of different nanoformulations."

"With the advances and availability of cost-effective in vivo imaging techniques, recent studies often assess the drug biodistribution profiles by whole-body imaging only."

"The case of CFZ and other covalent-modifier drugs may require careful consideration in assessing the drug biodistribution profiles."

"With continuing efforts to develop more effective and safe PI drugs and nanoformulations, it will be possible to realize the full therapeutic potential of PI drugs in MM, other types of cancer, and beyond (e.g., neurodegenerative and inflammatory diseases)."

Pharmacokinetic and pharmacodynamic analysis of neutropenia following nab-paclitaxel administration in Japanese patients with metastatic solid cancer [169] This is a machine-generated summary of:

Tsushima, Takahiro; Kasai, Hidefumi; Tanigawara, Yusuke: Pharmacokinetic and pharmacodynamic analysis of neutropenia following nab-paclitaxel administration in Japanese patients with metastatic solid cancer [169]

Published in: Cancer Chemotherapy and Pharmacology (2020)

Link to original: https://doi.org/10.1007/s00280-020-04140-x

If you want to cite the papers, please refer to the original.

For technical reasons we could not place the page where the original quote is coming from.

Abstract-Summary "To develop a pharmacokinetic (PK) and pharmacodynamic (PD) model for neutropenia following nab-paclitaxel administration and identify factors associated with drug disposition and changes in neutrophil counts in patients with solid cancer."

"The observed paclitaxel concentrations in whole blood and neutrophil counts in the first cycle were used for PK/PD analysis."

"Covariate analysis was performed to identify factors affecting PK and the decrease in neutrophil counts."

"Covariate factors affecting neutrophil counts were age and serum albumin level."

"Simulation based on the developed PK/PD model showed a substantial impact of age and serum albumin level on the time course of neutrophil counts after nab-paclitaxel administration."

"We have developed a novel PK/PD model for nab-paclitaxel in which age and serum albumin level were considered clinically important covariate factors."

Introduction

"A randomised phase III study in which nab-paclitaxel was compared with solvent-based paclitaxel (sb-paclitaxel) which contains Cremophor EL and dehydrated alcohol as solubilising agents, in a 3-week cycle among patients with metastatic breast cancer showed a higher response rate in the nab-paclitaxel arm with less incidence of grade 3 or 4 neutropenia than in the sb-paclitaxel arm [170]."

"Another randomised phase III study showed non-inferiority of nab-paclitaxel to sb-paclitaxel when administered weekly in metastatic gastric cancer patients [171]."

"Although nab-paclitaxel is widely used for the treatment of pancreatic, breast and non-small cell lung cancers, very limited pharmacokinetic (PK) and/or pharmacodynamic (PD) data of nab-paclitaxel are available to date, and clinical risk factors affecting severe neutropenia have not yet been fully elucidated."

"This study aimed to develop a PK/PD model for nab-paclitaxel that describes drug disposition in solid cancer patients and drug-induced neutropenia."

Materials and methods

"Taiho Pharmaceutical Co. provided paclitaxel concentration (324 samples, 12 samples per patient) and neutrophil count data (511 observations in total) together with related patient demographic data after all personal information was anonymised in accordance with privacy protection laws."

"Individual estimates of paclitaxel PK parameters obtained from the population PK model were used as input variables to predict drug concentrations."

"Once the developed PK/PD model was confirmed by the methods described above, we performed simulations to investigate the effect of significant covariates on neutrophil counts."

"A similar relationship for the percent decrease in the neutrophil count at nadir was investigated using our data as follows:where x was the time (h) when paclitaxel concentrations exceeded 0.05 μM during the dosing interval, y was the percent decrease in neutrophil count, and T_{50} was the time showing a 50% decrease in neutrophils."

Results

"A non-linear model with saturable distribution to compartment 2 was preferred over a linear 3-compartment model (ΔAIC = 37), while a non-linear saturable elimination model showed no improvement compared with the linear 3-compartment model (ΔAIC = – 4)."

"Because the shrinkage of ηCL and ηV3 were small enough (6.6% and 21%, respectively) [172], individual PK parameters were fixed to the POSTHOC estimates for subsequent PD model building."

"We developed a population PD model describing the changes in neutrophil count induced by nab-paclitaxel treatment."

"An optimal number of transit compartments was tested by changing from 3 to 11, and different models were compared by AIC."

"We modified the model by excluding η_{Kout} but keeping the other η values for Kprol, MTT and Slope."

"Incorporation of neither regimen nor SEX improved the model and serious imbalance of sex distribution between the regimens was observed: the proportion of female included in weekly regimen group was greater than that in tri-weekly regimen group (60% vs. 17%)."

Discussion

"We have developed a PK/PD model of nab-paclitaxel describing drug disposition in cancer patients and chemotherapy-induced neutropenia using clinical data following treatment of nab-paclitaxel monotherapy."

"In our analysis, we found that factors relating to the PK of nab-paclitaxel was BSA, and factors affecting the time-dependent change in neutrophil counts were age and serum albumin level."

"We found several patient factors that can affect the PK/PD of nab-paclitaxel, but the findings are preliminary, and these factors need to be further confirmed using large data."

"The present results describe a novel finding that the serum albumin level and increasing age are potential factors for nab-paclitaxel-induced neutropenia."

"We have developed a population PK/PD model describing nab-paclitaxel disposition and chemotherapy-induced changes in neutrophil counts in a nab-paclitaxel monotherapy setting."

"We found that BSA affected the PK profile of nab-paclitaxel, and that age and albumin level affected PD profiles for neutrophil counts."

5.4 Targeted Delivery to the Eye

Introduction by the Editor

Ocular drug delivery targets a range of molecular components within the eye to address various eye diseases and conditions. Some prominent molecular targets include vascular endothelial growth factor (VEGF), targeted in conditions like age-related macular degeneration (AMD) and diabetic retinopathy to inhibit abnormal blood vessel growth. Integrins, such as $\alpha v\beta 3$ integrin, are targeted for corneal neovascularization. Neurotrophic factors like brain-derived neurotrophic factor (BDNF) are of interest for neuroprotection in diseases like glaucoma. Additionally, ion channels, cytokines, and cell adhesion molecules are explored as potential targets for different ocular disorders. The specificity of these molecular targets offers a precision-driven approach to designing drug delivery strategies that can directly modulate disease pathways in the eye. Numerous nanoparticles have been designed to target specific molecular pathways in the eye for ocular drug delivery. For instance, in the treatment of neovascular age-related macular degeneration (AMD), anti-VEGF drugs are encapsulated in nanoparticles to inhibit abnormal blood vessel growth. Gold nanoparticles functionalized with antibodies target integrins in corneal neovascularization. Neuroprotective nanoparticles deliver neurotrophic factors to address conditions like glaucoma. Nanoparticles also target specific cell types; for instance, retinal pigment epithelium (RPE)-targeting nanoparticles deliver therapies for retinal diseases. By engineering nanoparticles with ligands that interact with these molecular targets, such as surface-conjugated antibodies or peptides, researchers are striving to enhance the precision and efficacy of ocular drug delivery while minimizing side effects.

Machine generated summaries

Disclaimer: The summaries in this chapter were generated from Springer Nature publications using extractive AI auto-summarization: An extraction-based summarizer aims to identify the most important sentences of a text using an algorithm and uses those original sentences to create the auto-summary (unlike generative AI). As the constituted sentences are machine selected, they may not fully reflect the body of the work, so we strongly advise that the original content is read and cited. The auto generated summaries were curated by the editor to meet Springer Nature publication standards.

To cite this content, please refer to the original papers.

Machine generated keywords: eye, ocular, animal model, paper, consideration, understand, animal, comprehensive, fill, improve patient, intraocular, morbidity, pressure, quality life, burden

Targeted Ocular Drug Delivery with Pharmacokinetic/Pharmacodynamic Considerations [173] This is a machine-generated summary of:

Shen, Jie; Lu, Guang Wei; Hughes, Patrick: Targeted Ocular Drug Delivery with Pharmacokinetic/Pharmacodynamic Considerations [173]

Published in: Pharmaceutical Research (2018)

Link to original: https://doi.org/10.1007/s11095-018-2498-y

If you want to cite the papers, please refer to the original.

For technical reasons we could not place the page where the original quote is coming from.

Abstract-Summary "The development of ophthalmic drug delivery systems is a long and comprehensive process including research, nonclinical, and clinical development stages."

"This paper reviews the constraints to various routes of ocular drug delivery and discusses the respective pharmacokinetic considerations, to lay the foundation for formulation approaches pharmaceutical scientists can use to maximize successful drug delivery for each route."

"The overall goal is to give both researchers and drug developers a better understanding of ocular drug delivery and offer tools to successfully develop new medicines that will fulfil unmet medical needs and improve patients' quality of life."

Ophthalmic Disease and Ocular Drug Delivery

"Other diseases contributing substantially to ocular morbidity include infection, inflammation and dry eye."

"Anatomically, physiologically and from a drug delivery and disposition standpoint the eye can be separated into the anterior and posterior segments."

"Successful formulation and drug delivery approaches must take into consideration both the route of administration as well as the disease to be treated."

"The development of ophthalmic drug delivery systems is a long and comprehensive process including research, nonclinical and clinical development stages."

"This will lay the foundation for formulation approaches pharmaceutical scientists can use to maximize successful drug delivery for each route."

"The overall goal of this paper is to give both researchers and developers a better understanding of ocular drug delivery and offer tools to successfully develop new medicines that in the end will fulfil unmet medical needs and improve our patients' quality of life."

Delivery of Pharmacueticals to the Anterior Segment

"To increasing permeability of the drug, the formulator can also improve precorneal retention time."

"Since no additional information is available on the release profile for each of these formulations tested in the clinic, it is difficult to assess whether lack of efficacy relative to topical eye drops of the drugs is due to suboptimal release rate or insufficient drug diffusion to target ocular tissue given the focal point of drug release in the eye offered by the plugs."

"Pehlivan and others reported the in vivo evaluation of CsA-loaded ocular PCL implants after subconjunctival placement in a dry eye mouse model and demonstrated efficacy against control animals by 90 days post dosing, while in vitro release studies showed continued release of drug from the implants through 60 days [174]."

Drug Delivery to the Posterior Segment

"For ranibizumab, despite the lack of concentration data in the dosing compartment of vitreous humor in patients, a population approach was taken utilizing serum drug concentration data pooled from 5 Phase 1 to Phase 3 studies following single or multiple intravitreal injections ranging in doses from 0.3 to 2 mg/eye in AMD patients to model pharmacokinetic profile both systemically and in the eye [175]."

"Most erodible drug delivery systems are matrix type systems with a diffusion controlled release."

"This is especially true in the posterior segment of the eye where there is limited space for the drug delivery systems."

"This has two benefits over diffusion controlled systems: a zero order system can be achieved and the polymers comprising the delivery system will be eliminated concurrent with drug release."

"Non erodible implants represent the first delivery systems in the clinic for the treatment of posterior segment disease."

Conclusion

"The successful development of ophthalmic medications requires a knowledge of the anatomy and physiology of the eye as well as the disease states to be treated."

"We attempted to integrate the disease state, the ocular pharmacokinetics and pharmacodynamics of the drug in the disease state, and the physicochemical properties of the drug into a holistic approach to formulation development."

"The Pharmaceutical Scientists need to be acutely aware of the different constraints to drug delivery for ocular surface diseases and diseases of the anterior and posterior segments."

"New classes of drugs, small molecules and macromolecules, are being identified to treat ocular diseases."

"It is our hope that this paper gives ophthalmic researchers and drug developers a better understanding of ocular drug delivery and offers tools to successfully develop new medicines that in the end will fulfil unmet medical needs and improve our patients' quality of life."

5.5 Targeted Delivery of Biomolecules and Gene Therapy

Introduction by the Editor

Targeted delivery of biomolecules and gene therapy offers a revolutionary approach to treat diseases at the molecular level. Lipid nanoparticles (LNPs) have been utilized to deliver messenger RNA (mRNA) vaccines like the Pfizer-BioNTech and Moderna COVID-19 vaccines, enabling the body to produce viral proteins for an immune response. Similarly, adeno-associated virus (AAV) vectors are employed in gene therapy to deliver corrective genes for genetic disorders. Nanoparticles encapsulating small interfering RNA (siRNA) specifically silence disease-associated genes, as seen in the treatment of hereditary transthyretin amyloidosis. Antibody-conjugated nanoparticles target cancer cells and deliver siRNA to silence oncogenes. These examples underscore how targeted delivery strategies harness the power of nanoparticles to precisely transport therapeutic biomolecules and gene therapies to their intended cellular destinations, offering potential cures for previously challenging diseases.

Machine generated summaries

Disclaimer: The summaries in this chapter were generated from Springer Nature publications using extractive AI auto-summarization: An extraction-based summarizer aims to identify the most important sentences of a text using an algorithm and uses those original sentences to create the auto-summary (unlike generative AI). As the constituted sentences are machine selected, they may not fully reflect the body of the work, so we strongly advise that the original content is read and cited. The auto generated summaries were curated by the editor to meet Springer Nature publication standards.

To cite this content, please refer to the original papers.

Machine generated keywords: biomolecule, gene, dna, transdermal, enhancer, chemical, transport, monoclonal antibody, kinetic, enzymatic, brain, alzheimer, alzheimer disease, monoclonal, device

Transdermal Delivery Systems for Biomolecules [176] This is a machine-generated summary of:

Peña-Juárez, Ma. Concepción; Guadarrama-Escobar, Omar Rodrigo; Escobar-Chávez, José Juan: Transdermal Delivery Systems for Biomolecules [176]

Published in: Journal of Pharmaceutical Innovation (2021)

Link to original: https://doi.org/10.1007/s12247-020-09525-2

If you want to cite the papers, please refer to the original.

For technical reasons we could not place the page where the original quote is coming from.

Abstract-Summary "The present review article focuses on highlighting the main technologies used as tools that improve the delivery of transdermal biomolecules, addressing them from the point of view of research in the development of transdermal systems that use physical and chemical permeation enhancers and nanocarrier systems or a combination of them."

"Transdermal drug delivery systems have increased in importance since the late 1970s when their use was approved by the Food and Drug Administration (FDA)."

"The first transdermal drug delivery system used for biomolecules was for the treatment of hormonal disorders."

"The latest technologies that have used such transdermal biomolecule transporters include electrical methods (physical penetration enhancers), some chemical penetration enhancers and nanocarriers."

Introduction

"The transdermal drug delivery (TDD) system reduces potential exposure to side effects in the host and improves protein stability by also reducing enzymatic degradation, which is very common to occur when a drug takes the most conventional routes to enter the body, such as oral and subcutaneous routes."

"For transdermal protein-drug delivery systems, specific methods or devices are used to allow or enhance intact passage through the stratum corneum (SC)."

"The most suitable systems for the release of biomolecules are the third-generation transdermal delivery systems, whose effects are directed to the SC."

"This article highlights the main technologies used as tools that improve transdermal protein-drug delivery systems, addressing them from the point of view of research in the development of transdermal systems where many technologies and devices have a high possibility of use in conventional drug treatments, such as nanotechnology devices."

Skin Anatomy

"The skin is a three-layer laminate that is made up of the SC, the epidermis, and the dermis [177]."

"The SC is the outermost and impermeable skin layer that prevents the entry of most chemicals [178]."

"This barrier is formed by around 10 to 25 lines of dead keratinocytes (corneocytes) that have united with each other with corneodesmosomes forming a wall embedded in a lipidic matrix."

"The SC acts as a barrier, preventing desiccation and pathogen entry and protecting against ultraviolet rays."

Transdermal Delivery Route

"The transdermal drug delivery systems also allow homogeneous absorption and have good pharmaceutical efficiency, reducing the adverse effects that drugs could cause in the host and reduce the drug risk to suffer alterations caused by enzymatic factors and the immune system present in the intrinsic environment."

"The first step is penetration, which occurs when the drug is deposited onto the specific skin layer."

"This type of drug delivery system has advantages over traditional systems because it can allow the storage of small amounts of drugs in the transport and delivery structure, which are developed by nanotechnology using mainly polymeric nanomaterials made with biopolymers, microneedles, and nanoparticles."

"With these devices, it is possible to improve drug delivery and its bioavailability."

Biomolecule Medical Application: Possible Target for Transdermal Protein Delivery Systems

"One of the most important applications is cancer treatment, which is a specific treatment that employs cytokines, a kind of protein produced by the immunity system that helps the patient's immune system eliminate cancer cells and prevents the replication of cancer cells."

"The US FDA approved the use of TDD systems and their use has been increasing in relevance for protein drugs because for some molecules these delivery systems improve the solubility of the compound, in addition to allowing it to pass efficiently through the skin."

"For immunotherapy, it might be possible to employ monoclonal antibody-loaded TDD systems and tumor agnostic therapies, non-specific immunotherapies (use of interferons and interleukins), and cancer theranostic treatment where SC could be interrupted by a nanogel delivering a drug through the skin and allowing it to accumulate at the tumor site [179] or these systems may even function by delivering vaccine antigens (cancer vaccines) using nanoparticles and liposomes as antigenic carriers loaded on nanogels to pass the antigen through the SC by iontophoresis [180]."

Passage of Biomolecules through the Skin

"These molecules are charged in a physiological pH and are hydrophilic, which complicates their permeability across the skin."

"Some drugs have a lower or higher affinity with the lipophilic condition of the SC, which determines the quantity of the molecule that can cross this layer of the skin naturally."

"It is better to use a technique that improves passage to the skin, using or not natural permeation molecular routes avoiding the degradation of peptides or proteins and ensuring passage and integrity during the transport and release processes [181]."

"In the skin, the SC and lipid layers have a barrier function, as both prevent the loss of water and avoid the entry of any substances, molecules, or microorganisms from the environment or microbiome."

"These layers play a crucial role in the passive passage of proteins and other chemicals into the skin."

Biomolecule Percutaneous Administration Systems

"In the drug delivery systems, some technologies are employed to permit the transport and delivery of certain molecules, such as peptides and proteins, decreasing the risk of degradation during their passage through the skin."

"The MNs are a minimally invasive delivery method for the recipient and offer an excellent option for delivery systems in which the user must use a hypodermic metal needle to introduce the protein drug as in the case of insulin treatment [182]."

"They must have an appropriate structure that allows penetrating the SC and must retain it during the necessary time for protein-drug delivery."

"It is possible to combine two or more polymers to generate MNs with a particular structure that allows better perforation of the skin that improves the passage of protein drug through the skin."

Conclusions

"A great number of therapies to resolve the problems associated with drug release have been implicated to probe numerous kinds of devices and pharmaceutical innovations to solve many diseases or applications in various clinical trials for the diagnosis of diseases or treatments of diseases."

"Many research studies have reported numerous results that indicate the most appropriate technologies to use in each case of a drug delivery system."

"In the coming years, we will observe an increase in their design, development, characterization, and clinical trials for the treatment of many important and chronic diseases."

Gene Therapy: A Pharmacokinetic/Pharmacodynamic Modelling Overview [183] This is a machine-generated summary of:

Parra-Guillén, Zinnia P.; González-Aseguinolaza, Gloria; Berraondo, Pedro; Trocóniz, Iñaki F.: Gene Therapy: A Pharmacokinetic/Pharmacodynamic Modelling Overview [183]

Published in: Pharmaceutical Research (2010)
Link to original: https://doi.org/10.1007/s11095-010-0136-4

If you want to cite the papers, please refer to the original.

For technical reasons we could not place the page where the original quote is coming from.

Abstract-Summary "Nonviral vectors present interesting properties for their clinical application, but their efficiency in vivo is relatively low, and further improvements in these vectors are needed."

"Model-based approach is a powerful tool to understand and describe the different processes that gene transfer systems should overcome inside the body."

"Model-based approach allows for proposing and predicting the effect of parameter changes on the overall gene therapy response, as well as the known application of the pharmacokinetic/pharmacodynamic modelling in conventional therapies."

INTRODUCTION

"Gene therapy can be defined as the transfer of genetic material (DNA or RNA) to somatic cells in order to obtain a therapeutic effect, by either (i) correcting genetic defects, (ii) over-expressing proteins that are therapeutically useful, or (iii) inhibiting the production of harmful proteins [184]."

"It can be expected that if models have been proved to be beneficial in understanding and predicting the effects of "traditional" drugs, the same would be the case for modern therapies such gene therapy [185]. (Semi-)mechanistic PK/PD models have to take into consideration the following: (i) formulation characteristics, (ii) biopharmaceutic aspects, (iii) pharmacokinetic and pharmacodynamic properties, and (iv) system behaviour."

"The apparent kinetics of the gene therapy product would be reflected by the time course of the synthesised therapeutic protein and represented by process ix above."

SYSTEMIC AND ORGAN PHARMACOKINETICS

"Less invasive alternatives as the development of viral and nonviral delivery systems that protect pDNA from systemic elimination and facilitate tissue uptake represent currently one of the major research focus in gene therapy."

"As in the case of the systemic pharmacokinetics, and despite that in several articles levels of the genetic material administered are measured and reported in different organs at different times after administration, not many efforts have been made to relate the time course in plasma with the corresponding in tissues."

"Of tissue pharmacokinetics, the models developed by Nomura and others [186] and recently by Mok and others [187] represent an interesting and valuable approach to understand the fate of the gene therapy systems in a more controlled environment compared to systemic (iv or oral) administration."

INTRACELLULAR PHARMACOKINETICS

"Mathematical models to characterize and quantitatively describe the intracellular processes of nonviral or viral gene expression systems can be developed to identify the main events controlling the desired transgene expression."

"Applying the model developed, the authors studied the effects of different intrinsic kinetic parameters in the expression level and demonstrated that it was not only a function of the promoter strength and the efficiency of gene transfer into the cell [188] (transfection), but also the intrinsic stability of the DNA, RNA and protein express."

"A more complex model was developed and validated by Varga and others [189] using data from previous works as well as their own experimental studies to describe the intracellular processes, differentiating between the cytoplasm and the nucleus compartments."

"The development of these intracellular models allows for quantitatively elucidating the rate-limiting steps for a variety of vector/cell systems and to predict the effect of parameter changes on gene expression."

PHARMACODYNAMICS

"Further studies of the same group observed that the number of DNA copies in the nucleus of one cell appeared to be fundamental in determining the gene expression levels, and suggested that DNA can work synergistically inside the nucleus, explaining this non-linear relationship [190]."

"The authors observed that, although intracellular PK was similar between these two systems, to achieve a comparable transgene expression, LFN required three more orders of intranuclear DNA copies than Ad due to the higher transcription efficiency of Ad (8100 times higher)."

"Taking as well the model of Kamiya and others [191] as a starting-point, intranuclear pharmacokinetic processes were inferred from the luciferase activity measures: in the nucleus, free DNA could be degraded ($k_{D\_DNAFree}$) or stabilized by transforming to bound DNA ($k_{DNABound}$); therefore, two DNA forms were proposed to coexist in the nucleus, although with different transcriptional rates ($k_{SFree\_mRNALUC}$ and $k_{SBound\_mRNALUC}$), and finally the mRNA was eliminated from the cell ($k_{D\_mRNALUC}$)."

SUMMARY

"Although it has been shown how PK and PD can help in the optimization and rational development of nonviral gene therapies, to our knowledge, non-integrated PK/PD model has not been developed yet."

"From our perspective, there are two fundamental aspects that have to be taken into consideration when integrating the modelling approach with the goal of optimizing and understanding gene therapy response."

"Given the complexity of the techniques employed and the knowledge required to accomplish the in vitro/in vivo experiments, along with the expertise needed to develop a mathematical model, it is necessary to establish multidisciplinary collaboration between experimental and modelling areas to be able to integrate the concepts of PK/PD modelling in gene therapy."

"These types of collaborative approaches are worthy not only for a better understanding of the biological processes regulating gene therapy responses, but also for optimizing preclinical and clinical phases in the development of new therapeutic agents."

Using nanotechnology to deliver biomolecules from nose to brain — peptides, proteins, monoclonal antibodies and RNA [192] This is a machine-generated summary of:

Borrajo, Mireya L.; Alonso, María José: Using nanotechnology to deliver biomolecules from nose to brain — peptides, proteins, monoclonal antibodies and RNA [192]

Published in: Drug Delivery and Translational Research (2021)

Link to original: https://doi.org/10.1007/s13346-021-01086-2

If you want to cite the papers, please refer to the original.

For technical reasons we could not place the page where the original quote is coming from.

Abstract-Summary "Nose-to-brain (N-to-B) delivery is now being investigated as a potential option for the direct transport of biomolecules from the nasal cavity to different brain areas."

"We discuss how different technological approaches enhance this N-to-B transport, with emphasis on those that have shown a potential for clinical translation."

"We also analyse how the physicochemical properties of nanocarriers and their modification with cell-penetrating peptides (CPPs) and targeting ligands affect their efficacy as N-to-B carriers for biomolecules."

Introduction

"As such, this defence mechanism represents an extraordinary barrier for the transport of drugs to the brain [193, 194]."

"An alternative approach to reach the brain that is gaining increasing attention makes use of the nose-to-brain (N-to-B) route."

"From then on, significant knowledge on the mechanisms of transport of molecules across the N-to-B barriers has been generated."

"In the past few decades, nanotechnology has been positioned as a promising strategy for enhancing the N-to-B transport of therapeutic biomolecules [194–201]."

"Background information, researchers, including members of our group, have published a number of review articles covering different N-to-B delivery strategies [194, 200–203]."

"We critically analyse the N-to-B drug delivery options for biologicals with emphasis on those based on nanotechnology."

"We disclose our understanding of the critical features for nanosystems to function as carriers to overcome the N-to-B barrier."

Challenges and barriers for the nose-to-brain delivery of biomolecules

"The olfactory epithelium is located on the upper region of the nasal cavity, separated from the CNS by the cribriform plate and the lamina propria and is also protected by a mucus layer [204–206]."

"Drugs administered intranasally can reach the brain indirectly, upon systemic absorption through the respiratory epithelium and subsequent transport across the BBB, or directly, across the olfactory epithelium and the olfactory nerves."

"Three mechanisms have been described for the N-to-B transport: direct internalization into the olfactory nerve, leading to axonal transport; paracellular transport between epithelial cells and across channels near olfactory nerves; and transcellular transport across cells of the olfactory epithelium [206–210]."

"The factors that influence the N-to-B transport of biomolecules include the physicochemical attributes of the biologicals themselves, and the intrinsic anatomical and physiological barriers of the nasal cavity [203, 211]."

"The olfactory epithelium presents long non-mobile cilia that, in combination with the mucus secretion and the presence of metabolic enzymes, hinders the N-to-B transport [195, 212–215]."

Clinical scenario of nose-to-brain transport of biomolecules

"Among the 196 clinical trials analysed, more than 67% refer to the administration of biomolecules in the form of a simple aqueous solution."

"The intranasal administration of an insulin aqueous solution has also been widely studied in up to Phase 2/3 clinical trials, as a therapy for Alzheimer's disease, mild cognitive impairment, diabetes, insulin resistance, and Parkinson's disease, among other conditions."

"Subsequent clinical trials were not conclusive for this indication [216, 217]."

"Additional clinical trials are currently being conducted to elucidate the correlation between administering intranasal insulin and both Alzheimer's disease and mild cognitive impairment."

"Other highly prevalent diseases extensively studied in clinical trials are diabetes and insulin resistance."

Technological approaches for nose-to-brain delivery of biomolecules

"Another mucoadhesive polymer that has been investigated for its potential to increase the N-to-brain access to biomolecules is chitosan, a polysaccharide extensively studied as mucoadhesive and cell penetration agent for N-to-B delivery of small molecules [218, 219]."

"In a separate study, different sizes of polyethylene glycol-polylactic acid (PEG-PLA) NPs were developed; according to their results, mean diameters of 100 nm showed greater brain accumulation and, in consequence, enhanced therapeutic effect, than nanosystems of 500 nm, in an epilepsy rat model after intranasal administration [220]."

"Although the interest on increasing the bioadhesion or mucoadhesion of nanocarriers to enhance the N-to-B delivery of biomolecules is still being elucidated, examples of the incorporation of adhesive regents into nanosystems to facilitate N-to-B transport of biomolecules are highlighted in the literature."

"Solanum tubersum lectin (STL), a glycoprotein that binds to N-acetyl-D-glucosamine of the nasal cavity epithelium [221, 222], has been reported to enhance the transport of PEG-PLGA NPs loaded with a fluorescent probe to the brain."

Illustrative nanotechnologies for nose-to-brain delivery of biomolecules

"The NR2BPc peptide, of potential interest for the treatment of ischemia and prevention of strokes, was loaded into WGA-functionalized PLA-PEG NPs (140 nm, and negative surface charge)."

"BFGF associated to functionalized STL-PEG-PLGA NPs (120 nm and negative surface charge) was administered intranasally in an AD mice model."

"The results showed high protein levels in different areas of the brain, including the olfactory bulb and striatum, and enhancement of their therapeutic effect after intranasal administration in a PD rat model, as compared with free protein and intravenous administration of the nanoencapsulated protein [223]."

"SiRNA Gal-1-loaded chitosan NPs (140 nm and positive surface charge) were administered intranasally in a glioblastoma mice model."

"SiRNA TNF-α-loaded Tat- modified PEG-PCL nanomicelles (62 nm and positive surface charge) were administered intranasally in an ischemic stroke mice model, and the result of this treatment was a reduction of the infarcted area [224]."

Conclusion and future perspectives

"Among the different possibilities to successfully access the CNS, the main difficulty in this line of study, the use of nanocarriers has shown some potential for the direct N-to-B transport."

"These approaches could be particularly beneficial for the N-to-B delivery of biomolecules, given that their access to the brain following systemic administration has proved extremely difficult so far."

"In order to develop potential nanocarriers for the N-to-B transport of biomolecules, it is fundamental to take into consideration the different barriers that must be overcome and the way that nanocarriers can be specifically designed to overcome them."

"Despite these promising findings and studies, none of these technologies intended for N-to-B delivery of biomolecules has reached the clinical development phase."

"N-to-B transport of biomolecules has been shown to be a potential alternative for the delivery of drug-loaded nanocarriers in sufficient amount to depict a therapeutic effect for the treatment of different neurological conditions."

Bibliography

1. Zhang W, Mehta A, Tong Z, Esser L, Voelcker NH (2021) Development of polymeric nanoparticles for blood–brain barrier transfer—strategies and challenges. Adv Sci 8(10):2003937
2. Nguyen TT, Maeng HJ (2022) Pharmacokinetics and pharmacodynamics of intranasal solid lipid nanoparticles and nanostructured lipid carriers for nose-to-brain delivery. Pharmaceutics 14(3):572
3. Pinheiro RG, Coutinho AJ, Pinheiro M, Neves AR (2021) Nanoparticles for targeted brain drug delivery: what do we know? Int J Mol Sci 22(21):11654
4. Yao Y, Zhou Y, Liu L, Xu Y, Chen Q, Wang Y, Wu S, Deng Y, Zhang J, Shao A (2020) Nanoparticle-based drug delivery in cancer therapy and its role in overcoming drug resistance. Front Mol Biosci 7:193
5. Sharifi E, Bigham A, Yousefiasl S, Trovato M, Ghomi M, Esmaeili Y, Samadi P, Zarrabi A, Ashrafizadeh M, Sharifi S, Sartorius R (2022) Mesoporous bioactive glasses in cancer diagnosis and therapy: stimuli-responsive, toxicity, immunogenicity, and clinical translation. Adv Sci 9(2):2102678
6. Cheng S, Nethi SK, Al-Kofahi M, Prabha S (2021) Pharmacokinetic—pharmacodynamic modeling of tumor targeted drug delivery using nano-engineered mesenchymal stem cells. Pharmaceutics 13(1):92
7. Shen AM, Minko T (2020) Pharmacokinetics of inhaled nanotherapeutics for pulmonary delivery. J Controlled Release 326:222–244
8. Li J, Zhang K, Wu D, Ren L, Chu X, Qin C, Han X, Hang T, Xu Y, Yang L, Yin L (2021) Liposomal remdesivir inhalation solution for targeted lung delivery as a novel therapeutic approach for COVID-19. Asian J Pharm Sci 16(6):772–783
9. Okuda T, Chan HK (2021) Formulation and pharmacokinetic challenges associated with targeted pulmonary drug delivery. In: Drug delivery approaches: perspectives from pharmacokinetics and pharmacodynamics
10. Swetledge S, Jung JP, Carter R, Sabliov C (2021) Distribution of polymeric nanoparticles in the eye: implications in ocular disease therapy. J Nanobiotechnol 19(1):1–9
11. Chandasana H, Prasad YD, Chhonker YS, Chaitanya TK, Mishra NN, Mitra K, Shukla PK, Bhatta RS (2014) Corneal targeted nanoparticles for sustained natamycin delivery and their PK/PD indices: an approach to reduce dose and dosing frequency. Int J Pharm 477(1-2):317–325
12. Li Z, Liu M, Ke L, Wang LJ, Wu C, Li C, Li Z, Wu YL (2021) Flexible polymeric nanosized micelles for ophthalmic drug delivery: research progress in the last three years. Nanoscale Adv 3(18):5240–5254
13. Ghezzi M, Ferraboschi I, Delledonne A, Pescina S, Padula C, Santi P, Sissa C, Terenziani F, Nicoli S (2022) Cyclosporine-loaded micelles for ocular delivery: Investigating the penetration mechanisms. J Controlled Release 349:744–755
14. Murata Y, Neuhoff S, Rostami-Hodjegan A, Takita H, Al-Majdoub ZM, Ogungbenro K (2022) In vitro to in vivo extrapolation linked to physiologically based pharmacokinetic models for assessing the brain drug disposition. AAPS J 24(1):28. https://doi.org/10.1208/s12248-021-00675-w
15. Abbott NJ, Friedman A (2012) Overview and introduction: the blood-brain barrier in health and disease. Epilepsia 53(Suppl 6):1–6
16. Maurer MH (2010) Proteomics of brain extracellular fluid (ECF) and cerebrospinal fluid (CSF). Mass Spectrom Rev 29(1):17–28
17. Saleh MAA, de Lange ECM (2021) Impact of CNS diseases on drug delivery to brain extracellular and intracellular target sites in human: a "WHAT-IF" simulation study. Pharmaceutics 13(1):95
18. Maurer MH, Berger C, Wolf M, Futterer CD, Feldmann RE Jr, Schwab S et al (2003) The proteome of human brain microdialysate. Proteome Sci 1(1):7
19. Yamamoto Y, Danhof M, de Lange ECM (2017) Microdialysis: the key to physiologically based model prediction of human CNS target site concentrations. AAPS J 19(4):891–909

20. Kakee A, Terasaki T, Sugiyama Y (1996) Brain efflux index as a novel method of analyzing efflux transport at the blood-brain barrier. J Pharmacol Exp Ther 227(3):1550–1559
21. Fridén M (2010) Development of methods for assessing unbound drug exposure in the brain. In vivo, in vitro and in silico. PhD Thesis Summary, Faculty of Pharmacy, Uppsala Universitet, Sweden, pp 1–57
22. Friden M, Ducrozet F, Middleton B, Antonsson M, Bredberg U, Hammarlund-Udenaes M (2009) Development of a high-throughput brain slice method for studying drug distribution in the central nervous system. Drug Metab Dispos 37(6):1226–1233
23. Westerhout J, Ploeger B, Smeets J, Danhof M, de Lange EC (2012) Physiologically based pharmacokinetic modeling to investigate regional brain distribution kinetics in rats. AAPS J 14(3):543–553
24. Yamamoto Y, Valitalo PA, Wong YC, Huntjens DR, Proost JH, Vermeulen A et al (2018) Prediction of human CNS pharmacokinetics using a physiologically-based pharmacokinetic modeling approach. Eur J Pharm Sci 112:168–179
25. Yamamoto Y, Välitalo PA, van den Berg D-J, Hartman R, van den Brink W, Wong YC, Huntjens DR, Proost JH, Vermeulen A, Krauwinkel W, Bakshi S, Aranzana-Climent V, Marchand S, Dahyot-Fizelier C, Couet W, Danhof M, van Hasselt JGC, de Lange ECM (2017) A generic multi-compartmental CNS distribution model structure for 9 drugs allows prediction of human brain target site concentrations. Pharm Res 34(2):333–351
26. Ball K, Bouzom F, Scherrmann JM, Walther B, Declèves X (2012) Development of a physiologically based pharmacokinetic model for the rat central nervous system and determination of an in vitro-in vivo scaling methodology for the blood-brain barrier permeability of two transporter substrates, morphine and oxycodone. J Pharm Sci 101(11):4277–4292
27. Ball K, Bouzom F, Scherrmann JM, Walther B, Decleves X (2014) A physiologically based modeling strategy during preclinical CNS drug development. Mol Pharm 11(3):836–848
28. Salem HF, Kharshoum RM, Abou-Taleb HA, Naguib DM (2019) Nanosized transferosome-based intranasal in situ gel for brain targeting of resveratrol: formulation, optimization, in vitro evaluation, and in vivo pharmacokinetic study. AAPS PharmSciTech 20:1–4. https://doi.org/10.1208/s12249-019-1353-8
29. Mura P, Mennini N, Nativi C, Richichi B (2018) In situ mucoadhesive-thermosensitive liposomal gel as a novel vehicle for nasal extended delivery of opiorphin. Eur J Pharm Biopharm 122:54–61. https://doi.org/10.1016/j.ejpb.2017.10.008
30. Sonvico F, Clementino A, Buttini F, Colombo G, Pescina S, Staniscuaski Guterres S et al (2018) Surface-modified nanocarriers for nose-to-brain delivery: from bioadhesion to targeting. Pharmaceutics 10(1):34. https://doi.org/10.3390/pharmaceutics10010034
31. Espinoza LC, Vacacela M, Clares B, Garcia ML, Fabrega MJCA (2018) Development of a nasal donepezil-loaded microemulsion for the treatment of Alzheimer's disease: in vitro and ex vivo characterization. CNS Neurol Disord Drug Targets 17(1):43–53
32. Gartziandia O, Herrán E, Ruiz-Ortega JA, Miguelez C, Igartua M, Lafuente JV et al (2016) Intranasal administration of chitosan-coated nanostructured lipid carriers loaded with GDNF improves behavioral and histological recovery in a partial lesion model of Parkinson's disease. J Biomed Nanotechnol 12(12):2220–2230
33. Wang Y, Li M, Qian S, Zhang Q, Zhou L, Zuo Z et al (2016) Zolpidem mucoadhesive formulations for intranasal delivery: characterization, in vitro permeability, pharmacokinetics, and nasal ciliotoxicity in rats. J Pharm Sci 105(9):2840–2847
34. Piazza J, Hoare T, Molinaro L, Terpstra K, Bhandari J, Selvaganapathy PR et al (2014) Haloperidol-loaded intranasally administered lectin functionalized poly(ethylene glycol)-block-poly(D,L)-lactic-co-glycolic acid (PEG-PLGA) nanoparticles for the treatment of schizophrenia. Eur J Pharm Biopharm 87(1):30–39. https://doi.org/10.1016/j.ejpb.2014.02.007
35. Karavasili C, Fatouros DG (2015) Smart materials: in situ gel-forming systems for nasal delivery. Drug Discov Today 21(1):157–166. https://doi.org/10.1016/j.drudis.2015.10.016

36. Callens C, Ceulemans J, Ludwig A, Foreman P, Remon JP (2003) Rheological study on mucoadhesivity of some nasal powder formulations. Eur J Pharm Biopharm 55:323–328
37. Manconi M, Caddeo C, Sinico C, Valenti D, Cristina M, Biggio G et al (2011) Ex vivo skin delivery of diclofenac by transcutol containing liposomes and suggested mechanism of vesicle–skin interaction. Eur J Pharm Biopharm 78(1):27–35. https://doi.org/10.1016/j.ejpb.2010.12.010
38. Sharma AK, Gupta L, Sahu H, Qayum A, Singh SK, Nakhate KT, Ajazuddin GU (2018) Chitosan engineered PAMAM dendrimers as nanoconstructs for the enhanced anti-cancer potential and improved in vivo brain pharmacokinetics of temozolomide. Pharm Res 35:1–4. https://doi.org/10.1007/s11095-017-2324-y
39. Louis DN, Ohgaki H, Wiestler OD, Cavenee WK, Burger PC, Jouvet A et al (2007) The WHO classification of tumours of the central nervous system. Acta Neuropathol 114:97
40. Sarin H (2009) Recent progress towards development of effective systemic chemotherapy for the treatment of malignant brain tumors. J Trans Med 7:77
41. Ruigrok MJ, de Lange EC (2015) Emerging insights for translational pharmacokinetic and pharmacokinetic-pharmacodynamic studies: towards prediction of nose-to-brain transport in humans. AAPS J 17:493–505. https://doi.org/10.1208/s12248-015-9724-x
42. Merkus FWHM, van den Berg MP (2007) Can nasal drug delivery bypass the??blood-brain barrier?: questioning the direct transport theory. Drugs R D 8(3):133–144
43. Harkema JR, Carey SA, Wagner JG (2006) The nose revisited: a brief review of the comparative structure, function, and toxicologic pathology of the nasal epithelium. Toxicol Pathol 34(3):252–269
44. Dhuria SV, Hanson LR, Frey WH (2010) Intranasal delivery to the central nervous system: mechanisms and experimental considerations. J Pharm Sci 99(4):1654–1673
45. Stevens J, Suidgeest E, van der Graaf PH, Danhof M, de Lange ECM (2009) A new minimal-stress freely-moving rat model for preclinical studies on intranasal administration of CNS drugs. Pharm Res 26(8):1911–1917
46. Haque S, Md S, Sahni JK, Ali J, Baboota S (2014) Development and evaluation of brain targeted intranasal alginate nanoparticles for treatment of depression. J Psychiatr Res 48(1):1–12
47. Born J, Lange T, Kern W, McGregor G, Bickel P, Fehm U et al (2002) Sniffing neuropeptides: a transnasal approach to the human brain. Nat Neurosci 5(6):514
48. Illum L (2004) Is nose-to-brain transport of drugs in man a reality? J Pharm Pharmacol 56(1):3–17
49. Kozlovskaya L, Abou-Kaoud M, Stepensky D (2014) Quantitative analysis of drug delivery to the brain via nasal route. J Control Release 189:133–140. https://doi.org/10.1016/j.jconrel.2014.06.053
50. Merkus P, Guchelaar HJ, Bosch DA, Merkus FWHM (2003) Direct access of drugs to the human brain after intranasal drug administration? Neurology 60(10):1669–1671
51. Stevens J, Ploeger BA, van der Graaf PH, Danhof M, de Lange ECM (2011) Systemic and direct nose-to-brain transport pharmacokinetic model for remoxipride after intravenous and intranasal administration. Drug Metab Dispos Biol Fate Chem 39(12):2275–2282
52. Stevens J, Ploeger BA, Hammarlund-Udenaes M, Osswald G, van der Graaf PH, Danhof M et al (2012) Mechanism-based PK-PD model for the prolactin biological system response following an acute dopamine inhibition challenge: quantitative extrapolation to humans. J Pharmacokinet Pharmacodyn 39(5):463–477
53. Movin-Osswald G, Hammarlund-Udenaes M (1995) Prolactin release after remoxipride by an integrated pharmacokinetic-pharmacodynamic model with intra- and interindividual aspects. J Pharm Exp Ther 274(2):921–927
54. Hammarlund-Udenaes M (2014) Pharmacokinetic concepts in brain drug delivery in drug delivery to the brain. In: Hammarlund-Udenaes M et al (eds) Physiological concepts, methodologies and approaches. Springer, New York, pp 127–161
55. Watson J, Wright S, Lucas A, Clarke KL, Viggers J, Cheetham S et al (2009) Receptor occupancy and brain free fraction. Drug Metab Dispos 37:753–760

56. Rompicherla SK, Arumugam K, Bojja SL, Kumar N, Rao CM (2021) Pharmacokinetic and pharmacodynamic evaluation of nasal liposome and nanoparticle based rivastigmine formulations in acute and chronic models of Alzheimer's disease. Naunyn-Schmiedeberg's Arch Pharmacol 394(8):1737–1755. https://doi.org/10.1007/s00210-021-02096-0
57. Sharma K (2019) Cholinesterase inhibitors as Alzheimer's therapeutics (review). Mol Med Rep 20:1479–1487
58. Vieira DB, Gamarra LF (2016) Getting into the brain: liposome-based strategies for effective drug delivery across the blood–brain barrier. Int J Nanomed 11:5381–5414
59. Li H et al (2009) Enhancement of gastrointestinal absorption of quercetin by solid lipid nanoparticles. J Control Release 133:238–244
60. Emerich DF, Walsh TJ (1990) Cholinergic cell loss and cognitive impairments following intraventricular or intradentate injection of colchicine. Brain Res 517:157–167
61. More S, Pawar A (2023) Brain targeted curcumin loaded turmeric oil microemulsion protects against trimethyltin induced neurodegeneration in adult zebrafish: a pharmacokinetic and pharmacodynamic insight. Pharm Res 40(3):675–687. https://doi.org/10.1007/s11095-022-03467-9
62. Maheshwari RK, Singh AK, Gaddipati J, Srimal RC (2006) Multiple biological activities of curcumin: a short review. Life Sci 78:2081–2087. Available from: https://doi.org/10.1016/j.lfs.2005.12.007
63. Mythri RB, Srinivas Bharath MM (2012) Curcumin: a potential neuroprotective agent in parkinson's disease. Curr Pharm Des 18:91–99
64. Perrone D, Ardito F, Giannatempo G, Dioguardi M, Troiano G, Lo Russo L et al (2015) Biological and therapeutic activities, and anticancer properties of curcumin (review). Exp Ther Med 10:1615–1623
65. Shehzad A, Islam SU, Lee YS (2019) Curcumin and inflammatory brain diseases. Curcumin Neurol Psychiatr Disord:437–458
66. Hamaguchi T, Ono K, Yamada M (2010) Curcumin and Alzheimer's disease. CNS Neurosci Ther 16:285–297
67. Darvesh AS, Carroll RT, Bishayee A, Novotny NA, Geldenhuys WJ, Van der Schyf CJ (2012) Curcumin and neurodegenerative diseases: a perspective. Expert Opin Investig Drugs 21:1123–1140. Available from: http://www.tandfonline.com/doi/full/10.1517/13543784.2012.693479
68. More SK, Pawar AP (2020) Preparation, optimization and preliminary pharmacokinetic study of curcumin encapsulated turmeric oil microemulsion in zebra fish. Eur J Pharm Sci 155:105539. Available from: https://doi.org/10.1016/j.ejps.2020.105539
69. Kudarha RR, Sawant KK (2021) Chondroitin sulfate conjugation facilitates tumor cell internalization of albumin nanoparticles for brain-targeted delivery of temozolomide via CD44 receptor-mediated targeting. Drug Deliv Transl Res 11:1994–2008. https://doi.org/10.1007/s13346-020-00861-x
70. Elzoghby AO, Samy WM, Elgindy NA (2012) Albumin-based nanoparticles as potential controlled release drug delivery systems. J Control Release 157:168–182
71. Yewale C, Baradia D, Vhora I, Misra A (2013) Proteins: emerging carrier for delivery of cancer therapeutics. Expert Opin Drug Deliv 10:1429–1448
72. Kudarha RR, Sawant KK (2017) Albumin based versatile multifunctional nanocarriers for cancer therapy: fabrication, surface modification, multimodal therapeutics and imaging approaches. Mater Sci Eng C Mater Biol Appl 81:607–626
73. Agrawal S, Dwivedi M, Ahmad H, Chadchan SB, Arya A, Sikandar R et al (2018) CD44 targeting hyaluronic acid coated lapatinib nanocrystals foster the efficacy against triple-negative breast cancer. Nanomedicine 14:327–337
74. Liu P, Chen N, Yan L, Gao F, Ji D, Zhang S et al (2019) Preparation, characterisation and in vitro and in vivo evaluation of CD44-targeted chondroitin sulphate-conjugated doxorubicin PLGA nanoparticles. Carbohydr Polym 213:17–26

75. Lo Y-L, Chou H-L, Liao Z-X, Huang S-J, Ke J-H, Liu Y-S et al (2015) Chondroitin sulfate-polyethylenimine copolymer-coated superparamagnetic iron oxide nanoparticles as an efficient magneto-gene carrier for microRNA-encoding plasmid DNA delivery. Nanoscale 7:8554–8565
76. Pandey A, Singh K, Patel S, Singh R, Patel K, Sawant K (2019) Hyaluronic acid tethered pH-responsive alloy-drug nanoconjugates for multimodal therapy of glioblastoma: an intranasal route approach. Mater Sci Eng C Mater Biol Appl 98:419–436
77. Baskin J, Jeon JE, Lewis SJ (2021) Nanoparticles for drug delivery in Parkinson's disease. J Neurol 268(5):1981–1994. https://doi.org/10.1007/s00415-020-10291-x
78. Jesus S et al (2019) Hazard assessment of polymeric nanobiomaterials for drug delivery: what can we learn from literature so far. Front Bioeng Biotechnol 7:261
79. Bobo D et al (2016) Nanoparticle-based medicines: a review of FDA-approved materials and clinical trials to date. Pharm Res 33(10):2373–2387
80. Anselmo AC, Mitragotri S (2019) Nanoparticles in the clinic: an update. Bioeng Transl Med 4(3):e10143
81. Saraiva C et al (2016) Nanoparticle-mediated brain drug delivery: overcoming blood-brain barrier to treat neurodegenerative diseases. J Control Release 235:34–47
82. Paul A, Yadav KS (2020) Parkinson's disease: Current drug therapy and unraveling the prospects of nanoparticles. J Drug Deliv Sci Technol 58:101790. https://doi.org/10.1016/j.jddst.2020.101790
83. Gambaryan PY et al (2014) Increasing the efficiency of Parkinson's Disease treatment using a poly(lactic-co-glycolic acid) (PLGA) based L-DOPA delivery system. Exp Neurobiol 23(3):246–252
84. Md S et al (2013) Bromocriptine loaded chitosan nanoparticles intended for direct nose to brain delivery: pharmacodynamic, pharmacokinetic and scintigraphy study in mice model. Eur J Pharm Sci 48(3):393–405
85. Sharma S, Lohan S, Murthy RS (2014) Formulation and characterization of intranasal muco-adhesive nanoparticulates and thermo-reversible gel of levodopa for brain delivery. Drug Dev Ind Pharm 40(7):869–878
86. Raj R et al (2018) Pramipexole dihydrochloride loaded chitosan nanoparticles for nose to brain delivery: Development, characterization and in vivo anti-Parkinson activity. Int J Biol Macromol 109:27–35
87. Tzeyung AS et al (2019) Fabrication, optimization, and evaluation of rotigotine-loaded chitosan nanoparticles for nose-to-brain delivery. Pharmaceutics 11:26. https://doi.org/10.3390/pharmaceutics11010026
88. Pardeshi CV, Belgamwar VS (2019) Improved brain pharmacokinetics following intranasal administration of N, N, N-trimethyl chitosan tailored mucoadhesive NLCs. Mater Technol 35(5):249–266
89. Wen Z et al (2011) Odorranalectin-conjugated nanoparticles: preparation, brain delivery and pharmacodynamic study on Parkinson's disease following intranasal administration. J Control Release 151(2):131–138
90. Bi C et al (2016) Intranasal delivery of rotigotine to the brain with lactoferrin-modified PEG-PLGA nanoparticles for Parkinson's disease treatment. Int J Nanomed 11:6547–6559
91. Yan X et al (2018) Lactoferrin-modified rotigotine nanoparticles for enhanced nose-to-brain delivery: LESA-MS/MS-based drug biodistribution, pharmacodynamics, and neuroprotective effects. Int J Nanomed 13:273–281
92. Hu K et al (2011) Lactoferrin conjugated PEG-PLGA nanoparticles for brain delivery: preparation, characterization and efficacy in Parkinson's disease. Int J Pharm 415(1–2):273–283
93. Ray S et al (2018) Polysorbate 80 coated crosslinked chitosan nanoparticles of ropinirole hydrochloride for brain targeting. J Drug Deliv Sci Technol 48:21–29
94. Qu M et al (2018a) A brain targeting functionalized liposomes of the dopamine derivative N-3,4-bis(pivaloyloxy)-dopamine for treatment of Parkinson's disease. J Control Release 277:173–182

95. Tan JPK et al (2019) Effective encapsulation of apomorphine into biodegradable polymeric nanoparticles through a reversible chemical bond for delivery across the blood-brain barrier. Nanomedicine 17:236–245
96. Wolfram J, Ferrari M (2019) Clinical cancer nanomedicine. Nano Today 25:85–98
97. Celsion corporation to continue following patients in Phase III OPTIMA Study for Overall Survival. 2020; Available from: https://investor.celsion.com/news-releases/news-release-details/celsion-corporation-continue-following-patients-phase-iii-optima
98. Gao G et al (2019) Gold nanoclusters for Parkinson's disease treatment. Biomaterials 194:36–46
99. Srivastava AK, Roy Choudhury S, Karmakar S (2020) Melatonin/polydopamine nanostructures for collective neuroprotection-based Parkinson's disease therapy. Biomater Sci 8(5):1345–1363
100. You L et al (2018) Targeted brain delivery of rabies virus glycoprotein 29-modified deferoxamine-loaded nanoparticles reverses functional deficits in Parkinsonian mice. ACS Nano 12(5):4123–4139
101. Huang R et al (2013) Angiopep-conjugated nanoparticles for targeted long-term gene therapy of Parkinson's disease. Pharm Res 30(10):2549–2559
102. Ali MM, Shoukri RA, Yousry C (2023) Thin film hydration versus modified spraying technique to fabricate intranasal spanlastic nanovesicles for rasagiline mesylate brain delivery: Characterization, statistical optimization, and in vivo pharmacokinetic evaluation. Drug Deliv Transl Res 13(4):1153–1168. https://doi.org/10.1007/s13346-022-01285-5
103. Kashyap K, Shukla R (2019) Drug delivery and targeting to the brain through nasal route: mechanisms, applications and challenges. Curr Drug Deliv 16(10):887–901. https://doi.org/10.2174/1567201816666191029122740
104. Johnson NJ, Hanson LR, Frey WH (2010) Trigeminal pathways deliver a low molecular weight drug from the nose to the brain and orofacial structures. Mol Pharm 7(3):884–893. https://doi.org/10.1021/mp100029t
105. Cunha S, Amaral MH, Lobo JMS, Silva AC (2017) Lipid nanoparticles for nasal/intranasal drug delivery. Crit Rev Ther Drug Carrier Syst 34(3):257–282. https://doi.org/10.1615/CritRevTherDrugCarrierSyst.2017018693
106. Zhao Y-Z, Lu C-T (2008) Increasing the entrapment of protein-loaded liposomes with a modified freeze–thaw technique: a preliminary experimental study. Drug Dev Ind Pharm 35:165–171. https://doi.org/10.1080/03639040802220300
107. Ramalingam P, Ganesan P, Prabakaran DS, Gupta PK, Jonnalagadda S, Govindarajan K, Vishnu R, Sivalingam K, Sodha S, Choi DK, Ko YT (2020) Lipid nanoparticles improve the uptake of α-asarone into the brain parenchyma: formulation, characterization, in vivo pharmacokinetics, and brain delivery. AAPS PharmSciTech 21:1–1. https://doi.org/10.1208/s12249-020-01832-8
108. Chellian R, Pandy V, Mohamed Z (2018) Alpha-asarone attenuates depression-like behavior in nicotine-withdrawn mice: evidence for the modulation of hippocampal pCREB levels during nicotine-withdrawal. Eur J Pharmacol 818:10–16
109. Chellian R, Pandy V, Mohamed Z (2017) Pharmacology and toxicology of α-and β-asarone: a review of preclinical evidence. Phytomedicine 32:41–58
110. Wang DK, Shi ZH, Liu L, Wang XY, Zhang CX, Zhao P (2006) Development of self-microemulsifying drug delivery systems for oral bioavailability enhancement of α-asarone in beagle dogs. PDA J Pharm Sci Technol 60(6):343–349
111. Deng L, Wang Y, Gong T, Sun X, Zhang Z-R (2017) Dissolution and bioavailability enhancement of alpha-asarone by solid dispersions via oral administration. Drug Dev Ind Pharm 43(11):1817–1826
112. Wang D, Wang X, Li X, Ye L (2008) Preparation and characterization of solid lipid nanoparticles loaded with α-asarone. PDA J Pharm Sci Technol 62(1):56–65

113. Maghsoudi A, Fakharzadeh S, Hafizi M, Abbasi M, Kohram F, Sardab S et al (2015) Neuroprotective effects of three different sizes nanochelating based nano complexes in MPP (+) induced neurotoxicity. Apoptosis 20(3):298–309
114. Sachdeva AK, Misra S, Kaur IP, Chopra K (2015) Neuroprotective potential of sesamol and its loaded solid lipid nanoparticles in ICV-STZ-induced cognitive deficits: behavioral and biochemical evidence. Eur J Pharmacol 747:132–140
115. Ramalingam P, Ganesan P, Choi D-K, Ko YT (2018) Development of a selective and sensitive LC–MS/MS method for the quantification of α-asarone in mouse plasma and its application to pharmacokinetic studies. J Pharm Biomed Anal 151:284–290
116. Lu J, Guo L-W, Fu T-M, Zhu G-L, Dai Z-N, Zhan G-J et al (2017) Pharmacokinetics of α-asarone after intranasal and intravenous administration with PLA-α-asarone nanoparticles. China J Chin Mater Med 42(12):2366–2372
117. Lu J, Fu T, Qian Y, Zhang Q, Zhu H, Pan L et al (2014) Distribution of α-asarone in brain following three different routes of administration in rats. Eur J Pharm Sci 63:63–70
118. Huang Z, Kłodzińska SN, Wan F, Nielsen HM (2021) Nanoparticle-mediated pulmonary drug delivery: state of the art towards efficient treatment of recalcitrant respiratory tract bacterial infections. Drug Deliv Transl Res 11:1634–1654. https://doi.org/10.1007/s13346-021-00954-1
119. Smyth AR, Bell SC, Bojcin S, Bryon M, Duff A, Flume P et al (2014) European cystic fibrosis society standards of care: best practice guidelines. J Cyst Fibros 13(Suppl 1):S23–S42. https://doi.org/10.1016/j.jcf.2014.03.010
120. Sethi S (2010) Infection as a comorbidity of COPD. Eur Respir J 35(6):1209–1215. https://doi.org/10.1183/09031936.00081409
121. Murphy TF, Brauer AL, Eschberger K, Lobbins P, Grove L, Cai X et al (2008) Pseudomonas aeruginosa in chronic obstructive pulmonary disease. Am J Respir Crit Care Med 177(8):853–860. https://doi.org/10.1164/rccm.200709-1413OC
122. Martinez-Solano L, Macia MD, Fajardo A, Oliver A, Martinez JL (2008) Chronic Pseudomonas aeruginosa infection in chronic obstructive pulmonary disease. Clin Infect Dis 47(12):1526–1533. https://doi.org/10.1086/593186
123. Hunter JD (2012) Ventilator associated pneumonia. BMJ 344(1):e3325-e. https://doi.org/10.1136/bmj.e3325
124. Fahy JV, Dickey BF (2010) Airway mucus function and dysfunction. N Engl J Med 363(10):2233–2247
125. Ladaviere C, Gref R (2015) Toward an optimized treatment of intracellular bacterial infections: input of nanoparticulate drug delivery systems. Nanomedicine 10(19):3033–3055. https://doi.org/10.2217/nnm.15.128
126. Radovic-Moreno AF, Lu TK, Puscasu VA, Yoon CJ, Langer R, Farokhzad OC (2012) Surface charge-switching polymeric nanoparticles for bacterial cell wall-targeted delivery of antibiotics. ACS Nano 6(5):4279–4287
127. Popov A, Schopf L, Bourassa J, Chen H (2016) Enhanced pulmonary delivery of fluticasone propionate in rodents by mucus-penetrating nanoparticles. Int J Pharm 502(1–2):188–197. https://doi.org/10.1016/j.ijpharm.2016.02.031
128. Raval M, Patel P, Airao V, Bhatt V, Sheth N (2021) Novel silibinin loaded chitosan-coated PLGA/PCL nanoparticles based inhalation formulations with improved cytotoxicity and bioavailability for lung cancer. Bionanoscience 11:67–83. https://doi.org/10.1007/s12668-020-00797-z
129. Rowinsky EK, Onetto N, Canetta RM et al (1992) Taxol: the first of the taxanes, an important new class of antitumor agents. Semin Oncol 19(6):646–662
130. Jiang L, Li X, Liu L, Zhang Q (2013) Thiolated chitosan-modified PLA-PCL-TPGS nanoparticles for oral chemotherapy of lung cancer. Nanoscale Res Lett 8(1):66
131. Stockmann C, Roberts JK, Yellepeddi VK, Sherwin CM (2015) Clinical pharmacokinetics of inhaled antimicrobials. Clin Pharmacokinet 54:473–492. https://doi.org/10.1007/s40262-015-0250-x

132. Heyder J (2004) Deposition of inhaled particles in the human respiratory tract and consequences for regional targeting in respiratory drug delivery. Proc Am Thorac Soc 1(4):315–320
133. Groneberg DA, Witt C, Wagner U, Chung KF, Fischer A (2003) Fundamentals of pulmonary drug delivery. Respir Med 97(4):382–387
134. Byron PR, Phillips EM (1990) Absorption, clearance and dissolution in the lung. In: Respiratory drug delivery I, vol 1, Chap 5, pp 107–141
135. Labiris NR, Dolovich MB (2003) Pulmonary drug delivery. Part I: physiological factors affecting therapeutic effectiveness of aerosolized medications. Br J Clin Pharmacol 56(6):588–599
136. Zhou QT, Leung SS, Tang P, Parumasivam T, Loh ZH, Chan HK (2014) Inhaled formulations and pulmonary drug delivery systems for respiratory infections. Adv Drug Deliv Rev 85:83–99. https://doi.org/10.1016/j.addr.2014.10.0224
137. Wilson J, Moorehead L, Montgomery B (2008) A phase 1 placebo-controlled, double-blind, randomized trial evaluating the safety and pharmacokinetics of three escalating doses of fosfomycin/tobramycin for inhalation (FTI) in healthy volunteers [abstract]. Pediatr Pulmonol 43:321
138. Clancy JP, Dupont L, Konstan MW et al (2013) Phase II studies of nebulised Arikace in CF patients with Pseudomonas aeruginosa infection. Thorax 68(9):818–825
139. McCoy KS, Quittner AL, Oermann CM, Gibson RL, Retsch-Bogart GZ, Montgomery AB (2008) Inhaled aztreonam lysine for chronic airway Pseudomonas aeruginosa in cystic fibrosis. Am J Respir Crit Care Med 178(9):921–928
140. Gibson RL, Retsch-Bogart GZ, Oermann C et al (2006) Microbiology, safety, and pharmacokinetics of aztreonam lysinate for inhalation in patients with cystic fibrosis. Pediatr Pulmonol 41(7):656–665
141. Retsch-Bogart GZ, Burns JL, Otto KL et al (2008) A phase 2 study of aztreonam lysine for inhalation to treat patients with cystic fibrosis and Pseudomonas aeruginosa infection. Pediatr Pulmonol 43(1):47–58
142. Cass LM, Efthymiopoulos C, Bye A (1999) Pharmacokinetics of zanamivir after intravenous, oral, inhaled or intranasal administration to healthy volunteers. Clin Pharmacokinet 36(Suppl 1):1–11
143. Watanabe A, Chang SC, Kim MJ, Chu DW, Ohashi Y (2010) Long-acting neuraminidase inhibitor laninamivir octanoate versus oseltamivir for treatment of influenza: a double-blind, randomized, noninferiority clinical trial. Clin Infect Dis 51(10):1167–1175
144. Hayden F (2009) Developing new antiviral agents for influenza treatment: what does the future hold? Clin Infect Dis 48(Suppl 1):S3–S13
145. Ishizuka H, Yoshiba S, Okabe H, Yoshihara K (2010) Clinical pharmacokinetics of laninamivir, a novel long-acting neuraminidase inhibitor, after single and multiple inhaled doses of its prodrug, CS-8958, in healthy male volunteers. J Clin Pharmacol 50(11):1319–1329
146. Ishizuka H, Yoshiba S, Yoshihara K, Okabe H (2011) Assessment of the effects of renal impairment on the pharmacokinetic profile of laninamivir, a novel neuraminidase inhibitor, after a single inhaled dose of its prodrug, CS-8958. J Clin Pharmacol 51(2):243–251
147. Je B (2001) Antifungal agents. In: Hardman GELL (ed) Goodman and Gilman's the pharmacological basis of therapeutics, 10th edn. McGraw-Hill, New York, pp 1295–1312
148. Osawa R, Alexander BD, Forrest GN et al (2010) Geographic differences in disease expression of cryptococcosis in solid organ transplant recipients in the United States. Ann Transplant 15(4):77–83
149. Monforte V, Ussetti P, Lopez R et al (2009) Nebulized liposomal amphotericin B prophylaxis for Aspergillus infection in lung transplantation: pharmacokinetics and safety. J Heart Lung Transplant 28(2):170–175
150. O'Callaghan C, Barry PW (1997) The science of nebulised drug delivery. Thorax 52(Suppl 2):S31–S44
151. Coates AL, Fink J, Chantrel G, Diot P, Vecellio L (2006) In vivo justification of a physiological insiratory:expiratory ratio to predict deposition of a novel valved spacer for liquid aerosol [abstract]. Am J Respir Crit Care Med 3:A84

152. Bennett WD, Smaldone GC (1987) Human variation in the peripheral air-space deposition of inhaled particles. J Appl Physiol 62(4):1603–1610
153. Geller DE, Konstan MW, Smith J, Noonberg SB, Conrad C (2007) Novel tobramycin inhalation powder in cystic fibrosis subjects: pharmacokinetics and safety. Pediatr Pulmonol 42(4):307–313
154. Varshosaz J, Ghaffari S, Khoshayand MR, Atyabi F, Dehkordi AJ, Kobarfard F (2012) Optimization of freeze-drying condition of amikacin solid lipid nanoparticles using D-optimal experimental design. Pharm Dev Technol 17(2):187–194
155. Hayes D Jr, Murphy BS, Mullett TW, Feola DJ (2010) Aerosolized vancomycin for the treatment of MRSA after lung transplantation. Respirology 15(1):184–186
156. Kwon S, Kim KB, Yeo Y, Lee W (2021) Pharmacokinetic aspects of the clinically used proteasome inhibitor drugs and efforts toward nanoparticulate delivery systems. J Pharm Investig 51(4):483–502. https://doi.org/10.1007/s40005-021-00532-0
157. Di Costanzo A, Del Gaudio N, Conte L, Altucci L (2020) The ubiquitin proteasome system in hematological malignancies: new insight into its functional role and therapeutic options. Cancers 12:1898
158. Huang Z, Wu Y, Zhou X, Xu J, Zhu W, Shu Y, Liu P (2014) Efficacy of therapy with bortezomib in solid tumors: a review based on 32 clinical trials. Future Oncol 10:1795–1807
159. Hanahan D, Weinberg RA (2011) Hallmarks of cancer: the next generation. Cell 144:646–674
160. Oakes SA (2017) Endoplasmic reticulum proteostasis: a key checkpoint in cancer. Am J Physiol Cell Physiol 312:C93–C102
161. Sanchez-Serrano I (2006) Success in translational research: lessons from the development of bortezomib. Nat Rev Drug Discov 5:107–114
162. Arastu-Kapur S, Anderl JL, Kraus M, Parlati F, Shenk KD, Lee SJ, Muchamuel T, Bennett MK, Driessen C, Ball AJ, Kirk CJ (2011) Nonproteasomal targets of the proteasome inhibitors bortezomib and carfilzomib: a link to clinical adverse events. Clin Cancer Res 17:2734–2743
163. Kupperman E, Lee EC, Cao Y, Bannerman B, Fitzgerald M, Berger A, Yu J, Yang Y, Hales P, Bruzzese F, Liu J, Blank J, Garcia K, Tsu C, Dick L, Fleming P, Yu L, Manfredi M, Rolfe M, Bolen J (2010) Evaluation of the proteasome inhibitor MLN9708 in preclinical models of human cancer. Cancer Res 70:1970–1980
164. Ashley JD, Stefanick JF, Schroeder VA, Suckow MA, Alves NJ, Suzuki R, Kikuchi S, Hideshima T, Anderson KC, Kiziltepe T, Bilgicer B (2014a) Liposomal carfilzomib nanoparticles effectively target multiple myeloma cells and demonstrate enhanced efficacy in vivo. J Control Release 196:113–121
165. Ashley JD, Stefanick JF, Schroeder VA, Suckow MA, Kiziltepe T, Bilgicer B (2014b) Liposomal bortezomib nanoparticles via boronic ester prodrug formulation for improved therapeutic efficacy in vivo. J Med Chem 57:5282–5292
166. Swami A, Reagan MR, Basto P, Mishima Y, Kamaly N, Glavey S, Zhang S, Moschetta M, Seevaratnam D, Zhang Y, Liu J, Memarzadeh M, Wu J, Manier S, Shi J, Bertrand N, Lu ZN, Nagano K, Baron R, Sacco A, Roccaro AM, Farokhzad OC, Ghobrial IM (2014) Engineered nanomedicine for myeloma and bone microenvironment targeting. Proc Natl Acad Sci USA 111:10287–10292
167. Shen S, Du XJ, Liu J, Sun R, Zhu YH, Wang J (2015) Delivery of bortezomib with nanoparticles for basal-like triple-negative breast cancer therapy. J Control Release 208:14–24
168. Park JE, Chun SE, Reichel D, Min JS, Lee SC, Han S, Ryoo G, Oh Y, Park SH, Ryu HM, Kim KB, Lee HY, Bae SK, Bae Y, Lee W (2017) Polymer micelle formulation for the proteasome inhibitor drug carfilzomib: anticancer efficacy and pharmacokinetic studies in mice. PLoS One 12:e0173247
169. Tsushima T, Kasai H, Tanigawara Y (2020) Pharmacokinetic and pharmacodynamic analysis of neutropenia following nab-paclitaxel administration in Japanese patients with metastatic solid cancer. Cancer Chemother Pharmacol 86:487–495. https://doi.org/10.1007/s00280-020-04140-x

170. Gradishar WJ, Tjulandin S, Davidson N, Shaw H, Desai N, Bhar P, Hawkins M, O'Shaughnessy J (2005) Phase III trial of nanoparticle albumin-bound paclitaxel compared with polyethylated castor oil-based paclitaxel in women with breast cancer. J Clin Oncol 23(31):7794–7803. https://doi.org/10.1200/JCO.2005.04.937
171. Shitara K, Takashima A, Fujitani K, Koeda K, Hara H, Nakayama N, Hironaka S, Nishikawa K, Makari Y, Amagai K, Ueda S, Yoshida K, Shimodaira H, Nishina T, Tsuda M, Kurokawa Y, Tamura T, Sasaki Y, Morita S, Koizumi W (2017) Nab-paclitaxel versus solvent-based paclitaxel in patients with previously treated advanced gastric cancer (ABSOLUTE): an open-label, randomised, non-inferiority, phase 3 trial. Lancet Gastroenterol Hepatol 2(4):277–287. https://doi.org/10.1016/S2468-1253(16)30219-9
172. Savic RM, Karlsson MO (2009) Importance of shrinkage in empirical bayes estimates for diagnostics: problems and solutions. AAPS J 11(3):558–569
173. Shen J, Lu GW, Hughes P (2018) Targeted ocular drug delivery with pharmacokinetic/pharmacodynamic considerations. Pharm Res 35(11):217. https://doi.org/10.1007/s11095-018-2498-y
174. Pehlivan SB, Yavuz B, Calamak S, Ulubayram K, Kaffashi A, Vural I et al (2015) Preparation and in vitro/in vivo evaluation of cyclosporin A-loaded nanodecorated ocular implants for subconjunctival application. J Pharm Sci 104(5):1709–1720
175. Xu L, Lu T, Tuomi L, Jumbe N, Lu J, Eppler S et al (2013) Pharmacokinetics of ranibizumab in patients with neovascular age-related macular degeneration: a population approach. Invest Ophthalmol Vis Sci 54(3):1616–1624
176. Peña-Juárez MC, Guadarrama-Escobar OR, Escobar-Chávez JJ (2021) Transdermal delivery systems for biomolecules. J Pharm Innov:1–4. https://doi.org/10.1007/s12247-020-09525-2
177. Scheuplein RJ (1967) Mechanism of percutaneous absorption. II. Transient diffusion and the relative importance of various routes of skin penetration. J Investig Dermatol 48:79–88
178. Wiechers JW (1989) The barrier function of the skin in relation to percutaneous absorption of drugs. Pharm Weekbl Sci Ed 11(6):185–198
179. Tran PHL, Duan W, Lee BJ, Tran TTD (2019) Nanogels for skin cancer therapy via transdermal delivery: current designs. Curr Drug Metab 20(7):575–582
180. Toyoda M, Hama S, Ikeda Y, Nagasaki Y, Kogure K (2015) Anti-cancer vaccination by transdermal delivery of antigen peptide-loaded nanogels via iontophoresis. Int J Pharm 483(1–2):110–114
181. Blattner CM, Coman G, Blickenstaff NR, Maibach HI (2014) Percutaneous absorption of water in skin: a review. Rev Environ Health 29(3):175–180
182. Xu B, Jiang G, Yu W, Liu D, Zhang Y, Zhou J et al (2017) H2O2-responsive mesoporous silica nanoparticles integrated with microneedle patches for the glucose-monitored transdermal delivery of insulin. J Mater Chem B 5(41):8200–8208
183. Parra-Guillén ZP, González-Aseguinolaza G, Berraondo P, Trocóniz IF (2010) Gene therapy: a pharmacokinetic/pharmacodynamic modelling overview. Pharm Res 27:1487–1497. https://doi.org/10.1007/s11095-010-0136-4
184. Gad SC (ed) (2007) Handbook of pharmaceutical biotechnology. Wiley, New York
185. Breimer D (2008) PK/PD modelling and beyond: impact on drug development. Pharm Res 25:2720–2722
186. Nomura T, Nakajima S, Kawabata K, Yamashita F, Takakura Y, Hashida M (1997) Intratumoral pharmacokinetics and in vivo gene expression of naked plasmid DNA and its cationic liposome complexes after direct gene transfer. Cancer Res 57:2681–2686
187. Mok W, Stylianopoulos T, Boucher Y, Jain RK (2009) Mathematical modeling of herpes simplex virus distribution in solid tumors: implications for cancer gene therapy. Clin Cancer Res 15:2352–2360
188. Nishikawa M, Hashida M (2002) Nonviral approaches satisfying various requirements for effective in vivo gene therapy. Biol Pharm Bull 25:275–283
189. Varga CM, Hong K, Lauffenburger DA (2001) Quantitative analysis of synthetic gene delivery vector design properties. Mol Ther 4:438–446

190. Moriguchi R, Kogure K, Harashima H (2008) Non-linear pharmacodynamics in the transfection efficiency of a non-viral gene delivery system. Int J Pharm 363:192–198
191. Kamiya H, Tsuchiya H, Yamazaki J, Harashima H (2001) Intracellular trafficking and transgene expression of viral and non-viral gene vectors. Adv Drug Deliver Rev 52:153–164
192. Borrajo ML, Alonso MJ (2022) Using nanotechnology to deliver biomolecules from nose to brain—peptides, proteins, monoclonal antibodies and RNA. Drug Deliv Transl Res 1:1–9. https://doi.org/10.1007/s13346-021-01086-2
193. Engelhardt B, Sorokin L (2009) The blood-brain and the blood-cerebrospinal fluid barriers: function and dysfunction. Semin Immunopathol 31:497–511
194. Lochhead JJ, Thorne RG (2012) Intranasal delivery of biologics to the central nervous system. Adv Drug Deliv Rev 64:614–628
195. Warnken ZN, Smyth HDC, Watts AB, Weitman S, Kuhn JG, Williams RO (2016) Formulation and device design to increase nose to brain drug delivery. J Drug Deliv Sci Technol 35:213–222
196. Dhuria SV, Hanson LR, Frey WH II. (2010) Intranasal delivery to the central nervous sytem: mechanisms and experimental consideration. J Pharm Sci Sci 99:2386–2398
197. Landis MS, Boyden T, Pegg S (2012) Nasal-to-CNS drug delivery: where are we now and where are we heading? An industrial perspective. Ther Deliv 3:195–208
198. Pardeshi CV, Belgamwar VS (2013) Direct nose to brain drug delivery via integrated nerve pathways bypassing the blood–brain barrier: an excellent platform for brain targeting. Expert Opin Drug Deliv 10:957–972
199. Mittal D, Ali A, Md S, Baboota S, Sahni JK, Ali J (2014) Insights into direct nose to brain delivery: current status and future perspective. Drug Deliv 21:75–86
200. Samaridou E, Alonso MJ (2018) Nose-to-brain peptide delivery – the potential of nanotechnology. Bioorganic Med Chem 26:2888–2905
201. Bourganis V, Kammona O, Alexopoulos A, Kiparissides C (2018) Recent advances in carrier mediated nose-to-brain delivery of pharmaceutics. Eur J Pharm Biopharm 128:337–362
202. Illum L (2000) Transport of drugs from the nasal cavity to the central nervous system. Eur J Pharm Sci 11:1–18
203. Kumar A, Pandey AN, Jain SK (2016) Nasal-nanotechnology: revolution for efficient therapeutics delivery. Drug Deliv 23:681–693
204. Moran DT, Rowley JC, Jaferk BW, Lovell MA (1982) The fine structure of the olfactory mucosa in man. J Neurocytol 11:721–746
205. Gizurarson S (2012) Anatomical and histological factors affecting intranasal drug and vaccine delivery. Curr Drug Deliv 9:566–582
206. Mistry A, Stolnik S, Illum L (2009) Nanoparticles for direct nose-to-brain delivery of drugs. Int J Pharm 379:146–157
207. Wu H, Hu K, Jiang X (2008) From nose to brain: understanding transport capacity and transport rate of drugs. Expert Opin Drug Deliv 5:1159–1168
208. Djupesland PG, Messina JC, Mahmoud RA (2014) The nasal approach to delivering treatment for brain diseases: an anatomic, physiologic, and delivery technology overview. Ther Deliv 5:709–733
209. Illum L (2004) Is nose-to-brain transport of drugs in man a reality? J Pharm Pharmacol 56:3–17
210. Hanson LR, Frey WH (2008) Intranasal delivery bypasses the blood-brain barrier to target therapeutic agents to the central nervous system and treat neurodegenerative disease. BMC Neurosci 9:1–4
211. Keller LA, Merkel O, Popp A (2021) Intranasal drug delivery: opportunities and toxicologic challenges during drug development. Drug Deliv Transl Res:1–23
212. Harkema JR, Carey SA, Wagner JG (2006) The nose revisited: a brief review of the comparative structure, function, and toxicologic pathology of the nasal epithelium. Toxicol Pathol 34:252–269
213. Cuschieri A (1974) Enzyme histochemistry of the olfactory mucosa and vomeronasal organ in the mouse. J Anat 118:477–489

214. Lee VHL, Yamamoto A (1989) Penetration and enzymatic barriers to peptide and protein absorption. Adv Drug Deliv Rev 4:171–207
215. Hu J, Sheng L, Li L, Zhou X, Xie F, D'Agostino J et al (2014) Essential role of the cytochrome P450 enzyme CYP2A5 in olfactory mucosal toxicity of naphthalene. Drug Metab Dispos 42:23–27
216. Craft S, Raman R, Chow TW, Rafii MS, Sun CK, Rissman RA et al (2020) Safety, efficacy, and feasibility of intranasal insulin for the treatment of mild cognitive impairment and Alzheimer disease dementia: a randomized clinical trial. JAMA Neurol 77:1099–1109
217. Rosenbloom M, Barclay TR, Kashyap B, Hage L, O'Keefe LR, Svitak A et al (2021) A phase II, single-center, randomized, double-blind, placebo-controlled study of the safety and therapeutic efficacy of intranasal glulisine in amnestic mild cognitive impairment and probable mild Alzheimer's disease. Drugs Aging 38:407–415
218. Vllasaliu D, Exposito-Harris R, Heras A, Casettari L, Garnett M, Illum L et al (2010) Tight junction modulation by chitosan nanoparticles: comparison with chitosan solution. Int J Pharm 400:183–193
219. Casettari L, Illum L (2014) Chitosan in nasal delivery systems for therapeutic drugs. J Control Release 190:189–200
220. Kubek MJ, Domb AJ, Veronesi MC (2009) Attenuation of kindled seizures by intranasal delivery of neuropeptide-loaded nanoparticles. Neurotherapeutics 6:359–371
221. Wirth M, Hamilton G, Gabor F (1998) Lectin-mediated drug targeting: quantification of binding and internalization of wheat germ agglutinin and Solanum tuberosum lectin using Caco-2 and HT-29 cells. J Drug Target 6:95–104
222. Lundh B, Brockstedt U, Kristensson K (1989) Lectin-binding pattern of neuroepithelial and respiratory epithelial cells in the mouse nasal cavity. Histochem J 21:33–43
223. Zhao YZ, Li X, Lu CT, Lin M, Chen LJ, Xiang Q et al (2014) Gelatin nanostructured lipid carriers-mediated intranasal delivery of basic fibroblast growth factor enhances functional recovery in hemiparkinsonian rats. Nanomed Nanotechnol Biol Med 10:755–764
224. Kanazawa T, Kurano T, Ibaraki H, Takashima Y, Suzuki T, Seta Y (2019) Therapeutic effects in a transientmiddle cerebral artery occlusion ratmodel by nose-to-brain delivery of anti-TNF-alpha siRNA with cell-penetrating peptide-modified polymer micelles. Pharmaceutics 11:478
225. Zhang M, Lu L, Ying M, Ruan H, Wang X, Wang H, Chai Z, Wang S, Zhan C, Pan J, Lu W (2018a) Enhanced glioblastoma targeting ability of carfilzomib enabled by a (d)A7R-modified lipid nanodisk. Mol Pharm 15:2437–2447
226. Zhang X, Yuan T, Dong H, Xu J, Wang D, Tong H, Ji X, Sun B, Zhu M, Jiang X (2018b) Novel block glycopolymers prepared as delivery nanocarriers for controlled release of bortezomib. Colloid Polym Sci 296:1827–1839
227. Taha MS, Cresswell GM, Park J, Lee W, Ratliff TL, Yeo Y (2019) Sustained delivery of carfilzomib by tannic acid-based nanocapsules helps develop antitumor immunity. Nano Lett 19:8333–8341

Chapter 6
Artificial Intelligence and Machine Learning in Pharmacokinetics and Pharmacodynamic Studies

Sankalp A. Gharat, Munira M. Momin, and Tabassum Khan

Introduction by the Editor

Artificial intelligence (AI) and machine learning (ML) have emerged as powerful tools in pharmacokinetics (PK) and pharmacodynamics (PD) studies, revolutionizing drug development and personalized medicine [1]. AI and ML algorithms analyse massive datasets, integrate complicated variables, and uncover patterns and correlations that traditional approaches lack [2]. In PK studies, AI and ML models can predict drug absorption, distribution, metabolism, and excretion parameters, enabling more accurate dosage recommendations and individualized treatment plans [3]. Moreover, in PD studies, these techniques can elucidate intricate relationship between drug concentration and pharmacological effect, aiding in dose optimization, response prediction, and the identification of potential biomarkers [4, 5]. AI and ML have the potential to accelerate the drug discovery process, optimize therapeutic regimen, and contribute to precision medicine by providing valuable insight into the complex interaction between drugs, diverse patient pool and diseases [6–8]. There are several software tools and programming libraries available that can be used for applying AI and ML techniques to PK and PD studies of nanoparticles. Some commonly used AI/ML tools are mentioned in Table 6.1.

When choosing a software tool, several factors are taken into account: the researcher's experience with the programming language, the distinctive features

S. A. Gharat · M. M. Momin (✉)
Department of Pharmaceutics, SVKM's Dr. Bhanuben Nanavati College of Pharmacy, Mumbai, Maharashtra, India
e-mail: sankalp.gharat@bncp.ac.in; munira.momin@bncp.ac.in

T. Khan
Department of Pharmaceutical Chemistry and Quality Assurance, SVKM's Dr. Bhanuben Nanavati College of Pharmacy, Mumbai, Maharashtra, India
e-mail: tabassum.khan@bncp.ac.in

S. A. Gharat et al. (eds.), *Pharmacokinetics and Pharmacodynamics of Novel Drug Delivery Systems: From Basic Concepts to Applications*,
https://doi.org/10.1007/978-981-99-7858-8_6

Table 6.1 Overview of various AI/ML tools for PK-PD studies of nanoparticles [9–15]

Sr. no	AI-ML tools	Description
1	Python	Python is a popular programming language with a rich ecosystem of libraries for AI and ML. Libraries like NumPy, Pandas, Scikit-learn, TensorFlow, and Keras provide tools for data preprocessing, modeling, and analysis. Python's versatility makes it a preferred choice for many researchers in pharmacokinetics and pharmacodynamics studies.
2	R	R is another programming language commonly used for data analysis and statistical modeling. It offers a wide range of packages for data manipulation, visualization, and modeling, such as caret and glmnet. **Caret:** Provides tools for training and evaluating ML models. **Glmnet:** Useful for regression tasks, such as predicting PK/PD outcomes.
3	MATLAB	MATLAB provides a user-friendly environment for data analysis, modeling, and simulation. It offers various toolboxes and functions for AI and ML, allowing researchers to develop and implement complex PK-PD models. **Statistics and Machine Learning Toolbox:** Offers functions for regression, classification, clustering, and more. **Neural Network Toolbox:** Useful for building and training neural networks.
4	KNIME	KNIME is an open-source platform for data analytics, reporting, and integration. It offers a graphical interface for building data pipelines and workflows, making it accessible to researchers without extensive programming skills.
5	Weka	Weka is a collection of machine learning algorithms for data mining tasks. It provides a user-friendly graphical interface for data preprocessing, classification, regression, and clustering
6	Orange	Orange is an open-source data visualization and analysis tool that includes components for AI and ML. It offers a visual programming interface for building and evaluating models.
7	Azure Machine Learning	Microsoft's Azure Machine Learning platform provides cloud-based tools for building, training, and deploying machine learning models. It offers various services for data preprocessing, feature engineering, and model deployment
8	Google Colab	Google Colab is a free cloud-based platform that provides Jupyter notebooks with built-in GPU support. It's a great option for running AI and ML experiments without the need for powerful local hardware.
9	DeepChem	DeepChem is an open-source library specifically designed for drug discovery and cheminformatics. It offers tools for molecular modeling, structure-activity relationship prediction, and more
10	PyBioMed	PyBioMed is a collection of Python tools for various bioinformatics and cheminformatics tasks. It might be useful for analyzing complex interactions involving nanoparticles
11	NONMEM	Nonlinear Mixed Effects Modeling is a widely used software tool for population pharmacokinetic (PK) and pharmacodynamic (PD) modeling. It's particularly designed for analyzing complex data from clinical trials and can be applied to various drug formulations, including nanoparticles. While NONMEM is not typically considered an AI/ML tool, it is a powerful platform for advanced PK/PD modeling. NONMEM allows researchers to develop population-based models that account for inter-individual variability, modeling the effects of covariates, and describing complex time-course data. It's widely used in pharmaceutical research for optimizing dosing regimens, understanding drug interactions, and making informed decisions during drug development.

required for PK-PD nanoparticle investigations, and the accessible libraries and resources that support the study aims. Furthermore, the choice of software often depends on the specific methodologies and approaches that the researcher intends to apply in AI and ML investigations. Applying AI/ML techniques to pharmacokinetics (PK) and pharmacodynamics (PD) studies presents numerous opportunities, but also comes with its own set of challenges and considerations. Some of the challenges are as follows:

Data Quality: Accurate and reliable data are essential for building effective AI models. Ensuring data quality and standardization is crucial.

Model Interpretability: Many AI models, especially deep learning ones, can be complex and difficult to interpret. Efforts are being made to develop models that are more transparent and explainable.

Validation and Generalization: AI models need to be validated using diverse datasets to ensure they generalize well across different patient populations and settings.

Regulatory Acceptance: While AI and ML offer powerful tools, regulatory agencies require transparency, validation, and evidence of reliability before these models can impact clinical decision-making.

Incorporating AI and ML into PK-PD studies requires interdisciplinary collaboration between pharmacologists, statisticians, computer scientists, and clinicians. These technologies hold the promise of accelerating drug development, optimizing therapy, and improving patient outcomes in the field of pharmacokinetics and pharmacodynamics. This chapter focuses on applications of in-silico and AI/ML models for predicting PK/PD of nanoparticles.

6.1 In-Silico Approach for Predicting PK/PD: In Vitro-In Vivo Extrapolation (IVIVE)

Introduction by the Editor

In silico approaches have emerged as powerful tools in pharmaceutical development to predict the pharmacokinetic (PK) and pharmacodynamic (PD) properties of drugs, offering insights that streamline drug discovery and optimization processes. One such approach is in vitro-in vivo extrapolation (IVIVE), which bridges the gap between laboratory experiments and real-world clinical outcomes. IVIVE involves the integration of in vitro data, often obtained from cell cultures or tissue studies, with mathematical models to predict how drugs will behave in vivo. This predictive modeling provides valuable information about drug absorption, distribution, metabolism, and elimination (ADME) in the human body. The IVIVE process typically begins with the generation of in vitro data, such as drug metabolism kinetics, cellular uptake rates, and binding affinities. These data are then used to develop mathematical models that simulate drug behaviour in vivo, factoring in physiological parameters, such as blood flow rates, organ sizes, and metabolic enzyme activities. These models aim to replicate the complexity of the human body, allowing researchers to predict how drugs will be processed and interact with target receptors or enzymes. The significance of IVIVE lies in its ability to guide decision-making throughout the drug development pipeline. It aids in optimizing drug dosing regimens, predicting potential drug-drug interactions, and identifying key factors influencing drug efficacy and safety. IVIVE is especially valuable when clinical trials are limited, expensive, or ethically challenging, as it enables researchers to gain insights into drug behaviour without directly involving human subjects. Accurate predictions rely on the availability of comprehensive and high-quality in vitro data, as well as the development of robust mathematical models that accurately reflect in vivo conditions. Variability among individuals, such as genetics and disease states, can impact the accuracy of predictions. In conclusion, in silico approaches like IVIVE hold immense potential to transform drug development and regulatory processes. By combining in vitro data with predictive models, IVIVE empowers researchers to anticipate drug behaviour in humans, enhancing decision-making and ultimately expediting the delivery of safe and effective therapeutics to patients.

Machine generated summaries

Disclaimer: The summaries in this chapter were generated from Springer Nature publications using extractive AI auto-summarization: An extraction-based summarizer aims to identify the most important sentences of a text using an algorithm and uses those original sentences to create the auto-summary (unlike generative AI). As the constituted sentences are machine selected, they may not fully reflect the body of the work, so we strongly advise that the original content is read and cited. The auto generated summaries were curated by the editor to meet Springer Nature publication standards.

To cite this content, please refer to the original papers.

Machine generated keywords: concentrationtime profile, extrapolation, pbpk, concentrationtime, vitroin, vitroin vivo, vitro, modeling, kinetic, quantitative, human, construct, testing, vivo, increasingly

Importance of in vitro conditions for modeling the in vivo dose in humans by in vitro–in vivo extrapolation (IVIVE) [16] This is a machine-generated summary of:

Algharably, Engi Abdel Hady; Kreutz, Reinhold; Gundert-Remy, Ursula: Importance of in vitro conditions for modeling the in vivo dose in humans by in vitro–in vivo extrapolation (IVIVE) [16]

Published in: Archives of Toxicology (2019)

Link to original: https://doi.org/10.1007/s00204-018-2382-x

If you want to cite the papers, please refer to the original.

For technical reasons we could not place the page where the original quote is coming from.

Abstract-Summary "We set out to apply physiologically based pharmacokinetic (PBPK) modeling to predict the in vivo dose of amiodarone that leads to the same concentration–time profile in the supernatant and the cell lysate of cultured primary human hepatic cells (PHH)."

"A PBPK human model was constructed based on the structure and tissue distribution of amiodarone in a rat model and using physiological human parameters."

"Using the validated kinetic model, we subsequently described the in vitro concentration–time data of amiodarone in PHH culture."

"This could be only appropriately modeled under conditions of zero protein binding and the very low clearance of the in vitro system in PHH culture."

"Our results reveal that, for meaningful quantitative extrapolation from in vitro to in vivo conditions in PBPK studies, it is essential to avoid non-intended differences between these conditions."

Introduction

"Concentration–response curves generated from in vitro data are not directly applicable for human risk and safety assessment, because concentration–response curves for effects on cells in culture do not define safe exposure levels in terms of in vivo doses (Louisse and others 1)."

"Using modeling approaches, e.g., physiologically based pharmacokinetic (PBPK) modeling, in vitro-derived data can be integrated to allow for 'reverse dosimetry' or quantitative in vitro–in vivo extrapolation (QIVIVE)."

"We used available in vitro biokinetic data on amiodarone (Pomponio and others 2) for an in vitro-in vivo extrapolation."

"Our aim was to predict the in vivo amiodarone dose in man leading to the same concentration–time profile as in the in vitro study using the QIVIVE approach which was validated by comparing the results of the modeling with experimental kinetic data in humans."

Methods

"The output of the oral model was validated by comparing plasma concentration–time data with the experimental profiles after oral administration of amiodarone in humans (Kannan and others 3)."

"For the evaluation of the data in the supernatant and in the liver cells obtained from the in vitro hepatocyte model (Pomponio and others 2), we used the kinetic parameter which they had calculated from the in vitro data (in vitro biokinetics) for a human model and compared the output with the experimental in vivo concentration–time profile."

"To determine the in vivo dose leading to the same concentration–time profile as in the in vitro study in both the supernatant and the liver cells, the concentration–time profile was simulated in an iterative process with different doses using both the in vitro metabolic parameters from the in vitro PHH study (Pomponio and others 2) and the in vitro metabolic parameters obtained from human liver microsomes (Trivier and others 4)."

Results

"In the evaluation step, using the clearance value based on the in vitro data from human liver microsomes (Trivier and others 4), we predicted the plasma concentration–time profile after i.v."

"To determine the dose that would lead to the same in vivo concentrations in plasma and liver in humans, and in the in vitro experiment in the supernatant and the hepatic cells, respectively, we employed dose simulations using the kinetic model that produced results in conformity with the in vivo data."

"According to this model simulation, the in vivo dose necessary to reach the in vitro concentrations in the supernatant was 7 mg/kg bw, whereas the concentration–time course in the liver cells could only be reached with an in vivo dose at least tenfold higher."

"We varied the dose between 5 and 40 mg/kg given into the human system and compared the SSQs of the resulting concentration–time simulation in the liver cells for these doses."

Discussion

"The current paper describes the reverse dosimetry for amiodarone for which measured concentration–time (biokinetic) data were available from an in vitro study in a human hepatic cell model system."

"Performing reverse dosimetry using the validated model to simulate the in vitro concentration–time profile in both the supernatant and liver cells—the latter representing one of the most relevant sites of amiodarone toxicity—the obtained simulation trials were difficult to fit to the in vitro data."

"A second important factor to consider is the in vitro clearance of amiodarone in PHH cells which was substantially lower [0.002 L/h (Pomponio and others 2)] than the reported in vivo clearance (~ 7 L/h), whereas the clearance resulting from the microsomal data used for the modeling was 5.1 L/h. We also observed a decline in the clearance in the in vitro system with respect to time."

6.2 AI/ML Based Software/Tools for Predicting PK/PD

Introduction by the Editor

AI/ML-based software and tools have revolutionized the field of pharmacokinetics (PK) and pharmacodynamics (PD) prediction of nanoparticles, offering powerful computational methods to streamline drug development. These tools leverage the immense data processing capabilities of artificial intelligence and machine learning algorithms to model and predict the complex interactions between nanoparticles and biological systems. These software and tools employ algorithms that analyse diverse physicochemical properties of nanoparticles, such as size, surface charge, and composition, alongside biological factors like tissue permeability, protein binding, and cellular uptake. By integrating this information, AI/ML models predict nanoparticle biodistribution, accumulation, and clearance dynamics. Machine learning algorithms utilize large datasets of in vitro and in vivo experiments to identify patterns and correlations that would be challenging to discern manually. They learn from these datasets to create predictive models that estimate key PK parameters like half-life, distribution volume, and clearance rate, as well as PD responses and toxicity. AI/ML-based tools enhance efficiency in nanoparticle development by expediting preclinical evaluations. These tools can aid in optimizing nanoparticle design, identifying ideal drug release kinetics, and minimizing off-target effects. Additionally, they facilitate the translation of laboratory findings into clinical predictions, offering insights into dosing regimens and potential therapeutic outcomes. Despite their immense potential, AI/ML-based tools require accurate and comprehensive datasets for training to ensure reliable predictions. Variability in experimental conditions, limited availability of specific nanoparticle datasets, and the need for data validation remain challenges. However, as these tools continue to evolve, they hold promise for accelerating nanoparticle-based drug development, improving treatment strategies, and fostering personalized medicine approaches in the realm of nanomedicine.

Machine generated summaries

Disclaimer: The summaries in this chapter were generated from Springer Nature publications using extractive AI auto-summarization: An extraction-based summarizer aims to identify the most important sentences of a text using an algorithm and uses those original sentences to create the auto-summary (unlike generative AI). As the constituted sentences are machine selected, they may not fully reflect the body of the work, so we strongly advise that the original content is read and cited. The auto generated summaries were curated by the editor to meet Springer Nature publication standards.

To cite this content, please refer to the original papers.

Machine generated keywords: nonmem, estimation, pkpd model, pkpd, model, modeling, likelihood, conditional, algorithm, covariate, run, effect model, implement, scenario, parameter

Evaluation of FOCEI and SAEM Estimation Methods in Population Pharmacokinetic Analysis Using NONMEM® Across Rich, Medium, and Sparse Sampling Data [17] This is a machine-generated summary of:

Sukarnjanaset, Waroonrat; Wattanavijitkul, Thitima; Jarurattanasirikul, Sutep: Evaluation of FOCEI and SAEM Estimation Methods in Population Pharmacokinetic Analysis Using NONMEM® Across Rich, Medium, and Sparse Sampling Data [17]

Published in: European Journal of Drug Metabolism and Pharmacokinetics (2018)

Link to original: https://doi.org/10.1007/s13318-018-0484-8

If you want to cite the papers, please refer to the original.

For technical reasons we could not place the page where the original quote is coming from.

Abstract-Summary "First-order conditional estimation with interaction (FOCEI) is one of the most commonly used estimation methods in nonlinear mixed effects modeling, while the stochastic approximation expectation maximization (SAEM) is the newer estimation algorithm."

"This work aimed to compare the performance of FOCEI and SAEM methods when using NONMEM® with the classical one- and two-compartment models across rich, medium, and sparse data."

"One- and two-compartment models of the previous studies were used to simulate data in three scenarios: rich, medium, and sparse data."

"Every data set was estimated with both FOCEI and SAEM methods."

"Both FOCEI and SAEM methods provided comparable completion rates, median %RERs (ranged from −9.03 to 3.27% for FOCEI and −9.17 to 3.27% for SAEM) and RMSEs (ranged from 0.0004 to 1.244 for FOCEI and 0.0004 to 1.131 for SAEM) for most parameters in both models across three scenarios."

"For the classical one- and two-compartment models, FOCEI method exhibited comparable performance similar to SAEM method but with significantly shorter runtimes across rich, medium, and sparse sampling scenarios."

Introduction

"Johansson and others [18] investigated performance of different estimation methods available in NONMEM® 7 for a various set of PD models."

"Liu and others [19] evaluated three EM methods [SAEM, Monte Carlo importance sampling parametric EM (IMP), quasi-random parametric EM (QREM)] and FOCE for their accuracy and speed when solving complex population physiologically based pharmacokinetic models."

"Most of the published population PK analyses in clinical studies, both rich and sparse data, still use FOCEI and the pharmacokinetic behavior of most drugs can be described by the classical one- or two-compartment model."

"We are interested in comparing the performance (accuracy, precision, completed estimations, and runtimes) of FOCEI and SAEM estimation methods in population PK analysis using NONMEM® when implemented with the classical one- and two-compartment models across rich, medium, and sparse sampling data."

Methods

"We simulated concentration–time profiles in three different scenarios for each model: rich sampling data (12 samples per subject for one-compartment model; 21 samples per subject for two-compartment model as previously described [20]), medium sampling data (7 samples per subject), and sparse sampling data (3 samples per subject)."

"Every simulated data set was separately estimated with both FOCEI and SAEM estimation methods with default options to obtain parameter estimates (see Electronic Supplementary material for the control stream used)."

"The completion rates were calculated from the proportion of data sets that produced parameter estimates with each algorithm, however; when the completion rates were less than 50%, the accuracy and precision were not evaluated."

"Runtimes were recorded as the computation (CPU) time needed to estimate parameters for each data set with each algorithm."

Results

"Full completion rate (100%) was obtained for all scenarios except sparse data sets from two-compartment model (21% completion rate), and therefore, the accuracy and precision were not assessed for this scenario because of failing to meet the 50% completion criterion."

"For the one-compartment model, both FOCEI and SAEM methods comparably estimated all five parameters across three scenarios."

"They provided accurate parameter estimates, although a few marked deviations were found in some random effect parameters particularly with the sparse data."

"For the two-compartment model, both FOCEI and SAEM exhibited comparable estimations from the medium or rich data."

Discussion

"This study evaluated the performance of FOCEI and SAEM estimation methods in population PK analysis when implemented with the classical one- and two-compartment models across the rich, medium, and sparse data."

"FOCEI and SAEM provided full completion rate (100%) for most scenario except when implemented with the sparse data of two-compartment model, the completion rates were equally low (21%)."

"Both methods provided comparably accurate and precise parameter estimates of the one- and two-compartment models for all studied scenarios."

"Similar to a previous study with rich sampling data, fixed effect parameter estimates of both FOCEI and SAEM methods with naïve options were similarly close to their true values, while random effect parameter estimates had more deviation [21]."

"Although SAEM, using exact likelihood algorithm, has been documented that it would be a useful method for the sparse data, the results of this study demonstrated that SAEM did not show higher accuracy and precision than FOCEI."

Conclusions

"The performance of FOCEI and SAEM methods was evaluated using simulated data from previously published population pharmacokinetics models of piperacillin in patients."

"For the classical one- and two-compartment models with default options, FOCEI exhibited comparable performance similar to SAEM but with significantly shorter runtimes across sparse, medium, and rich data scenarios."

"Each of them showed noticeable, but comparable, bias in some parameters of both models."

"FOCEI would be the appropriate method for the classical one- and two-compartment models."

Computing optimal drug dosing with OptiDose: implementation in NONMEM [22] This is a machine-generated summary of:

Bachmann, Freya; Koch, Gilbert; Bauer, Robert J.; Steffens, Britta; Szinnai, Gabor; Pfister, Marc; Schropp, Johannes: Computing optimal drug dosing with OptiDose: implementation in NONMEM [22]

Published in: Journal of Pharmacokinetics and Pharmacodynamics (2023)

Link to original: https://doi.org/10.1007/s10928-022-09840-w

If you want to cite the papers, please refer to the original.

For technical reasons we could not place the page where the original quote is coming from.

Abstract-Summary "An optimal dosing algorithm (OptiDose) was developed to compute the optimal doses for any pharmacometrics/PKPD model for a given dosing scenario."

"To demonstrate the potential of the OptiDose implementation in NONMEM, four relevant but substantially different optimal dosing tasks are solved."

"The impact of different dosing scenarios as well as the choice of the therapeutic goal on the computed optimal doses are discussed."

Introduction

"The optimal dosing algorithm OptiDose [23, 24] solves an optimal control problem (OCP) to compute the doses which bring the associated response of the PKPD model as closely as possible to a desired therapeutic goal."

"OptiDose was specifically designed for optimal dosing tasks in PKPD modeling since it optimizes the drug doses, i.e., the control variables, generating the drug level according to the dosing scenario which then drives the model response, and it is robust with respect to jumps in the model solution occurring for intravenous (IV) bolus, oral and subcutaneous dose administration."

"In the Motivation section, the optimal dosing task is illustrated based on an indirect response model [25, 26]."

"In the Results section, the OptiDose implementation in NONMEM is applied to four relevant but substantially different PKPD models with different optimal dosing tasks."

Motivation

"As a motivating example, we consider an elevated biomarker B, characterized by an indirect response model, which shall return safely to the healthy level, e.g., a reference range."

"A clinically realistic dosing scenario for non-hospitalized patients is a daily administration of doses that may change on a weekly basis."

"The OFV is illustrated by the shaded area between the two curves."

Theoretical

"The general case of an optimal dosing task is presented."

"For given dosing scenario and therapeutic goal, the optimal control problem of OptiDose is formulated."

"Its reformulation as an optimization problem, which allows an implementation in NONMEM, is discussed."

"The reformulated problem is an optimization problem and can be interpreted as a classical nonlinear least squares problem with one data point arising from an observation at the end point T."

Methods

"In the optimization context, AMT will serve only as a factor multiplied with the dose, therefore, we set AMT = 1.0."

"Point, the model output Y is evaluated as defined in the $ERROR block of the control stream to compute the OFV."

"That F2 is the corresponding scale factor for a dose administration into the second compartment, etc In our optimization context, we choose AMT = 1.0 in the data file such that F1*AMT = F1 serves as the dose."

"For IV infusion, the infusion duration needs to be specified, e.g., by D1 if the dosing compartment is the first."

"We run the first order estimation to compute the optimal doses THETA via The -2LL option indicates the minimization of the objective function as well as that the OFV is given by the model output Y as specified in the $ERROR block."

Results

"We present four examples of PKPD models covering a broad variety of optimal dosing tasks arising in drug development and clinical pharmacology."

"Both choices are problem-specific and depend on the desired accuracy of the optimal doses, the OFV and the gradient."

"The five investigated optimal dosing tasks for the motivating example provide different optimal doses."

"These illustrate the impact of specific choices, such as the reference function or the number of repeated dose administrations, in the problem setup of the optimal dosing task."

"The OFV can be always improved by increasing the degree of freedom, i.e., the number of doses to be optimized, such as when considering scenario B instead of A. Reducing the degree of freedom usually stabilizes the underlying optimization problem and provides clinically more realistic solutions with less frequent dose changes."

Discussion

"For a given dosing scenario, the optimal doses, which bring the model response as closely as possible to a desired therapeutic goal, can be computed efficiently in NONMEM utilizing standard commands."

"User-defined items such as the maximal tolerable doses, the number of repeated dose administrations and the reference function characterizing the therapeutic goal have an impact on the computed optimal doses."

"An indispensable extension to the method will be to compute optimal doses for a target population not solely based on the model parameters of a typical individual, but to include inter-individual variability."

"In [27], the dose optimization for a target population is addressed based on a target variable and a risk (i.e., objective) function."

"Further interesting and important issues to address in future work are to perform a sensitivity analysis for the computed optimal doses to account for variability and uncertainty in the model parameters, and to include state constraints in the optimization problem."

Handling underlying discrete variables with bivariate mixed hidden Markov models in NONMEM [28] This is a machine-generated summary of:

Brekkan, A.; Jönsson, S.; Karlsson, M. O.; Plan, E. L.: Handling underlying discrete variables with bivariate mixed hidden Markov models in NONMEM [28]

Published in: Journal of Pharmacokinetics and Pharmacodynamics (2019)

Link to original: https://doi.org/10.1007/s10928-019-09658-z

If you want to cite the papers, please refer to the original.

For technical reasons we could not place the page where the original quote is coming from.

Abstract-Summary "Non-linear mixed effects models typically deal with stochasticity in observed processes but models accounting for only observed processes may not be the most appropriate for all data."

"Hidden Markov models (HMMs) characterize the relationship between observed and hidden variables where the hidden variables can represent an underlying and unmeasurable disease status for example."

"The two hidden states included in the model were remission and exacerbation and two observation sources were considered, patient reported outcomes (PROs) and forced expiratory volume (FEV1)."

"Estimation properties in the software NONMEM of model parameters were investigated with and without random and covariate effect parameters."

"The influence of including random and covariate effects of varying magnitudes on the parameters in the model was quantified and a power analysis was performed to compare the power of a single bivariate MHMM with two separate univariate MHMMs."

"A drug effect was included on the transition rate probability and the precision of the drug effect parameter improved with increasing magnitude of the parameter."

"The power to detect the drug effect was improved by utilizing a bivariate MHMM model over the univariate MHMM models where the number of subject required for 80% power was 25 with the bivariate MHMM model versus 63 in the univariate MHMM FEV1 model and > 100 in the univariate MHMM PRO model."

"The results advocates for the use of bivariate MHMM models when implementation is possible."

Introduction

"Non-linear mixed effects models (NLMEs) are typically restrained to handle stochastic processes in observed variables; in contrast, hidden Markov models (HMMs) are a class of statistical models that can be used to characterize relationships between observed variables and unobserved stochastic processes."

"Incorporating the observed variables, FEV1 and PRO, simultaneously in the analysis of COPD data may give insight on the hidden patient disease status, i.e. whether patients are in remission or whether they are experiencing an exacerbation."

"Specific questions were: (i) how do random effect- and covariate (including drug effect) relationship magnitudes affect parameter estimation accuracy and precision, (ii) how well is the correlation between the two observed variables estimated with the bivariate model and what is the impact of ignoring it, and (iii) what is the power to detect a drug effect with a bivariate MHMM incorporating both observation sources simultaneously compared to two separate univariate MHMMs?"

Methods

"The stationary distribution, governing the probability to start in one state or another, and the transition probabilities, describing probabilities for a patient's disease status to move from remission to exacerbation, were modelled using a logit function, to constrain the function between 0 and 1 in case random or covariate effects were included."

"Focus was on the parameters influencing the transition probabilities (including the magnitude of drug effect and IIV) and on the correlation in the bivariate model, as estimation of these parameters may influence other parameters in the model as well."

"To determine the effect of ignoring the correlation when present in the simulations, a separate SSE analysis was performed, where the simulation model included a correlation while a reduced estimation model without correlation, in addition to the full simulation model, was employed."

"According to the general MCMP steps, a large simulated dataset (5000 individuals) was simulated from a model with a treatment effect reducing the probability to transition from remission to exacerbation."

Results

"Doubling the transition probabilities, π_{RE} and π_{ER}, from 0.05 and 0.15 to 0.1 and 0.3, respectively, moderately improved the estimation of all parameters in all scenarios, reducing the average RRMSE across all parameters by 5%."

"In scenario 4, where the drug effect was 0.5, the RRMSE of SLP was 17.6% and improved with increasing parameter magnitude of SLP (RRMSE = 10.0% and 6.7% for scenarios 1 and 3, respectively)."

"Doubling the transition probabilities (scenario 5) decreased the RRMSE of SLP in the investigations of the influence of drug effect magnitude by between 2.1 and 5.2 percentage points."

"Estimation of data simulated with ρ_R and $\rho_E = -0.33$ with a reduced model (both parameters set to zero) and a full model (ρ's estimated) resulted in an average ΔOFV of 2993 when the simulation model was based on scenario 1."

Discussion

"We included random effects on the modes (IIV), variances of the observed variable distributions (residual error) and assumed distinct distributions of PRO and FEV1 for each disease state."

"The dependence on number of transitions likely depends on the difference in the observed variables in the two modelled states, where larger differences should enable more precise estimation of "hidden state parameters"."

"The drug effect in this model influences just one of the transition rates and differences in the observed variables propagate from there."

"The number of subjects required for detection of a drug effect in this analysis was relatively low and is expected to be larger for observed variables associated with larger uncertainty but the results show that the use of a bivariate model to detect drug effects is advocated, when possible, over using single variable models."

Conclusion

"A bivariate MHMM was developed for simulating and analysing correlated continuous observations connected to hidden states."

"The data generated consisted of PROs and FEV1 measurements in COPD patients conditional on latent/hidden exacerbation/remission disease states."

"Parameters associated with the "observable" portion of the model were in general more precisely estimated than those associated with the "hidden" portion; in addition, precision depended on the magnitude of parameters such as the transition probabilities, the drug effect on the transition probabilities, IIV of the transition probability and correlation."

"The power to detect a hypothetical drug effect was consistently highest with the bivariate model compared with univariate models."

gPKPDSim: a SimBiology®-based GUI application for PKPD modeling in drug development [29] This is a machine-generated summary of:

Hosseini, Iraj; Gajjala, Anita; Bumbaca Yadav, Daniela; Sukumaran, Siddharth; Ramanujan, Saroja; Paxson, Ricardo; Gadkar, Kapil: gPKPDSim: a SimBiology®-based GUI application for PKPD modeling in drug development [29]

Published in: Journal of Pharmacokinetics and Pharmacodynamics (2018)

Link to original: https://doi.org/10.1007/s10928-017-9562-9

If you want to cite the papers, please refer to the original.

For technical reasons we could not place the page where the original quote is coming from.

Abstract-Summary "We have developed a versatile graphical user interface (GUI) application to enable easy use of any model constructed in SimBiology® to execute various common PKPD analyses."

"The MATLAB®-based GUI application, called gPKPDSim, has a single screen interface and provides functionalities including simulation, data fitting (parameter estimation), population simulation (exploring the impact of parameter variability on the outputs of interest), and non-compartmental PK analysis."

"gPKPDSim was designed primarily for use in preclinical and translational drug development, although broader applications exist."

"We illustrate the use and features of gPKPDSim using multiple PKPD models to demonstrate the wide applications of this tool in pharmaceutical sciences."

"GPKPDSim provides an integrated, multi-purpose user-friendly GUI application to enable efficient use of PKPD models by scientists from various disciplines, regardless of their modeling expertise."

Introduction

"While model development might require such expertise, certain model-based exploration and applications could be conducted by scientists from other disciplines (e.g., pharmacology, toxicology, biomarker discovery & development, bioanalytical sciences, and clinical sciences) if provided with validated models in user-friendly, graphical-user-interface (GUI) based tools."

"A variety of software and applications are available to perform PKPD analysis, modeling, and simulation."

"A simpler multi-purpose tool with a user-friendly interface can promote the broader use of PKPD models in various stages of drug development, by scientists from diverse disciplines."

"Germani and others [30] introduced the A4S Simulator, developed in MATLAB®, that uses a library of models and provides different dosing options; however, features such as non-compartmental analysis, interactive visualization and project management are limited in this program."

"We have developed a GUI-based application called gPKPDSim (Genentech PKPD Simulator) to broaden utilization of preclinical and translational PKPD models in drug development."

Methods

"The expert modeler builds the model in SimBiology® and configures it into a "Session" file, which contains specification of the SimBiology® model Variants,

Doses, Species, Parameters and Settings for use within the user interface (see Supplementary Method S1 for how to create a session file)."

"Enables simulation of the model encapsulated in the session file with different parameter values, variants, dosing regimens and simulation time."

"This functionality allows the end-user to estimate model parameter values, providing a best fit of model simulation to data."

"After the data fitting algorithm is complete, the user can save estimated parameter values as a new variant (which will appear in the list of variants and the tag column will show "Custom") within the current session, enabling use of the parameter estimates for subsequent simulation or population simulation functionality (Supplementary Method S3)."

Results

"The units for antibody amounts and concentrations are μg/kg and μg/mL. Parameters V_1 and V_2 represent the volume of central and peripheral compartments, CL represents the clearance from the central compartment, CL_d denotes the distribution clearance between the central and peripheral compartments, k_{abs} represents the absorption rate from the extravascular compartment and f_{bio} is the bioavailability, that is the fraction of drug available in the central compartment after extravascular dosing (see Supplementary File casestudy1_TwoCompPK_equations.pdf for the units of parameters and species in the model)."

"The units for antibody amounts, antibody concentrations, and target concentrations are μg/Kg, μg/mL, and ng/mL. Compared to the model presented in Case Study #1, this model has additional parameters including k_{on}, K_D, MW_{Ab}, MW_T, $t_{1/2,T}$, $T_{c,init(nM)}$, and α_{CL} representing the drug-to-target association rate constant, equilibrium dissociation constant, molecular weight of drug, molecular weight of target, half-life of target, initial concentration of target, and the ratio of the clearance of the bound drug-target complex to free drug clearance."

Discussion

"With the increasing application of PKPD modeling in pharmaceutical research and development, one key challenge is to efficiently deploy models built by expert modelers and enable scientists and collaborators with limited M&S experience to correctly use these models to answer research questions."

"To do so, we have developed gPKPDSim, a multi-purpose MATLAB® application with a user-friendly interface for deploying SimBiology® models and allowing scientists to perform various PKPD analyses and model-based tasks, regardless of their modeling expertise."

"The case studies presented in this paper demonstrate how scientists familiar with the concepts of PKPD modeling can efficiently use gPKPDSim to analyze PKPD data and address their drug research and development questions."

"We have found that by empowering pharmaceutical scientists and drug development team members to conduct straightforward model-based tasks, gPKPDSim encourages broader use of PKPD models by collaborators across multiple functions to support drug development, optimize study design, reduce animal usage and help with decision-makings."

Population pharmacokinetic reanalysis of a Diazepam PBPK model: a comparison of Stan and GNU MCSim [31] This is a machine-generated summary of:

Tsiros, Periklis; Bois, Frederic Y.; Dokoumetzidis, Aristides; Tsiliki, Georgia; Sarimveis, Haralambos: Population pharmacokinetic reanalysis of a Diazepam PBPK model: a comparison of Stan and GNU MCSim [31]

Published in: Journal of Pharmacokinetics and Pharmacodynamics (2019)

Link to original: https://doi.org/10.1007/s10928-019-09630-x

If you want to cite the papers, please refer to the original.

For technical reasons we could not place the page where the original quote is coming from.

Abstract-Summary "The aim of this study is to benchmark two Bayesian software tools, namely Stan and GNU MCSim, that use different Markov chain Monte Carlo (MCMC) methods for the estimation of physiologically based pharmacokinetic (PBPK) model parameters."

"Both tools produced very good fits at the individual and population levels, despite the fact that GNU MCSim is not able to consider multivariate distributions."

"GNU MCSim exhibited much faster convergence and performed better in terms of effective samples produced per unit of time."

Introduction

"Besides these drug-independent physiological parameters, PBPK models incorporate information about the drug as well, through the drug-related parameters."

"Adding to that, the complex structure that includes both drug-related and physiological parameters permits the creation of more informative models, where data from independent sources (especially for the physiological parameters) can be integrated together with the drug specific experimental results to create a model with more robust inference capabilities [32]."

"An alternative approach is to use again a covariate model for explaining a part of the inter-individual variability, with physiological parameters as fixed effects and drug-related parameters as random effects [33]."

"The two software platforms are used to deploy a hierarchical Bayesian model and estimate the drug-related parameters of an existing WBPBPK model, based on published clinical data, extending the comparison between NONMEM and Winbugs performed by Langdon and others [34]."

Materials and methods

"The first Bayesian tool used for obtaining the posterior distribution of the parameters was Stan, which is a statistical modeling language that can be used through many interfaces such as R, Python and Matlab and includes MCMC sampling, variational inference and penalized maximum likelihood estimation with optimization."

"The nature of the HMC algorithm allows Stan to offer some very helpful diagnostic tools, which generate alerts in the process of model fitting that are reported at the end of each run."

"If any of these tools suggests that there are pathologies, the user can refit the model by changing the parameters of NUTS or by reparameterizing the model."

"The model that was built with GNU MCSim had a warm-up period (initial samples that are discarded, also termed burn-in period) of 1000 iterations, while the models in Stan used 400 warm-up iterations."

Results

"The same procedure was followed in the GNU MCSim VPC plot, with the difference that both the population and individual parameters were drawn from univariate distributions."

"To parameter estimation results, the Stan and the GNU MCSim VPC plots are almost identical, which implies that the population structure is correctly modeled when an independence assumption is made at the population level."

"Increased computational time for the non-centered model was due to the extra parameters at the individual level."

"The parameter estimates of the centered and non-centered models were all the same with the exception of fKp, which differed in both the population and individual level."

"The non-centered model considered a higher fKp, which resulted in slightly worse individual plots and VPC plots."

Discussion

"The small number of selected parameters is also directly connected to the structural and statistical identifiability problems potentially affecting complex PBPK models [35, 36]."

"We added the rest of the body Kp to the parameters to infer on (prior information about it is weak, as there was no such parameter in the rat model), but we gained no insight about its true value (its prior remained unchanged)."

"The shape parameter of the LKJ prior had a substantial impact on the posterior of the correlation matrix, which, however, did not affect the posterior of any other parameter and, therefore, the goodness-of-fit of the model on both individual and population levels."

"For the specific combination of model, data and priors, the posterior of the centered model proved to be easily explored."

"It seems superfluous to use multivariate distributions for the specific combination of structural model and information (data and priors)."

Conclusions

"Two Bayesian software inference tools namely, Stan and GNU MCSim, were compared in this work on updating the parameters of a Diazepam population PBPK model."

"The two software tools produced almost identical parameter estimates and both exhibited very good individual and population fits."

"The goodness-of-fit of the model was not affected by the change in the posterior results of the correlation matrix, which was completely dependent on the prior, indicating the lack of information about parameter correlation in the data."

"In a future work, we are planning to investigate how the addition of more parameters affects the results of the two software tools."

Comparing the performance of first-order conditional estimation (FOCE) and different expectation–maximization (EM) methods in NONMEM: real data experience with complex nonlinear parent-metabolite pharmacokinetic model [37] This is a machine-generated summary of:

Bach, Thanh; An, Guohua: Comparing the performance of first-order conditional estimation (FOCE) and different expectation–maximization (EM) methods in NONMEM: real data experience with complex nonlinear parent-metabolite pharmacokinetic model [37]

Published in: Journal of Pharmacokinetics and Pharmacodynamics (2021)

Link to original: https://doi.org/10.1007/s10928-021-09753-0

If you want to cite the papers, please refer to the original.

For technical reasons we could not place the page where the original quote is coming from.

Abstract-Summary "First-order conditional estimation (FOCE) has been the most frequently used estimation method in NONMEM, a leading program for population pharmacokinetic/pharmacodynamic modeling."

"In NONMEM 7, expectation–maximization (EM) estimation methods and FOCE with FAST option (FOCE FAST) were introduced."

"We compared the performance of FOCE, FOCE FAST, and two EM methods, namely importance sampling (IMP) and stochastic approximation expectation–maximization (SAEM), utilizing the rich pharmacokinetic data of oxfendazole and its two metabolites obtained from the first-in-human single ascending dose study in healthy adults."

"All methods yielded similar parameter estimates, but great differences were observed in parameter precision and modeling time."

"For simpler models (i.e., models of oxfendazole and/or oxfendazole sulfone), FOCE and FOCE FAST were more efficient than EM methods with shorter run time and comparable parameter precision."

"For the most complex model (i.e., model of all three analytes, one of which having high level of data below quantification limit), FOCE failed to reliably assess parameter precision, while parameter precision obtained by IMP and SAEM was similar with SAEM being the faster method."

Introduction

"Many simulation-based studies comparing the performance of estimation methods in NONMEM as well as in other popPK/PD modeling programs have been published in the last decade [18, 19, 21, 38–41]."

"It is, therefore, the objective of our study to compare the performance of different estimation methods during popPK model development using a complicated parent-metabolite data set obtained from a first-in-human (FIH) single ascending dose (SAD) study of oxfendazole in healthy adults [42]."

"Among the various estimation methods in NONMEM, we focused on FOCE, FOCE with FAST option (henceforth referred to as FOCE FAST), IMP, and SAEM in our study because the other methods are not suitable for either highly variable/ sparse data (e.g., FO and ITS) or parameter estimation (e.g., MCMC Bayesian analysis whose main goal is to obtain sample distribution of population parameters and would need large sample size in order to provide reliable parameter estimates)."

Theoretical

"Integral of the joint density is very difficult to evaluate deterministically, thus various approaches of approximation corresponding to different estimation algorithms have been proposed."

"In FOCE, the Hessian matrix is approximated by the expected value of the second derivative of the joint density."

"The specific Hessian matrix used by NONMEM FOCE was discussed by Almquist and others[43]."

"Standard errors of parameter estimates are generated based on the variance–covariance matrix."

"S matrix is the cross-product of the gradient and its transpose, where gradient is the first derivative of the objective function evaluated at the final parameter estimates."

"R matrix is the Hessian matrix (or exact second derivative matrix) of the objective function evaluated at the final estimates."

Method

"Objective function value for hypothesis testing can be obtained by performing IMP with option EONLY = 1, which evaluates the objective function without updating the population parameters, following SAEM estimation; the values of population parameters used in this IMP step are the final values estimated by SAEM [44]."

"For that reason, the performance of FOCE, FOCE FAST, IMP, and SAEM in terms of convergence, parameter estimates, precision, and time were evaluated at each stage of structural model exploration."

"Parameter precision were assessed through relative standard error (%RSE) calculated using the equation: Parallelization efficiency was evaluated by comparing time to convergence of each estimation method when the final model (i.e., model of oxfendazole and both metabolites) optimization was performed on 1, 2, 4, 8, 12 and 16 CPUs."

Results

"Comparison of population parameter estimates and precision across three estimation methods, FOCE, IMP, and SAEM, each with two sets of estimation options showed that in general, there was good agreement in population parameter estimates among estimation methods."

"Goodness-of-fit plots for models optimized by FOCE, IMP and SAEM were similar, which was expected considering that all methods yielded comparable final estimates of population parameters."

"Because FOCE failed to provide reliable assessment of the covariance matrix and IMP was sensitive to model misspecification, model parameters optimized by SAEM were considered the final results."

"Model estimation speed of FOCE, FOCE FAST, IMP, and SAEM with ADVAN13 subroutine were evaluated separately for the three models presented above using a parallel platform with 16 CPUs and NONMEM recommended estimation options."

"Model optimization time decreased with increasing number of CPUs for all estimation methods with FOCE being the slowest method and SAEM being the fastest method at all levels of parallelization."

Discussion

"Regarding the estimation time of IMP and SAEM, Liu and Wang [19] and Gibiansky and others [21] showed that IMP was as fast or faster than SAEM for all models evaluated."

"To fill in this literature gap, the popPK model of oxfendazole and popPK model of oxfendazole and oxfendazole sulfone were used to demonstrate the relative performance of FOCE FAST to FOCE, IMP, and SAEM."

"FOCE FAST was also faster than IMP and SAEM for the two models evaluated."

"Terminated FOCE FAST analysis could not be resolved for the oxfendazole and oxfendazole sulfone model even after 5 updates of initial estimates."

"Gibiansky and others [21] reported an almost proportional increase in model optimization speed with increasing number of CPUs from 1 to 12 with IMP and SAEM, and to a slightly lesser extent with FOCE."

Conclusion

"If FOCE/FOCE FAST fails to reliably evaluate parameter precision, provided that other potential causes (e.g., model misspecification, data quality) are eliminated, IMP and SAEM might be tested."

"The relative speed of IMP and SAEM certainly depends on model complexity but there is no general rule of which method is faster under which scenario."

"IMP should still be prioritized because we evaluated the performance of IMP and SAEM with popPK/PD lifespan models (data not presented) and noticed that IMP was faster than SAEM most of the time."

Prior information for population pharmacokinetic and pharmacokinetic/pharmacodynamic analysis: overview and guidance with a focus on the NONMEM PRIOR subroutine [45] This is a machine-generated summary of:

Chan Kwong, Anna H.-X. P.; Calvier, Elisa A. M.; Fabre, David; Gattacceca, Florence; Khier, Sonia: Prior information for population pharmacokinetic and pharmacokinetic/pharmacodynamic analysis: overview and guidance with a focus on the NONMEM PRIOR subroutine [45]

Published in: Journal of Pharmacokinetics and Pharmacodynamics (2020)

Link to original: https://doi.org/10.1007/s10928-020-09695-z

If you want to cite the papers, please refer to the original.

For technical reasons we could not place the page where the original quote is coming from.

Abstract-Summary "Due to data sparseness issues, available datasets often do not allow the estimation of all parameters of the suitable model."

"The PRIOR subroutine in NONMEM supports the estimation of some or all parameters with values from previous models, as an alternative to fixing them or adding data to the dataset."

"The guidance provides general advice on how to select the most appropriate reference model when there are several previous models available, and to implement and weight the selected parameter values in the PRIOR function."

"On the model built with PRIOR, the similarity of estimates with the ones of the reference model and the sensitivity of the model to the PRIOR values should be checked."

"Covariates could be implemented a priori (from the reference model) or a posteriori, only on parameters estimated without prior (search for new covariates)."

Introduction

"When data are not sufficient to build a model, one may use prior information to stabilize the estimation of some parameters of the model."

"In population pharmacokinetics (popPK), there are two alternatives to stabilize poorly estimated parameters with prior information: either to fix them to their previous estimated values or to "inform" them thanks to their previous estimated values. """

"To "inform" poorly estimated parameters, the PRIOR subroutine in NONMEM can be used, regardless of the estimation method."

"Adding a prior to a Maximum Likelihood Estimation would technically convert these into a mode a posteriori (MAP) estimation of the population parameters, even though this term does not show up on the NONMEM report."

"The OFV is the sum of the OFV on the sparse data (O^S) and the penalty function (O^P), which reflects the deviation of the iterated parameters from their previous estimate value [46]."

Literature review

"Literature was screened for articles reporting the use of the PRIOR subroutine in NONMEM, in a four-step approach, as described in Online Resource 1."

"In each step, the full text of eligible articles was checked to retain articles actually reporting the use of the PRIOR subroutine in NONMEM."

"A total of 33 articles reporting the use of the PRIOR subroutine in NONMEM was found in literature [34, 46–77]."

"Eight articles used the PRIOR subroutine in NONMEM to inform mechanistic popPK models [34, 50, 53, 57, 64, 74–76]."

"The PRIOR subroutine is an alternative to fixing the parameters to their previous estimates or to pooling the new data with the previous rich data (when available)."

"To analyze sparse pediatric data when adult data are available, two articles concluded that it was better to build first the adult model (with allometric scaling) and then use it as prior for the pediatric model than to build a model on pooled data [54, 55]."

Guidance

"If a model with full informative priors has much greater estimates of interindividual variability as compared to the prior value, it might stem from the strength of the prior values for the corresponding fixed effects together with a potentially different population parameter estimate in the new population."

"In cases where the previous data are available, the similarity in PK parameter distribution between populations can be assessed by comparing the results of models estimated without prior on the pooled data stratified by two different approaches [69]."

"Specific considerations should be kept in mind when searching for covariates on parameters that are estimated with priors: with prior information on THETA, the typical values of the parameters are constrained to be close to the one of the reference model: if the covariates were not similarly distributed in the previous and the new population, the covariate should be centered around its median in the previous dataset."

Conclusion

"The PRIOR subroutine is a valuable approach to analyze sparse/rare data or estimate mechanistic-based models in an easy way and acceptable run times."

"In order to specify the prior weight, the usual approach is to retain the model with priors on the least parameters and with the lowest informativeness on some parameters, but that still allows for a good estimation."

"The sensitivity of the model parameters to the prior specification should also be evaluated."

"It is tricky to identify new covariates or to confirm previously existing parameter/covariate relationships with a model built with prior."

Delay differential equations based models in NONMEM [78] This is a machine-generated summary of:

Yan, Xiaoyu; Bauer, Robert; Koch, Gilbert; Schropp, Johannes; Perez Ruixo, Juan Jose; Krzyzanski, Wojciech: Delay differential equations based models in NONMEM [78]

Published in: Journal of Pharmacokinetics and Pharmacodynamics (2021)

Link to original: https://doi.org/10.1007/s10928-021-09770-z

Copyright of the summarized publication:

If you want to cite the papers, please refer to the original.

For technical reasons we could not place the page where the original quote is coming from.

Abstract-Summary "Several DDE solvers have been implemented in NONMEM 7.5 for the first time."

"The purpose of this tutorial is to introduce basic concepts underlying DDE based models and to show how they can be developed using NONMEM."

"We evaluated the accuracy of NONMEM DDE solvers, their ability to handle stiff problems, and their performance in parameter estimation using both first-order conditional estimation (FOCE) and the expectation–maximization (EM) method."

"All DDE solvers provide accurate and precise solutions with the number of significant digits controlled by the error tolerance parameters."

"For estimation of population parameters, the EM method is more stable than FOCE regardless of the DDE solver."

Introduction

"The delays in drug response observed in biological systems become an additional complexity to take into account in developing mechanism-based PK/PD models."

"The ADVAN16 algorithm in NONMEM utilizes RADAR5, a delay differential equation solver developed by Guglielmi and Hairer [79]."

"The ADVAN18 algorithm in NONMEM utilizes DDE_SOLVER, a delay differential equation solver developed by Thompson and Shampine [80]."

"DDE_SOLVER uses a modified explicit Runge–Kutta (RK) order [18, 19] method [80, 81], and solves non-stiff differential equations with state-dependent delays."

"DDE extension algorithms have been added to the general ODE solvers (in the manner of Petzoldt and Soetaert for the R DeSolve package [82]) ADVAN9, 13, 14, and 15 in NONMEM version 7.5."

"The objective of this tutorial is to introduce basic concepts underlying DDE based models and to show how they can be developed using ADVAN13, ADVAN16 and ADVAN18 in NONMEM."

Methods

"The DDE solvers are implemented in NONMEM as ADVAN9, 13, 14, 15, 16, and 18."

"TAU1 may be defined in the $PK record with a fixed effect THETA and random effect ETA: The second variable type AD_x_y, the state variable A(x) delayed by time TAUy, is used in $DES."

"Consider the following DDE model: where k_g and y_{ss} are model parameters, τ is the delay, y(t) is the state variable and y(t-τ) is the delay state."

"The presence of a delay variable requires evaluation of delay states AD_x_y in the past (times before the initial time)."

"A past variable is defined as AP_x_y in NONMEM record section $DES, for state variable x, delayed by time TAUy."

"The -dde option is required so additional lines are added to the user's control stream to define the AD_x_y variables, and to make special calls to the DDE delay algorithm."

"We further compare ADVAN13, ADVAN16, ADVAN18 with the MATLAB dde23 solver using the delay logistic growth model."

"The comparison of ADVAN13, ADVAN16, ADVAN18, implemented in NONMEM, and dde23, implemented in MATLAB, was done by calculating the absolute difference of solution of delayed logistic growth model between ADVANs and dde23."

"The drug concentration c(t) is expressed as A(t)/V. The equation describes GM-CSF dynamics G(t) as follows:where k_1, k_2, k_3, a, b are cytokine growth related parameters, and Emax model is used to describe the drug effect."

"The purpose of this example is to demonstrate the estimation of population parameters for the LIDR model including the inter-individual variability on the lifespan parameter (i.e. delay)."

"The predicted serum concentrations were used as a driving force for the PD modeling, where PD parameters (including both random and fixed effect parameters) were estimated based on the PD data for RETs, mature RBCs (MRBCs), and hemoglobin (Hgb)."

Discussion

"Implementation of DDE based models in NONMEM is similar to the implementation of ODE based models with two additional requirements specific to DDEs: (1) one needs to introduce delay times and delay variables; (2) one needs to define history for delay states."

"MOS was used to implement DDE based models in NONMEM prior to the release of version 7.5 [83–85]."

"The ddexpand utility that is called when the -dde option is placed on the command line is capable of augmenting NMTRAN code to provide additional differential equations for the MOS method (when; MOS comment is inserted in the beginning of the NMTRAN code), or provide additional function call code to implement a DDE solver's inherent delay differential equation abilities (when; DDE comment is inserted at the beginning of the NMTRAN code)."

"One uses existing DDE solvers and adds to the NONMEM library on par with ODE solvers (e.g., ADVAN16 and ADVAN18)."

Development of a genetic algorithm and NONMEM workbench for automating and improving population pharmacokinetic/pharmacodynamic model selection [86] This is a machine-generated summary of:

Ismail, Mohamed; Sale, Mark; Yu, Yifan; Pillai, Nikhil; Liu, Sihang; Pflug, Beth; Bies, Robert: Development of a genetic algorithm and NONMEM workbench for automating and improving population pharmacokinetic/pharmacodynamic model selection [86]

Published in: Journal of Pharmacokinetics and Pharmacodynamics (2021)

Link to original: https://doi.org/10.1007/s10928-021-09782-9

If you want to cite the papers, please refer to the original.

For technical reasons we could not place the page where the original quote is coming from.

Abstract-Summary "The current approach to selection of a population PK/PD model is inherently flawed as it fails to account for interactions between structural, covariate, and statistical parameters."

"GAs are general, powerful, robust algorithms and can be used to find global optimal solutions for difficult problems even in the presence of non-differentiable functions, as is the case in the discrete nature of including/excluding model components in search of the best performing mixed-effects PK/PD model."

"A genetic algorithm implemented in an R-based NONMEM workbench for identification of near optimal models is presented."

"This approach will further facilitate the scientist to shift efforts to focus on model evaluation, hypotheses generation, and interpretation and applications of resulting models."

Introduction

"The current approach to population PK/PD model selection is formally known as downhill search and belongs to a class of optimization methods known as local search [87]."

"Within the set of optimization algorithms, downhill search is the most efficient (completing with the fewest evaluations of model quality), but the least robust."

"Exhaustive search entails examining all possible combinations of structural, statistical, and covariate model features and would give rise to millions of potential models for even a standard population pharmacokinetic model search space."

"Out of necessity, the traditional approach to population PK/PD modeling greatly reduces the sample space of models by proceeding in a stepwise manner, with the user first identifying the best base structural model, and then searching for significant covariate relationships and statistical models."

"While a uniform random search will sample models with uniform probability, genetic algorithms exploit historical information from previous iterations to direct the search towards better solutions."

Theory

"To understand and appreciate the mechanisms of the genetic algorithm, it is important to have an understanding of evolution and in particular the concept of survival of the fittest."

"The phrase survival of the fittest was coined by biologist Herbert Spencer in his book Principles of Biology [88]."

"Evolution then, driven by survival of the fittest, becomes a trial-and-error iterative procedure towards establishing 'fitter' organisms as generations pass."

"This is essentially a naturally occurring optimization algorithm, where efficiency is likely important to the continuation of the species."

"This is the driving principle in the genetic algorithm applied in this setting."

Model components commonly included in population pharmacokinetic model search space

"Lag time: presence or absence of a lag time Transit compartments: the number of absorption site transit compartments, if any Absorption rate: zeroth or first order absorption, or both."

"Residual error on dependent variable/s: additive, proportional, or combined Inter-individual variability (IIV) on structural parameters: lognormal or absent With a linear relationship, the typical value of a parameter changes linearly for each unit change in the covariate value."

"The traditional approach to selecting the optimal popPK/PD model proceeds in a stepwise manner."

"Model fitness and the selection of an optimal model is guided by an objective function (usually the − 2LL) as well as various graphical indicators such as residual plots and visual predictive checks."

"The current approach is a local search and is not guaranteed to find the globally optimal combination of model components."

Methods

"The genetic algorithm begins by randomly selecting, without replacement, an initial population of models to be run."

"Due to the uniqueness of each modeling assignment the following penalty terms can be added to the fitness function: User defined penalty for unsuccessful convergence User defined penalty for unsuccessful covariance step User defined penalty for highly correlated parameters User defined penalty for large eigenvalues After the fitness of the models are determined, the genetic operators described in the next sections apply selective pressure and sporadic random changes (i.e., mutations) to determine the next generation of models."

"After crossover and mutation are complete, the elitism operator randomly selects one model from the new generation to be replaced with the best performing model from the previous generation unaltered."

"Genetic algorithms will perform better if the most fit model is always included in the crossover pool [89], and elitism ensures this."

Results

"Beginning from a one compartment base model NONMEM control stream with no covariate effects, the workflow implementing the genetic algorithm on a PK dataset will be shown."

"Once all components to be tested have been added, the user can choose to view all possible models and run them manually or go directly to initiating the genetic algorithm."

"Covariate effects of phenotype (e.g. poor metabolizer) and Weight (power model) were included on clearance."

"The genetic algorithm approach considered the following search scope: (1) one, two or three compartment model; (2) inter-individual variability present/absent as additive or exponentiated forms on all parameters; (3) continuous covariate relationships on any of the parameters as multiplicative, additive and exponentiated; (4) categorical covariate relationships on any of the parameters as present/absent with an exponentiated form (switching from a unique effect for each category)."

Discussion

"The application of GA to population pk model selection requires the separation of the hypothesis generation process (human intelligence/experience required) from the hypothesis testing process (can be automated)."

"Initial hypotheses are generated, models are run, diagnostics are generated, hypotheses are tested, then additional hypotheses are generated, and the process repeat many times."

"Experience suggests that modeling using GA still requires cycling through iterations of hypothesis generation followed by hypothesis testing."

"While the GA selected the same number of compartments as the traditional and the "true" model, the estimates for Cl and central volume were quite different."

"The PK model identified by the stepwise user (0.05 forward and 0.01 reverse), resulted in a model that was quite different from the one used for the simulations, but with a lower objective function value than that discovered using the GA (with a 10 point per parameter penalty)."

6.3 Applications of AI/ ML in PK/PD

Introduction by the Editor

Artificial intelligence (AI) and machine learning (ML) are reshaping the landscape of pharmacokinetics (PK) and pharmacodynamics (PD) assessment for novel drug delivery systems, presenting a paradigm shift in drug development and personalized medicine. These technologies offer multifaceted applications that expedite decision-making, optimize drug formulations, and enhance therapeutic outcomes. AI/ ML-driven predictive modeling enables efficient exploration of complex PK/PD relationships. By assimilating diverse data sources, including the physicochemical properties of drug delivery systems, biological interactions, and patient characteristics, AI/ML algorithms create accurate models that forecast drug behaviour within the body. These models facilitate the design of optimal drug delivery systems tailored to specific diseases and patient populations. Virtual screening powered by AI/ ML expedites the identification of promising drug candidates and delivery platforms. Computational simulations analyse molecular structures, drug interactions, and ADME properties, aiding in selecting the most viable options. This accelerates the screening process and reduces experimental costs, ensuring the most suitable drug delivery systems are pursued. AI/ML-driven optimization refines drug release kinetics and dosing regimens. These technologies analyse complex datasets to determine the ideal parameters for sustained or targeted drug release. Consequently, drug delivery systems can be precisely modified to maintain therapeutic concentrations, minimizing fluctuations and enhancing efficacy while mitigating side effects. In the context of clinical trials, AI/ML guides trial design and patient recruitment.

By analysing historical data and patient characteristics, these tools aid in selecting appropriate patient populations for trials, enhancing trial success rates and reducing costs. Moreover, AI/ML enables real-time analysis of clinical data to adjust dosing regimens and optimize therapeutic outcomes on an individual basis. This concept of adaptive dosing holds immense promise in developing treatment strategies for each patient's unique response. However, successful implementation of AI/ML in PK/PD modeling necessitates high-quality, diverse datasets for training and validation, as well as rigorous model validation. Overcoming these challenges can empower researchers and the pharmaceutical industry to harness the potential of AI/ML for novel drug delivery systems, ultimately revolutionizing drug development and ushering in a new era of personalized medicine.

Machine generated summaries

Disclaimer: The summaries in this chapter were generated from Springer Nature publications using extractive AI auto-summarization: An extraction-based summarizer aims to identify the most important sentences of a text using an algorithm and uses those original sentences to create the auto-summary (unlike generative AI). As the constituted sentences are machine selected, they may not fully reflect the body of the work, so we strongly advise that the original content is read and cited. The auto generated summaries were curated by the editor to meet Springer Nature publication standards.

To cite this content, please refer to the original papers.

Machine generated keywords: machine, algorithm, bind, population pharmacokinetic, combine, simulation, dynamic, prediction, dataset, individual, molecular, insight, excipient, transcription, population

Drug Clearance in Neonates: A Combination of Population Pharmacokinetic Modelling and Machine Learning Approaches to Improve Individual Prediction [90] This is a machine-generated summary of:

Tang, Bo-Hao; Guan, Zheng; Allegaert, Karel; Wu, Yue-E.; Manolis, Efthymios; Leroux, Stephanie; Yao, Bu-Fan; Shi, Hai-Yan; Li, Xiao; Huang, Xin; Wang, Wen-Qi; Shen, A.-Dong; Wang, Xiao-Ling; Wang, Tian-You; Kou, Chen; Xu, Hai-Yan; Zhou, Yue; Zheng, Yi; Hao, Guo-Xiang; Xu, Bao-Ping; Thomson, Alison H.; Capparelli, Edmund V.; Biran, Valerie; Simon, Nicolas; Meibohm, Bernd; Lo, Yoke-Lin; Marques, Remedios; Peris, Jose-Esteban; Lutsar, Irja; Saito, Jumpei; Burggraaf, Jacobus; Jacqz-Aigrain, Evelyne; van den Anker, John; Zhao, Wei: Drug Clearance in Neonates: A Combination of Population Pharmacokinetic Modelling and Machine Learning Approaches to Improve Individual Prediction [90]

Published in: Clinical Pharmacokinetics (2021)

Link to original: https://doi.org/10.1007/s40262-021-01033-x

Copyright of the summarized publication:

If you want to cite the papers, please refer to the original.

For technical reasons we could not place the page where the original quote is coming from.

Abstract-Summary "Population pharmacokinetic evaluations have been widely used in neonatal pharmacokinetic studies, while machine learning has become a popular approach to solving complex problems in the current era of big data."

"The aim of this proof-of-concept study was to evaluate whether combining population pharmacokinetic and machine learning approaches could provide a more accurate prediction of the clearance of renally eliminated drugs in individual neonates."

"Individual estimates of clearance obtained from population pharmacokinetic models were used as reference clearances, and diverse machine learning methods and nested cross-validation were adopted and evaluated against these reference clearances."

"Using the combined method, more than 95% of predictions for all six drugs had a relative error of < 50% and the mean relative error was reduced by an average of 44.3% and 71.3% compared with the other two predictive methods."

"A combined population pharmacokinetic and machine learning approach provided improved predictions of individual clearances of renally cleared drugs in neonates."

"For a new patient treated in clinical practice, individual clearance can be predicted a priori using our model code combined with demographic data."

Introduction

"Population pharmacokinetic analysis has been widely used in neonatal pharmacology and optimal drug dosages based on model-based simulation techniques have been proposed [91–93]."

"Combining population pharmacokinetics with machine learning approaches may result in more computationally powerful data science tools that could enhance the achievement of precision medicine in this vulnerable, neonatal population [94, 95]."

"For renally eliminated drugs, clearance in neonates is often expressed as a function of growth (size), maturation (gestational, postnatal, or postmenstrual age [PMA]) and kidney function, all based on developmental population pharmacokinetic analyses [96]."

"In this proof-of-concept study, we hypothesized that we could predict individual clearance values by combining population pharmacokinetics with machine learning approaches."

"The objective of this study was to evaluate whether a combination of the two methods could accurately predict the individual clearances of renally eliminated drugs in neonates."

Methods

"This study consisted of three steps: population pharmacokinetic analysis, machine learning analysis, and predictive performance comparison."

"NeCV has been accepted widely in the machine learning community as 'state-of-the-art', as it has been found to be an (almost) unbiased model assessment method when estimating the true error [97, 98]."

"The outer test set was then used to evaluate the best machine learning model. (through each training of the machine learning (ML) algorithm, each outer test set of reference clearances was not used for model development)."

"Graphical and statistical criteria were used to select the optimal machine learning approach and validate the performance of the final model."

"For predictive method 2, F_{size}, F_{age} and F_{renal} were estimated according to different population pharmacokinetic models (step 1)."

"Five data sets (the five outer test data sets in the machine learning analysis step) were used as the evaluation data sets for the three methods."

Results

"ETR was the optimal machine learning approach for latamoxef, amoxicillin and ceftazidime, and although it was not the best approach, it also performed well for vancomycin, cefepime and azlocillin."

"The mean relative errors for the combined predictive method (method 1) were 15.4%, 2.2%, 2.8%, 16.9%, 10.1% and 2.0% for vancomycin, cefepime, latamoxef, amoxicillin, azlocillin and ceftazidime, respectively."

"With the exception of method 2 for azlocillin (9.9%), all the mean relative errors were higher with methods 2 and 3 than with method 1."

"The overall mean relative error of the combined method was 8.24%, which was lower by an average of 44.3% and 71.3% than the other two predictive methods (14.8% and 28.7%), respectively."

"For all six drugs, method 1 achieved more than 95% of predictions for all antibiotics within a relative error of 50%."

Discussion

"This is the first study to demonstrate an innovative method that uses a combination of population pharmacokinetic models and machine learning approaches to predict individual clearances of renally eliminated drugs in neonates."

"The models can be used to predict individual clearances of each drug in neonates, based on the patient's demographic data (e.g. BW, CW, GA, PNA, PMA, CREA)."

"Building complex machine learning models from sparse data always carries a risk of data memorization and this study was not designed to identify the number of neonates required to accurately predict the clearance of renally eliminated drugs using this approach."

"Future studies using a combined population pharmacokinetic and machine learning analysis approach should evaluate the impact of these covariates on the prediction of individual clearances, examine drug clearance following non-intravenous routes of administration, and identify predictors for drugs that are also metabolized."

Conclusion

"A combined population pharmacokinetic and machine learning approach provided consistent descriptions of individual clearances of renally cleared drugs in neonates."

"For new neonatal patients treated in clinical practice, individual clearances can be predicted in advance using the model code and demographic data, and can be used to individualize the initial dosing regimen."

Population pharmacokinetic model selection assisted by machine learning [99] This is a machine-generated summary of:

Sibieude, Emeric; Khandelwal, Akash; Girard, Pascal; Hesthaven, Jan S.; Terranova, Nadia: Population pharmacokinetic model selection assisted by machine learning [99]

Published in: Journal of Pharmacokinetics and Pharmacodynamics (2021)

Link to original: https://doi.org/10.1007/s10928-021-09793-6

If you want to cite the papers, please refer to the original.

For technical reasons we could not place the page where the original quote is coming from.

Abstract-Summary "We compared the classical pharmacometric approach with two machine learning methods, genetic algorithm and neural networks, in different scenarios based on simulated pharmacokinetic data."

"Machine learning provided a selection based only on statistical rules and achieved accurate selection."

"Neural network regression tasks were less precise than neural network classification and genetic algorithm methods."

"The computational gain obtained by using machine learning was substantial, especially in the case of neural networks."

"We demonstrated that machine learning methods can greatly increase the efficiency of pharmacokinetic population model selection in case of large datasets or complex models requiring long run-times."

"Our results suggest that machine learning approaches can achieve a first fast selection of models which can be followed by more conventional pharmacometric approaches."

Introduction

"Nonlinear mixed effects modeling (NONMEM) was historically developed to build population PK models around a first order approximation of the random effect."

"In the age of digital medicine, where large amounts of data are available, models have become increasingly complex, and model selection can be further optimized using recent state-of-the-art developments in artificial intelligence algorithms."

"To existing approaches to model selection, machine learning (ML) and deep learning (DL) offer numerous algorithms which can be applied to medicine [100–102] and drug development [103–109] when built on statistical rules."

"One existing ML approach for model selection is genetic algorithm (GA), an optimization process that tries to mimic Darwinian natural selection [110]."

"Neural networks (NNs) are another existing ML approach for model selection belonging to a group of supervised learning algorithms."

"We investigated the accuracy and computational costs for ML approaches (GA and NNs) and classical pharmacometric (PMX) approaches in the context of population PK model selection."

Methods

"Statistical assumptions and relationships (e.g., parameter distributions, covariance matrix structure for random effects, and residual error model) also differed across models."

"Work, the process was implemented in R version 3.5.1 by automatically executing different structural and statistical models through Monolix, then comparing results to commonly pre-defined model selection criteria also considered in the GA fitness function."

"The application of GA to PK model selection problems follows the same rationale [111], with populations made of PK models selected in subsequent generations according to a fitness function, based on pre-defined PMX criteria."

"This method consists of running several "tournaments" among a few models, assessing each of them against a random opponent and selecting the winner (the one with the best fitness) for crossover in the next generation."

"The hybrid component performed an exhaustive local search around the best models every N generation (a parameter [integer] set by the user for the GA)."

Results

"GA assessment on the five true models was performed according to the two defined fitness functions, with and without the hybrid component and with two different generation sizes."

"For the first dataset, the best model across the four selections (with and without the hybrid component, for the two different fitness functions) was the 1 compartment model with 1 order absorption and linear elimination."

"The transitory compartment present in the true model was selected for some models of the last generation for the second fitness function."

"The selected error model was correctly predicted for three selections (fitness function 1 with and without hybrid component and fitness function 2 without hybrid component)."

"With the second fitness function, the true fifth model was always correctly selected, except for the number of compartments (1 instead of 3)."

Discussion

"Four different approaches were used to recover the true models: PMX, GA, and NN with classification and regression."

"Model selection based on classical PMX approach was partly done manually while it was fully automated for ML (GA and NN) approaches."

"We investigated GA models and classic PMX model approaches."

"Even when the fitness of the selected model was less than the fitness criterion of the true model, GA tended to select models that were too simple, suggesting the penalties set for a new parameter may have been too high in this study or the simulated design not informative enough to enable selection of more complex models."

"While the optimization of covariate screening by using ML approaches can be addressed separately [112], both these model building steps could be combined by expanding GA and NN approaches to include the assessment of relationships between parameters and covariates."

Conclusions

"In this new digital era, where increasing amounts of data are collected, integrating ML with PMX processes could confer great benefits within this discipline, including reduced computational costs and the ability to handle different data types without losing interpretability."

"The results of this study demonstrated that ML methods can greatly increase the efficiency of population model selection in case of large datasets or complex models requiring long run-times."

"Our results suggest that ML approaches can achieve a first fast selection which can be followed by more conventional PMX approaches."

"Conventional PMX methods could take several days to weeks, depending on previous knowledge."

Molecular dynamics simulations on RORγt: insights into its functional agonism and inverse agonism [113] This is a machine-generated summary of:

Yuan, Cong-min; Chen, Hai-hong; Sun, Nan-nan; Ma, Xiao-jun; Xu, Jun; Fu, Wei: Molecular dynamics simulations on RORγt: insights into its functional agonism and inverse agonism [113]

Published in: Acta Pharmacologica Sinica (2019)

Link to original: https://doi.org/10.1038/s41401-019-0259-z

If you want to cite the papers, please refer to the original.

For technical reasons we could not place the page where the original quote is coming from.

Abstract-Summary "We performed molecular dynamics simulations on four different RORγt systems, i.e., the apo protein, protein bound with agonist, protein bound with inverse agonist in the orthosteric-binding pocket, and protein bound with inverse agonist in the allosteric-binding pocket."

"The tracked data from MD simulations supported that RORγt could be activated by an agonist binding at the orthosteric-binding pocket, because the bound agonist helped to enhance the triplet His479–Tyr502–Phe506 interactions and stabilized H12 structure."

"The stabilized H12 helped RORγt to form the protein-binding site, and therefore made the receptor ready to recruit a coactivator molecule."

"We also showed that transcription function of RORγt could be interrupted by the binding of inverse agonist at the orthosteric-binding pocket or at the allosteric-binding site."

"After the inverse agonist was bound, H12 either structurally collapsed, or reorientated to a different position, at which the presumed protein-binding site was not able to be formed."

Introduction

"After the binding of an agonist at the orthosteric-binding pocket, H12 at the LBD is stabilized in a conformation that is ready to interact with a coactivator, such as steroid receptor activator 2 (SRC2), at the protein-binding site."

"H12 can also be destabilized or completely reorientated after an inverse agonist is bound at the orthosteric-binding pocket or at the new allosteric site, thus repressing the gene transcription function of RORγt [114–117]."

"Although this structural information has provided profound insights into how RORγt binds with an agonist or inverse agonist and where H12 of the LBD could interact with or reorient relative to the neighboring structural components, the mechanisms of these conformational changes resulting from ligand binding remain ambiguous."

"We selected a system of RORγt without ligand binding (apo-form) and three typical systems of RORγt bound with different modulators: agonist, inverse agonist bound at the orthosteric-binding pocket, and inverse agonist bound at the allosteric-binding site."

Materials and methods

"Each of the selected protein structures was visually checked and prepared by using the Protein Preparation Wizard encoded in the Schrodinger 3.5 software package [118]."

"Each system of protein structures was solvated in a cubic SPC water box with at least 10 Å from the box boundary to any residue."

"Based on each of the prepared systems, energy minimization was performed first for the solvent molecules, including ions, and then on the protein and ligand molecule with constraints on protein backbone atoms [119]."

"The representative structure of each system was derived from the largest conformational cluster based on the MD trajectories [120]."

"To reveal the most important internal motion of each simulated system, we performed principal component analysis (PCA) [121–123] on the MD trajectories."

"We used the starting system as the reference structure and performed a least square fitting of the MD trajectory to the reference structure [124, 125]."

Results and discussion

"To confirm that the apo-form RORγt is active and ready to bind coactivator if provided, we performed PCA and DSSP [126] calculations to track the secondary structure of H11' and H12 based on the MD trajectory."

"After agonist 921 is bound inside the orthosteric-binding pocket, H12 is stabilized by the triplet His479–Tyr502–Phe506 interactions and maintains its structure as a typical helix."

"Due to such intercalation of the inverse agonist 92A into the triplet His479–Tyr502–Phe506 of the RORγt LBD structure, it could be reasonably expected that the local structures around H11, H11', and H12 were seriously perturbed or distorted, and thus, H12 was destabilized."

"We performed MD simulations on the 5C4O system for the RORγt LBD bound to the inverse agonist 4F1 [115] to test the dynamics of the binding structure and stability of the reorientated H12."

Conclusion

"Once an agonist is bound at the orthosteric-binding pocket in the RORγt 5VB7 system, the helical structure of H12 is stabilized by strong triplet His479–Tyr502–Phe506 intramolecular interactions."

"The transcription function of RORγt can be interrupted by the binding of an inverse agonist at the orthosteric-binding pocket."

"The inverse agonist 92A, which is bound at the orthosteric-binding pocket in the RORγt 5VB5 system, clashed sterically with the side chain of His479, destabilized the triplet His479–Tyr502–Phe506, and completely collapsed the structure of H11' and H12."

"The tracked intermolecular and intramolecular interactions from the MD simulations explicitly demonstrate the conformational dynamics and molecular mechanisms of how the RORγt receptor is activated by binding an agonist at the orthosteric-binding pocket and how the RORγt is inversely agonized by destabilizing H12 or reorientating H12 to a different position."

Molecular Dynamics Simulations and Experimental Results Provide Insight into Clinical Performance Differences between Sandimmune® and Neoral® Lipid-Based Formulations [127] This is a machine-generated summary of:

Warren, Dallas B.; Haque, Shadabul; McInerney, Mitchell P.; Corbett, Karen M.; Kastrati, Endri; Ford, Leigh; Williams, Hywel D.; Jannin, Vincent; Benameur, Hassan; Porter, Christopher J.H.; Chalmers, David K.; Pouton, Colin W.: Molecular Dynamics Simulations and Experimental Results Provide Insight into Clinical Performance Differences between Sandimmune® and Neoral® Lipid-Based Formulations [127]

Published in: Pharmaceutical Research (2021)

Link to original: https://doi.org/10.1007/s11095-021-03099-5

If you want to cite the papers, please refer to the original.

For technical reasons we could not place the page where the original quote is coming from.

Abstract-Summary "Molecular dynamics (MD) simulations provide an in silico method to study the structure of lipid-based formulations (LBFs) and the incorporation of poorly water-soluble drugs within such formulations."

"In order to validate the ability of MD to effectively model the properties of LBFs, this work investigates the well-known cyclosporine A formulations, Sandimmune® and Neoral®."

"MD simulations were performed of both LBFs to investigate the differences observed in fasted and fed conditions."

"These conditions were also tested using an in vitro experimental model of dispersion and digestion."

"The current data suggests that MD simulations are a potential method to model the fate of LBFs in the gastrointestinal tract, predict their dispersion and digestion, investigate behaviour of APIs within the formulations, and provide insights into the clinical performance of LBFs."

Introduction

"Sandimmune® and Neoral® are LBFs developed for oral delivery of cyclosporine A and represent a well-known story within the formulation field, being the gold standard for LBF development and performance."

"A clinical study by Mueller and others [128] reported pharmacokinetic data for Sandimmune® and Neoral® formulations under fasted and fed conditions, observed in 24 healthy patients."

"The two formulations, Sandimmune® and Neoral®, are well characterized systems that have significant differences in performance."

"We use MD simulations to model the dissolution behaviour of Sandimmune® and Neoral® in water and in models of the fasted and fed gastrointestinal environments."

"MD simulations of neat formulation and with small amounts of water are then presented to determine whether the simulations provide evidence that the formulation is likely to undergo phase separation upon dispersion."

"The influence of the formulation digestion process on the dispersion behaviour of the formulations is then studied using MD simulations."

Methods

"Recovery of CsA from formulation digestion experiments by ACN extraction was assessed to ensure measurement of total CsA in the various phases."

"This was conducted by spiking dispersion media, aqueous digestion media, and oil-phase digestion media; all produced by digestion of blank Sandimmune® or Neoral® formulations (900 mg of each over 60 min) with known concentrations of CsA similar to those expected in experiments."

"Sandimmune® displayed poor dispersion in both FaSSIF and FeSSIF and the concentration of the oil phase was therefore calculated based on the theoretical total CsA minus the aqueous and pellet phase at each time point."

"In order to negate the matrix effect of unknown components in the formulation, the QC values and aqueous concentrations were normalized to 100% i.e. the theoretical total CsA. Dynamic light scattering measurements were performed using a Malvern Zetasizer Nano Series (Malvern, UK) running DTS software and operating a 4 mW He-Ne laser at 633 nm."

Results

"Sandimmune® and Neoral® dispersion within 90% w/w water, fasted and fed phases was then studied to indicate the performance of the formulations upon dispersion."

"To investigate how the formulation structure changes upon dispersion within the gastrointestinal tract, MD simulations were conducted of Sandimmune® and Neoral® dispersed in 90% water, 90% fasted buffer, and 90% fed buffer conditions."

"Sandimmune® formed water pools at these low water concentrations, but indicated a propensity to phase separate at higher water concentrations, and Neoral® formed hydrophilic channels indicating a propensity to disperse into a fine micellar system when more water is present."

"Fed conditions were modelled using higher concentrations of bile species in the dispersing medium and, in addition, the digestible molecules (di- and tri-glycerides) within the formulation were broken down into their more water-soluble digestion products (free fatty acids and mono-glycerides)."

"Both Sandimmune® and Neoral® provide consistent solubilization of the CsA within the neat formulation and upon dispersion in water."

Discussion

"MD simulations provide molecular scale information about the physicochemical behaviour of complex, colloidal systems, such as LBFs, including information about the dispersibility of different molecular mixtures."

"The MD simulations provide useful information about interactions between of drug molecules within the formulation under different conditions."

"The simulations also show that the degree of association does not change to any great degree in each of the conditions tested (neat, in water, fasted or fed environments), showing that both formulations effectively prevent drug aggregation as they are dispersed in the GI tract."

"The size of the simulation and the formulation water content limit the size of the colloidal structures that can be observed."

"The study indicates that MD simulations can be used as a tool to assist in selection of lipid-based formulations, in particular to avoid differences in bioavailability after oral administration in the fed and fasted states."

Conclusions

"We have performed in vitro experimental and in silico computational molecular dynamics studies of the physicochemical behaviour of the clinically important CsA formulations Sandimmune® and Neoral®."

"Consistent with previous reports [129, 130] these in vitro experimental studies show that Sandimmune® is less water-dispersible than Neoral®, is digested more slowly and is less able to maintain the API, CsA, in the aqueous phase."

"Concordant with the experimental studies, changes in the dispersing medium (water, fasted, or fed) do not affect the structures formed by Neoral® as much as they do for Sandimmune®."

"Further, the simulations performed in this study are unique in that they include multicomponent MD models of formulation excipients and include a cyclic peptide API."

"This study highlights the power of MD to model the dispersion and digestion of LBFs, to trace the location and behaviour of the API within the formulation, and to provide insights into the experimental and clinical performance of these formulations."

A Hybrid Algorithm Combining Population Pharmacokinetic and Machine Learning for Isavuconazole Exposure Prediction [131] This is a machine-generated summary of:

Destere, Alexandre; Marquet, Pierre; Labriffe, Marc; Drici, Milou-Daniel; Woillard, Jean-Baptiste: A Hybrid Algorithm Combining Population Pharmacokinetic and Machine Learning for Isavuconazole Exposure Prediction [131]

Published in: Pharmaceutical Research (2023)

Link to original: https://doi.org/10.1007/s11095-023-03507-y

If you want to cite the papers, please refer to the original.

For technical reasons we could not place the page where the original quote is coming from.

Abstract-Summary "The aim of this study was to confirm the previous results by developing a hybrid algorithm combining POPPK, MAP-BE and ML that accurately predicts isavuconazole clearance."

"A total of 1727 isavuconazole rich PK profiles were simulated using a POPPK model from the literature, and MAP-BE was used to estimate the clearance based on: (i) the full PK profiles (refCL); and (ii) C24h only (C24h-CL)."

"C24h-CL as well as ML-corrected C24h-CL were evaluated in a testing dataset (25%) and then in a set of PK profiles simulated using another published POPPK model."

"A strong decrease in mean predictive error (MPE%), imprecision (RMSE%) and the number of profiles outside ± 20% MPE% (n-out20%) was observed with the hybrid algorithm (decreased in MPE% by 95.8% and 85.6%; RMSE% by 69.5%

and 69.0%; n-out20% by 97.4% and 100% in the training and testing sets, respectively."

"The hybrid model proposed significantly improved isavuconazole AUC estimation over MAP-BE based on the sole C24h and may improve dose adjustment."

Introduction

"It relies on population pharmacokinetics (POPPK) models and Bayesian estimators (MAP-BE) [132, 133] based on priors and a limited sampling strategy (LSS) and allows to estimate individual pharmacokinetics parameters (e.g. distribution volume or clearance…) and/or exposure indices as the area under the curve (AUC)."

"We propose an hybrid algorithm that use combine a POPPK model and a machine learning algorithm [134]."

"Based on simulations from a literature POPPK model of iohexol [135], a ML algorithm was trained to predict the bias between the CL obtained from simulated full PK profiles and the CL obtained after MAP-BE based on a three-points LSS."

"The aim of this work was to develop a hybrid model based on POPPK and ML algorithm to correct the CL estimation bias (and the isavuconazole estimation of exposure) in the MAP-BE & LSS based on 1 sample."

Material and Methods

"The variables used as predictors were: (i) the PK parameters (CL and V2) obtained using the MAP-BE on C24h, (ii) the simulated C24h (SC_{24h}), (iii) the MAP-BE estimated C24h (EC_{24h}), (iv) the simulated sex and BMI and (v) the η of MAP-BE CL and V2 (corresponding to the estimated deviation (in %) relative to the typical population CL and V2)."

"The predicted error was added to C24h-CL to obtain the corrected CL (corCL): The performances of the Xgboost algorithm developed to correct ISA CL in another ten fold cross-validation set (to check out overfitting) from the training set and in the testing dataset were evaluated using the MPE%, the relative RMSE% and the percentage of profiles with MPE% out of the ± 20% MPE interval between corCL and refCL."

Results

"After excluding the 1% extreme percentiles in CL and V2 (n = 233) and thirty-five PK profiles that failed to be estimated by MAPBAYR, 1727 (out of the 2000) simulated profiles remained."

"The refCL and C24h-CL were (mean ± SD [min–max]) 4.9 ± 3.3[1.0–19.2] L/h and 4.4 ± 2.5[1.1–14.8], respectively."

"The shrinkage of C24h-CL was 45.6% and 44.6% in the training and testing datasets, respectively."

"The hybrid model decreased the shrinkage of ISA CL estimation by 54.0% and 58.5% respectively (training: 21.2% and testing: 18.5%)."

"The hybrid model decreased the MPE% by approximately 96% and the RMSE% by 68% as compared to MAP-BE alone."

Discussion

"We developed a hybrid algorithm combining MAP-BE and ML, able to estimate very accurately ISA CL based on the steady-state C24h."

"The algorithm developed yields very accurate estimation of ISA CL, hence of its AUC (CL = dose/AUC)."

"We did not try to develop a ML algorithm to directly predict the CL of isavuconazole based on features and available concentrations as previously done by our group for other drugs [136, 137]."

"In a previous collaborative work, we showed that these approaches had similar (or slightly better) performances in comparison to the ML algorithm that directly estimated the CL of vancomycin [137]."

"Is the works of Hughes et al who developed a ML algorithm which learned from the error between observations and predictions and was able to select patients in whom it would be better to use flat priors rather than conventional MAP-BE for vancomycin [138]."

Machine learning-driven multifunctional peptide engineering for sustained ocular drug delivery [139] This is a machine-generated summary of:

Hsueh, Henry T.; Chou, Renee Ti; Rai, Usha; Liyanage, Wathsala; Kim, Yoo Chun; Appell, Matthew B.; Pejavar, Jahnavi; Leo, Kirby T.; Davison, Charlotte; Kolodziejski, Patricia; Mozzer, Ann; Kwon, HyeYoung; Sista, Maanasa; Anders, Nicole M.; Hemingway, Avelina; Rompicharla, Sri Vishnu Kiran; Edwards, Malia; Pitha, Ian; Hanes, Justin; Cummings, Michael P.; Ensign, Laura M.: Machine learning-driven multifunctional peptide engineering for sustained ocular drug delivery [139]

Published in: Nature Communications (2023)

Link to original: https://doi.org/10.1038/s41467-023-38056-w

If you want to cite the papers, please refer to the original.

For technical reasons we could not place the page where the original quote is coming from.

Abstract-Summary "For many chronic ocular diseases, patient adherence to eye drop dosing regimens and the need for frequent intraocular injections are significant barriers to effective disease management."

"We utilize peptide engineering to impart melanin binding properties to peptide-drug conjugates to act as a sustained-release depot in the eye."

"When the lead multifunctional peptide (HR97) is conjugated to brimonidine, an intraocular pressure lowering drug that is prescribed for three times per day topical dosing, intraocular pressure reduction is observed for up to 18 days after a single intracameral injection in rabbits."

"Engineered multifunctional peptide-drug conjugates are a promising approach for providing sustained therapeutic delivery in the eye and beyond."

Introduction

"One approach for circumventing the issues associated with sustained release devices is to impart enhanced retention time and therapeutic effect to drugs upon administration to the eye without the need for an excipient matrix/implant."

"To impart beneficial melanin-binding properties to drugs, one approach is to engineer peptides with high melanin binding that could be conjugated to small molecule drugs through a reducible linker."

"The peptide would provide enhanced retention time, while the linker would ensure that drug could be released and exert its therapeutic action in a sustained manner."

"There are available databases describing how peptide sequence affects cell-penetration [140, 141], and separately cytotoxicity [142], enabling the potential for engineering multifunctional peptides that can be chemically conjugated to drugs."

"We describe the development of engineered peptides informed by machine learning, which have three properties: high binding to melanin, cell-penetration (to enter cells and access melanin in the melanosomes), and low cytotoxicity."

Results

"We used the trained random forest model to predict melanin binding for ~630,000 randomly generated peptides, and those classified as melanin binding were selected."

"units) from the 119-peptide microarray, demonstrating the enrichment of melanin binding properties from training the random forest model."

"Similar to melanin binding, adversarial controls had decreased generalization performances, where the MCC, F_1, and balanced accuracy were -0.002 ± 0.05, 0.52 ± 0.03, and 0.50 ± 0.03 for cell-penetration, and 0.001 ± 0.01, 0.05 ± 0.02, and 0.62 ± 0.04 for cytotoxicity."

"Among the 127 peptide candidates, 113 peptides were classified as cell-penetrating and 117 peptides were predicted as non-toxic."

"We next characterized how the predicted cell-penetrating properties of the peptides affected cell uptake in a retinal pigment epithelium cell line (ARPE-19)."

Discussion

"We developed a machine learning-based methodology to engineer tri-functional peptides that displayed melanin binding, cell-penetration, and non-toxic properties."

"The peptide sequence that provided the optimal combination of high melanin binding, high cell-penetration, and low cytotoxicity, HR97, was then conjugated to brimonidine as a proof-of-principle."

"The second peptide microarray designed using the initial machine learning model provided more potent melanin binding peptides compared to the first peptide microarray, demonstrating the rapid improvement in design by machine learning model refinement."

"Engineered HR97 peptide demonstrated increased cell-penetrating properties compared to known cell-penetrating peptides, such as TAT, and simultaneously possessed high melanin binding capacity and low cytotoxicity."

"In the current context, utilizing short peptide sequences that impart melanin binding to a drug conjugate may provide an avenue for creating safe and effective implant-free sustained intraocular drug release systems."

Methods

"Peptides from the crude peptide library were further purified by being first dissolved in 50% acetonitrile (ACN) with 0.1% TFA at 10 mg/mL. Shimadzu LC20 high-performance liquid chromatography (HPLC) system with Phenomenex reverse-phase preparative HPLC column (Gemini® 10 µm C18 110 Å, LC Column 250 × 21.2 mm, AXIA™ Packed) were used to separate and collect the peptides with an elution gradient of 5/5/90/90/5/5% solvent B (TFA 0.05% in ACN) at 0/2/10/12/13.5/15 min with a flow rate of 5 mL/min with monitoring at 220 nm."

"To better characterize variable contributions to peptide property predictions, models trained on the outer loop training sets with the same hyperparameters as the final predictive model were used to calculate SHAP values using the corresponding test sets."

"Explanations of HR97 multifunctional peptide predictions were computed using the final models trained on the whole machine learning data sets."

Bibliography

1. Keutzer L, You H, Farnoud A, Nyberg J, Wicha SG, Maher-Edwards G, Vlasakakis G, Moghaddam GK, Svensson EM, Menden MP, Simonsson US (2022) Machine learning and pharmacometrics for prediction of pharmacokinetic data: differences, similarities and challenges illustrated with rifampicin. Pharmaceutics 14(8):1530
2. Vora LK, Gholap AD, Jetha K, Thakur RR, Solanki HK, Chavda VP (2023) Artificial intelligence in pharmaceutical technology and drug delivery design. Pharmaceutics 15(7):1916
3. Singh AV, Ansari MH, Rosenkranz D, Maharjan RS, Kriegel FL, Gandhi K, Kanase A, Singh R, Laux P, Luch A (2020) Artificial intelligence and machine learning in computational nanotoxicology: unlocking and empowering nanomedicine. Adv Healthc Mater 9(17):1901862
4. Singh AV, Chandrasekar V, Janapareddy P, Mathews DE, Laux P, Luch A, Yang Y, Garcia-Canibano B, Balakrishnan S, Abinahed J, Al AA (2021) Emerging application of nanorobotics and artificial intelligence to cross the BBB: advances in design, controlled maneuvering, and targeting of the barriers. ACS Chem Neurosci 12(11):1835–1853
5. Singh AV, Maharjan RS, Kanase A, Siewert K, Rosenkranz D, Singh R, Laux P, Luch A (2020) Machine-learning-based approach to decode the influence of nanomaterial properties on their interaction with cells. ACS Appl Mater Interf 13(1):1943–1955
6. Winkler DA (2020) Role of artificial intelligence and machine learning in nanosafety. Small 16(36):2001883
7. Scott-Fordsmand JJ, Amorim MJ (2023) Using machine learning to make nanomaterials sustainable. Sci Total Environ 859:160303
8. Villa Nova M, Lin TP, Shanehsazzadeh S, Jain K, Ng SC, Wacker R, Chichakly K, Wacker MG (2022) Nanomedicine ex machina: between model-informed development and artificial intelligence. Front Digit Health 4:799341
9. Yaghini E, Tacconi E, Pilling A, Rahman P, Broughton J, Naasani I, Keshtgar MR, MacRobert AJ, Della PO (2021) Population pharmacokinetic modelling of indium-based quantum dot nanoparticles: preclinical in vivo studies. Eur J Pharm Sci 157:105639

10. Yang S, Kar S (2023) Application of artificial intelligence and machine learning in early detection of adverse drug reactions (ADRs) and drug-induced toxicity. Artif Intell Chem 10:100011
11. Kutumova EO, Akberdin IR, Kiselev IN, Sharipov RN, Egorova VS, Syrocheva AO, Parodi A, Zamyatnin AA Jr, Kolpakov FA (2022) Physiologically based pharmacokinetic modeling of nanoparticle biodistribution: a review of existing models, simulation software, and data analysis tools. Int J Mol Sci 23(20):12560
12. Leonis G, Melagraki G, Afantitis A (2017) Open source chemoinformatics software including KNIME analytics. In: Handbook of computational chemistry, p 2201
13. Vrontaki E, Mavromoustakos T, Melagraki G, Afantitis A (2015) Quantitative nanostructure-activity relationship models for the risk assessment of nanomaterials. In: Roy K (ed) Quantitative structure-activity relationships in drug design, predictive toxicology, and risk assessment. IGI Global, Hershey, pp 537–561
14. Winkler DA, Burden FR, Yan B, Weissleder R, Tassa C, Shaw S, Epa VC (2014) Modelling and predicting the biological effects of nanomaterials. SAR QSAR Environ Res 25(2):161–172
15. Speck-Planche A, Kleandrova VV, Luan F, Cordeiro MN (2015) Computational modeling in nanomedicine: prediction of multiple antibacterial profiles of nanoparticles using a quantitative structure-activity relationship perturbation model. Nanomedicine 10(2):193–204
16. Algharably EAH, Kreutz R, Gundert-Remy U (2019) Importance of in vitro conditions for modeling the in vivo dose in humans by in vitro–in vivo extrapolation (IVIVE). Arch Toxicol 93:615. https://doi.org/10.1007/s00204-018-2382-x
17. Sukarnjanaset W, Wattanavijitkul T, Jarurattanasirikul S (2018) Evaluation of FOCEI and SAEM estimation methods in population pharmacokinetic analysis using NONMEM® across rich, medium, and sparse sampling data. Eur J Drug Metab Pharmacokinet 43:729. https://doi.org/10.1007/s13318-018-0484-8
18. Johansson AM, Ueckert S, Plan EL, Hooker AC, Karlsson MO (2014) Evaluation of bias, precision, robustness and runtime for estimation methods in NONMEM 7. J Pharmacokinet Pharmacodyn 41(3):223–238
19. Liu X, Wang Y (2016) Comparing the performance of FOCE and different expectation-maximization methods in handling complex population physiologically-based pharmacokinetic models. J Pharmacokinet Pharmacodyn 43(4):359–370
20. Bulitta JB, Duffull SB, Kinzig-Schippers M, Holzgrabe U, Stephan U, Drusano GL et al (2007) Systematic comparison of the population pharmacokinetics and pharmacodynamics of piperacillin in cystic fibrosis patients and healthy volunteers. Antimicrob Agents Chemother 51(7):2497–2507
21. Gibiansky L, Gibiansky E, Bauer R (2012) Comparison of Nonmem 7.2 estimation methods and parallel processing efficiency on a target-mediated drug disposition model. J Pharmacokinet Pharmacodyn 39(1):17–35
22. Bachmann F, Koch G, Bauer RJ, Steffens B, Szinnai G, Pfister M, Schropp J (2023) Computing optimal drug dosing with OptiDose: implementation in NONMEM. J Pharmacokinet Pharmacodyn 50:173. https://doi.org/10.1007/s10928-022-09840-w
23. Bachmann F, Koch G, Pfister M, Szinnai G, Schropp J (2021) OptiDose: computing the individualized optimal drug dosing regimen using optimal control. J Optim Theory Appl 189:46–65
24. Bachmann F, Koch G, Pfister M, Szinnai G, Schropp J (2022) A sensitivity analysis of the optimal drug dosing algorithm OptiDose. Pure Appl Funct Anal 7(4):1127–1140
25. Dayneka NL, Garg V, Jusko WJ (1993) Comparison of four basic models of indirect pharmacodynamic responses. J Pharmacokinet Biopharm 21(4):457–478
26. Koch G, Schropp J (2018) Delayed indirect response models: realization of oscillating behaviour. J Pharmacokinet Pharmacodyn 45(1):49–58
27. Jönsson S, Karlsson MO (2003) A rational approach for selection of optimal covariate-based dosing strategies. Clin Pharmacol Ther 73:7–19
28. Brekkan A, Jönsson S, Karlsson MO, Plan EL (2019) Handling underlying discrete variables with bivariate mixed hidden Markov models in NONMEM. J Pharmacokinet Pharmacodyn 46:591. https://doi.org/10.1007/s10928-019-09658-z

29. Hosseini I, Gajjala A, Bumbaca Yadav D, Sukumaran S, Ramanujan S, Paxson R, Gadkar K (2018) gPKPDSim: a SimBiology®-based GUI application for PKPD modeling in drug development. J Pharmacokinet Pharmacodyn 45:259. https://doi.org/10.1007/s10928-017-9562-9
30. Germani M, Del Bene F, Rocchetti M, Van Der Graaf PH (2013) A4S: a user-friendly graphical tool for pharmacokinetic and pharmacodynamic (PK/PD) simulation. Comput Methods Programs Biomed 110(2):203–214. https://doi.org/10.1016/j.cmpb.2012.10.006
31. Tsiros P, Bois FY, Dokoumetzidis A, Tsiliki G, Sarimveis H (2019) Population pharmacokinetic reanalysis of a Diazepam PBPK model: a comparison of Stan and GNU MCSim. J Pharmacokinet Pharmacodyn 46:173. https://doi.org/10.1007/s10928-019-09630-x
32. Nestorov I (2003) Whole body pharmacokinetic models. Clin Pharmacokinet 42(10):883–908. https://doi.org/10.2165/00003088-200342100-00002
33. Wendling T, Dumitras S, Ogungbenro K, Aarons L (2015) Application of a Bayesian approach to physiological modelling of mavoglurant population pharmacokinetics. J Pharmacokinet Pharmacodyn 42(6):639657. https://doi.org/10.1007/s10928-015-9430-4
34. Langdon G, Gueorguieva I, Aarons L, Karlsson M (2007) Linking preclinical and clinical whole-body physiologically based pharmacokinetic models with prior distributions in NONMEM. Eur J Clin Pharmacol 63(5):485–498. https://doi.org/10.1007/s00228-007-0264-x
35. Garcia RI, Ibrahim JG, Wambaugh JF, Kenyon EM, Setzer RW (2015) Identifiability of PBPK models with applications to dimethylarsinic acid exposure. J Pharmacokinet Pharmacodyn 42(6):591–609. https://doi.org/10.1007/s10928-015-9424-2
36. Yates JWT (2006) Structural identifiability of physiologically based pharmacokinetic models. J Pharmacokinet Pharmacodyn 33(4):421–439. https://doi.org/10.1007/s10928-006-9011-7
37. Bach T, An G (2021) Comparing the performance of first-order conditional estimation (FOCE) and different expectation–maximization (EM) methods in NONMEM: real data experience with complex nonlinear parent-metabolite pharmacokinetic model. J Pharmacokinet Pharmacodyn 48:581. https://doi.org/10.1007/s10928-021-09753-0
38. Bauer RJ, Guzy S, Ng C (2007) A survey of population analysis methods and software for complex pharmacokinetic and pharmacodynamic models with examples. AAPS J 9(1):E60–E83
39. Duffull SB, Kirkpatrick CM, Green B, Holford NH (2005) Analysis of population pharmacokinetic data using NONMEM and WinBUGS. J Biopharm Stat 15(1):53–73. https://doi.org/10.1081/bip-200040824
40. Plan EL, Maloney A, Mentré F, Karlsson MO, Bertrand J (2012) Performance comparison of various maximum likelihood nonlinear mixed-effects estimation methods for dose–response models. AAPS J 14(3):420–432
41. Chan PL, Jacqmin P, Lavielle M, McFadyen L, Weatherley B (2011) The use of the SAEM algorithm in MONOLIX software for estimation of population pharmacokinetic-pharmacodynamic-viral dynamics parameters of maraviroc in asymptomatic HIV subjects. J Pharmacokinet Pharmacodyn 38(1):41–61. https://doi.org/10.1007/s10928-010-9175-z
42. An G, Murry DJ, Gajurel K, Bach T, Deye G, Stebounova LV, Codd EE, Horton J, Gonzalez AE, Garcia HH, Ince D, Hodgson-Zingman D, Nomicos EYH, Conrad T, Kennedy J, Jones W, Gilman RH, Winokur P (2019) Pharmacokinetics, safety, and tolerability of oxfendazole in healthy volunteers: a randomized, placebo-controlled first-in-human single-dose escalation study. Antimicrob Agents Chemother 63:10–1128. https://doi.org/10.1128/AAC.02255-18
43. Almquist J, Leander J, Jirstrand M (2015) Using sensitivity equations for computing gradients of the FOCE and FOCEI approximations to the population likelihood. J Pharmacokinet Pharmacodyn 42(3):191–209. https://doi.org/10.1007/s10928-015-9409-1
44. Bauer RJ (2017) NONMEM users guide-Introduction to NONMEM 7.4.1. ICON Plc, Dublin
45. Kwong C, Anna H-XP, Calvier EAM, Fabre D, Gattacceca F, Khier S (2020) Prior information for population pharmacokinetic and pharmacokinetic/pharmacodynamic analysis: overview and guidance with a focus on the NONMEM PRIOR subroutine. J Pharmacokinet Pharmacodyn 47:431. https://doi.org/10.1007/s10928-020-09695-z
46. Gisleskog PO, Karlsson MO, Beal SL (2002) Use of prior information to stabilize a population data analysis. J Pharmacokinet Pharmacodyn 29:473–505. https://doi.org/10.1023/A:1022972420004

47. Saito M, Kaibara A, Kadokura T, Toyoshima J, Yoshida S, Kazuta K, Ueyama E (2019) Pharmacokinetic and pharmacodynamic modelling for renal function dependent urinary glucose excretion effect of ipragliflozin, a selective sodium–glucose cotransporter 2 inhibitor, both in healthy subjects and patients with type 2 diabetes mellitus. Br J Clin Pharmacol 85:1808–1819. https://doi.org/10.1111/bcp.13972
48. Nemoto A, Masaaki M, Yamaoka K (2017) A Bayesian approach for population pharmacokinetic modeling of alcohol in Japanese individuals. Curr Ther Res Clin Exp 84:42–49. https://doi.org/10.1016/j.curtheres.2017.04.001
49. Brill MJE, Svensson EM, Pandie M, Maartens G, Karlsson MO (2017) Confirming model-predicted pharmacokinetic interactions between bedaquiline and lopinavir/ritonavir or nevirapine in patients with HIV and drug-resistant tuberculosis. Int J Antimicrob Agents 49:212–217. https://doi.org/10.1016/j.ijantimicag.2016.10.020
50. Sadiq MW, Nielsen EI, Khachman D, Conil J-M, Georges B, Houin G, Laffont CM, Karlsson MO, Friberg LE (2017) A whole-body physiologically based pharmacokinetic (WB-PBPK) model of ciprofloxacin: a step towards predicting bacterial killing at sites of infection. J Pharmacokinet Pharmacodyn 44:69–79. https://doi.org/10.1007/s10928-016-9486-9
51. Knøsgaard KR, Foster DJR, Kreilgaard M, Sverrisdóttir E, Upton RN, van den Anker JN (2016) Pharmacokinetic models of morphine and its metabolites in neonates: Systematic comparisons of models from the literature, and development of a new meta-model. Eur J Pharm Sci 92:117–130. https://doi.org/10.1016/j.ejps.2016.06.026
52. Muto C, Shoji S, Tomono Y, Liu P (2015) Population pharmacokinetic analysis of voriconazole from a pharmacokinetic study with immunocompromised Japanese pediatric subjects. Antimicrob Agents Chemother 59:3216–3223. https://doi.org/10.1128/AAC.04993-14
53. Tsamandouras N, Dickinson G, Guo Y, Hall S, Rostami-Hodjegan A, Galetin A et al (2015) Development and application of a mechanistic pharmacokinetic model for simvastatin and its active metabolite simvastatin acid using an integrated population PBPK approach. Pharm Res 32(6):1864–1883
54. Knebel W, Gastonguay MR, Malhotra B, El-Tahtawy A, Jen F, Gandelman K (2013) Population pharmacokinetics of atorvastatin and its active metabolites in children and adolescents with heterozygous familial hypercholesterolemia: selective use of informative prior distributions from adults. J Clin Pharmacol 53:505–516. https://doi.org/10.1002/jcph.66
55. Robbie GJ, Zhao L, Mondick J, Losonsky G, Roskos LK (2012) Population pharmacokinetics of palivizumab, a humanized anti-respiratory syncytial virus monoclonal antibody, in adults and children. Antimicrob Agents Chemother 56:4927–4936. https://doi.org/10.1128/AAC.06446-11
56. Krogh-Madsen M, Bender B, Jensen MK, Nielsen OJ, Friberg LE, Honoré PH (2012) Population pharmacokinetics of cytarabine, etoposide, and daunorubicin in the treatment for acute myeloid leukemia. Cancer Chemother Pharmacol 69:1155–1163. https://doi.org/10.1007/s00280-011-1800-z
57. Marshall S, Macintyre F, James I, Krams M, Jonsson NE (2006) Role of mechanistically-based pharmacokinetic/pharmacodynamic models in drug development : a case study of a therapeutic protein. Clin Pharmacokinet 45:177–197. https://doi.org/10.2165/00003088-200645020-00004
58. Golubović B, Vučićević K, Radivojević D, Kovačević SV, Prostran M, Miljković B (2019) Exploring sirolimus pharmacokinetic variability using data available from the routine clinical care of renal transplant patients population pharmacokinetic approach. J Med Biochem 38:323. https://doi.org/10.2478/jomb-2018-0030
59. Naidoo A, Chirehwa M, Ramsuran V, McIlleron H, Naidoo K, Yende-Zuma N, Singh R, Ncgapu S, Adamson J, Govender K, Denti P, Padayatchi N (2019) Effects of genetic variability on rifampicin and isoniazid pharmacokinetics in South African patients with recurrent tuberculosis. Pharmacogenomics 20:225–240. https://doi.org/10.2217/pgs-2018-0166
60. Chotsiri P, Zongo I, Milligan P, Compaore YD, Somé AF, Chandramohan D, Hanpithakpong W, Nosten F, Greenwood B, Rosenthal PJ, White NJ, Ouédraogo J-B, Tarning J (2019)

Optimal dosing of dihydroartemisinin-piperaquine for seasonal malaria chemoprevention in young children. Nat Commun 10:480. https://doi.org/10.1038/s41467-019-08297-9
61. Lohy Das J, Rulisa S, de Vries PJ, Mens PF, Kaligirwa N, Agaba S, Tarning J, Karlsson MO, Dorlo TPC (2018) Population pharmacokinetics of artemether, dihydroartemisinin, and lumefantrine in Rwandese pregnant women treated for uncomplicated Plasmodium falciparum malaria. Antimicrob Agents Chemother 62:10–1128. https://doi.org/10.1128/AAC.00518-18
62. Ali AM, Penny MA, Smith TA, Workman L, Sasi P, Adjei GO, Aweeka F, Kiechel J-R, Jullien V, Rijken MJ, McGready R, Mwesigwa J, Kristensen K, Stepniewska K, Tarning J, Barnes KI, Denti P (2018) Population pharmacokinetics of the antimalarial amodiaquine: a pooled analysis to optimize dosing. Antimicrob Agents Chemother 62:10–1128. https://doi.org/10.1128/AAC.02193-17
63. Lohy Das JP, Kyaw MP, Nyunt MH, Chit K, Aye KH, Aye MM, Karlsson MO, Bergstrand M, Tarning J (2018) Population pharmacokinetic and pharmacodynamic properties of artesunate in patients with artemisinin sensitive and resistant infections in Southern Myanmar. Malar J 17:1–10. https://doi.org/10.1186/s12936-018-2278-5
64. Guiastrennec B, Sonne D, Hansen M, Bagger J, Lund A, Rehfeld J, Alskär O, Karlsson M, Vilsbøll T, Knop F, Bergstrand M (2016) Mechanism-based modeling of gastric emptying rate and gallbladder emptying in response to caloric intake. CPT Pharmacometrics Syst Pharmacol 5:692–700. https://doi.org/10.1002/psp4.12152
65. Milosheska D, Lorber B, Vovk T, Kastelic M, Dolžan V, Grabnar I (2016) Pharmacokinetics of lamotrigine and its metabolite N-2-glucuronide: influence of polymorphism of UDP-glucuronosyltransferases and drug transporters. Br J Clin Pharmacol 82:399–411. https://doi.org/10.1111/bcp.12984
66. Denti P, Jeremiah K, Chigutsa E, Faurholt-Jepsen D, PrayGod G, Range N, Castel S, Wiesner L, Hagen CM, Christiansen M, Changalucha J, McIlleron H, Friis H, Andersen AB (2015) Pharmacokinetics of isoniazid, pyrazinamide, and ethambutol in newly diagnosed pulmonary TB patients in Tanzania. PLoS One 10:e0141002. https://doi.org/10.1371/journal.pone.0141002
67. Cella M, Knibbe C, de Wildt SN, Van Gerven J, Danhof M, Della Pasqua O (2012) Scaling of pharmacokinetics across paediatric populations: the lack of interpolative power of allometric models. Br J Clin Pharmacol 74:525–535. https://doi.org/10.1111/j.1365-2125.2012.04206.x
68. Pérez-Ruixo JJ, Doshi S, Chow A (2011) Application of pharmacokinetic-pharmacodynamic modeling and simulation for erythropoietic stimulating agents. Clin Trial Simul:307–323. https://doi.org/10.1007/978-1-4419-7415-0_14
69. Cella M, de Vries FG, Burger D, Danhof M, Pasqua OD (2010) A model-based approach to dose selection in early pediatric development. Clin Pharmacol Ther 87:294–302. https://doi.org/10.1038/clpt.2009.234
70. Abdelwahab MT, Leisegang R, Dooley KE, Mathad JS, Wiesner L, McIlleron H, Martinson N, Waja Z, Letutu M, Chaisson RE, Denti P (2020) Population pharmacokinetics of isoniazid, pyrazinamide, and ethambutol in pregnant South African women with tuberculosis and HIV. Antimicrob Agents Chemother 64:10–1128. https://doi.org/10.1128/AAC.01978-19
71. Deng R, Gibiansky L, Lu T, Agarwal P, Ding H, Li X, Kshirsagar S, Lu D, Li C, Girish S, Wang J, Boyer M, Humphrey K, Freise KJ, Salem AH, Seymour JF, Kater AP, Miles D (2019) Bayesian population model of the pharmacokinetics of venetoclax in combination with rituximab in patients with relapsed/refractory chronic lymphocytic leukemia: results from the phase III MURANO study. Clin Pharmacokinet 58:1621–1634. https://doi.org/10.1007/s40262-019-00788-8
72. Magnusson MO, Samtani MN, Plan EL, Jonsson EN, Rossenu S, Vermeulen A, Russu A (2017) Population pharmacokinetics of a novel once-every 3 months intramuscular formulation of paliperidone palmitate in patients with schizophrenia. Clin Pharmacokinet 56:421–433. https://doi.org/10.1007/s40262-016-0459-3
73. Edlund H, Steenholdt C, Ainsworth MA, Goebgen E, Brynskov J, Thomsen OØ, Huisinga W, Kloft C (2017) Magnitude of increased infliximab clearance imposed by anti-infliximab anti-

bodies in crohn's disease is determined by their concentration. AAPS J 19:223–233. https://doi.org/10.1208/s12248-016-9989-8
74. Quartino AL, Karlsson MO, Lindman H, Friberg LE (2014) Characterization of endogenous G-CSF and the inverse correlation to chemotherapy-induced neutropenia in patients with breast cancer using population modeling. Pharm Res 31:3390–3403. https://doi.org/10.1007/s11095-014-1429-9
75. Lledó-García R, Mazer NA, Karlsson MO (2013) A semi-mechanistic model of the relationship between average glucose and HbA1c in healthy and diabetic subjects. J Pharmacokinet Pharmacodyn 40:129–142. https://doi.org/10.1007/s10928-012-9289-6
76. Stevens J, Ploeger BA, Hammarlund-Udenaes M, Osswald G, van der Graaf PH, Danhof M et al (2012) Mechanism-based PK-PD model for the prolactin biological system response following an acute dopamine inhibition challenge: quantitative extrapolation to humans. J Pharmacokinet Pharmacodyn 39(5):463–477
77. Kshirsagar SA, Blaschke TF, Sheiner LB, Krygowski M, Acosta EP, Verotta D (2007) Improving data reliability using a non-compliance detection method versus using pharmacokinetic criteria. J Pharmacokinet Pharmacodyn 34:35–55. https://doi.org/10.1007/s10928-006-9032-2
78. Yan X, Bauer R, Koch G, Schropp J, Ruixo P, Jose J, Krzyzanski W (2021) Delay differential equations based models in NONMEM. J Pharmacokinet Pharmacodyn 48:763. https://doi.org/10.1007/s10928-021-09770-z
79. Guglielmi N, Hairer E (2001) Implementing Radau IIA methods for stiff delay differential dquations. Computing 67(1):1–12
80. Thompson S, Shampine LF (2004) A friendly FORTRAN DDE solver. DDE_SOLVER User's Guide
81. Shampine LF, Thompson S (2001) Solving DDEs in Matlab. Appl Numer Math 37(4):441–458
82. Soetaert K, Petzoldt T, Setzer R (2010) Solving differential equations in R: package deSolve. J Stat Softw 33(9):1–25
83. Krzyzanski W, Perez-Ruixo JJ, Vermeulen A (2008) Basic pharmacodynamic models for agents that alter the lifespan distribution of natural cells. J Pharmacokinet Pharmacodyn 35(3):349–377
84. Krzyzanski W, Jusko WJ, Wacholtz MC, Minton N, Cheung WK (2005) Pharmacokinetic and pharmacodynamic modeling of recombinant human erythropoietin after multiple subcutaneous doses in healthy subjects. Eur J Pharm Sci 26:295–306
85. Ernst H, Nørsett S, Gerhard W (2000) Solving ordinary differential equations I, 2nd revised edn. Springer, Berlin
86. Ismail M, Sale M, Yu Y, Pillai N, Liu S, Pflug B, Bies R (2021) Development of a genetic algorithm and NONMEM workbench for automating and improving population pharmacokinetic/pharmacodynamic model selection. J Pharmacokinet Pharmacodyn:1–14. https://doi.org/10.1007/s10928-021-09782-9
87. Sale M, Sherer EA (2015) A genetic algorithm based global search strategy for population pharmacokinetic/pharmacodynamic model selection. Br J Clin Pharmacol 79(1):28–39
88. Jonsson EN, Wade JR, Karlsson MO (2000) Nonlinearity detection: advantages of nonlinear mixed-effects modeling. AAPS PharmSci 2(3):E32
89. Sharma P, Wadhwa A, Komal K (2014) Analysis of selection schemes for solving an optimization problem in genetic algorithm. Int J Comput Appl 93:11. https://doi.org/10.5120/16256-5714
90. Tang BH, Guan Z, Allegaert K, Wu YE, Manolis E, Leroux S, Yao BF, Shi HY, Li X, Huang X, Wang WQ, Shen AD, Wang XL, Wang TY, Kou C, Xu HY, Zhou Y, Zheng Y, Hao GX, Xu BP, Thomson AH, Capparelli EV, Biran V, Simon N, Meibohm B, Lo YL, Marques R, Peris JE, Lutsar I, Saito J, Burggraaf J, Jacqz-Aigrain E, van den Anker J, Zhao W (2021) Drug clearance in neonates: a combination of population pharmacokinetic modelling and machine learning approaches to improve individual prediction. Clin Pharmacokinet 60:1435. https://doi.org/10.1007/s40262-021-01033-x

91. Jacqz-Aigrain E, Leroux S, Thomson AH, Allegaert K, Capparelli EV, Biran V et al (2019) Population pharmacokinetic meta-analysis of individual data to design the first randomized efficacy trial of vancomycin in neonates and young infants. J Antimicrob Chemother 74(8):2128–2138. https://doi.org/10.1093/jac/dkz158
92. Tang BH, Wu YE, Kou C, Qi YJ, Qi H, Xu HY et al (2019) Population pharmacokinetics and dosing optimization of amoxicillin in neonates and young infants. Antimicrob Agents Chemother 63:10–1128. https://doi.org/10.1128/AAC.02336-18
93. Bradley JS, Sauberan JB, Ambrose PG, Bhavnani SM, Rasmussen MR, Capparelli EV (2008) Meropenem pharmacokinetics, pharmacodynamics, and Monte Carlo simulation in the neonate. Pediatr Infect Dis J 27(9):794–799. https://doi.org/10.1097/INF.0b013e318170f8d2
94. Zhu H, Huang SM, Madabushi R, Strauss DG, Wang Y, Zineh I (2019) Model-informed drug development: a regulatory perspective on progress. Clin Pharmacol Ther 106(1):91–93. https://doi.org/10.1002/cpt.1475
95. Goulooze SC, Zwep LB, Vogt JE, Krekels EHJ, Hankemeier T, van den Anker JN et al (2020) Beyond the randomized clinical trial: innovative data science to close the pediatric evidence gap. Clin Pharmacol Ther 107(4):786–795. https://doi.org/10.1002/cpt.1744
96. Wilbaux M, Fuchs A, Samardzic J, Rodieux F, Csajka C, Allegaert K et al (2016) Pharmacometric approaches to personalize use of primarily renally eliminated antibiotics in preterm and term neonates. J Clin Pharmacol 56(8):909–935. https://doi.org/10.1002/jcph.705
97. Krstajic D, Buturovic LJ, Leahy DE, Thomas S (2014) Cross-validation pitfalls when selecting and assessing regression and classification models. J Cheminform 6(1):10. https://doi.org/10.1186/1758-2946-6-10
98. Varma S, Simon R (2006) Bias in error estimation when using cross-validation for model selection. BMC Bioinform 7:91. https://doi.org/10.1186/1471-2105-7-91
99. Sibieude E, Khandelwal A, Girard P, Hesthaven JS, Terranova N (2021) Population pharmacokinetic model selection assisted by machine learning. J Pharmacokinet Pharmacodyn 49:257. https://doi.org/10.1007/s10928-021-09793-6
100. Darzi SA, Munz Y (2004) The impact of minimally invasive surgical techniques. Annu Rev Med 55:223–237
101. Hockstein NG, Gourin CG, Faust RA, Terris DJ (2007) A history of robots: from science fiction to surgical robotics. J Robot Surg 1(2):113–118
102. Liang H, Tsui BY, Ni H, Valentim CCS, Baxter SL, Liu G et al (2019) Evaluation and accurate diagnoses of pediatric diseases using artificial intelligence. Nat Med 25(3):433–438
103. Baker RE, Pena JM, Jayamohan J, Jerusalem A (2018) Mechanistic models versus machine learning, a fight worth fighting for the biological community? Biol Lett 14(5):20170660
104. Haghighatlari M, Hachmann J (2019) Advances of machine learning in molecular modeling and simulation. Curr Opin Chem Eng 23:51–57
105. Jeon J, Nim S, Teyra J, Datti A, Wrana JL, Sidhu SS et al (2014) A systematic approach to identify novel cancer drug targets using machine learning, inhibitor design and high-throughput screening. Genome Med 6(7):57
106. Khandelwal A, Bahadduri PM, Chang C, Polli JE, Swaan PW, Ekins S (2007) Computational models to assign biopharmaceutics drug disposition classification from molecular structure. Pharm Res 24(12):2249–2262
107. Khandelwal A, Krasowski MD, Reschly EJ, Sinz MW, Swaan PW, Ekins S (2008) Machine learning methods and docking for predicting human pregnane X receptor activation. Chem Res Toxicol 21(7):1457–1467
108. Liu Q, Zhu H, Liu C, Jean D, Huang SM, ElZarrad MK et al (2020) Application of machine learning in drug development and regulation: current status and future potential. Clin Pharmacol Ther 107(4):726–729
109. You W, Widmer N, de Micheli G (2011) Example-based support vector machine for drug concentration analysis. In: Paper in proceedings of the 33rd annual international conference of the IEEE engineering in medicine and biology society (EMBC 2011)

110. Holland JH (1975) Adaptation in natural and artificial systems. University of Michigan Press, Ann Arbor, p 100
111. Bies RJ, Muldoon MF, Pollock BG, Manuck S, Smith G, Sale ME (2006) A genetic algorithm-based, hybrid machine learning approach to model selection. J Pharmacokinet Pharmacodyn 33(2):195–221
112. Sibieude E, Khandelwal A, Girard P, Hesthaven JS, Terranova N (2021) Fast screening of covariates in population models empowered by machine learning. J Pharmacokinet Pharmacodyn 48:597–609
113. Yuan C-m, Chen H-h, Sun N-n, Ma X-j, Xu J, Fu W (2019) Molecular dynamics simulations on RORγt: insights into its functional agonism and inverse agonism. Acta Pharmacol Sin 40:1480. https://doi.org/10.1038/s41401-019-0259-z
114. Fauber B, Magnuson S (2014) Modulators of the nuclear receptor retinoic acid receptor-related orphan receptor-γ (RORγ or RORc). J Med Chem 57:5871–5892
115. Scheepstra M, Leysen S, van Almen GC, Miller JR, Piesvaux J, Kutilek V et al (2015) Identification of an allosteric binding site for RORγt inhibition. Nat Commun 6:8833
116. Olsson RI, Xue Y, von Berg S, Aagaard A, McPheat J, Hansson E et al (2016) Benzoxazepines achieve potent suppression of IL-17 release in human T-helper 17 (TH17) cells through an induced-fit binding mode to the nuclear receptor RORγ. ChemMedChem 11:207–216
117. Kallen J, Izaac A, Be C, Arista L, Orain D, Kaupmann K et al (2017) Structural states of RORγt: X-ray elucidation of molecular mechanisms and binding interactions for natural and synthetic compounds. ChemMedChem 12:1014–1021
118. Sastry GM, Adzhigirey M, Day T, Annabhimoju R, Sherman W (2013) Protein and ligand preparation: parameters, protocols, and influence on virtual screening enrichments. J Comput Aid Mol Des 27:221–234
119. Bian Y, He X, Jing Y, Wang L, Wang J, Xie X (2019) Computational systems pharmacology analysis of cannabidiol: a combination of chemogenomics-knowledgebase network analysis and integrated in silico modeling and simulation. Acta Pharmacol Sin. 40:374–386
120. Huang Z, Zhao J, Deng W, Chen Y, Shang J, Song K et al (2018) Identification of a cellular active SIRT6 allosteric activator. Nat Chem Biol 14:1118–1126
121. Hotelling H (1933) Analysis of a complex of statistical variables into principal components. J Educ Psychol 24:417–441
122. Amadei A, Linssen AB, Berendsen HJ (1993) Essential dynamics of proteins. Proteins 17:412–425
123. David CC, Jacobs DJ (2014) Principal component analysis: a method for determining the essential dynamics of proteins. Methods Mol Biol 1084:193–226
124. Jiang H, Deng R, Yang X, Shang J, Lu S, Zhao Y et al (2017) Peptidomimetic inhibitors of APC-Asef interaction block colorectal cancer migration. Nat Chem Biol 13:994–1001
125. Shen Q, Cheng F, Song H, Lu W, Zhao J, An X et al (2017) Proteome-scale investigation of protein allosteric regulation perturbed by somatic mutations in 7,000 cancer genomes. Am J Hum Genet 100:5–20
126. Wolfgang K, Christian S (1983) Dictionary of protein secondary structure: pattern recognition of hydrogen-bonded and geometrical features. Biopolymers 22:2577–2637
127. Warren DB, Haque S, McInerney MP, Corbett KM, Kastrati E, Ford L, Williams HD, Jannin V, Benameur H, Porter CJH, Chalmers DK, Pouton CW (2021) Molecular dynamics simulations and experimental results provide insight into clinical performance differences between Sandimmune® and Neoral® lipid-based formulations. Pharm Res 38:1531. https://doi.org/10.1007/s11095-021-03099-5
128. Mueller EA, Kovarik JM, van Bree JB, Grevel J, Lücker PW, Kutz K (1994) Influence of a fat-rich meal on the pharmacokinetics of a new Oral formulation of cyclosporine in a crossover comparison with the market formulation. Pharm Res 11(1):151–155
129. Fatouros DG, Karpf DM, Nielsen FS, Mullertz A (2007) Clinical studies with oral lipid based formulations of poorly soluble compounds. Ther Clin Risk Manag 3(4):591–604
130. Klyashchitsky BA, Owen AJ (1998) Drug delivery systems for cyclosporine: achievements and complications. J Drug Target 5(6):443–458

131. Destere A, Marquet P, Labriffe M, Drici M-D, Woillard J-B (2023) A hybrid algorithm combining population pharmacokinetic and machine learning for isavuconazole exposure prediction. Pharm Res 40:951–959. https://doi.org/10.1007/s11095-023-03507-y
132. Woillard J-B, Debord J, Benz-de-Bretagne I, Saint-Marcoux F, Turlure P, Girault S et al (2017) A time-dependent model describes methotrexate elimination and supports dynamic modification of MRP2/ABCC2 activity. Ther Drug Monit 39:12
133. Benkali K, Rostaing L, Premaud A, Woillard J-B, Saint-Marcoux F, Urien S et al (2010) Population pharmacokinetics and bayesian estimation of tacrolimus exposure in renal transplant recipients on a new once-daily formulation. Clin Pharmacokinet 49:683–692
134. Destere A, Marquet P, Gandonnière CS, Åsberg A, Loustaud-Ratti V, Carrier P et al (2022) A Hybrid model associating population pharmacokinetics with machine learning: a case study with Iohexol Clearance Estimation. Clin Pharmacokinet 61(8):1157–1165. https://doi.org/10.1007/s40262-022-01138-x
135. Destere A, Gandonnière CS, Åsberg A, Loustaud-Ratti V, Carrier P, Ehrmann S et al (2021) A single Bayesian estimator for iohexol clearance estimation in ICU, liver failure and renal transplant patients. Br J Clin Pharma 88:bcp.15197
136. Labriffe M, Woillard J, Debord J, Marquet P (2022) Machine learning algorithms to estimate everolimus exposure trained on simulated and patient pharmacokinetic profiles. CPT Pharmacometrics Syst Pharmacol 11:psp4.12810
137. Bououda M, Uster DW, Sidorov E, Labriffe M, Marquet P, Wicha SG et al (2022) A machine learning approach to predict interdose vancomycin exposure. Pharm Res 39:721–731
138. Hughes JH, Keizer RJ (2021) A hybrid machine learning/pharmacokinetic approach outperforms maximum a posteriori Bayesian estimation by selectively flattening model priors. CPT Pharmacometrics Syst Pharmacol 10:1150–1160
139. Hsueh HT, Chou RT, Rai U, Liyanage W, Kim YC, Appell MB, Pejavar J, Leo KT, Davison C, Kolodziejski P, Mozzer A, Kwon HY, Sista M, Anders NM, Hemingway A, Rompicharla SVK, Edwards M, Pitha I, Hanes J, Cummings MP, Ensign LM (2023) Machine learning-driven multifunctional peptide engineering for sustained ocular drug delivery. Nat Commun 14:2509. https://doi.org/10.1038/s41467-023-38056-w
140. Wei L, Tang J, Zou Q (2017) SkipCPP-Pred: an improved and promising sequence-based predictor for predicting cell-penetrating peptides. BMC Genom 18:742
141. Agrawal P et al (2016) CPPsite 2.0: a repository of experimentally validated cell-penetrating peptides. Nucleic Acids Res 44:D1098–D1103
142. Gupta S et al (2013) In silico approach for predicting toxicity of peptides and proteins. PLoS One 8:e73957
143. Louisse J, Beekmann K, Rietjens IMCM (2017) Use of physiologically based kinetic modeling-based reverse dosimetry to predict in vivo toxicity from in vitro data. Chem Res Toxicol 30:114–125. https://doi.org/10.1021/acs.chemrestox.6b00302
144. Pomponio G et al (2015) In vitro kinetics of amiodarone and its major metabolite in two human liver cell models after acute and repeated treatments. Toxicol In Vitro 30:36–51. https://doi.org/10.1016/j.tiv.2014.12.012
145. Kannan R, Nademanee K, Hendrickson JA, Rostami HJ, Singh BN (1982) Amiodarone kinetics after oral doses. Clin Pharmacol Ther 31:438–444
146. Trivier JM, Libersa C, Belloc C, Lhermitte M (1993) Amiodarone N-deethylation in human liver microsomes: involvement of cytochrome P450 3A enzymes (first report). Life Sci 52:PL91–PP96
147. Abraham MJ, Murtola T, Schulz R, Páll S, Smith JC, Hess B, Lindahl E (2015) GROMACS: high performance molecular simulations through multi-level parallelism from laptops to supercomputers. SoftwareX 1-2:19–25
148. Pall S, Abraham MJ, Kutzner C, Hess B, Lindahl E (2015) Tackling exascale software challenges in molecular dynamics simulations with GROMACS. In: Markidis S, Laure E (eds) Solving software challenges for exascale. Springer, Stockholm, pp 3–27

Printed in the USA
CPSIA information can be obtained
at www.ICGtesting.com
CBHW060327100924
14324CB00003BA/65